W0079726

Monopole '83

NATO ASI Series

Advanced Science Institutes Series

A series presenting the results of activities sponsored by the NATO Science Committee, which aims at the dissemination of advanced scientific and technological knowledge, with a view to strengthening links between scientific communities.

The series is published by an international board of publishers in conjunction with the NATO Scientific Affairs Division

A	**Life Sciences**	Plenum Publishing Corporation
B	**Physics**	New York and London
C	**Mathematical and Physical Sciences**	D. Reidel Publishing Company Fordrecht Boston, and Lancaster
D	**Behavioral and Social Sciences**	Martinus Nijhoff Publishers
E	**Engineering and Materials Sciences**	The Hague, Boston, and Lancaster
F	**Computer and Systems Sciences**	Springer-Verlag
G	**Ecological Sciences**	Berlin, Heidelberg, New York, and Tokyo

Recent Volumes in this Series

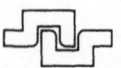

Series B: Physics

Monopole '83

Edited by

James L. Stone

University of Michigan
Ann Arbor, Michigan

Plenum Press
New York and London
Published in cooperation with NATO Scientific Affairs Division

Proceedings of a NATO Advanced Research Workshop entitled
Monopole '83, held October 6–9, 1983, in Ann Arbor, Michigan

Library of Congress Cataloging in Publication Data

Monopole '83 (1983: Ann Arbor, Mich.)
  Monopole '83.

  (NATO ASI series. Series B, Physics; v. 111)
  "Proceedings of a NATO advanced research workshop entitled Monopole '83,
held October 6–9, 1983, in Ann Arbor, Michigan"—Verso t.p.
  Includes bibliographies and index.
  1. Magnetic monopoles—Congresses. I. Stone, James L. II. Title. III. Series.
QC760.4.M33M66   1983   530.1   84-16047

ISBN 978-1-4757-0377-1   ISBN 978-1-4757-0375-7 (eBook)
DOI 10.1007/978-1-4757-0375-7

© 1984 Plenum Press, New York
Softcover reprint of the hardcover 1st edition 1984

A Division of Plenum Publishing Corporation
233 Spring Street, New York, N.Y. 10013

All rights reserved

No part of this book may be reproduced, stored in a retrieval system, or transmitted,
in any form or by any means, electronic, mechanical, photocopying, microfilming,
recording, or otherwise, without written permission from the Publisher

PREFACE

Ten years have passed since 't Hooft and Polyakov demonstrated that superheavy magnetic monopoles were a natural consequence of any Grand Unified Theory (GUT) in which the unifying group contains a U(1) factor as a subgroup. An analysis of these GUTs in an expanding, cooling universe yields a phase transition at an energy $\sim 10^{15}$ GeV and at a cosmic time $\sim 10^{-35}$ seconds after the big bang. The general consequences of GUTs and this phase transition are the prediction of proton decay, the production of superheavy magnetic monopoles, and an understanding of the observed excess of matter over anti-matter in the universe. Attempts to provide experimental verification of GUTs has led to valiant experimental efforts in recent years to observe nucleon decay in massive underground detectors. Experiments to search for superheavy monopoles may eventually require similar efforts. Since the unification scale is unreachable in the laboratory, monopole detectors must search for relics of the big bang.

Much theoretical groundwork has been accomplished in recent years with the development of GUTs. In Part I of this book, Erick Weinberg gives a theoretical overview of the role of magnetic monopoles in the various unification schemes. Monopoles in the context of the newly revived Kaluza-Klein theories are presented by several authors and are summarized by Qaisar Shafi.

Mike Turner begins Part II with a discussion of monopoles in standard big bang cosmology. Paul Steinhardt follows with his perspectives on the inflationary universe; C. Wetterich introduces Kaluza-Klein cosmology. Although there is little guidance from cosmology on monopole production in the early universe, Gene Parker presents an upper limit on the monopole flux based on the continued existence of a large scale galactic magnetic field. The "Parker limit" of 10^{-15} monopoles cm^{-2} sr^{-1} s^{-1} sets the scale for future search experiments.

One of the more exciting ideas to emerge from GUTs is the possibility that monopoles may "catalyze" nucleon decay with relatively large cross sections. Observation of a time sequence of

monopole catalyzed nucleon decays would be dramatic--the simultaneous discovery of monopoles and nucleon decay! In Part III of
this book, Curt Callan discusses the technical aspects of the
catalysis mechanism. Several theoretical papers on catalysis
follow which are summarized by Sally Dawson. Also in Part III,
Rocky Kolb reviews the astrophysical consequences of catalysis and
summarizes monopole flux limits based on catalysis in neutron
stars. Steve Errede's paper gives a complete summary of the
experimental situation and reports current results for monopole
catalysis searches using nucleon decay detectors.

The possibilities that monopoles may induce nuclear fission
of heavy elements, induce β-decay, attach to nuclei, or produce
Zeeman effect level splittings in passing through hydrogen or
helium atoms are presented in Part IV. Monopole detection
techniques relevant for building detectors are given in Part V.
The discussion by Ray Hagstrom on using narrow band gap infrared
phosphors as monopole detectors is particularly interesting.

Since the announcement by Blas Cabrera in early 1982 of the
possible observation of a monopole with a superconducting coil and
SQUID detector, several groups have built similar superconducting
detectors to search for monopoles. Detailed reports from these
efforts are presented in Part VI. Henry Frisch summarizes the
results from the induction experiments and reports that the
accumulation of 100 times the original Cabrera exposure has
revealed no additional monopole candidates. Some groups are now
considering the feasibility of constructing large ($\sim$100m^2) arrays.

The energy loss mechanism for heavy monopoles moving $< 10^{-3}$c
has become better understood. Steve Ahlen and Greg Tarle calculate
a conservative velocity threshold of 6×10^{-4}c for monopoles to
produce a measurable photon yield in plastic scintillators.
Status reports from monopole search experiments using ionization
and excitation detectors appear in Part VII. Some authors discuss
detectors of unprecedented collecting power (10^3-10^4 m^2) which
could challenge the Parker limit in a few years.

In Part VIII, Giorgio Giacomelli summarizes the overall
experimental situation and John Preskill gives the theoretical
status of monopoles in 1983.

Clearly, the subject of magnetic monopoles has grown into a
fascinating interdisciplinary field of physics. Experimental
techniques and interaction mechanisms encompass atomic, nuclear,
low temperature, and solid state physics. The existence or non-
existence of monopoles has fundamental implications in particle
physics, cosmology, and astrophysics.

James L. Stone
Ann Arbor, Michigan

ACKNOWLEDGMENTS

The MONOPOLE '83 Conference held in Ann Arbor, Michigan from October 6-9, 1983 was sponsored by the North Atlantic Treaty Organization, the U.S. Department of Energy, the National Science Foundation, and the University of Michigan. The program was organized with the assistance and advice of an International Advisory Committee consisting of

CHAIRMAN: James Stone (University of Michigan, USA)

ADVISORS: Richard Carrigan, Jr. (Fermilab, USA)
           John Ellis (CERN, Switzerland)
           Giorgio Giacomelli (Bologna, Italy)
           Alfred Goldhaber (State University of New York,USA)
           David Ritson (Stanford University, USA)
           Frank Wilczek (ITP, Santa Barbara, USA)

We acknowledge the valuable work of the topical workshop chairpersons: Steve Ahlen (Indiana) , Sally Dawson (LBL), Henry Frisch (Chicago), Roberto Peccei (MPI), and Qaisar Shafi (Bartol).

Appreciation is extended to the local organizers from the Michigan Physics Department: M. Einhorn, S. Errede, J. Frere, D. Gudehus, D. Heygi, L. Jones, G. Kane, I. Leedom, M. Longo, E. Shumard, L. Sulak, Y. Tomozawa, J. van der Velde, M. Veltman, and Y. Yao.

Special thanks go to Barb Nieman for handling all financial matters; Peter Ross for managing the distribution of over 30,000 pages of transparencies; and Karl Luttrell, Emil Hurst, and Dan Higby for daily operations.

Sue Streicher and JoAnne Sulak are to be commended and thanked for their expert organizational efforts and attention to details before and during the conference, and following the conference for their technical skills and devotion in preparing this manuscript.

Conference   photographs appearing in this manuscript   are by K. Luttrell and L. Lin.

CONTENTS

PART III: Monopole Catalysis of Nucleon Decay

PART VII: Ionization/Excitation Experiments

PART VIII: Conference Highlights and Summation

---

*Asteriks designate the authors who presented the papers.

MONOPOLES AND GRAND UNIFICATION

Erick J. Weinberg

Department of Physics
Columbia University
New York, New York 10027

INTRODUCTION

The study of magnetic monopoles may be divided into four eras. In the first, monopoles were unobserved objects whose existence was merely an interesting possibility. The second period began in 1931 with Dirac's[1] observation that the existence of a magnetic monopole would explain the observed quantization of electric charge. A third era was entered when it was realized that electric charge is naturally quantized in unified theories, where electromagnetism is imbedded in a spontaneously broken gauge theory based on a compact semi-simple group; monopoles did not appear to be needed. Finally, a fourth era began in 1974 with the realization[2] that such unified theories imply the existence of magnetic monopoles and that these monopoles have calculable properties. Before proceeding further, it may be useful to recall why, aside from the question of monopoles, certain of these unified theories have come to be of particular interest.

GRAND UNIFIED THEORIES

Although the "standard" $SU_3 \times SU_2 \times U_1$ model of the strong, weak and electromagnetic interactions fits the observed data rather well, it has a number of features which are somewhat unsatisfactory. It involves three independent gauge coupling constants, in addition to the large number of parameters needed to specify the Higgs potential and the fermion mass matrix. The multiplet structure of the observed particles appears to be somewhat random; for each generation (e.g., u, d, e, ν_e) of left-handed fermions we have

1

$$(3,2)^{1/6} + (\bar{3},1)^{-2/3} + (\bar{3},1)^{1/3} + (1,2)^{-1/2} + (1,1)^1$$

$$\tag{1}$$

$$(u,d)_L \qquad \bar{u}_L \qquad \bar{d}_L \qquad (\nu_e,e^-)_L \qquad e_L^+$$

Finally, there is no explanation for the fact that the weak hypercharge, which has been indicated by a superscript in Eq. (1), is always a rational number; this is equivalent to the problem of the quantization of electric charge.

One way to resolve these difficulties is to imbed the standard model in a grand unified theory (GUT)[3] based on a simple gauge group $G \supset SU_3 \times SU_2 \times U_1$. The fact that G is simple both implies that there is only a single gauge coupling constant and explains the quantization of weak hypercharge (and hence of electric charge). It also turns out to be possible to construct GUTs in which the observed fermions fit into a relatively simple multiplet structure): in SU_5 the particles listed in Eq. (1) can be accommodated in a $\bar{5}$ and a 10, with no new particles being needed, while in SO_{10} the same particles can be put into a single 16-dimensional representation, with only one new particle needed to complete the multiplet.

The symmetry breaking in GUTs may be viewed as occuring in stages

$$G \xrightarrow[\phi_1]{M_1} H_1 \xrightarrow[\phi_2]{M_2} H_2 \longrightarrow \cdots \xrightarrow[\phi_n]{M_n} SU_3 \times U_1 \tag{2}$$

(At each stage both the Higgs field responsible for the symmetry breaking and the vector meson mass which results are indicated.) For the simplest possible GUT,[4]

$$SU_5 \xrightarrow[\phi_{24}]{10^{15}\text{ GeV}} SU_3 \times SU_2 \times U_1 \xrightarrow[\phi_5]{10^2\text{ GeV}} SU_3 \times U_1 \tag{3}$$

while more complex sequences are possible with larger gauge groups, such as SO_{10} or E_6.

Even after specifying the sequence of symmetry breaking, there are still many variations possible. For example, one can make the theory supersymmetric. The main significance of this for our purpose is that it tends to raise the unification mass; in SU_5 this mass may be increased to 10^{16} GeV or more. Another possibility is to add new scalars in order to better account for the pattern of fermion masses; this too has relatively little effect on the GUT monopole.

There are several predictions which are common to all GUTs:
1) The sum of the electric charges of all particle species

vanishes. In fact, one does get zero if the top quark is included along with the observed particles.

2) The running gauge coupling constants (properly normalized) for the weak, electromagnetic, and strong interactions are all equal when evaluated at some sufficiently high unification mass; equivalently, one can use the fine structure constant and Λ_{QCD} to determine the unification mass and then evolve the Weinberg angle down to low energy to obtain a value consistent with experiment. This procedure appears to work; if it is assumed that there are no new particles between the electroweak and the unification scales (the "Desert Hypothesis") one obtains a unification mass M_{unif} of roughly 10^{15} GeV.

3) Baryon number is not absolutely conserved. A very rough guess for the proton decay rate would be $\Gamma \sim (m_p/M_{unif})^4 m_p$, which for $M_{unif} \sim 10^{15}$ GeV gives a lifetime of 10^{28} to 10^{29} years. While this is somewhat less than the present experimental limits[5] (at least for some decay modes), it is not hard to modify the theory so that the predicted lifetime is too long to be measured by any plausible experiment.

4) Magnetic monopoles should exist, with properties which will be described shortly. Observation of such monopoles may be one of the few possibilities for obtaining experimental support for the unification hypothesis.

QUANTIZATION OF ELECTRIC AND MAGNETIC CHARGES

Before discussing the specific properties of GUT monopoles, I will review the quantization conditions which all monopoles must satisfy. Consider first pure electromagnetism, with a monopole whose magnetic charge g is defined by

$$\vec{B} = g \frac{\hat{r}}{r^2} \tag{4}$$

(Note however that I am normalizing electric charge q so that $\vec{E} = q\hat{r}/(4\pi r^2)$.) If we try to find a vector potential such that

$$\vec{B} = \vec{\nabla} \times \vec{A} \tag{5}$$

we will always discover that $\vec{A}$ is singular. For example, one possible choice is

$$\vec{A} = g \frac{(1-\cos\theta)}{r \sin\theta} \hat{\phi} \tag{6}$$

which is singular along the negative z-axis. This vector potential may be viewed as arising from an infinitely thin solenoid running along the z-axis from the origin to $z = -\infty$ and carrying a flux $\Phi = 4\pi g$. Because of the Aharonov-Böhm effect[6] the solenoid can be detected by interference experiments using quantum mechanical particles of charge q unless

$$e^{iq\Phi} = e^{4\pi iqg} = 1 \tag{7}$$

or equivalently,

$$qg = 0, \pm 1/2, \pm 1, \cdots \tag{8}$$

(I have set $h/2\pi = c = 1$.) This and all such solenoids will be undetectable by any interference experiment if all electric and magnetic charges are integral multiples of q_{min} and g_{min}, respectively, where

$$g_{min} \equiv g_{Dirac} = \frac{1}{2q_{min}} \tag{9}$$

(These statements apply to particles carrying only electric or only magnetic charge; the quantization conditions on dyons are somewhat different.) Such an undetectable solenoid is a purely mathematical singularity, commonly referred to as a Dirac string. The orientation of the string is arbitrary, as long as it begins at the position of the point monopole.

However, it appears that the unbroken gauge symmetry of the real world is not simply that of the electromagnetism but instead $SU_3^{color} \times U_1^{EM}$. In this case we should write

$$\vec{B} = G \frac{\hat{r}}{r^2} \tag{10}$$

where G is a matrix in the (reducible) representation of $SU_3 \times U_1$ corresponding to the full set of particle species. G can be diagonalized by a gauge transformation to give

$$G = G_{EM} + G_{color}$$

$$= \frac{a}{2e_{EM}} Q_{Elec} + \frac{1}{2e_{color}} (b \ T_3^{color} + c \ Y^{color}) \tag{11}$$

An appropriate choice of vector potential is then

$$\vec{A} = G \frac{(1 - \cos\theta)}{r \sin\theta} \hat{\phi} \tag{12}$$

which as before has a string singularity along the negative z-axis. Now suppose that we perform an Aharonov-Böhm experiment using particles with "electric" charges $e_{EM}q$, $e_{color} t_3{}^c$, and $e_{color} y^c$. (The apparatus should be much smaller than 10^{-13} cm in order to avoid complications arising from confinement.) The condition for the string to be undetectable turns out to be[7]

$$\exp 4\pi i \left[\frac{a}{2} q + \frac{b}{2} t_3{}^c + \frac{c}{2} y^c\right] = 1 \quad . \tag{13}$$

In the standard model all "electrically" charged particles are either

   a)   color singlets, with $t_3{}^c = y^c = 0$ and $q = n$ ($n = 0, \pm 1, \pm 2, \dots$)

   b)   color 3's, with $t_3{}^c = \pm \frac{1}{2}$, $y^c = 1/3$ or $t_3{}^c = 0$, $y^c = -2/3$, and $q = n - 1/3$

   c)   color $\bar{3}$'s, with $t_3{}^c = \pm 1/2$, $y^c = -1/3$ or $t_3{}^c = 0$, $y^c = 2/3$, and $q = n + 1/3$

   or d) anything which can be made from combinations of a)-c).

Let us see which magnetic charges are consistent with this pattern of electric charges. Applying Eq. (13) to the color singlets implies that "a" must be an integer. From the color triplets we get the three conditions

$$
\begin{aligned}
b &= \text{integer} \\
c - \frac{1}{2} b &= \text{integer} \\
\frac{2}{3} c + \frac{1}{3} a &= \text{integer}
\end{aligned}
\tag{14}
$$

If the first two of these are satisfied, it is always possible to gauge transform G so that b is an even integer; the second condition then implies that c is also an integer. The third condition then requires that c = a (mod 3). Applying Eq. (13) to the remaining cases gives no new conditions.

   Not all G's which satisfy these quantization conditions can actually occur. Unlike G_{EM}, the color magnetic charge G_{color} is not conserved. G_{color} will adjust itself[8] so as to reduce as much as possible the energy in the magnetic field, which is proportional to

$$\text{tr } G^2 = \text{tr } G^2{}_{EM} + \text{tr } G^2{}_{color} \quad . \tag{15}$$

Together with the quantization conditions, this implies that the only possibilities for stable monopoles are

$$a = 3n + 1, \qquad b = 0, \ c = 1$$

$$a = 3n - 1, \qquad b = 0, \ c = -1 \qquad\qquad (16)$$

$$a = 3n, \qquad\qquad b = c = 0$$

where n is any integer. In particular, a monopole carrying the Dirac charge ($a = 1$) must have a color magnetic charge. (Note that these stability criteria must be modified in the Prasad-Sommerfield[9] limit.)

Of course, if we allowed fewer "electric" charges a wider variety of G's would have been permitted. For example, if we had only zero triality representations of $SU_3 {}^{color}$ with integral values for q, then Dirac charge monopoles with $G_{color} = 0$ would be permitted.

THE 'T HOOFT-POLYAKOV MONOPOLE

It may happen that a classical field theory has a stable solution in which the energy density is localized in space. Such a solution corresponds to a particle in the spectrum of the quantized version of the theory; to lowest order in perturbation theory, the mass of this particle is just the energy of the classical solution. In particular, certain spontaneously broken gauge theories have solutions which correspond to magnetic monopoles.

The simplest such solution[2] occurs in an SU_2 gauge theory with a triplet Higgs field $\vec{\phi}$. (Throughout this section vector notation will refer to SU_2 indices.) The Lagrangian is

$$\mathcal{L} = -\frac{1}{4} \vec{F}_{\mu\nu}^2 + \frac{1}{2}(D_\mu \vec{\phi})^2 - \frac{\lambda}{4}(\vec{\phi}^2 - v^2)^2 \qquad\qquad (17)$$

where

$$\vec{F}_{\mu\nu} = \partial_\mu \vec{A}_\nu - \partial_\nu \vec{A}_\mu - e\vec{A}_\mu \times \vec{A}_\nu$$
$$\qquad\qquad (18)$$
$$D_\mu \vec{\phi} = \partial_\mu \vec{\phi} - e \vec{A}_\mu \times \vec{\phi}$$

The Higgs potential is minimized by $\langle\vec{\phi}\rangle = v\hat{n}$, where $\hat{n}$ is an arbitrary unit vector; this Higgs vacuum expectation value breaks SU_2 down to a U_1 which I will refer to as electromagnetism. The particle spectrum includes a neutral massless vector, two vectors with mass $M_W = ev$ and electric charges $\pm e$, and a neutral scalar with mass $\sqrt{2\lambda}v$. A richer spectrum can be obtained by including additional matter field multiplets. Odd-dimensional multiplets lead to particles with electric charges 0, $\pm e$, $\pm 2e$, ..., while those with even-dimension lead to charges $\pm 1/2e$, $\pm 3/2e$, ...

We are interested in finding stable static solutions of the Euler-Lagrange equations of this theory; such solutions correspond to local minima of the energy. We want the solutions to be non-singular and of finite energy; the latter condition imposes the requirement that as $r \to \infty$, $D_\mu \vec{\phi}$ must vanish faster than $r^{-3/2}$ and $|\vec{\phi}|$ must approach v. Finally, the solution should be non-trivial and not simply the vacuum.

One strategy for finding such a solution begins by choosing any configuration of finite energy. One then proceeds to reduce the energy by making continued deformations of the configuration. When this is no longer possible, the configuration is a local minimum of the energy and thus a static solution.

For example, one might take the initial configuration to have $\vec{\phi} = vz$ and $\vec{A}_\mu = 0$ for $r > R$ and some choice of smoothly varying functions $\vec{\phi}(\vec{r})$ and $\vec{A}_\mu(\vec{r})$ for $r < R$. Reducing the energy by continuous deformation, one can eventually arrive at the uniform configuration $\vec{\phi} = v\hat{z}$, $\vec{A}_\mu = 0$. Unfortunately, this is just the vacuum.

A better choice of initial configuration is one where at large distances $\vec{\phi} = v\hat{r}$; i.e., the direction of $\vec{\phi}$ in internal space is correlated with the position in coordinate space. The important thing is that there is no way to smoothly transform this "hedgehog" configuration into one where the direction of $\vec{\phi}$ is uniform. Consequently, the solution which results from the procedure outlined above cannot be the vacuum but must instead be the non-trivial solution we want.

In order that the energy be finite,

$$D_\mu \vec{\phi} = \partial_\mu \vec{\phi} - e \vec{A}_\mu \times \vec{\phi}$$

must fall faster than $r^{-3/2}$ at large distances. However, with $\vec{\phi} \approx v\hat{r}$, the first term on the right hand side of Eq. (19) falls only as $1/r$. In order to cancel this term, there must be a vector potential falling as $1/r$ and thus a magnetic field falling as $1/r^2$. This long-range field must correspond to the massless particle; i.e., it must lie in the unbroken U_1 subgroup. We expect the large distance behavior of the solution to be spherically symmetric, implying a U_1 magnetic field

$$B_i^{EM} \approx g \frac{\hat{r}_i}{r^2} \quad . \tag{20}$$

We will see shortly that the magnetic charge $g = 1/e$. In the pure U_1 theory a monopole must have a string singularity; since our procedure leads to a non-singular solution, it must involve the other, massive, SU_2 gauge fields. These massive fields should fall off outside a core region of radius $R_{core} \sim M_W^{-1}$. The energy of the solution may be written as

$$E = E_{core} + E_{field} \tag{21}$$

where

$$E_{field} = \frac{1}{2} \int\limits_{r \, > \, R_{core}} dV \, \vec{B}^2$$

$$\approx \frac{2\pi}{e^2} M_W \quad . \tag{22}$$

If, as seems plausible, the two terms in Eq. (21) are comparable in magnitude, the monopole mass will be roughly

$$M_{mon} \approx \frac{4\pi}{e^2} M_W \quad . \tag{23}$$

These qualitative arguments can be supplemented by more explicit calculations. Consider the spherically symmetric ansatz[2]

$$\phi^a(\vec{r}) = v\hat{r}^a \, H(r)$$

$$A_i^a(\vec{r}) = \varepsilon_{iaj} \frac{\hat{r}_j}{er} [1-K(r)] \quad . \tag{24}$$

(Here a is an SU_2 index while i labels space components.) Substituting this ansatz into the field equations gives a pair of non-linear differential equations for the functions H and K. The boundary conditions are that $H(0) = K(\infty) = 0$ and $H(\infty) = K(0) = 1$. These equations can be solved numerically; one finds that for $r \gtrsim M_W^{-1}$ both H and K approach their asymptotic values exponentially fast. The magnetic field is

$$B_i^a = \frac{1}{e} \frac{\hat{r}_i}{r^2} \hat{r}_a + \frac{1}{e} \hat{r}_a \hat{r}_i (\frac{K'}{r} - \frac{K^2}{r^2}) - \frac{1}{e} \delta_{ai} \frac{K'}{r} \quad . \tag{25}$$

Only the first term survives at large distances. This long-range part of $\vec{B}$ is parallel to $\vec{\phi}$ in internal space, and thus truly electromagnetic. As expected, it has the magnetic monopole form, with the magnetic charge being $g = 1/e$. This is the Dirac charge if all possible representations of matter fields are included, but twice the Dirac charge if only odd-dimensional representations are allowed. The mass, which may be written as

$$M_{mon} = \frac{4\pi}{e^2} M_W \, f(\lambda/e^2) \tag{26}$$

is rather insensitive to the value of the scalar coupling; the function f varies between

$$f(0) = 1 \quad , \text{ Ref. 9} \tag{27}$$

and

$$f(\infty) = 1.787 \quad , \text{ Ref. 10} \quad . \tag{28}$$

It should be stressed that the use of a spherically symmetric ansatz is simply a guess; it is quite conceivable that the energy could be lowered by deforming the solution in some non-spherically symmetric manner. The important point is that it doesn't really matter. The qualitative arguments given above guarantee that there is a non-trivial solution with a mass given roughly by Eq. (23). (In fact, one can show[11] that $M_{mon} \geqslant (\frac{4\pi}{e^2})M_W$.) Further, since magnetic charge is quantized, any solution obtained by continuous deformation of the sperically symmetric one must also have $g = 1/e$.

The secret to finding this monpole was the recognition that there is a class of finite energy configurations which cannot be deformed into the vacuum. It is natural to ask whether there is some other class of configurations which cannot be deformed into either the vacuum or the 't Hooft-Polyakov solution. The answer is given by topology. Let $f_1(\theta,\phi)$ and $f_2(\theta,\phi)$ be continuous functions defined on a two-sphere and taking values in some manifold M. These will be said to belong to the same homotopy class if it is possible to smoothly deform one into the other. (For a more precise definition and fuller discussion, see Ref. 12.) These homotopy classes are elements of a group denoted by $\pi_2(M)$.

For the problem we are interested in, $f(\theta,\phi) = \vec{\phi}(r = \infty, \theta, \phi)$. M is the set of all three-vectors with length v; this set may also be identified as the sphere S^2 or as the quotient group SU_2/U_1. The possible finite energy configurations can therefore be classified by the elements of $\pi_2(SU_2/U_1)$, which turns out to be equal to Z, the additive group of the integers. Consequently, we can label any configuration by an integer (its topological charge) which is just equal to the magnetic charge in units of $1/e$. The group structure of $\pi_2(SU_2/U_1)$ implies that topological charges add in the usual manner.

Are there any monopole solutions with higher magnetic charge? Topology alone cannot answer this question. For example, the topological charge 2 configuration with least energy could be a localized solution corresponding to a monopole with magnetic charge $2/e$ or it could be two unit charge monopoles infinitely far apart. Determining which is the case is a difficult dynamical question which is still undecided, although the latter case seems the more likely. (There is a solution of the former type[13] in the mathematically useful but unphysical Prasad-Sommerfield[9] limit. However, it belongs to a continuous family of solutions which appear to correspond to two-monopole states rather than to a new type of particle.)

MONOPOLES IN GRAND UNIFIED THEORIES

Let us now turn to theories with larger gauge groups. Specifically, let us suppose that a gauge group G is broken to a subgroup H by some Higgs field ϕ. In any finite energy configuration the asymptotic value of ϕ in any direction must minimize the Higgs potential; the set of values which do this form the manifold G/H. There will be topologically non-trivial choices of $\phi(r = \infty, \theta, \phi)$, and thus monopole solutions, if the homotopy group $\pi_2(G/H)$ is non-trivial; i.e., if it has more than one element.

There is a mathematical theorem which is quite useful in determining $\pi_2(G/H)$. If G is compact, semi-simple, and simply connected, then

$$\pi_2(G/H) = \pi_1(H) \tag{29}$$

where $\pi_1(H)$ is the group whose elements are homotopy classes of functions from the circle S^1 to H(i.e., classes of closed loops in H).   For H = U_1, we have $\pi_1(U_1) = \pi_1(S^1) = Z$; together with Eq. (29), this implies that $\pi_2(SU_2/U_1) = Z$, as was asserted earlier. If H is semi-simple, $\pi_1(H)$ is a finite group; e.g., $\pi_1(SU_2) = 0$ and $\pi_1(SO_3) = Z_2$. (Note that one must pay more attention than usual to the global properties of the gauge group.) Finally, if $\hat{H}$ is any semi-simple group, then $\pi_1(U_1 \times \hat{H})$ is either Z or the product of Z with some finite group; in particular, $\pi_1(U_1 \times SU_3) = Z$.

In any GUT the gauge group G is both compact and semi-simple (in fact, by the definition of a GUT it must be simple). Further,

G can always be taken to be the covering group of the Lie algebra and thus simply connected. Equation (29) then applies, and $\pi_2(G/H) = \pi_1 (U_1^{EM} \times SU_3^{color}) = Z$. Thus all GUTs have monopoles.

Many of the properties of these monopoles can be deduced by considering the sequence of symmetry breaking stages, Eq. (2). The first step is to compute $\pi_2(G/H_1)$, $\pi_2(G/H_2)$, and so on until a non-trivial π_2 is encountered. For the moment let use assume that no finite groups occur in this sequence, so that the first non-trivial π_2 is $\pi_2(G/H_j) = \pi_2(H_{j-1}/H_j)$, where H_j is the first subgroup with an explicit U_1 factor. Now consider the theory with ϕ_{j+1}, $\phi_{j+2}, \ldots, \phi_n$ omitted and obtain a solution by finding the lowest energy configuration with unit topological charge. From this point the argument proceeds as in the SU_2 case. Since the solution is topologically non-trivial, ϕ_j must "twist" as one moves about the sphere at spatial infinity. Consequently its gradient falls only as $1/r$, so in order that its covariant derivative decreases rapidly enough there must be a $1/r$ vector potential and hence a magnetic field falling as $1/r^2$ and lying in the unbroken subgroup H_j. For the solution to be non-singular it must show evidence of arising from a spontaneously broken gauge theory. It must therefore involve the vector mesons which acquire a mass $M_X \sim M_j$ from ϕ_j and so should have a core with radius $R_{core} \sim M_j^{-1}$. Reasoning as before, we then obtain the estimate

$$M_{mon} \sim \frac{M_j}{\alpha_{GUT}} \; . \tag{30}$$

For most popular models $M_j \gtrsim 10^{15}$ GeV; however, it is possible[14] to construct GUTs which have monopoles of this sort with masses as low as 10^4 GeV.

The effects of the lighter Higgs scalars ϕ_{j+1}, $\phi_{j+2}, \ldots, \phi_n$ must still be included. There are two cases to be considered. The first occurs when the U_1 factor of the H_j contributes to the U_1 of electromagnetism (e.g., $SU_5 \rightarrow SU_3 \times SU_2 \times U_1 \rightarrow SU_3 \times U_1$). In this case the light scalars have little effect on the monopole. The U_1 part of the magnetic charge, G_{EM}, turns out to have the Dirac value. (More precisely, it is the Dirac charge if the minimum electric charge allowed by the group G actually occurs).

The other possibility is that the U_1 factor of H_j is broken completely at some later stage of symmetry breaking (e.g., $SO_{10} \rightarrow SU_5 \times U_1 \rightarrow S_5 \rightarrow \ldots$). In this case the H_j monopoles described above cannot exist as free particles, just as ordinary magnetic monopoles are forbidden in a superconductor. To find the GUT monopole one must return to the sequence of symmetry breaking stages and look for the next U_1 to appear.

As in the SU_2 example, a more concrete construction of the monopole can be obtained by using a spherically symmetric ansatz. A general prescription for obtaining such ansatz has been given by Wilkinson and Goldhaber[15]; this prescription can be applied in all GUTs of interest although, somewhat surprisingly, there are theories (e.g., SU_3 broken to the U_1 generated by λ_8) where the monopole cannot be spherically symmetric. The method is based on finding an SU_2 subgroup of G of which only a U_1 subgroup survives the symmetry breaking; in a certain sense the case of $G \rightarrow H$ is reduced to that of $SU_2 \rightarrow U_1$.

Consider for example the case of SU_5 broken to $SU_3 \times U_1$. The SU_5 basis may be fixed by giving the first generation of fermions

$$\overline{\psi_5} = (\overline{d}_1, \ \overline{d}_2, \ \overline{d}_3, \ e^-, \ \nu)_L$$

$$\psi_{10} = \frac{1}{\sqrt{2}} \begin{bmatrix} 0 & \overline{u}_3 & -\overline{u}_2 & u_1 & d_1 \\ -\overline{u}_3 & 0 & \overline{u}_1 & u_2 & d_2 \\ -u_2 & -\overline{u}_1 & 0 & u_3 & d_3 \\ -u_1 & -u_2 & -u_3 & 0 & e^+ \\ -d_1 & -d_2 & -d_3 & -e^+ & 0 \end{bmatrix} \tag{31}$$

(Mixing angles have been omitted in ψ_{10}.) With this basis an appropriate choice for the SU_2 subgroup is the one generated by

$$\begin{bmatrix} 0 & 0 & 0 & 0 & 0 \\ 0 & 0 & 0 & 0 & 0 \\ 0 & 0 & \left[\frac{1}{2}\,\tau^a\right] & 0 \\ 0 & 0 & & & 0 \\ 0 & 0 & 0 & 0 & 0 \end{bmatrix} \tag{32}$$

where the τ^a $(a = 1,2,3)$ are the usual Pauli matrices. This SU_2 leads to a spherically symmetric solution[16] with magnetic charge

$$G = \text{diag}(0,0, \ -1/2, \ 1/2,0)$$

$$= \frac{1}{2} Q_{Elec} + \frac{1}{2} Y^C \tag{33}$$

(i.e., the Dirac charge) and with mass

$$M_{mon} = \frac{3}{8} \frac{M_x}{\alpha_{GUT}} f \tag{34}$$

where f depends on the details of the Higgs potential and lies between 1 and 1.787.

Spherically symmetric solutions with two and three times the Dirac charge can be constructed in a similar fashion by imbedding the J=1 and J=3/2 representations of SU_2 in the fundamental SU_5 representation.[16] However, mass calculations[17] indicate that these solutions are unstable against decay into several singly charged monopoles. (This does not rule out the possibility of multiply charged monopoles. Gardner and Harvey[18] have shown that for a certain range of parameters the SU_5 model has solutions with magnetic charges 2,3,4, and 6 which are stable against decay into Dirac charge monopoles. These solutions cannot be constructed by the above procedure.)

The SU_2 subgroup used in the Wilkinson-Goldhaber construction is of significance for the Callan-Rubakov[19] effect. The details of the catalyzed proton decay depend on the transformation properties of the various particle species under this SU_2. Under the SU_2 of Eq. (31), the first generation of fermions decomposes into four doublets -- $(d_{\bar{3}}, e^-)_L$, $(u_1, \bar{u}_2)_L$, $(u_2, \bar{u}_1)_L$, and $(e^+, d_3)_L$ -- and seven singlets; only the former are involved in the Callan-Rubakov effect. Consequently, any proton decay thus catalyzed by the Dirac charge SU_5 monopole cannot involve a neutrino.

It has so far been assumed that the homotopy groups $\pi_2(G/H_j)$ were either trivial or else equal to Z. It can also happen that $\pi_2(G/H_j) = Z_n$, the finite group defined by addition of the integers modulo n; topological, and hence magnetic, charges are then defined only modulo n. Thus in addition to the usual possibility of monopole-antimonopole annihilation, one can also have n like sign monopoles combine to yield magnetically neutral products.

For example[20], in the symmetry breaking

$$SO_{10} \xrightarrow{M_1} SO_6 \times SO_4 \tag{35}$$

(or more precisely, $Spin_{10} \longrightarrow (Spin_6 \times Spin_4)/Z_2$) the relevant homotopy group is Z_2. There is a monople solution with unit topological charge and $M_{mon} \sim M_1/\alpha_{GUT}$. The Z_2 nature of the monopole becomes apparent if one tries to construct a configuration with topological charge 2 by having the Higgs field at spatial infinity "twist" twice as much as in the topological charge 1 monopole. It turns out that such a double twist can be deformed into no twist at all, so topological charge 2 is equivalent to topological charge 0. It follows that there is no distinction between a monopole and an antimonopole, and one can have monopole-monopole annihilation.

One might well object that this may be all very interesting, but that it is hardly relevant to the real world, which does not have an unbroken $SO_6 \times SO_4$ symmetry. Equation (35) must be supplemented by further symmetry breaking; one possibility is

$$SO_{10} \xrightarrow[\phi_1]{M_1} SO_6 \times SO_4 \xrightarrow[\phi_2]{M_2} SO_6 \times SU_2 \times U_1 \longrightarrow \cdots \xrightarrow[\phi_n]{M_n} SU_3 \times U_1 \quad (36)$$

$$\pi_2(G/H_1) = Z_2 \qquad \pi_2(G/H_2) = Z \qquad \pi_2(G/H_n) = Z$$

At the second the subsequent stages of symmetry breaking topological charge adds in the usual fashion, so one might think that the earlier Z_2 would be of no consequence. However, it affects the monopole spectrum. At the second stage of symmetry breaking the usual arguments lead to a monopole solution with unit topological charge in which ϕ_1 and ϕ_2 are twisted and whose mass must therefore be $\sim M_1/\alpha$. One could also construct configurations with topological charge 2 by having ϕ_1 and ϕ_2 twist twice as much. However, the double twist in ϕ_1 can be undone, thus leading to configurations in which only ϕ_2 twists and for which the mass is $\sim M_2/\alpha$. If $M_2 \ll M_1$, both the heavier singly charged and the lighter doubly charged solutions must be stable. (More generally[21], an original Z_n homotopy group can lead to a light monopole with topological charge n.)

At the final $SU_3 \times U_1$ level these two monopoles are rather different. The former is rather similar to the Dirac charge SU_5 monopole. It catalyzes proton decay by the Callan-Rubakov effect and has a mass $> 10^{16}$ GeV. The latter, doubly charged, monopole does not lead to Callan-Rubakov proton decay, as can be seen by examining the structure of the fermion multiplets.[22] (Note that this is just the monopole arising from $SO_6 \times SO_4 \rightarrow SO_6 \times SU_2 \times U_1$, and that the gauge bosons of $SO_6 \times SO_4$ do not mediate baryon number violation.) It may, however, catalyze proton decay by a much weaker mechanism.[23] Its mass, which is determined by the intermediate symmetry breaking scale M_2, could be anywhere from 10^{16} GeV down to 10^{10} GeV[24], or perhaps even lower[25].

CONCLUSION

To summarize briefly, there are two types of monopoles which, from the viewpoint of grand unification, are the most important to keep in mind when devising monopole searches or deducing astrophysical limits. The first is superheavy ($M > 10^{16}$ GeV), strongly catalyzes proton decay, and occurs with magnetic charge g(Dirac) = 1/2e or, possibly, with multiples of this charge. The second, which is absent in the SU_5 model but occurs in other popular GUTs, is lighter ($M \sim 10^{10}$ to 10^{16} GeV), does not catalyze proton decay by the Callan-Rubakov mechanism, and occurs only with multiple magnetic charge.

ACKNOWLEDGEMENTS

I would like to thank Curtis Callan, Jeffrey Harvey, and Qaisar Shafi for helpful comments. This research was supported in part by the U.S. Department of Energy.

REFERENCES

1. P.A.M. Dirac, Proc. Roy. Soc. A133, 60 (1931).
2. G. 't Hooft, Nucl. Phys. B79, 276 (1974); A.M. Polyakov, JETP Lett. 20, 194 (1974).
3. P. Langacker, Phys. Rep. 72, 185 (1981).
4. H. Georgi and S.L. Glashow, Phys. Rev. Lett. 32, 438 (1974).
5. R. Bionta et al., Phys. Rev. Lett. 51, 27 (1983); J. Bartelt et al., Phys. Rev. Lett. 50, 651 (1983); G. Battistoni et al., Phys. Lett. 118B, 461 (1982); M. Krishnaswamy et al., Phys. Lett. 115B, 349 (1982).
6. Y. Aharonov and D. Böhm, Phys. Rev. 115, 485 (1959).
7. F. Englert and P. Windey, Phys. Rev. D14, 2728 (1976); P. Goddard, J. Nuyts, and D. Olive, Nucl. Phys. B125, 1 (1977).
8. R. Brandt and F. Neri, Nucl. Phys. B161, 253 (1979); S. Coleman, "The Magnetic Monopole Fifty Years Later," in The Unity of the Fundamental Interactions, ed., A. Zichichi, (Plenum, 1983).
9. M. Prasad and C. Sommerfield, Phys. Rev. Lett. 35, 760 (1975).
10. T. Kirkman and C. Zachos, Phys. Rev. D24, 999 (1981).
11. E. Bogomol'nyi, Sov. J. Nucl. Phys. 24, 449 (1976); S. Coleman, S. Parke, A. Neveu, and C. Sommerfield, Phys. Rev. D15, 544 (1977).
12. S. Coleman, "Classical Lumps and Their Quantum Descendents," in New Phenomena in Subnuclear Physics, ed., A. Zichichi, (Plenum, 1977).
13. R. Ward, Comm. Math. Phys. 80, 563 (1981); Phys. Lett. B102, 136 (1981); P. Forgács, Z. Horváth, and L. Palla, Phys. Lett.

99B, 239 (1981).

14. J. Kim, Phys. Rev. D23, 2706 (1981).

15. D. Wilkinson and A. Goldhaber, Phys. Rev. D16, 1221 (1977).

16. C. Dokos and T. Tomaras, Phys. Rev. D21, 2940 (1980).

17. A. Schellekens and C. Zachos, Phys. Rev. Lett. 50, 1242 (1983); P. Eckert, D. Altschüler, T. Schücker, and G. Wanner, Univ. of Geneva preprint UGVA-DPT 1983/03-383 (1983); P. Eckert, D. Altschüler, and T. Schücker, Univ. of Geneva preprint UGVA-DPT 1983/07-402 (1983).

18. C. Gardner and J. Harvey, Princeton University preprint (1983).

19. V. Rubakov, JETP Lett. 33, 644 (1981); Nucl. Phys. B203, 311 (1982); C. Callan, Phys. Rev. D25, 2141 (1982); Phys. Rev. D26, 2058 (1982).

20. G. Lazarides and Q. Shafi, Phys. Lett. 94B, 149 (1980); G. Lazarides, M. Magg, and Q. Shafi, Phys. Lett. 97B, 87 (1980).

21. F.A. Bais, Phys. Lett. 98B, 437 (1981); E. Weinberg, D. London, and J. Rosner, Univ. of Chicago preprint EFI83-39 (1983).

22. S. Dawson and A. Schellekens, Phys. Rev. D27, 2119 (1983).

23. F. Wilczek, Phys. Rev. Lett. 48, 1146 (1982).

24. H. Georgi and D. Nanopoulos, Nucl. Phys. B159, 16 (1979); S. Dawson and H. Georgi, Nucl. Phys. B179, 477 (1981); R. Robinett and J. Rosner, Phys. Rev. D26, 2396 (1982); N. Deshpande and R. Johnson, Phys. Rev. D27, 1165, 1193 (1983).

25. T. Rizzo and G. Senjanovic, Phys. Rev. Lett. 46, 1315 (1981); Phys. Rev. D24, 704 (1981); Phys. Rev. D25, 235 (1982).

NONABELIAN MONOPOLES BREAK COLOR

A.P. Balachandran

Physics Department
Syracuse University
Syracuse, New York   13210

INTRODUCTION

Grand unified theories[1] (GUTs) are gauge theories based on a unifying group G which is spontaneously broken into a subgroup H. Such theories predict "magnetic" monopoles[2] as topological excitations. These monopoles are generalizations of the abelian Dirac monopoles, in particular some of them are sources of long range nonabelian magnetic fields.

Monopoles are known to be characterized by many remarkable properties of which we may recall the following: a) It was recognized quite early that a test particle with zero spin (say) in a monopole field may behave like a particle of spin half as a consequence of the unusual topology of the monopole, this resemblance to a fermion is now known to extend to its statistics as well.[3] b) Monopoles (or rather dyons) can be fractionally charged in the absence of zero mass fermions.[4]   c) Monopoles can catalyze proton decay, the decay lifetime is expected to be of the order of a typical strong interaction lifetime.[5]

In this lecture, I want to point out yet another bizarre property which is displayed by nonabelian monopoles.[6,7]   Such monopoles emit long range nonabelian (and possibly also abelian) magnetic fields, and are associated with a nonabelian "unbroken" subgroup H. We shall see that although the group H, which is the little group of the Higgs field at spatial infinity, is perfectly well defined as a set, still it is impossible to implement a generic transformation of H on the states due to global topological obstructions. There are, however, several subgroups H_0, H_0', ... of H which can be globally implemented, with the transformation laws of the observables differing from group to group in a novel way. When G is SU(5) and H is (locally) $SU(3)_C \times SU(2)_{WS} \times U(1)_Y$, a possible choice for H_0 is $SU(2)_C \times U(1) \times U(1) \times U(1)$. Here $SU(2)_C$ acts on the first two quarks (say) and the three U(1)'s and the rotations around the third axis in $SU(2)_C$ are generated by the Cartan subalgebra of SU(5). When G = SU(5) and H = $SU(3)_C \times U(1)_{EM}$, a possible choice for H_0 is $U(2)_C \times U(1)_{EM}$.

It is important to recognize that the results reported here show a fundamental structural property of any field theory which predicts nonabelian magnetic monopoles by means of a suitable Higgs mechanism: In the presence of these monopoles, the symmetry group of the theory is not the little group H of the Higgs field at spatial infinity, rather it is a very different group with a novel action on the fields.

The topological problems I shall discuss arise because any definition of a generic "transformation of H" at all points of a two-sphere surrounding the monopole necessarily contains (at least) a string singularity. If we perversely ignore the string and apply the transformation, then the monopole is mapped into a state of infinite energy. These problems are consequences of topology and are not associated with any energy scale. Further, they cannot be perceived in any local experiment which is involved with only a portion of a two-sphere surrounding the monopole. To put it differently, if the monopole is behind the moon, there is certainly no insurmountable difficulty in defining color multiplets in terrestrial experiments. However, even in such experiments, when the effects of the monopole are considered, there is no natural and unique definition of a generic transformation of H, while any one particular definition of this notion looks clumsy. But this sort of nonuniqueness is not serious if the solid angle subtended by the experimental set up at the monopole is small in comparison with 4π.

The theoretical difficulties in giving a meaning to the action of the "unbroken" subgroup on the states indicate that the concept of color (and of the standard model group $SU(3)_C \times SU(2)_{WS} \times U(1)_Y$) partially break down in the vicinity of nonabelian GUT monopoles. There is other evidence as well to support such a conclusion. Thus in the presence of an elementary (nonabelian) monopole which is created when SU(5) breaks to $SU(3)_C \times U(1)_{EM}$, the color triplet in $\bar{5}$ contains two fermions and one boson. A generic color transformation therefore mixes bosons with fermions, does not commute with angular momentum and is incompatible with superselection rules. We will also see later that color (or more generally H) singlets composed of color (H) nonsinglet constituents will disintegrate into color (H) non-singlets if scattered by such monopoles suggesting that the free existence of colored monopoles requires the observability of other colored states. All this shows the importance of investigating the mechanisms which can "screen" the topology of the nonabelian monopoles.[8]

THE MONOPOLE IN THE U-GAUGE

Let us first recall the description of the monopole sectors of the 't Hooft-Polyakov model in the U-gauge. The model is the SU(2) gauge theory of a triplet

$$\Phi = \tau_\alpha \, \phi_\alpha, \quad \tau_\alpha = \text{Pauli matrices} \tag{1}$$

of Higgs fields, wherein the gauged group $SU(2) \equiv G$ is spontaneously broken to a subgroup $U(1) \equiv H$. Such a symmetry breakdown is achieved by introducing a potential $V(\Phi)$ with a minimum located at a nonzero value of $\text{Tr}\Phi^2$. At large distances, Φ approaches a minimum of V, therefore the asymptotic behavior of Φ in the topologically trivial vacuum sector is

$$\Phi \underset{r \to \infty}{\to} \Phi^{(\infty)} = c\tau_3,$$

$$c = \text{constant}, \quad r^2 = \Sigma x_i x_i \tag{2}$$

The little group of $c\,\tau_3$ is of course U(1):

$$U(1) = H = \left\{ e^{i\eta\tau_3} \right\},$$

$$e^{i\eta\tau_3} \, c\tau_3 \, e^{-i\eta\tau_3} = c\tau_3 \qquad . \tag{3}$$

In the (topologically nontrivial) monopole sectors as well, we can adopt the U-gauge wherein the asymptotic form of Φ is as in Eq. (2). As we know, however, in particular from the work of Wu and Yang,[2] in such a gauge a charged field [with the charge of the $U(1) = H$ group] must be described by "sections" in the presence of a monopole. Thus we divide the sphere S_∞^2 at $r = \infty$ into the two patches

$$\Sigma_N = S_\infty^2 - S, \qquad S = (0, \ 0, \ -1),$$

$$\Sigma_S = S_\infty^2 - N, \qquad N = (0, \ 0, \ 1). \tag{4}$$

An SU(2) doublet field splits under U(1) into two fields of opposite charge. Such a field has two descriptors ("sections") ψ_N and ψ_S which are defined and smooth on Σ_N and Σ_S respectively. On the overlap $\Sigma_N \cap \Sigma_S$, the two descriptors are related to each other by a "transition function" $h(\phi) \in U(1)$ which is characteristic of the monopole:

$$\psi_S(\hat{x}) = h(\phi)\psi_N(\hat{x}), \quad \hat{x} \in \Sigma_N \cap \Sigma_S,$$

(5)

ϕ = Azimuthal angle on S_∞^2

For an elementary monopole [with the lowest value of the "magnetic charge"], the transition function is

$$h(\phi) = e^{i\tau_3\phi}$$

(6)

while for its antiparticle, it is

$$h(\phi) = e^{-i\tau_3\phi} \quad .$$

(7)

[Here we have confined our discussion to spatial infinity since it suffices for our purposes.]

The point to note is that the U(1) transformations we perform on ψ_N and ψ_S must respect the relation (5). For the 't Hooft-Polyakov model, since H is abelian, such a rule does not create difficulties. Thus if

$$\psi_N'(\hat{x}) = e^{i\tau_3\alpha} \psi_N(\hat{x}),$$

(8)

$$\psi_S'(\hat{x}) = e^{i\tau_3\alpha} \psi_S(\hat{x})$$

are the transformed fields, then clearly,

$$\psi_S'(\hat{x}) = h(\phi) \psi_N'(\hat{x}) \quad .$$

(9)

Therefore the U(1) transformations can be implemented on the fields.

Let us now consider nonabelian monopoles and for simplicity choose G to the be the GUT group SU(5) and H to be (locally) $SU(3)_C \times SU(2)_{WS} \times U(1)_Y$. The breakdown $G \rightarrow H$ is achieved by a 24 of Higgs. In the U-gauge, at $r = \infty$, it has the form

$$\phi(\infty) = C \begin{bmatrix} 1 & & & & \\ & 1 & & & \\ & & 1 & & \\ & & & -3/2 & \\ & & & & -3/2 \end{bmatrix} \tag{10}$$

It is manifestly invariant under $SU(3)_C \times SU(2)_{WS} \times U(1)_Y$.

In the presence of an elementary monopole, the $\bar{5}$ multiplet for instance has two descriptors ψ_N and ψ_S which are defined and smooth on Σ_N and Σ_S respectively. The transition function for such a monopole is

$$h(\phi) = e^{i\tau_3\phi},$$

$$\tau_3 = \begin{bmatrix} 0 & & & & 0 \\ & 0 & & & \\ & & 1 & & \\ & & & -1 & \\ 0 & & & & 0 \end{bmatrix} \tag{11}$$

so that

$$\psi_S(\hat{x}) = h(\phi)\psi_N(\hat{x}), \quad \hat{x} \in \Sigma_N \bigcap \Sigma_S \tag{12}$$

Consider now the transformation of the $\bar{5}$ by a generic element $s \in H$:

$$\psi'_N = s\psi_N \quad ,$$

$$\psi'_S = s\psi_S \quad . \tag{13}$$

The transformed descriptors fulfill, instead of Eq. (12),

$$\psi'_S = s\,h(\phi)s^{-1}\psi'_N \quad . \tag{14}$$

Since only those transformations which preserve Eq. (12) are legal, we see that only those transformations which commute with $h(\phi)$ can be implemented. This is the subgroup $SU(2)_C \times U(1) \times U(1) \times U(1)$.

If we had broken G all the way down to $SU(3)_C \times U(1)_{EM}$, since the transition function still is given by Eq. (11), the implementable subgroup would be $U(2)_C \times U(1)_{EM}$.

In this discussion we have assumed that s in Eq. (13) does not depend on $\hat{x}$. When we allow for such dependence, the situation is more involved[6] and several subgroups H_0, H_0', ... of globally implementable transformations emerge.

It may be shown that a transformation s of H which does not commute with $h(\phi)$ will as a rule map a state of finite energy into one of infinite energy. For instance the magnetic field strength on transformation by s acquires a singularity which by a gauge transformation can be positioned along the negative z-axis. If we do so, the singular term is proportional to

$$\frac{1}{r^2} \theta(-z)\delta(x)\delta(y) \qquad\qquad (15)$$

so that the term $1/(2g^2)\ TrF_{ij}F_{ij}$ in the energy density becomes infinite along the negative z-axis. Thus loosely put, there is an infinite potential barrier to generic H rotations in the presence of a nonabelian monopole.

CONCLUDING REMARKS

The contention of this lecture is that when a symmetry group G spontaneously breaks to H, and it creates nonabelian monopoles, then the concept of H as a transformation group partially breaks down in the vicinity of such monopoles, and only the transformations of certain distinguished subgroups $H_0, H_0', ...$ of H can be consistently implemented on the states.

Locally, in any region of space far from such a monopole which also does not enclose the monopole, it is possible to implement all the transformation in H. [For such a local region can always be taken to be contained in one patch, say Σ_N(times a suitable range of the radial variable r).] Thus locally a test particle in a monopole field can be classified into H-multiplets. We can thus inquire whether an H-singlet composed of such test particles will emerge as an H-singlet when scattered by such a monopole. The answer seems to be that it will not. For instance, when $G = SU(5)$ and $H = SU(3)_C \times U(1)_{EM}$, the $\bar{5}$ multiplet splits under H into a $\bar{3}$ [$(d_1^c, d_2^c, d_3^c)_L$] and two singlets [e_L^-, ν_e]. The elementary monopole interacts only with d_{3L}^c and e_L^-. If we now scatter an H-singlet formed from three of these $\bar{3}$'s, the constituent d_{1L}^c and d_{2L}^c are not scattered at all while d_{3L}^c is

scattered. This scrambles the phase relations between the constituents of the singlet so that the outgoing wave function is not going to be a singlet. Since color confining forces are color singlets, they cannot bind the emergent constituents into a singlet. Thus the free existence of a colored monopole implies the existence of other colored objects.

The effect discussed in the preceding paragraph will be there if the nonabelian monopole has definite transformation properties under H, it does not really show that an H transformation cannot be implemented at all on such a monopole. But there is another observation which lends credence to such a contention and that is the following. In the presence of a nonabelian monopole, an H multiplet is likely to contain particles of different spin, it may even consist of both bosons and fermions. Thus in the model $SU(5) \to SU(3)_C \times U(1)_{EM}$, only the third member of the color triplet interacts with the elementary monopole, its angular momentum is thereby changed by the addition of an extra $1/2$ unit of angular momentum. It is thus a boson whereas d_{1L}^c and d_{2L}^c continue to behave as fermions. In other words, the color triplet has two fermions and one boson. Consequently, generic color transformations even locally do not commute with angular momentum when the latter includes the influence of the monopole. Worse still, since they mix up bosons with fermions, they are not compatible with superselection rules. Since these transformations do not involve Grassmann variables, it is also not possible to interpret them as some sort of supersymmetry transformations. Problems of this sort do not arise for the distinguished subgroup $H_0 = U(2)_C \times U(1)_{EM}$ under which the triplet splits into a doublet $[(d_1^c, d_2^c)]$ and a singlet $[d_{3L}^c]$.

It is our opinion that these remarks on the incompatibility of generic H transformations and superselection rules provide strong support to our finding that the transformations of the "unbroken" subgroup H cannot all be implemented in the presence of a nonabelian monopole.

ACKNOWLEDGEMENTS

This work has been supported by the Department of Energy under contracts DE AC02-76ERO 3533 and DE-AS05-76ERO 3992, by the Swedish National Research Council under contract F310, and by INFN, Italy. A.P.B. thanks V.P. Nair and S.M. Roy for discussions.

REFERENCES

1. Cf. P. Langacker, Phys. Reports 72C, 185 (1981).
2. Cf. P. Goddard and D.I. Olive, Rep. Prog. Phys. 41, 1357 (1978).
3. A.S. Goldhaber, Phys. Rev. Lett. 36, 1122 (1976); J.L. Friedman and R.D. Sorkin, Comm. Math. Phys. (in press); R.D. Sorkin, Phys. Rev. D 27, 1787 (1983).
4. E. Witten, Phys. Lett. 86B, 283 (1979).
5. V. Rubakov, JETP Lett. 33, 644 (1981); Nucl. Phys. B 203, 311 (1982); C.G. Callan, Phys. Rev. D 26, 2058 (1982); F. Wilczek, Phys. Rev. Lett. 48, 1146 (1982); F. Bais, J. Ellis, D.V. Nanopoulos and K.A. Olive, CERN preprint (1982); V.P. Nair, Syracuse University preprints (1983).
6. A.P. Balachandran, G. Marmo, N. Mukunda, J.S. Nilson, E.C.G. Sudarshan and F. Zaccaria, Phys. Rev. Lett. 50, 1553 (1983); "Nonabelian Monopoles Break Color, I. Classical Mechanics", Syracuse preprint 4217-254 (1983), "Nonabelian Monopoles Break Color, II. Field Theory and Quantum Mechanics", Syracuse preprint 4217-268 (1983); A.P. Balachandran, "Color Breaking by Nonabelian Monopoles" (lectures delivered at the December 1982 Winter Institute at Bangalore), Syracuse preprint 4217-251 (1983).
7. P. Nelson and A. Manohar, Phys. Rev. Lett 50, 943 (1983). See also A. Abouelsaood, Harvard preprint HUTP-82 A058 (82) and Phys. Lett. 125B, 467 (1983).
8. F. Lizzi, "Confinement of Nonabelian Monopoles by MIT Bag", Syracuse University preprint 4217-255 (1983).

THE SCALES AND UPPER BOUNDS OF CLASSICAL MONOPOLE SOLUTIONS

Cosmas K. Zachos

High Energy Physics Division
Argonne National Laboratory
Argonne, Illinois 60439

I would like to report on some observations made together with T. Kirkman and A.N. Schellekens, on the mathematical structure of the classical monopole solutions. These observations may guide our intuition on the structure of the relevant physical objects. For lack of time, I will concentrate on results: the technical underpinnings of our formal argument are to be found in Refs. 1 and 2.

The equations of the Grand Unified Theories describing monopoles involve several mass scales which typify the various gauge bosons and Higgs scalars of these systems.[3,4,5] However, the monopole solutions to these equations depend <u>extremely weakly</u> on the masses of the Higgs particles and the electroweak breaking scale, regardless of their size. We may highlight this by finding the finite limits which bound monopole masses from above. What makes these upper bounds especially interesting is that they are quite close to the lower bounds (the corresponding Prasad-Sommerfield/Bogomolnyi[6] limits)--they are roughly twice these bounds.

To obtain the upper limits, we increase the concavity of the Higgs potential indefinitely in all group directions, maintaining the Higgs field vacuum values fixed. This essentially[7] freezes the scalar field variables to their vacuum values. In consequence, <u>all</u> effects of the scalar potential drop out, and the resulting drastic reduction of variables permits a straightforward numerical solution of the variational problem.[1,2]

The above limit of frozen Higgs variables is an upper bound, for the following reason: when the masses of the gauge bosons (equivalently, the v.e.v.'s of the scalars) are kept fixed, the mass of a monopole is a nondecreasing function of the Higgs masses (equivalently, the Higgs self-couplings).[1]   In the limit of infinite Higgs masses, all scalar variables freeze and their potential vanishes.   There is also an alternative way to rationalize this upper bound: For _any_ Higgs self-couplings, constraining scalar variables in the variational problem raises the energy and decouples the potential--naturally, if left unconstrained, these functions will vary so as to lower this energy.

The resulting upper bound for the SO(3) monopole ('t Hooft-Polyakov) is 1.7867 m, where m is the Prasad-Sommerfield/Bogomolnyi lower bound $m = M_w/\alpha$.   For the spherically symmetric[3,4] SU(5) monopoles we find[2] the following upper bounds:

1.7867 m   for the _single_ strength pole

2.0741 × 2m for the _double_ strength pole

2.3155 × 3m for the _triple_ strength pole

where m is the PSB lower bound $m = 3M_x/8\alpha$; the limit for the single SU(5) pole reduces to the 't Hooft-Polyakov problem, which leads to the identical numerical bound.   These numbers are purely empirical, and we have no fundamental understanding of them. Note that in this limit, the higher strength monopoles in this family are above threshold for decay into single strength monopoles.

Just as in the case of the lower limit (PSB), in this (upper) limit the specific form of the Higgs potential is irrelevant, as long as it constrains the vacuum to a unique absolute minimum. Several variables which couple to the rest of the system only through the potential drop out as a consequence.   In particular, in the SU(5) monopoles examined, the variable for the Higgs 5 completely decouples and there is no trace of the associated electroweak breaking scale M_w left in the problem.   We should not therefore expect a significant dependence on the electroweak scale for finite Higgs masses, regardless of the magnitude of this scale: it would turn out to be unimportant even if it happened to be enormous ($\approx$ GUT scale)!   The above argument then provides formal support for the conventional statement that the only essential scale for the GUT monopoles is set by just M_x, and it further justifies the use of the PSB limit in astrophysical estimates.

Eckert et al.[8] have recently provided some illustrations to the points just made. They solve for the single SU(5) monopole in a wide range of different finite Higgs masses and couplings. They find that the upper bound can in fact be approached close to saturation.   In addition, they verify the weak dependence of the

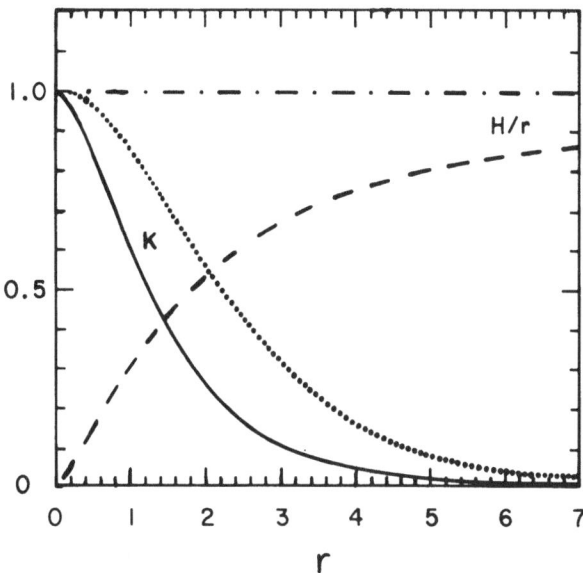

Fig. 1   The scalar and vector field radial functions for the 't
          Hooft-Polyakov monopole (Ref. 1) in the upper and lower
          (PSB) limits. r is measured in units of $1/M_w$. The vector
          function K in the lower limit is dotted line, and it moves
          monotonically with the Higgs mass to the solid line in the
          upper limit.   The scalar function H/r in the lower limit
          is the dashed line, and it moves to the dash-dot step
          function in the upper limit. The corresponding curves for
          all values of the Higgs mass lie inside these envelopes.

monopole mass and structure on $\sigma \equiv$ electroweak scale/GUT scale
($\approx 5M_w/\sqrt{2}M_x$):   as σ increases from 0 to 1, the monopole mass in
general decreases less than 5%.   Finally, they also observe that
all radial variables turn out to be monotonic in r throughout
their entire investigation.

      This technique is easy to generalize. For instance, for this
family of SU(N) spherically symmetric monopoles, the strength ℓ
pole relies on a spin $\ell/2$ representation of the crucial SU(2)
subgroup of SU(N) which controls the topology.   The corresponding
limit can be taken by freezing out all scalar functions and
solving for only ℓ vector variables numerically.

      Of course, the infinite coupling limit we have been taking is
not meant to be physical. However, it brackets any classical case
of interest conveniently, and likewise those situations which
involve only irrelevant quantum corrections to the effective
action (e.g. modifications of the Higgs potential which do not
affect the symmetry breaking pattern).

ACKNOWLEDGEMENTS

Work performed under the auspices of the United States Department
of Energy.

REFERENCES

1. T. Kirkman and C. Zachos, Phys. Rev. $\underline{D24}$, 999 (1981).
2. A.N. Schellekens and C. Zachos, Phys. Rev. Lett. $\underline{50}$, 1242
    (1983).
3. D. Wilkinson and A. Goldhaber, Phys. Rev. $\underline{D16}$, 1221 (1977); D.
    Wilkinson, Nucl. Phys. $\underline{B125}$, 423 (1977).
4. D.M. Scott, Nucl. Phys. $\underline{B125}$, 423 (1977); C. Dokos and T.
    Tomaras, Phys. Rev. $\underline{D21}$, 2940 (1980).
5. For a review , see S. Coleman's 1983 Erice Lectures, in The
    Unity of the Fundamental Interactions, A. Zichichi ed., Plenum
    Press (N.Y.), 1983.
6. M. Prasad and C. Sommerfield, Phys. Rev. Lett. $\underline{35}$, 760 (1975);
    E. Bogomolnyi, Sov. J. Nucl. Phys. $\underline{24}$, 449 (1976); S. Coleman
    et al., Phys. Rev. $\underline{D15}$, 544 (1977).
7. Actually, there is a slight technical complication involved
    here: certain Higgs radial variables must vanish at the origin
    for the monopole Ansätze to be meaningful, as dictated by the
    topology of the solutions. Nevertheless, simple asymptotics
    imply that these variables in fact drop to zero from their
    finite vacuum value in a region of radius $O(1/M_{Higgs})$; this
    lowers the energy by $O(1/M_{Higgs})$. Hence both the size of the
    transition region and the decrement of the energy vanish in the
    special limit of infinite Higgs masses.
8. P. Eckert et al., Nucl. Phys. B, in press; unpublished report
    UGVA-DPT 1983/07-402; unpublished report Ref.TH.3661-CERN, July
    1983.

KALUZA-KLEIN THEORIES AND THE DIRAC MONOPOLE

Malcolm J. Perry

Department of Physics
Princeton University
Princeton, New Jersey 08544

Conventional grand unified theories unify the strong interactions with electroweak interactions by postulating that at
energy scales of $\sim 10^{15}$ GeV the two interactions become different
facets of the same. In a typical scheme of this type, a gauge
group G becomes spontaneously broken down to H = SU(3)×SU(2)×U(1)
at energies less than the grand unification scale. G is often
supposed to be SU(5), although it can in principle be any compact
gauge group which contains H. At these very high energies, it
seems to be rather difficult to imagine any direct tests of such
theories. Nevertheless, there are at least two indirect tests of
these ideas. The first is that they predict the existence of
baryon-number violating interactions.[1] These would give rise to
an observable decay of the proton on timescales of the order of
$10^{31\pm2}$ years. Such decays have not been observed. One should,
however, point out that the observability of large amounts of
antimatter in the universe is superficially incompatible with the
standard big-bang model (or inflationary models) of the universe
unless such baryon-number non-conserving reactions took place in
the early universe.[2] These would guarantee that only matter would
be present in the universe at its current epoch. A second test is
that they predict the existence of magnetic monopoles with a mass
of $\sim 10^{16}$ GeV, provided at least that G is not of the form $\tilde{G}$×U(1).[3]
Magnetic monopoles have not been conclusively observed. However,
there seems no compelling reason to suppose that they will not be
seen once detectors become sufficiently sensitive to see a flux
lower than that of the Parker bound.[4]

This type of standard scheme seems too good to be true.
Indeed, it is hard to reconcile it with various cosmological
problems.[5] Perhaps the time is ripe to ask whether it is possible

29

to invent different realistic unification schemes. The unification scale of 10^{15} GeV is very close to the Planck scale of 10^{19} GeV, where quantum gravitational effects cannot be ignored. Perhaps one cannot really have a grand unified theory unless gravitation is also taken into account. One way to do this is via supergravity, however, it seems hard to build realistic theories in this fashion.[6] Another route is via the Kaluza-Klein approach, in which spacetime is supposed to have a dimensionality higher than four. In fact, supergravity theories seem most naturally formulated in dimensions greater than four, so maybe the evidence all points toward investigations of theories in higher spacetime dimensions. Supergravity, however, cannot be formulated in dimensions greater than eleven. Even supergravity does not cure all our problems. It faces the same problems that has beset all theories which contain gravity (at least Einstein gravity and not fourth-order theories which have their own problems). This is the problem of renormalizability; all these theories are, by power-counting, unrenormalizable. In contrast, superstring theory (which is not a quantum field theory) is finite,[7] can only be formulated in ten spacetime dimensions, and in some limit reduces to an N = 2 supergravity theory in ten dimensions. Some feel that this is the best candidate for a solution to the problems of unification. Anyway, it would seem that the idea of spacetime having more than four dimensions is not to be immediately dismissed. Instead, we should ask what the experimental or observational consequences could be of spacetime dimensions greater than four.

Spacetime appears to be four-dimensional on distance scales down to those which can be probed at the highest energy scales available today, which is around $\sim 10^{-16}$ cm. In all Kaluza-Klein models with d spacetime dimensions, it is assumed that spacetime is locally the product of four-dimensional Minkowski space M with a compact space Σ.[8] Σ has (d-4)-dimensions and linear dimensions similar to the Planck length scale of 10^{-33} cm. Distances of this scale cannot be probed directly at reasonable energies. However, they do have certain very interesting indirect consequences. Let us suppose that gravitation is described by the Einstein-Hilbert Lagrangian in d-dimensions, together with possible supersymmetric extensions. Thus, the action is

$$I = \frac{1}{16\pi\Gamma} \int R(-g)^{1/2} \, d^d x + \text{supersymmetric terms} \tag{1}$$

R is the Ricci scalar of the d-dimensional metric tensor g_{ab}, and is invariant under the group of general co-ordinate transformation in d-dimensions. g can be decomposed into pieces where both indices lie in Σ, both in M, or one in each.

$$
g_{ab} = \begin{bmatrix} \overset{M}{g_{\alpha\beta}} & \overset{\Sigma}{g_{i\beta}} \\ g_{\alpha j} & g_{ij} \end{bmatrix} \begin{matrix} M \\ \Sigma \end{matrix} \tag{2}
$$

$g_{\alpha\beta}$ is interpreted as the metric tensor representing the four-dimensional gravitational field. g_{ij} and $g_{i\beta}$ are four-dimensional scalars and vectors respectively. If Σ has an isometry group I, these fields can be interpreted as the gauge fields with gauge group I. To be precise, I is generated by a set of n Killing vectors on Σ, $K^{(p)}_i$. These Killing vectors have commutation rules

$$
[K^{(p)}, K^{(q)}] = c^{pqr} K^{(r)} \tag{3}
$$

where c^{pqr} are the structure constants of I. Notice that if I acts effectively on Σ, as will be assumed throughout, then $d < n < \frac{1}{2} d(d+1)$. The set of n gauge vectors is given by

$$
A^{(p)}_\alpha = K^{(p)i} g_{i\alpha} \tag{4}
$$

and the set of $\frac{1}{2} n(n+1)$ scalars by

$$
\phi^{(pq)} = K^{(p)i} K^{(q)j} g_{ij} \quad . \tag{5}
$$

These vector and scalar fields have gauge transformation properties, rather like in Yang-Mills theories, but which arise from the possibility of general co-ordinate transformations in M. General co-ordinate transformations in Σ simply perform global transformations of I into itself. I will thus manifest itself indirectly as a low energy (at least relative to the Planck energy) gauge group. By a judicious choice of Σ one can make $\tilde{I}$ be more or less whatever one wishes. Many examples are provided by homogeneous spaces listed below.

Σ	Gauge group
S^n	$SO(n+1)$
CP^n	$SU(n+1)$
QP^n	$Sp(n+1)$

If one wishes to find a space which has an isometry group of the form $\tilde{I} \times U(1)$, then Σ can be chosen to be of the form $\tilde{\Sigma} \times S^1$. To obtain an isometry group of $SU(3) \times SU(2) \times U(1)$, Σ can be one of the seven-dimensional spaces described by Witten,[6] the simplest of which is topologically $S^5 \times S^3 / S^1$.

In Kaluza-Klein theories, like grand unified theories, soliton-type excitations are classified by $\pi^1(I)$.[3] Thus, if $G = \tilde{G} \times U(1)$, there will be excitations classified by the integers. In this case $U(1)$ describes electromagnetic interactions, and these solitons are magnetic monopoles. We will now, for simplicity, describe the simplest possible Kaluza-Klein theory. This is a five-dimensional theory where Σ is S^1. In fact, this is precisely the model proposed by Kaluza in 1922 as a way of unifying the only interactions known at that time, electromagnetism and gravitation. The S^1, or Kaluza-Klein circle, is generated by a Killing vector $\partial/\partial\psi$. The co-ordinate ψ lies on a circle, and thus is identified with period $2\pi R$ where $2\pi R$ is the radius of the Kaluza-Klein circle. A convenient form for the metric tensor is then given by the line element

$$ds^2 = V^2(d\psi + A_\alpha dx^\alpha)^2 + g_{\alpha\beta}dx^\alpha dx^\beta \tag{6}$$

$x^\alpha (\alpha = 1,2,3,4)$ are a set of spacetime co-ordinates, $g_{\alpha\beta}$ is the metric tensor of spacetime. A_α plays the role of the electromagnetic vector potential. V is a scalar field. The Kaluza-Klein vacuum is described by

$$V = 1, \quad A_\alpha = 0, \quad g_{\alpha\beta} = n_{\alpha\beta} \quad , \tag{7}$$

where $n_{\alpha\beta}$ is the metric of Minkowski space.

To see the physical content of this theory, we can compute the action as specified in Eq. (1). In this case, no auxiliary fields are needed to produce a consistent picture. Since $\partial/\partial\psi$ is generated by a Killing vector, then we can perform the ψ-integral in Eq. (1) and obtain

$$I \sim \int g^{1/2} \, d^4x \left[VR + \frac{1}{4} V^3 F_{\alpha\beta}F^{\alpha\beta} - 2 \,\Box V \right] \tag{8}$$

where $\hat{R}$ is the Ricci scalar of the metric $g_{\alpha\beta}$, all tensorial operations are with respect to the metric $g_{\alpha\beta}$, and $F_{\alpha\beta}$ is defined by

$$F_{\alpha\beta} = \nabla_\alpha A_\beta - \nabla_\beta A_\alpha. \tag{9}$$

Thus the field content of this theory is a massless spin-2 graviton, a massless spin-1 photon, and Brans-Dicke type scalar field. The equations of motion can be easily found and they are

$$R_{\alpha\beta} - \frac{1}{2}Rg_{\alpha\beta} = \frac{V2}{2} \left[F_{\alpha\gamma}F_\beta^{\ \gamma} - \frac{1}{4}g_{\alpha\beta}F_{\gamma\delta}F^{\gamma\delta} \right] - \frac{\nabla_\alpha\nabla_\beta V}{V} + \frac{g_{\alpha\beta}\Box V}{V} \tag{10a}$$

$$\nabla_\alpha(V^3 F_\beta^{\ \alpha}) = 0. \tag{10b}$$

These are the analogs of the Einstein and Maxwell equations respectively. The equation of motion for V can be found by taking the divergence of Eq. (10a). Note, however, that the normalization of $F_{\alpha\beta}$, the Maxwell tensor, and hence A_α, differs from the conventional one by a factor of $(16\pi)^{1/2}$.

The issue of the existence of solitons in this picture was first examined by Einstein and Pauli.[9] They erroneously concluded that no solitons could exist. In the context of a Kaluza-Klein theory a soliton means a stable, localized lump in the fields, which is not singular. The magnetic monopole in the Kaluza-Klein theory[10,11] corresponds to

$$g_{\alpha\beta}dx^\alpha dx^\beta = \frac{1}{V^2}(dr^2 + r^2 d\theta^2 + r^2 \sin^2\theta d\phi^2) - dt^2 \tag{11}$$

$$A_\alpha dx^\alpha = 8m\sin^2\frac{1}{2}\theta\,d\phi \tag{12}$$

$$V = (1+4m/r)^{-1} \tag{13}$$

r, θ, ϕ are spherical polar co-ordinates on R^3, t is a time coordinate, and m is an arbitrary constant of integration.

This configuration looks singular at r = 0, and along the axis $\theta = 0$. However, these are coordinate singularities provided ψ is identified with period $16\pi m$. Thus, there is a relation between the radius of the Kaluza-Klein circle and m, namely

$$m = \frac{R}{8}\,. \tag{14}$$

In addition, the vector field A_α does not vanish, and its strength is controlled by m. We can compute the electric and magnetic fields associated with A_α. The conventionally normalized electric field

$$E_i = F_{0i}/\sqrt{16\pi} \tag{15}$$

(i = 1,2,3) vanishes. The magnetic field is given by

$$B_i = \frac{1}{2}\varepsilon_{ijk}F^{jk}/\sqrt{16\pi} \tag{16}$$

where ε_{ijk} is the three-dimensional alternating symbol. B is a vector which points in the radial direction only, and is given by

$$B_r = \frac{4m}{r^2}\frac{1}{\sqrt{16\pi}} \tag{17}$$

Thus, this object behaves like a magnetic monopole of strength $P = m/\sqrt{\pi}$.

The radius of the Kaluza-Klein circle, R, can be interpreted as determining the fundamental unit electric charge, e. To see this, we recall that a scalar field Φ is single-valued. Since ψ is a periodic variable with period $2\pi R$, Φ must be given by functions of the form

$$\Phi \sim \exp\left(\frac{in\psi}{R}\right) \times \hat{\Phi} \qquad (18)$$

where n is some integer, and $\hat{\Phi}$ is independent of ψ. If we examine the kinetic part of the action for such a field it must be given

$$\int \frac{1}{2} (\partial_a \Phi)^*(\partial_b \Phi) g^{ab}(-g)^{1/2} d^5x \qquad (19)$$

Making the 4+1-dimensional decomposition of Eq. (6) together with Eq. (18), we find that the action is, up to an overall constant

$$\int \left[\frac{1}{2} \left(\frac{1}{v^2} + A_\gamma A_\delta g^{\gamma\delta} \right) \frac{n^2}{R^2} \hat{\Phi}^2 - \frac{in}{R} A^\alpha \partial_\alpha \hat{\Phi} \cdot \hat{\Phi} \right.$$
$$\left. + \frac{1}{2} (\partial_\alpha \hat{\Phi})(\partial_\beta \hat{\Phi}) g^{\alpha\beta} \right] V (-g)^{1/2} d^4x \qquad (20)$$

By inspection of Eq. (20), we can find the value of the electric charge of the field of Eq. (18) by looking at the coefficient of the $(A^\alpha \partial_\alpha \hat{\Phi})\hat{\Phi}$ term. It is in/R. Hence the fundamental unit of electric charge e is given by

$$e = \frac{\sqrt{16\pi}}{R} = \sqrt{\pi/4} \cdot m^{-1} \qquad (21)$$

Given that $e^2 \cong 4\pi/137$, we can find R and m. They are, in conventional units

$$R = 3.7 \times 10^{-32} \text{ cm}$$

$$m = 6.4 \times 10^{-5} \text{ gm} = 3.6 \times 10^{19} \text{ GeV} \qquad (22)$$

The magnetic charge P can also be related to these fundamental quantities,

$$P = \frac{m}{\sqrt{\pi}} = \frac{1}{2e} \quad . \qquad (23)$$

Thus, the magnetic charge obeys the Dirac condition, it has one Dirac unit.

Earlier m was said to the mass of the object. Certain qualifications need to be made concerning this statement. In theories of gravitation, there are two distinct types of mass: the gravitational mass, and the inertial mass. Normally, it is assumed that these are identical. This was first observed by Galileo, and subsequently verified by Dicke and collaborators[12] to an accuracy of $\sim 10^{-12}$. The inertial mass of a system is determined by the behavior of the spatial components of the metric tensor for large r. The gravitational mass is determined by the time components of the metric. In ordinary four-dimensional relativity these are the same. In Kaluza-Klein theories, they are not necessarily the same. The inertial mass of the monopole is m. The gravitational mass vanishes.[11] It should be pointed out however that this does not violate any experimentally established principle since no one has ever observed a magnetic monopole.

There are also multi-magnetic monopole solutions in this theory. They are given by

$$g_{\alpha\beta}dx^{\alpha}dx^{\beta} = \frac{1}{V^2} (dx^2 + dy^2 + dz^2) - dt^2 \qquad (24)$$

$$V^2 = \left(1 + 4m \sum_{i=1}^{N} \frac{1}{|\vec{x}-\vec{x}_i|}\right) \qquad (25)$$

$$\text{grad } V^2 = \text{curl } A \quad . \qquad (26)$$

Here $\vec{x} = (x,y,z)$ and the operations grad and curl are the familiar flat space differential operators. $\vec{x}_i$ is the spatial location of the ith monopole. $|\vec{x}-\vec{x}_i|$ means the Euclidian distance between the points $\vec{x}$ and $\vec{x}_i$, i.e.

$$|\vec{x}-\vec{x}_i| = \left[(x-x_i)^2 + (y-y_i)^2 - (z-z_i)^2\right]^{1/2} \qquad (27)$$

The vectors $\vec{x}_i$ are completely arbitrary.

Each of the N magnetic monopoles has the same mass and charge. Exactly as before, if ψ is identified with period $16\pi m$, the system has no singularities in it. If the sign is changed in Eq. (26) so that

$$\text{grad } V^2 = -\text{curl } A \qquad (28)$$

the solution represents N antimonopoles. It should be noticed that the vectors $\vec{x}_i$ are completely arbitrary. This is because the monopoles do not interact with each other. The electromagnetic forces between them are cancelled by the other interactions in the theory.

It seems highly probable that most of these properties will be carried over into any realistic theory. Simply on topological grounds can one guarantee that this overall picture must be correct although precise details may not be the same.

One can also perform similar constructions in non-abelian versions of the theory.[13] Suppose that one starts out with a six-dimensional theory. Let us take two dimensions to be wrapped up to form a two-sphere. In this case, one would have a theory with an SO(3) gauge group. In this theory, there is a magnetic monopole which carries a Z_2 quantum number. Z_2 is the group realized by the elements +1, -1 under multiplication. Thus a superposition of two monopoles has the quantum numbers of the vacuum. We can illustrate this explicitly. To arrange for the compactification down to S^2, we introduce an auxiliary abelian antisymmetric gauge field F_{abcd}, and a cosmological constant Λ. Thus the action is

$$I = \int g^{1/2} \, d^6x \left[\frac{1}{16\pi} (R-2\Lambda) + \frac{1}{8} F_{abcd} F^{abcd} \right] \tag{29}$$

The metric of the vacuum is given by

$$ds^2 = -dt^2 + dr^2 r^2 d\theta^2 + r^2\sin^2\theta d\phi^2 + \frac{1}{2\Lambda} \left[d\chi^2 + \sin^2\chi d\psi^2 \right] . \tag{30}$$

This is the direct product of Minkowski space with a sphere of radius $(2\Lambda)^{-1/2}$. The auxiliary field F_{abcd} is given by

$$F_{tr\theta\phi} = \sqrt{\Lambda/24\pi} \ r^2\sin\theta \quad . \tag{31}$$

The Z_2-monopole solution is given by

$$ds^2 = -V^2 dt^2 + \frac{1}{V^2} \, dr^2 + r^2(d\theta^2 + \sin^2\theta d\phi^2)$$

$$+ \frac{1}{2\Lambda} \left[d\psi^2 + \sin^2\psi \left(d\chi + \frac{1}{2} \cos\theta d\phi - \frac{1}{2r} \, dt \right)^2 \right] \tag{32}$$

where

$$V^2 = 1 - \frac{1}{2\sqrt{\Lambda}r} + \frac{1}{16\Lambda r^2} \quad . \tag{33}$$

The non-compact part of the spacetime is geometrically equivalent to the extreme Reissner-Nordstrom solution. It thus looks like a black hole with zero temperature. One presumes that this object is stable. It can also be verified that the spacetime contains no singularities.

The field F_{abcd} required to find this is rather complicated, and will not be discussed further. Many features of this solution are similar to what happened in the previous case. Once the size of the Kaluza-Klein sphere has been fixed by the cosmological constant, the mass of this object is also fixed, and is equal to

$$M = \frac{1}{4\sqrt{\Lambda}} \tag{34}$$

In this case, however, the gravitational and inertial masses are the same. Also, the gauge fields are fixed. In the gauge we have written the solution in, the gauge field points in a single direction in isospin space. This is usually labelled by A^3, and corresponds to fields projected out by one of the three Killing vectors on the sphere $\partial/\partial\chi$. The corresponding electric and magnetic fields are entirely radial and given by

$$E^3 = \frac{1}{4\Lambda r^2}$$

$$B^3 = \frac{1}{4\Lambda r^2} \tag{35}$$

Thus this solution is really a dyon with equal electric charge and magnetic charge. Further details of this solution can be found in Ref. 13.

In summary, Kaluza-Klein theories provide a viable alternative to conventional unification schemes. Furthermore, if the electromagnetic interactions arise this way, then there will be magnetic monopoles with masses around the Planck mass. Perhaps one day such objects will be discovered which would then give us some kind of observable tests of current ideas about unification and quantum gravity.

ACKNOWLEDGEMENTS

Research supported by the National Science Foundation PHY80-19754 and the Alfred P. Sloan Foundation.

REFERENCES

1. J. Ellis, M.K. Gaillard, D.V. Nanopoulos and S. Rudaz, Nucl. Phys. B 176, 61 (1980).
2. E. Kolb and S. Wolfram, Nucl. Phys. B 172, 224 (1980).
3. S. Coleman, "The Magnetic Monopole Fifty Years Later," Harvard preprint (1982).
4. E.N. Parker, Ap. J. 160, 383 (1970).
5. J. Preskill, Phys. Rev. Lett. 43, 1365 (1979); A.H. Guth, Phys. Rev. D 23, 347 (1981); A. Albrecht and P. Steinhardt, Phys. Rev. Lett 48, 1220 (1982).
6. E. Witten, Nucl. Phys. B 186, 412 (1981) and Princeton preprint 1983.
7. J.H. Schwarz, Phys. Rep. 89, 223 (1982).
8. T. Kaluza, Sitzungber Preuss Akad. Wiss. Phys. Math. Kl, 996 (1921); O. Klein, Z. Phys. 37, 895 (1982); A. Einstein and P. Bergman, Ann. Math. 39, 683 (1938); P. Jordan, Ann. Phys. (Leipzig) 1, 219 (1947); Y. Thiry, Comptes Rendus Acad. Sci. (Paris), 226, 216 (1948); 226, 881, (1948); A. Lichnérowicz, Théories Relativistes de la gravitation et de l'electro-magnetisme, A. Masson et Cie, Paris (1955); O. Klein, New Theories in Physics, Martinus Nijshoff, den Haag (1939); B.S. deWitt, Dynamical theory of groups and fields, Gordon and Breach, London (1965); R. Kerner, Ann. Inst. H. Poincaré 9, 143 (1968); A. Trautman, Rep. Math. Phys. 1, 291 (1970); Y.M. Cho and P.G.O. Freund, Phys. Rev. D 12, 1711 (1972); Y.M. Cho, J. Math. Phys. 16, 2029 (1975); E. Cremmer and J.H. Schwarz, Phys. Lett. 57B, 463 (1975).
9. A. Einstein and W. Pauli, Ann. Math. 44, 131 (1943).
10. R. Sorkin, Phys. Rev. Lett. 51, 87 (1983).
11. D.J. Gross and M.J. Perry, Nucl. Phys. B 226, 29 (1983).
12. P.G. Roll, R. Krotkov and R.H. Dicke, Ann. Phys. (NY) 26, 442 (1967).
13. M.J. Perry, Phys. Lett. B (in press).

A 5-DIMENSIONAL MONOPOLE

Rafael Sorkin

Department of Physics
Syracuse University
Syracuse, New York   13210

The Kaluza-Klein idea assimilates gauge fields to gravity by comprehending gauge-transformations as diffeomorphism ("coordinate transformations").   Although this can be done for any choice of gauge group, the original U(1)-theory has the distinction of requiring only a single fundamental field: it is nothing but 5-dimensional gravity with a specific choice of metric and topology to represent the vacuum.   Since you have already heard this theory described, I will sketch it only in enough detail to fix notation.

The 5-dimensional takes the form

$$S = S_0 + S_\infty$$

$$= \int {}^5L \, d^5v + \oint_\infty , \tag{1}$$

$${}^5L = (1/2k) \, {}^5R \tag{2}$$

where ${}^5R$ is the 5-dimensional scalar curvature and d^5v the 5-dimensional element of spacetime volume.   The integral at spatial infinity would be needed only if we wanted to compute our monopole's mass from first principles; and I will henceforth ignore it.

By assumption the "ground state" has a U(1)-symmetry, i.e., a Killing vector, K^A, which we can take to have period 2π:

$$\underset{K}{£} \, g_{AB} \equiv \nabla_A K_B + \nabla_B K_A = 0, \tag{3}$$

$$\exp (2\pi K) = 1 \quad .$$

I am using capital Roman letters A,B,... for 5-dimensional tensor
indices which can assume the values 0,1,2,3,5. In such coordinates the Lie derivative $\pounds_K$ will reduce to differentiation with
respect to x^5, and Eq. (3) will express that g_{AB} is independent of
the fifth coordinate. The "vacuum metric" can be written

$$ds^2 = -dt^2 + dx^2 + dy^2 + dz^2 + \lambda_\infty^2 \, d\psi^2 \tag{4}$$

where $\psi = x_5$ runs periodically from 0 to 2π. Thus $\ell \equiv 2\pi \lambda_\infty$ is
the circumference of the Kaluza-Klein circle, or in other words
the extension of spacetime in its "internal," or fifth, dimension.
For the vacuum Killing vector we have, as I just said,

$$K^A = (0,0,0,0,1) = \partial/\partial\psi \tag{5}$$

Since we are interested in a one-particle solution, we must
require the metric to approach its vacuum form near spatial
infinity, but in the interior region not only the metric but even
the topology can be chosen freely. Within the broad class of
field configurations, consider those which retain an exact Killing
vector, K^A. Such U(1)-symmetrical configurations are the only
ones that can be excited at low energies, and they also include
the monopole solution which interests us. Because of its special
symmetry such a 5-dimensional metric g_{AB} is equivalent to a set of
three four-dimensional fields: a scalar, a gauge-variant "vector-
potential," and a tensor. The scalar

$$\lambda = (g_{AB} \, K^A K^B)^{1/2}$$

just gives the radius of the local Kaluza-Klein circle. The
vector-potential is

$$A_A = g_{AB} \, K^B/\lambda^2, \tag{6}$$

or in other words just K itself rescaled so that $K \cdot A = 1$. The
tensor is the projection of g_{AB} orthogonal to K:

$$\gamma_{AB} = g_{AB} - \lambda^2 A_A A_B, \tag{7}$$

and plays the role of 4-dimensional spacetime metric.

In coordinates x^A the decomposition of g_{AB} takes the form

$$g_{\mu\nu} = \gamma_{\mu\nu} + \lambda^2 A_\mu A_\nu, \quad g_{5\mu} = g_{\mu5} = \lambda^2 A\mu, \quad g_{55} = \lambda^2,$$

where $\mu = 0,1,2,3$ is an index for the 4-dimensional quotient
spacetime. Conversely, from $\gamma_{\mu\nu}$, A_μ and λ we can determine g_{AB}
because by defraction $A_5 = 1$, $\gamma_{\mu5} = 0$. A gauge-transformation is
a diffeomorphism $x^5 \to x^5 + \theta(x^\mu)$, or more invariantly $\exp(\theta K)$,

where θ is any function such that $K^A \partial_A \theta = 0$. It is easy to confirm that $\gamma_{\mu\nu}$ and λ are invariant under such a diffeomorphism, whereas $A_\mu \to A_\mu - \partial_\mu \theta$. We can also define

$$F_{AB} = \partial_A A_B - \partial_B A_A, \tag{8}$$

which is gauge-invariant and purely 4-dimensional in the sense that $F_{5A} = 0$.

In this four-dimensional language the action S_0 becomes

$$S_0 = (2\pi/k) \int (\frac{\lambda}{2} R - \frac{\lambda^3}{8} F_{\mu\nu}F^{\mu\nu} - \Box\lambda)d^4V \tag{9}$$

where now $\gamma_{\mu\nu}$ is treated as the spacetime metric in forming R, $\Box$ and d^4V and in raising and lowering indices, etc. Comparing the first term with the known gravitational action $(2\kappa)^{-1} \int R \, d^4V$ ($\kappa \equiv 8\pi G$) shows that

$$\kappa^{-1} = 2\pi\lambda_\infty/k \equiv \ell/k, \text{ or}$$
$$k = \ell\kappa. \tag{10}$$

Thus, the effective 4-dimensional gravitational constant is the quotient of the fundamental 5-dimensional constant k by the circumference of the vacuum ℓ. Similarly the comparison

$$- \frac{h^2}{16\pi^2 e^2} = - \frac{2\pi\lambda_\infty^3}{8k}$$

shows (in the presence of our normalization $K \cdot A = 1$) that the 4-dimensional unit of electric charge is given (in "Heaviside" or "rationalized" units) by

$$e/h = \sqrt{2\kappa}/\ell \tag{11}$$

In particular, ℓ will be of Planckian dimensions if e is of order one (which presumably should mean "of order one at Planck energies" in the sense of the renormalization group.)

The fact that spacetime is now five-dimensional allows us to seek stationary solutions with the very simple ("ultrastatic") form

$$ds^2 = -dt^2 + g_{ab} \, dx^a \, dx^b \tag{12}$$

where the spatial indices a,b can assume the values 1,2,3,5. For such a metric the field equation $R_{AB}=0$ reduces to the purely spatial form $R_{ab}=0$, where R_{ab} is the contracted curvature tensor of the positive-definite metric g_{ab}. In a 3+1-dimensional theory, the corresponding Ansatz would be useless since the only 3-dimensional solutions of $R_{ab}=0$ are flat; but of course that is far from true in 4-dimensions, since ordinary relativity is not a trivial theory.

The simplest solution describing a monopole can be given as

$$g_{ab} \ dx^a \ dx^b = \left(1 + \frac{\lambda_\infty}{4\rho}\right)^2 \ dr^2 + \left(\frac{r\lambda_\infty}{\rho+\lambda_\infty/4}\right)^2 \left(\frac{d\sigma_3}{2}\right)^2 + r^2 d\Omega^2 \qquad (13)$$

where $\rho^2 = r^2 + \lambda_\infty^2/16$, $-d\sigma_3$ denotes

$$2d\psi + \cos\theta \ d\phi, \qquad\qquad\qquad\qquad\qquad\qquad (14)$$

and as usual $d\Omega^2$ is short for $d\theta^2 + \sin^2\theta \ d\phi^2$. The parameters r, θ, ϕ, ψ coordinatize the manifold R^4 with $r \in [0,\infty]$ being the radial coordinate and $(\theta,\phi,2\psi)$ being Euler coordinates for S^3. Thus $\theta \in [0,\pi]$, $\phi,\psi \in [0,2\pi]$, and one makes the identifications

$$(\theta,\phi,\psi) = (\theta,\phi + \Delta, \ \psi \pm \Delta/2) \qquad\qquad\qquad (15)$$

at $\theta = \pm\pi$. The metric Eq. (13), [one of the "NUT" family] can alternately be written in terms of Cartesian coordinates w,x,y,z for R^4 by setting

$$re^{1/2\phi} \ e^{1/2\theta j} \ e^{1/2\psi k} = w + ix + jy + kz, \qquad\qquad (16)$$

where i,j,k are the unit quaternions; and then $d\sigma_3$ is just the k-part of $-2r^{-2} \ q* \ dq$. Using these coordinates one can verify that g_{ab} is regular at r = 0, which is not obvious from Eq. (13). In fact, in w,x,y,z coordinates Eq. (13) reduces near r=0 to

$$ds^2 \approx -dt^2 + 4(dx^2 + dy^2 + dz^2 + dw^2)$$

which is manifestly free from singularity.

Thus Eq. (13) with Eq. (12) defines a 5-dimensional vacuum solution with topology R^5, and the only question remaining is whether it admits interpretation in the Kaluza-Klein fashion. To answer we may note first of all that in t,r,θ,ϕ,ψ coordinates the metric is independent of ψ, which means that $K = \partial/\partial\psi$ is again a Killing vector. Moreover, as $r \to \infty$, $\rho \to r$ and our solution reduces (modulo terms of order $1/r$, such as dr^2/r and $d\phi d\psi$) to

$$-dt^2 + dr^2 + \lambda_\infty^2 \ d\psi^2 + r^2 \ d\Omega^2,$$

which indeed is Eq. (4), only in spherical coordinates. A further glance at Eqs. (13) and (14) now allows us to identify $\lambda = \lambda(r)$ as

$$\lambda = \frac{r\lambda_\infty}{\rho + \lambda_\infty/4} \qquad . \tag{17}$$

Similarly we can recognize from Eq. (7) that

$$A_A = 2 \, \partial_A \psi + \cos\theta \, \partial_A \phi \tag{18}$$

and that

$$g_{\mu\nu} dx^\mu dx^\nu = -dt^2 + \left(1 + \frac{\lambda_\infty}{4\rho}\right)^2 dr^2 + r^2 d\Omega^2 \qquad . \tag{19}$$

We have established that Eq. (13) + (12) defines a soliton solution to the U(1) Kaluza-Klein equations, but what sort of particle is this soliton? From the 4d standpoint there are, as we know, 3 massless fields in our theory ($g_{\mu\nu}$, $F_{\mu\nu}$, λ), and a soliton can therefore carry several kinds of charge, namely spin, mass, and electric, magnetic, and scalar charge. At this conference it is of course the magnetic charge Q_B which is most interesting. To compute it we need merely form

$$F = \text{curl } A = -\sin\theta \, \partial\theta \times \partial\phi,$$

whose integral over any 2-sphere linking the world line r=0 is -2π. In our units the magnetic charge is therefore $-h/e$, which is the minimum possible or "Dirac" value. (In Gaussian units one has $eQ_B = h/4\pi$). Observe that the 1-form A_A is completely regular for $r \neq 0$, even though the "μ-component" of Eq. (18) has apparent string singularities arising from the singularity of the θ, ϕ, ψ coordinate system at $\theta = 0, \pi$. In the same way F blows up at r=0, even though g_{AB} does not, the reason being that the decomposition in terms of which A_A is defined ceases to exist when $\lambda = 0$. Geometrically the Kaluza-Klein circle shrinks to a point at r=0 and the local 4+1 "product structure" of spacetime breaks down.

The other non-zero charges are the scalar charge Q_S and the mass M, which can be found from Eqs. (17) and (19). The normalization of Q_S, and especially of M, is somewhat subtle. The simplest way seems to be to rewrite the 4-action in terms of the conformally rescaled metric $\bar{g}_{\mu\nu} = (\lambda/\lambda_\infty)\gamma_{\mu\nu}$, whose effect is to remove the λ-dependence from the "gravitational" term in Eq. (9). Then the values of M and Q_S extracted from the 1/r pieces of $\bar{g}_{\mu\nu}$ and λ will enter into the corresponding inter-soliton Coulomb force laws with their usual normalizations. The final result (in rationalized units) is

$$M = \frac{\ell}{2\kappa}, \qquad |Q_B| = \frac{\ell}{\sqrt{2}\kappa}, \qquad Q_S = \sqrt{3/8} \ \frac{\ell}{\sqrt{\kappa}} \qquad .$$

In particular both M and Q_S are of Planckian scale, as one would anticipate on dimensional grounds.

In the few minutes remaining I would like to offer a few comments on this monopole solution, and also to pose a couple of questions. The first thing to be stressed is that the solution is completely regular; it has neither event horizon (being ultra-static) nor any type of singularity. Even its topology is unremarkable, being just that of flat spacetime, R^5. However, this regularity is obtained only in five dimensions, since, as we have seen, the equivalent 4-dimensional potential and freed, $F_{\mu\nu}$, are both singular. For such a metric "dimensional reduction" is untenable and the reality of the fifth dimension cannot be escaped.

The fact that our monopole is thus a "counterexample to dimensional reduction" should be no surprise if you recall that Charlie Misner once described the Taub-NUT metric as a "counter-example to almost anything." In its Kaluza-Klein incarnation it continues to live up to that description. For a second example, consider the bound--valid for coupled gravity and electromagnetism in 4 dimensions--that $Q^2 \leqslant \kappa M^2/2$ where $Q^2 = Q_E^2 + Q_B^2$. For the monopole, we have $Q^2 = Q_B^2 = (1/2)(\ell^2/\kappa)$ but $\kappa M^2/2 = (1/8)(\ell^2/\kappa)$. This violation is bound up with the existence of the extra scalar field λ, and also with the fact that 2 identical monopoles will exert no long range force on each other:

$$\frac{M^2\kappa}{8\pi r^2} + \frac{Q_S^2}{4\pi r^2} - \frac{Q_B^2}{4\pi r^2} = \frac{\ell^2}{r^2\kappa}\left(\frac{1}{8} + \frac{3}{8} - \frac{1}{2}\right) = 0 \quad .$$

Another theorem which might seem to rule out a Kaluza-Klein soliton is the Gannon singularity theorem[1] which infers a singularity from the existence of a non-simply connected hypersurface at t=0 (technically a Cauchy hypersurface). The metric, Eqs. (12)-(13), evades this theorem in several ways, one of which is that the t=0 metric, Eq. (13), is not actually non-simply connected (being just R^4) although it appears so locally.

Notice incidentally that--in contrast to the "Grand Unified Theories"--the electromagnetic U(1) is here embedded in no larger (broken) gauge group such as SU(5). Correspondingly our monopole solution has no interior region where the U(1)-symmetry breaks down; hence it possesses no "internal U(1)-degree of freedom" such as characterizes the 't Hooft-Polyakov monopoles.

Finally, our monopole furnishes a further disproof of the idea that one cannot obtain spin-1/2 without introducing at least one elementary spinor field. Here there are no such fields, and yet (because the monopole carries charge h/e) a 2-particle state containing both the monopole and one of the charge-e particles of the theory (corresponding to excitations of the internal dimension) would necessarily possess half-integral angular momentum.

The questions I wanted to pose are of course only two among several that might be asked at this point. The first question is the natural one whether there exist analogous--stationary and non-singular--particle-like solutions for higher-dimensional Kaluza-Klein theories, which would represent monopole sources of non-abelian gauge-fields. Here the situation is complicated by the presence of fields other than just the higher dimensional metric, without which the presumed vacuum would not actually be a solution (at least classically). Nevertheless, prospects look good that such solutions exist.

The second question seems more urgent but unfortunately also harder to answer. One would of course like to know the probability to produce Kaluza-Klein monopoles under conditions such as those thought to characterize the early universe. But the creation of monopole-anti-monopole pairs entails a change in the topology of spacetime itself. To compute the amplitude for such a process is to solve a problem in 5-dimensional quantum gravity. One might try to find an instanton mediating such pair creation, but at present we neither know of one nor possess convincing rules for interpreting one if it should be found. The moral then is that before we really understand Kaluza-Klein monopoles, we will have to understand better the quantum processes by which the topology of spacetime undergoes change.

REFERENCE

1. D. Gannon, J. Math. Phys. 16, 2364 (1975).

MAGNETIC MONOPOLES IN GRAND UNIFIED AND KALUZA-KLEIN THEORIES

Qaisar Shafi

Bartol Research Foundation
University of Delaware
Newark, Delaware   19716

The apparent reluctance of the proton to decay into $e^+\pi^\circ$[1] suggests that the superheavy gauge bosons that mediate this process must be heavier than the value predicted by minimal SU(5). Some ways of curing this difficulty within the SU(5) framework have recently been discussed[2,3]. Alternatively, one may resort to gauge groups larger than SU(5). Indeed, it was pointed out some years ago[4] that in grand unified theories (GUTs) with intermediate mass scales, the decay rate for $p \rightarrow e^+\pi^\circ$ (and for other processes mediated by the superheavy gauge bosons) can be considerably lower than the value(s) given by SU(5).

Grand unified theories based on gauge groups larger than SU(5) present novel possibilities for magnetic monopole charges and their masses.[5] In the SO(10) model, for example, the lightest monopole can carry two units of the (Dirac) magnetic charge and weigh about 10^{13} GeV (or even less). In an E_6 model, the lightest monopole carries three units of magnetic charge and weighs as much as $10^{18}-10^{19}$ GeV. These monopoles need not catalyze nucleon decay with strong cross sections, and constraints on their number density are consequently much less severe than on the SU(5) monopole. The flux of the SO(10) monopole, for instance, can be as large as 10^{-16} cm^{-2} s^{-1} sr^{-1}, while the flux of the E_6 monopole may be even larger, perhaps as much as 10^{-14} cm^{-2} s^{-1} sr^{-1}. The appearance of intermediate mass scales in these theories provides a mechanism for suppressing the number density of the above monopoles to cosmologically acceptable, yet experimentally accessible levels.

The resurge of interest lately in the study of Kaluza-Klein (KK) theories stems from the belief that these theories may provide an elegant approach for unifying gravity and the gauge interactions. The original KK model was based on five dimensional Einstein's gravity, and provided unification of four dimensional gravity with electrodynamics. Attempts to extend this approach to include the strong and electroweak interactions are presently under way. Incorporation of known quarks and leptons poses problems for these theories[6] which we shall not pursue here.

Let us consider how monopoles appear in KK theories.[7] The simplest example is provided by the original 5-dimensional model of Kaluza. The ground state of this theory is taken to be $M^4 \times S^1$, the product of four dimensional Minkowski space-time with a one dimensional circle. The symmetry group of S^1 is U(1) which can be identified with the electromagnetic gauge group. Magnetic monopoles in this theory are characterized as non-trivial S^1 bundles over S^2, where the latter forms the boundary of 3-space. It turns out that there are an infinite number of such bundles, each labelled by an integer. From this one infers the existence of topologically stable magnetic monopoles.

The lightest monopole has mass on the order of M_p/q, where q is the elementary electric charge. We expect q to be 1/3 e, the charge of the d-quark. Thus, the monopole mass is $\simeq 10^{20}$ GeV, and it carries three Dirac units of magnetic charge. Its flux may be as large 10^{-13} cm^{-2} s^{-1} sr^{-1}. On account of its huge mass and consequently low speed ($\lesssim 10^{-4}$ c), such a monopole is best looked for in a Cabrera type experiment.

The original KK model, of course, is not fully realistic since it does not incorporate the strong and the weak inter-actions. The gauge group that describes the presently observed particle interactions is SU(3) × SU(2) × U(1)/Z_3 × Z_2 (here Z_3 × Z_2 denotes the product group of the centers of SU(3) and SU(2)). The KK idea can be extended to incorporate this group. Topologically stable magnetic monopoles are once again expected to appear. However, the lightest monopole with mass on the order of 10^{19}-10^{20} GeV may now carry one unit of Dirac charge, provided the underlying group of symmetry transformation is simply connected. It would then also carry color magnetic fields. Whether or not these monopoles catalyze nucleon decay with strong cross section would depend, to a large extent, on the underlying higher dimensional KK theory. If one assumes that the underlying theory is eleven dimensional supergravity, then it is likely that strong catalysis of $\Delta B \neq 0$ processes would occur in the presence of the monopoles (the underlying theory has no global symmetry that could be interpreted as baryon number). The flux of these monopoles would consequently be extremely tiny and presumably not accessible to present experiments.

ACKNOWLEDGEMENTS

This work is supported in part by the Department of Energy Contract No. DE-AC02-78ER05007.

REFERENCES

1. According to the IMB experiment $\tau_{p \to e^+\pi^\circ} > 2.3 \times 10^{32}$ yr.
2. Q. Shafi, C. Wetterich, Bartol preprint BA-83-43 (1983). Also see the discsusion by C. Hill, Fermi Lab. preprint (1983).
3. P. Frampton, S.L. Glashow, Phys. Lett. 131B, 340 (1983).
4. Q. Shafi, C. Wetterich, Phys. Lett. 85B, 52 (1979); H. Georgi, D. Nanopoulous, Nucl. Phys. B159, 16 (1979); Also see R. Robinett, J.L. Rosner, Phys. Rev., D25, 3036 (1982); Y.Tosa, G. Branco and R. Marshak, Phys. Rev. D28, 1731 (1983).
5. For further details and other relevant references, see Q. Shafi and C. Wetterich, Bartol preprint BA-83-45.
6. E. Witten, Princeton preprint (1983) and references therein.
7. See the articles by R. Sorkin and M. Perry in this volume.

BLACK HOLES IN COMPACTIFIED SUPERGRAVITY

F. Alexander Bais

Instituut voor Theoretische Fysica
Rijksuniversiteit Utrecht
The Netherlands

ABSTRACT

Some consequences of dimensional compactification for black holes in the context of $d = 11$ supergravity were discussed.[1,2,3] The new solutions exhibit novel features. Some of them have tightly dressed singularities, by which we denote a situation where the horizon coincides with a (light like) singularity. It leads to a new type of Penrose diagram from which one can see that there is only a single universe connected to the singularity. The fact that it is not possible to analytically extend the co-ordinates over the horizon implies that the Hawking temperature is undetermined. Within the set of ansätze which were studied, namely the time independent $SO_3 \times SO_8$ symmetric metrics, no soliton type solutions were found.

REFERENCES

1. P. van Baal, F.A. Bais and P. van Nieuwenhuizen, Nucl. Phys. B (to be published).
2. P. van Baal and F.A. Bais, Phys. Lett. (to be published).
3. F.A. Bais, Invited talk at the 7th John Hopkins Workshop, Bad Honnef, Germany, 1983.

MAGNETIC MONOPOLES IN PURELY FERMIONIC MODELS

M. Arik, M. Hortacsu and J. Kalayci

Physics Department
Bogazici University
Istanbul, Turkey

Magnetic monopoles[1] have received a great deal of renewed attention over the past few years. Indeed any nonabelian gauge theory[2] has a magnetic monopole solution. In its simplest form this solution is the Wu-Yang monopole[3] with a singularity at the site of the monopole. The mechanism of spontaneous symmetry breaking can be utilized[4] to smear this singularity and make these solutions have finite energy and topological stability provided that one starts with a simple gauge group and unbroken gauge symmetries including an electromagnetic U(1). On the other hand, the fermions in such a theory play no role in these features. As an alternative it would be interesting to look for magnetic monopoles constructed in terms of fermion fields.

Recently, it has been proposed[5] that nonabelian gauge bosons may not be fundamental particles but arise as composite fields in a class of purely fermionic models with fractional fermion self-interaction. Here we would like to present the monopole solutions in such a model. Our starting point is the Lagrangian density

$$\mathcal{L} = \bar{\psi} i \not{\partial} \psi + g_1 (\bar{\psi}\gamma_\mu \tau_a \psi \bar{\psi}\gamma_\mu \tau_a \psi)^{2/3}$$
$$- g_2 (\bar{\psi}\gamma_\mu \gamma_5 \tau_a \psi \bar{\psi}\gamma_\mu \gamma_5 \tau_a \psi)^{2/3} \tag{1}$$

where fermion field ψ belongs to the doublet representation of the internal symmetry SU(2) group. The fractional powers are chosen such that the couplings g_1 and g_2 are dimensionless. The intrepretation of fractional powers, especially in a quantum theory is likely to cause problems. However, the fractional powers can be put into polynomial form provided one introduces two auxiliary fields for each interaction and does the replacement

$$(\bar{\psi}\gamma_\mu\tau_a\psi\bar{\psi}\gamma_\mu\tau_a\psi)^{2/3} \rightarrow \lambda_{a\mu}(\bar{\psi}\gamma_\mu\tau_a\psi-V^2V_{a\mu}) + \bar{\psi}\gamma_\mu\tau_a\psi V_{a\mu} \qquad (2)$$

$$(\bar{\psi}\gamma_\mu\gamma_5\tau_a\psi\bar{\psi}\gamma_\mu\gamma_5\tau_a\psi)^{2/3} \rightarrow \rho_{a\mu}(\bar{\psi}\gamma_\mu\gamma_5\tau_a\psi-A^2A_{a\mu}) + \bar{\psi}\gamma_\mu\gamma_5\tau_a\psi A_{a\mu}$$

In path integral language, performing the path integral over the auxiliary fields gives the nonpolynomial pure spinor interaction. On the other hand, performing the path integral over the fermion fields gives the effective action for the auxiliary fields. One can show that for each pair of auxiliary fields one of the auxiliary fields propagates[5,6]

$$\frac{\delta^2 S_{eff}}{\delta V_{a\mu}\delta V_{b\nu}} = \frac{\delta^2 S_{eff}}{\delta A_{a\mu}\delta A_{b\nu}} = \delta_{ab}(q^2 g_{\mu\nu} - q_\mu q_\nu) \qquad (3)$$

provided that

$$g_1 = g_2 \sim (\ln \Lambda)^{-2/3} \qquad (4)$$

where Λ is the cutoff. Hence the model, when quantized, contains a composite massless vector boson and a composite massless pseudovector boson. Since the bare coupling goes to zero as the cutoff goes to infinity the interaction is asymptotically free as in the CP^n models and as in QCD.

The composite vector and pseudovector bosons in terms of the spinor fields are given by

$$V^2V_{a\mu} = \bar{\psi}\gamma_\mu\tau_a\psi$$

$$\qquad (5)$$

$$A^2A_{a\mu} = \bar{\psi}\gamma_\mu\gamma_5\tau_a\psi \quad .$$

At the classical level it is more convenient to work with the pure spinor form of the Lagrangian as given by Eq. (1). We look for a classical solution such that $V_{a\mu}$ is of the form of a Wu-Yang monopole and the spinor equation of motion

$$i\!\!\not{\partial}\psi + \frac{4}{3} g V_{a\mu}\gamma_\mu\tau_a\psi - \frac{4}{3} g A_{a\mu}\gamma_\mu\gamma_5\tau_a\psi = 0 \qquad (6)$$

is satisfied. The solution for the spinor field is

$$\psi = C \, r^{-3/2} \begin{pmatrix} (\sqrt{3} + i \, \dfrac{\vec{x}\cdot\vec{\sigma}}{r})\sigma_2 \\[2ex] (-\sqrt{3} + i \, \dfrac{\vec{x}\cdot\vec{\sigma}}{r})\sigma_2 \end{pmatrix} \qquad (7)$$

where

$$|C|^2 = 3^{11/2} \times 2^{-42} \times g^{-3}.$$

Here, ψ is a 4×2 matrix whose rows are labelled by the spinor indices and whose columns are labelled by the internal symmetry SU(2) indices. The composite vector and pseudovector fields as given by Eq. (5) are

$$V_{ao} = A_{ao} = 0$$

$$V_{ai} = \frac{1}{e} \frac{\varepsilon_{iaj} x^j}{r^2} \tag{8}$$

$$A_{ai} = -\frac{1}{\sqrt{3}e} \left(\frac{\delta_{ia}}{r} + \frac{x_i x_a}{r^3} \right)$$

where

$$e = 2^{40/3} \times 3^{-2} \times g \quad .$$

The expression for V is precisely that of a Wu-Yang monopole, as advertised. Note that finding such a solution is by no means a trivial task. For example, with the axial vector interaction term absent such a solution does not exist. It can also be checked that there is no solution where the pseudovector field rather than the vector field is of monopole form. In fact, in the general spirit of the model[7] one can consider adding fractional inter-action terms with dimensionless couplings formed by considering scalar-pseudoscalar combinations of the form $\bar{\psi}(A+B\gamma_5)\tau_a\psi$ or vector-pseudovector combinations of the form $\bar{\psi}\gamma_\mu(A+B\gamma_5)\tau_a\psi$ to the Lagrangian together with the vector interaction term. Again the monopole solution exists only if the additional term is a pure axial vector.

As might be expected the monopole solution is invariant under the diagonal SU(2) group of

$$SU(2)_{rotations} \times SU(2)_{internal \; symmetry}.$$

A rotation in space together with a rotation of the internal symmetry leaves the solution unchanged. To show this note that ψ in Eq. (7) can be expressed as

$$\psi \sim \begin{pmatrix} \chi\sigma_2 \\ -\chi^*\sigma_2 \end{pmatrix} \tag{9}$$

where

$$\chi = e^{\,i\,\frac{\vec{\sigma}\cdot\vec{x}}{r}\,\theta} \quad , \qquad \theta = 30° \quad .$$

The spinor rotation matrices Σ are block diagonal and one has

$$R(R^{-1} \chi R \, \sigma_2)R^T = \chi\sigma_2 \tag{10}$$

where R is an SU(2) rotation matrix. The R on the left rotates the spinor indices, the R^{-1} and R on the two sides of χ rotate the x-dependence of the solution and the R^T on the right is an internal symmetry transformation.

The transformation of χ and its x-dependence is reminiscent of a chiral soliton.[8] It would be interesting to search for a modification of the model such that the solution would have an r dependent θ. Such a solution could smear the singularity at the origin and make the monopole topologically stable just as in chiral model solitons.

REFERENCES

1. P.A.M. Dirac, Proc. Roy. Soc. Lond. A 133, 60 (1931).
2. C.N. Yang and R. Mills, Phys. Rev. 96, 191 (1954).
3. T.T. Wu and C.N. Yang, in Properties of Matter Under Unusual Conditions, edited by H. Mark and S. Fernback (Interscience, New York) (1968).
4. G. 't Hooft, Nucl. Phys., B 35, 276 (1974); A.M. Polyakov, Phys. Lett. B 59, 82 (1975).
5. K.G. Akdeniz et al., Phys. Lett. B 116, 41 (1982); ibid, B 124, 79 (1983).
6. Ibid, B 116, 34 (1982).
7. F. Gursey, Nuovo Cimento 3, 988 (1956).
8. T.H.R. Skyrme, Proc. Roy. Soc. A 260, 127 (1961).

SUMMARY OF DISCUSSION ON GUTS AND KALUZA-KLEIN THEORIES[1]

Qaisar Shafi

Bartol Research Foundation
University of Delaware
Newark, Delaware 19716

The simplest 't Hooft-Polyakov monopoles only carry the long range magnetic field and are known (in semiclassical quantization) to possess a spectrum of excitation called dyons. Dyons carry not only a magnetic charge but also a definite electric charge. Consequently they transform according to definite representations of the electromagnetic gauge group.

Grand unified theories with conventional charge assignments for quarks and leptons predict the existence of a topologically stable monopole that carries one unit of Dirac magnetic charge and also carries screened color magnetic fields.

One may wonder whether the GUT monopole (which emits chromo-magnetic flux) also is accompanied by excitations (chromodyons) which transform according to definite representations of color SU(3). This question was addressed by Balachandran who showed that in the presence of a monopole the full set of SU(3) transformations could not be implemented. Only a subgroup SU(2) × U(1) of SU(3) can be implemented on a sphere that contains the monopole. He suggested that this may imply a breaking of the color SU(3) symmetry to SU(2) × U(1) in the presence of a monopole. In any case it implies that chromodyons transforming according to definite representations of color SU(3) do not exist.

Note that had the above chromodyons existed, one would arrive at some striking predictions that would apparently conflict with selection rules imposed by the underlying GUT theory. For instance, combining a "triplet" chromodyon state with an anti-triplet antiquark state leads to a color singlet state that

carries fractional electric charge.   Such states   are not allowed
by the underlying GUT theory (at least not in the zero topological
charge sector).

During the past five years or so the cosmological monopole
problem has attracted a great deal of attention. Many interesting
mechanisms for suppressing the monopole number density to cosmologically acceptable levels have been proposed.   One such attempt
was described by Olive and was based on local supersymmetric GUTS.
The underlying idea is to make the SU(5) phase transition during
the inflationary era.   Alas, the prediction is that there should
be no monopoles in our present universe!

There is an old idea of Heisenberg that the fundamental
theory of the world should only involve fermions.   Such theories,
of course, are unrenormalizable. Attempts have recently been made
to obtain effective gauge theories starting from a pure spinor
theory.   Arik discussed the possibility that magnetic monopoles
also may arise in theories of this kind.   Although one does find
classical solutions that may possibly be labelled as monopoles,
much remains to be done before such monopoles are on a par with
the 't Hooft-Polyakov monopoles.

Great hope rests on a brilliant idea put forward by Kaluza
more than sixty years ago showing that unification of ordinary
gravity and the gauge interactions can be achieved by dimensional
reduction of a higher dimensional theory of gravity.   The idea has
languished in the past due to lack of testable prediction.
Recently, however, things have started to change.   The talks by
Sorkin and Perry dealt with magnetic monopoles that appear in
Kaluza-Klein theories, especially in the original five dimensional
model of Kaluza. The fundamental monopole of the latter theory is
topologically stable and weighs about 10^{20} GeV.   It differs in
other respects from the SU(5) monopole (see the contribution of Q.
Shafi in these proceedings).   It does not carry color magnetic
fields nor is it expected to catalyze $\Delta B \neq 0$ processes strongly.
However, the underlying five dimensional theory is not fully
realistic and construction of a new non-GUT monopole solution in
more realistic models has not yet been done.   Also their production mechanism in the very early universe needs to be explored.

Before the last remark can be reasonably answered one needs
to understand, qualitatively at least, the phase transition from a
4+n dimensional universe to the observed four dimensional
universe. This problem was discussed by Wetterich. He considered
a scenario in the context of a pure 4+n dimensional model of
gravity which admits spontaneous compactifiction of n dimensions.
Not only did he propose a scenario which led to the standard
cosmology, he also discussed the possibility that the phase
transition from 4+n dimensions to four dimensions involves an

inflationary phase.

One Kaluza-Klein theory that has attracted a great deal of attention lately is D = 11 Supergravity. This theory unifies the graviton with the gauge and matter multiplets, and there are speculations that it may provide a unified description of the four observed interactions. In Bais' talk the possibility of finding soliton solutions (including black holes) in these theories was discussed.

REFERENCES

1. See papers by M. Perry, R. Sorkin, C. Wetterich, A. Balachandran, M. Arik, F. Bais, and Q. Shafi in this volume.

SUPERHEAVY MAGNETIC MONOPOLES AND THE STANDARD COSMOLOGY

Michael S. Turner

Theoretical Astrophysics Group
Fermi National Accelerator Laboratory
Batavia, Illinois 60510

and

Enrico Fermi Institute
The University of Chicago
Chicago, Illinois 60637

ABSTRACT

The superheavy magnetic monopoles predicted to exist in grand unified theories (GUTs) are very interesting objects, both from the point of view of particle physics, as well as from astrophysics and cosmology. Astrophysical and cosmological considerations have proved to be invaluable in studying the properties of GUT monopoles. Because of the glut of monopoles predicted in the standard cosmology for the simplest GUTs (so many that the Universe should have reached a temperature of 3 K at the tender age of $\simeq 10,000$ yrs), the simplest GUTs and the standard cosmology are <u>not</u> compatible. This is a very important piece of information about physics at unification energies (E $\gtrsim 10^{14}$ GeV) and about the earliest moments (t $\lesssim 10^{-34}$ s) of the Universe. In this talk I review the cosmological consequences of GUT monopoles within the context of the standard hot big bang model.

INTRODUCTION

In the past five years or so progress in both elementary particle physics and in cosmology has become increasingly dependent upon the interplay between the two disciplines. On the particle physics side, the $SU(3)_C \times SU(2)_L \times U(1)_Y$ model seems to very accurately describe the interactions of quarks and leptons at energies below, say, 10^3 GeV. At the very least the so-called standard model is a satisfactory, effective low energy theory. The frontiers of particle physics now involve energies of much greater than 10^3 GeV--energies which are not now available in terrestrial accelerators, nor are ever likely to be available in terrestrial accelerators. For this reason particle physicists have turned both to the early Universe with its essentially unlimited energy budget (up to 10^{19} GeV) and high particle fluxes (up to 10^{107} cm^{-2} s^{-1}), and to various unique, contemporary astrophysical environments (centers of main sequence stars where temperatures reach 10^8 K, neutron stars where densities reach 10^{14}-10^{15} g cm^{-3}, our galaxy whose magnetic field can impart 10^{11} GeV to a Dirac magnetic charge, etc.) as non-traditional laboratories for studying physics at very high energies and very short distances.

On the cosmological side, the hot big bang model, the so called standard model of cosmology, seems to provide an accurate accounting of the history of the Universe from about 10^{-2} s after 'the bang' when the temperature was about 10 MeV, until today, some 10-20 billion years after 'the bang' and temperature of about 3 K ($\simeq 3 \times 10^{-13}$ GeV). Extending our understanding further back, to earlier times and higher temperatures, requires knowledge about the fundamental particles (presumably quarks and leptons) and their interactions at very high energies. For this reason, progress in cosmology has become linked to progress in elementary particle physics.

Grand unification provides a particularly good example of the importance of the interplay between particle physics and cosmology. GUTs give us only a 'few windows' from our low energy world to the physics of unification energies, the two most familar and important ones being baryon number nonconservation and superheavy magnetic monopoles. Both of these predictions also have profound implications for cosmology and astrophysics. Baryon nonconservation provides for the first time a framework for understanding the 'baryon asymmetry of the Universe' (for a recent review see Ref. 1). Baryogenesis is a spectacular and unqualified success for the marriage of cosmology and GUTs, and a useful window to very high energies and the earliest history of the Universe. Monopoles, on the other hand, have been a cosmological disaster. In the standard cosmology so many monopoles should have been produced (at least for the simplest GUTs) that the Universe should have reached a temperature of 3 K while still in its infancy (t $\simeq$ 10,000 yrs). This is the so-called 'Monopole Problem': the simplest GUTs and the standard cosmology are not compatible (at least at times as early as 10^{-34} s). Although somewhat discouraging, this too is an important piece of information about physics at very high energies and the earliest history

of the Universe. Monopoles have been a real boon for astrophysics. Because of their macroscopic masses [$\simeq 10^{-8}$ g in SU(5)],
hefty magnetic charge (integer multiple of $\simeq 69$ e), and their
remarkable ability to catalyze nucleon decay at a prodigious
rate, monopoles if they exist, should today be doing very
interesting things--contributing mass density, 'shorting out'
astrophysical magnetic fields, and gobbling up nucleons (releasing
1 GeV per gobble) to mention but three. Astrophysical considerations have resulted in very stringent bounds on the flux of
relic monopoles. I have summarized these bounds in Fig. 1, and
they will be discussed in great detail by other speakers.

In this talk I will first briefly review the standard
cosmology. Next I will discuss monopole production in the very
early Universe, both as topological defects (the 'Kibble
mechanism') and by energetic particle collisions. The glut which
results from the Kibble mechanism in the standard cosmology is the
root of the monopole problem. On the other hand, if Kibble
production can be suppressed, we are apparently left with a

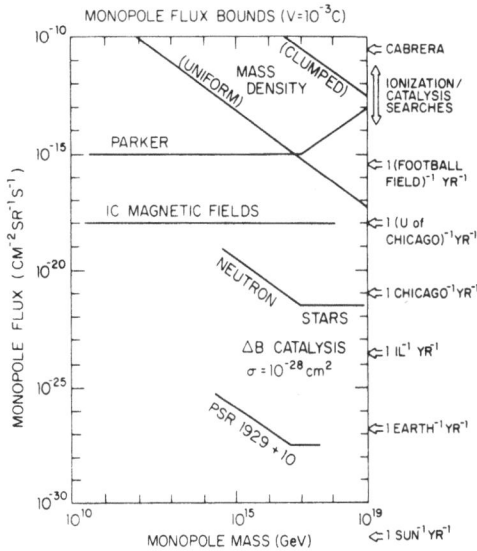

Fig. 1   Summary of the astrophysical and cosmological constraints
         on the monopole flux as a function of monopole mass.
         Wherever necessary the monopole velocity was taken to be
         $\simeq 10^{-3}$ c. The magnetic field bounds are discussed in Refs.
         2-5; the bounds based upon monopole catalysis of nucleon
         decay in neutron stars are discussed in Refs. 6-11. The
         most stringent bound (based upon monopole catalysis of
         nucleon decay in PSR 1929 + 10) was obtained also by
         taking into account the monopoles that the progenitor of
         this pulsar captured while it was on the main sequence.[8]

monopole famine, due to the dearth of monopoles produced in
energetic particle collisions (although the uncertainties here are
exponential). I will then trace the history of monopoles from
their birth during the earliest moments of the Universe, through
their adolescence, until today, with the aim of answering the
important questions like, where should one expect to find monopoles today?, and with what velocities should they be moving? I
will finish with some concluding remarks.

THE STANDARD COSMOLOGY[12]

The hot big bang model nicely accounts for the universal
(Hubble) expansion, the 2.7 K cosmic microwave background radiation, and through primordial nucleosynthesis, the abundances of
D, ^{4}He and perhaps also ^{3}He and ^{7}Li. Light received from the most
distant objects observed (QSOs at redshifts $\simeq$ 3.5) left these
objects when the Universe was only a few billion years old. Thus
observations of galaxies allow us to directly probe the history of
the Universe to within a few billion years of 'the bang'. The
surface of last scattering for the microwave background is the
Universe about 100,000 yrs. after the bang when the temperature
was about 1/3 eV. Thus the microwave background is a fossil
record of the Universe at that very early epoch. In the standard
cosmology the epoch of big bang nucleosynthesis takes place from
$t \simeq 10^{-2}$ s - 10^2 s when the temperature was $\simeq$ 10 MeV - 0.1 MeV.
The light elements synthesized, primarily D, ^{3}He, ^{4}He, and ^{7}Li,
are relics from this early epoch, and thus comparing their
predicted big bang abundances with their inferred primordial
abundances is the most stringent test of the standard cosmology we
have at present. [Note that I must say inferred primordial
abundance because contemporary astrophysical processes can affect
the abundance of these light isotopes, e.g., stars very efficiently burn D, and produce ^{4}He.] At present the predicted
abundances of D, ^{3}He, ^{4}He, and ^{7}Li are all simultaneously
consistent with their inferred primordial abundances so long as
the number of light ($\lesssim$ 1 MeV) neutrino species is less than or
equal to 4, and the baryon-to-photon ratio, η, is in the range[13]
(see Figs. 2 and 3):

$$\eta \simeq (4 - 7) \times 10^{-10} \qquad (1)$$

The baryon-to-photon ratio is related to the fraction of critical
density contributed by baryons by,

$$\eta = 2.83 \times 10^{-8} \ \Omega_b h^2 \ (2.7 \ K/T)^3, \qquad (2)$$

where the Hubble parameter H_0 = 100 h kms^{-1} Mpc^{-1}, and T is the
present temperature of the cosmic microwave background. Observations strongly suggest that: 1/2 < h < 1 and 2.7 K < T < 3.0 K,

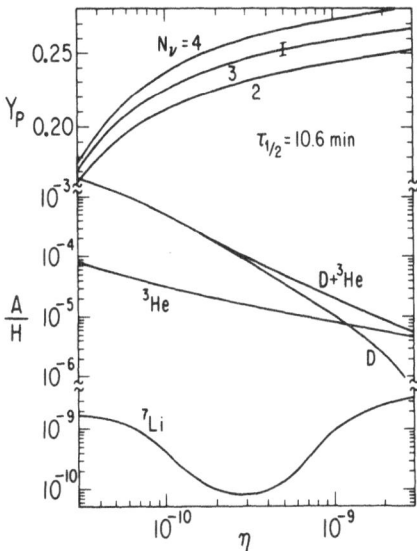

Fig. 2   The predicted primordial abundances of D, ^{3}He, ^{4}He, and
^{7}Li [$\tau_{1/2}(n)$ = 10.6 min was used; error bar shows
$\Delta\tau_{1/2}$ = ± 0.2 min; Y_p = mass fraction of ^{4}He].   Inferred
primordial abundances: $Y_p \simeq 0.23 - 0.25$; $(D/H)_p \gtrsim 10^{-5}$;
$(D + {}^3He)_p/H \lesssim 10^{-4}$; $(^7Li/H)_p \simeq (1.1 \pm 0.4) \times 10^{-10}$.   Consistency of the predicted abundances with observations can
only be achieved for $\eta \simeq (4 - 7) \times 10^{-10}$ (= baryon-to-
photon ratio) and $N_\nu \lesssim 4$ (= number of light neutrino
species).   For $4 < \eta/10^{-10} < 7$, $0.014 < \Omega_b < 0.15$.   See
Ref. 13 for more details.

so that the concordant range for η implies

$$0.014h^{-2} (T/2.7 \text{ K})^3 < \Omega_b < 0.034h^{-2} (T/3.0 \text{ K})^3, \tag{3a}$$

$$0.014 < \Omega_b < 0.15; \tag{3b}$$

implying that baryons alone cannot provide the closure density.

Note that other information we have about Ω_b (e.g., lower
bound based upon the amount of luminous matter in the Universe,
the total amount of matter associated with a galaxy) is consistent
with this range.   The concordance of the predictions and observations of D and ^{4}He are particularly compelling evidence because
there is no known contemporary astrophysical site where the
observed amounts of ^{4}He ($\simeq$ 25% by mass) and D (D/H $\simeq$ few $\times 10^{-5}$)
can be produced.   It is the successful predictions of big bang
nucleosynthesis that gives us confidence in the standard model
back to $\simeq 10^{-2}$ s after 'the bang'.

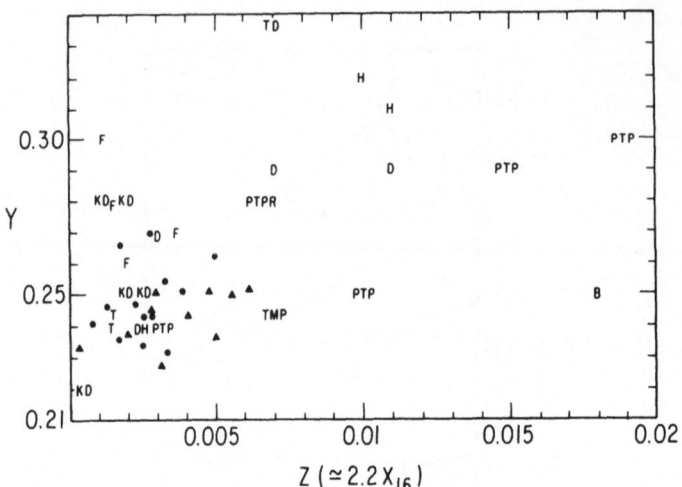

Fig. 3   Summary of determinations of ^{4}He mass fraction (Y) in HII
         regions   as   a function   of   the   metal   abundance Z (more
         precisely, 2.2 times the mass fraction of ^{16}0).   Where the
         metal abundance is lowest,   one expects the stellar con-
         tribution to Y to be the smallest.   The data exhibit this
         trend and clearly show the existence of a primordial ^{4}He
         component of about 0.23-0.25 (by mass).   For more details
         see Ref. 13.

     On the large scale (>> 100 Mpc), the Universe is isotropic
and homogeneous, and so it can accurately be described by the
Robertson-Walker line element[12]

$$ds^2 = -dt^2 + R(t)^2 \ [dr^2/(1-kr^2) + r^2 \ d\theta^2 + r^2 \ \sin^2 \ \theta d\phi^2], \quad (4)$$

where ds^2 is the proper separation between two spacetime events,
k = 1, 0, or -1 is the curvature signature, and R(t) is the cosmic
scale factor.   The expansion of the Universe is embodied in
R(t)--as R(t) increases all proper (i.e., measured by meter
sticks) distances scale with R(t), e.g., the distance between two
galaxies comoving with the expansion (i.e., fixed r, θ, ϕ), or the
wavelength of a freely-propagating photon ($\lambda \propto R$).   The k > 0
spacetime has positive spatial curvature and is finite in extent;
the k < 0 spacetime has negative spatial curvature and is infinite
in extent; the k = 0 spacetime is spatially flat and is also
infinite in extent.

     The evolution of the cosmic scale factor is determined by the
Friedmann equations:

$$H^2 \equiv (\dot{R}/R)^2 = 8\pi G\rho/3 - k/R^2, \qquad\qquad (5a)$$

$$d(\rho R^3) = -p \ d(R^3), \qquad\qquad (5b)$$

where ρ is the total energy density and p is the pressure. The expansion rate H (also called the Hubble parameter) sets the characteristic time for the growth of $R(t)$; $H^{-1} \simeq$ e-folding time for R. The present value of H is 100 h kms^{-1} Mpc$^{-1} \simeq$ h $(10^{10}$ yr$)^{-1}$ $(1/2 < h < 1)$. As can be seen from Eqn. 5a model Universes with $k < 0$ expand forever, while a model Universe with $k > 0$ must eventually recollapse. The sign of k (and hence the geometry of spacetime) can be determined from measurements of ρ and H:

$$k/H^2R^2 = \rho/(3H^2/8\pi G) - 1, \tag{6}$$

$$\equiv \Omega - 1,$$

where $\Omega = \rho/\rho_{crit}$ and $\rho_{crit} = 3H^2/8\pi G \simeq 1.88$ h$^2 \times 10^{-29}$ gcm^{-3}. From primordial nucleosynthesis we know that $\Omega \gtrsim \Omega_b \gtrsim 0.014$. The best upper limit to Ω follows by considering the age of the Universe:

$$t_U = 10^{10} \text{ yr } \left(h^{-1} f(\Omega)\right), \tag{7}$$

where $f(\Omega) < 1$ and is monotonically decreasing. The ages of the oldest stars (in globular clusters) strongly suggest that $t_U \gtrsim 10^{10}$ yr; combining this with Eqn. 7 implies that: $\Omega f^2 (\Omega) \gtrsim \Omega h^2$. The function Ωf^2 is monotonically increasing and asymptotically approaches $(\pi/2)^2$, implying that independent of h, $\Omega h^2 \lesssim 2.5$. Restricting h to the interval $(1/2, 1)$ it follows that: $\Omega h^2 \lesssim 0.8$ and $\Omega \lesssim 3.2$.

The energy density contributed by nonrelativistic matter varies as $R(t)^{-3}$--due to the fact that the number density of particles is diluted by the increase in the proper (or physical) volume of the Universe as it expands. For a relativistic species the energy density varies as $R(t)^{-4}$, the extra factor of R due to the redshifting of the particle's momentum $\left(\text{recall } \lambda \propto R(t)\right)$. The energy density contributed by a relativistic species (T $\gg$ m) at temperature T is

$$\rho = g_{eff}\pi^2T^4/30, \tag{8}$$

where g_{eff} is the number of degrees of freedom for a bosonic species, and 7/8 that number for a fermionic species. Note that $T \propto R(t)^{-1}$. Here and throughout I have taken $h/2\pi = c = k_B = 1$, so that 1 GeV $= (1.97 \times 10^{-14}$ cm$)^{-1} = (1.16 \times 10^{13}$ K$) = (6.57 \times 10^{-25}$ s$)^{-1}$, $G = m_{pl}^{-2}$ $(m_{pl} = 1.22 \times 10^{19}$ GeV), and 1 GeV$^4 = 2.32 \times 10^{17}$ g cm^{-3}.

Today, the energy density contributed by relativistic particles (photons and 3 neutrino species) is negligible: $\Omega_{rel} \simeq 4 \times 10^{-5}$ h^{-2} (T/2.7 K)^4. However, since $\rho_{rel} \propto R^{-4}$, while

$\rho_{nonrel} \propto R^{-3}$, early on relativistic species will dominate the energy density. For $R/R_{today} < 4 \times 10^{-5}$ $(\Omega h^2)^{-1}$ $(T/2.7$ K$)^4$, which corresponds to $t < 4 \times 10^{10}$ s $(\Omega h^2)^{-2}$ $(T/2.7$ K$)^6$ and $T > 6$ eV $(\Omega h^2)(2.7$ K$/T)^3$, the energy density of the Universe was dominated by relativistic particles. Since the curvature term varies as $R(t)^{-2}$, it too will be small compared to the energy density contributed by relativistic particles, and Eqn. 5a simplifies to:

$$H \equiv (\dot{R}/R) \simeq (4\pi^3 \ g_\ast/45)^{1/2} \ T^2/m_{pl},$$
$$\simeq 1.66 \ g_\ast^{1/2} \ T^2/m_{pl}, \tag{9}$$

(valid for $t \lesssim 10^{10}$ s, $T \gtrsim 10$ eV).

Here $g_\ast$ counts the total number of effective degrees of freedom of all the relativistic particles (i.e., those species with mass $\ll T$):

$$g_\ast = \sum_{Bose} g_i(T_i/T)^4 + 7/8 \sum_{Fermi} g_i(T_i/T)^4 \tag{10}$$

and T is the photon temperature. For example: $g_\ast(3$ K$) \simeq 3.36$ (γ, $3\nu\bar{\nu}$); $g_\ast$(few MeV) $\simeq 10.75$ (γ, $e^\pm$, 3 $\nu\nu$); $g_\ast$(few 100 GeV) $\simeq 110$ (γ, $W^\pm$ Z°, 8 gluons, 3 families of quarks and leptons, and 1 Higgs doublet).

If thermal equilibrium is maintained, then the second Friedmann equation, Eqn. 5b - conservation of energy, implies that the entropy per comoving volume (a volume with fixed r, θ, ϕ coordinates) $S \propto sR^3$ remains constant. Here s is the entropy density, which is dominated by the contribution from relativistic particles, and

$$s = (\rho + p)/T \simeq 2\pi^2 \ g_\ast \ T^3/45, \tag{11}$$

which is proportional to the number density of relativistic particles. So long as the expansion is adiabatic (i.e., in the absence of entropy production) S(and s) will prove to be useful fiducials. For example, at low energies ($E \ll 10^{14}$ GeV) baryon number is effectively conserved, and so the net baryon number per comoving volume $N_B \propto n_B(\equiv n_b - n_{\bar{b}})$ R^3 remains constant, implying that the ratio n_B/s is a constant of the expansion. Today $s \simeq 7n_\gamma$, so that

$$n_B/s \simeq \eta/7 \simeq (6 - 10) \times 10^{-11}. \tag{12}$$

Once monopole-antimonopole annihilations are no longer important, the number of monopoles per comoving volume is also conserved, $N_M \propto n_M \ R^3$. Comparing this to the baryon number and entropy per comoving volume we get two ratios which remain constant (so long as annihilations are not important, and entropy and baryon number are conserved) and are related to the present average flux of monopoles in the Universe by:

$$\langle F_M \rangle \simeq 10^{10} \ (n_M/s)(v_M/10^{-3} \ c)cm^{-2} \ sr^{-1} \ s^{-1}, \tag{13a}$$

$$\simeq (n_M/n_B)(v_M/10^{-3} \ c)cm^{-2} \ sr^{-1} \ s^{-1}, \tag{13b}$$

where v_M is the typical monopole velocity (which will be discussed at length later). The fraction of critical density contributed by monopoles (Ω_M) is:

$$\Omega_M \simeq 10^{24} \ (n_M/s)(m_M/10^{16} \ GeV), \tag{14a}$$

$$\simeq 10^{14} \ \langle F_M \rangle (10^{-3} \ c/v_M)cm^2 \ sr \ s. \tag{14b}$$

Whenever $g_* \simeq$ constant, the constancy of the entropy per comoving volume implies that $T \propto R^{-1}$; together with Eq. 9 this gives

$$R(t) = R(t_o)(t/t_o)^{1/2}, \tag{15a}$$

$$t \simeq 0.3 \ g_*^{-1/2} \ m_{pl}/T^2,$$

$$\simeq 2.4 \times 10^{-6} \ s \ g_*^{-1/2} \ (T/GeV)^{-2}, \tag{15b}$$

valid for $t \lesssim 10^{10}$ s and $T \gtrsim 10$ eV.

Finally, let me mention one more important feature of the standard cosmology, the existence of particle horizons. The distance that a light signal could have propagated since the bang is finite, and easy to compute. Photons travel on paths characterized by $ds^2 = 0$; for simplicity (and without loss of generality) consider a trajectory with $d\theta = d\phi = 0$. The coordinate distance covered by this photon since 'the bang' is just $\int_o^t dt'/R(t')$, corresponding to a physical distance (measured at time t) of

$$d_H(t) = R(t) \int_o^t dt'/R(t') \tag{16}$$

$$= t/(1 - n) \ [for \ R \propto t^n, \ n < 1]$$

If $R \propto t^n$, then this distance is finite and $\simeq t \simeq H^{-1}$. Note that even if $d_H(t)$ diverges and there is no finite particle horizon (e.g., if $R \propto t^n$, $n > 1$), the Hubble radius H^{-1} still sets the scale for the 'physics horizon'. Since all physical length scales roughly e-fold in a time H^{-1}, causally-coherent microphysical processes can only operate over times (also distances) $\lesssim H^{-1}$.

During the radiation-dominated epoch $n = 1/2$ and $d_H = 2t$; the baryon number and entropy within the horizon at time t are easily computed:

$$S_{HOR} = (4\pi/3)t^3 s,$$

$$\simeq 0.05 \ g_*^{-1/2} \ (m_{pl}/T)^3; \tag{17a}$$

$$N_{B-HOR} = (n_B/s) \times S_{HOR},$$

$$\simeq 10^{-12} \ (m_{pl}/T)^3; \tag{17b}$$

note that I have implicitly assumed the constancy of the baryon-to-entropy ratio in computing N_{B-HOR}.

Although our verifiable knowledge of the early history of the Universe only takes us back to $t \simeq 10^{-2}$ s, and $T \simeq 10$ MeV, nothing in our present understanding of the laws of physics suggests that it is unreasonable to extrapolate back to times as early as $\simeq 10^{-43}$ s and temperatures as high as $\simeq 10^{19}$ GeV. At high energies the interactions of quarks and leptons are asymptotically free (and/or weak) justifying the dilute gas approximation made in Eqn. 8, and at energies below 10^{19} GeV quantum corrections to general relativity are expected to be small. I hardly need to remind the reader that 'not unreasonable' does not necessarily mean 'correct'. Making this extrapolation, I have summarized 'The Complete History of the Universe' in Fig. 4.

BIRTH: GLUT OR FAMINE

In 1931 Dirac[14] showed that if magnetic monopoles exist, then the single-valuedness of quantum mechanical wavefunctions requires the magnetic charge of a monopole to satisfy the quantization condition

$$g = ng_D, \ n = 0, \ \pm 1, \ \pm 2 \ \ldots$$

$$g_D = 1/2e \simeq 69e.$$

However, one is not required to have Dirac monopoles in the theory--you can take 'em or leave 'em! In 1974 't Hooft[15] and Polyakov[16] independently made a remarkable discovery. They showed that monopoles are obligatory in the low-energy theory whenever a semi-simple group G, e.g., SU(5); breaks down to a group G' $\times$ U(1) which contains a U(1) factor [e.g., SU(3) $\times$ SU(2) $\times$ U(1)]; this, of course, is the goal of unification. These monopoles are associated with nontrivial topology in the Higgs field responsible for SSB, topological knots if you will, have a mass $m_M \simeq O(M/\alpha)$ [$\simeq 10^{16}$ GeV in SU(5); M = scale of SSB], and have a magnetic

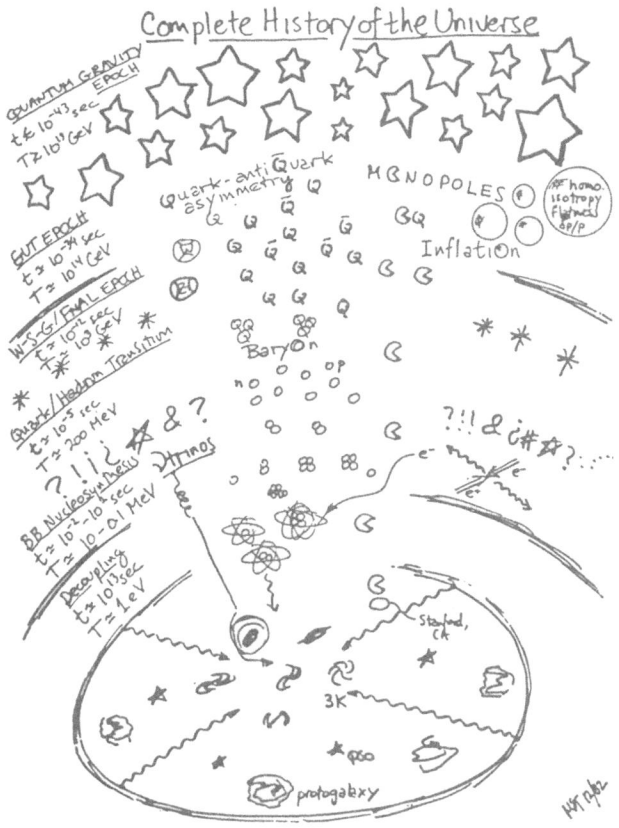

Fig. 4 'The Complete History of the Universe'. Highlights in-
        clude: <u>decoupling</u> (t ≃ 10^{13} s, T ≃ 1/3 eV) - the surface
        of last scattering for the cosmic microwave background,
        epoch after which matter and radiation cease to interact
        and matter 'recombines' into neutral atoms (D, ^{3}He, ^{4}He,
        ^{7}Li); also marks the beginning of the formation of
        structure; <u>primordial nucleosynthesis</u> (t ≃ 10^{-2} s,
        T ≃ 10 MeV) - epoch during which all of the free neutrons
        and some of the free protons are synthesized into D, ^{3}He,
        ^{4}He, and ^{7}Li, and the surface of last scattering for the
        cosmic neutrino backgrounds; <u>quark/hadron transition</u>
        (t ≃ 10^{-5} s, T ≃ few 100 MeV) - epoch of 'quark enslave-
        ment' [confinement transition in SU(3)]; <u>W-S-G epoch</u>
        (t ≃ 10^{-12} s, T ≃ 10^3 GeV) - SSB phase transition as-
        sociated with electroweak breaking, SU(2) × U(1) → U(1);
        <u>GUT epoch</u> (?? t ≃ 10^{-34} s, T ≃ 10^{14} GeV ??) - SSB of the
        GUT, during which the baryon asymmetry of the Universe
        evolves, monopoles are produced, and 'inflation' may
        occur; the <u>Quantum Gravity Wall</u> (t≃10^{-43} s, T ≃ 10^{19} GeV).

charge which is a multiple of the Dirac charge.

Since there exist no contemporary sites for producing particles of mass even approaching 10^{16} GeV, the only plausible
production site is the early Universe, about 10^{-34} s after 'the
bang' when the temperature was $\simeq O(10^{14}$ GeV). There are two ways
in which monopoles can be produced: (1) as topological defects
during the SSB of the unified group G; (2) in monopole-
antimonopole pairs by energetic particle collisions. The first
process has been studied by Kibble[17], Preskill[18], and Zel'dovich
and Khlopov[19], and I will review their important conclusions
here.

The magnitude of the Higgs field responsible for the SSB of
the unified group G is determined by the minimization of the free
energy. However, this does not uniquely specify the direction of
the Higgs field in group space. A monopole corresponds to a
configuration in which the direction of the Higgs field in group
space at different points in physical space is topologically
distinct from the configuration in which the Higgs field points in
the same direction (in group space) everywhere in physical space
(which corresponds to no monopole):

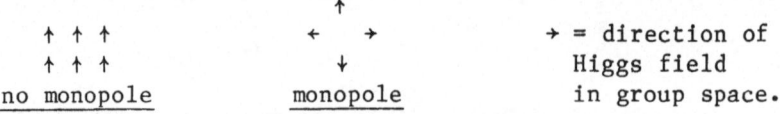

| no monopole | monopole | $\rightarrow$ = direction of
Higgs field
in group space. |

Clearly monopole configurations cannot exist until the SSB
$[G \rightarrow G' \times U(1)]$ transition takes place. When spontaneous symmetry
breaking occurs, the Higgs field can only be smoothly oriented
(i.e., the no monopole configuration) on scales smaller than some
characteristic correlation length ξ. On the microphysical side,
the inverse Higgs mass at the Ginzburg temperature (T_G) sets such
a scale: $\xi \simeq m_H^{-1}(T_G)$ (in a second-order phase transition)[20].
[The Ginzburg temperature is the temperature at which it becomes
improbable for the Higgs field to fluctuate between the SSB
minimum and $\phi = 0$.] Cosmological considerations set an absolute
upper bound: $\xi \lesssim d_H (\simeq t$ in the standard cosmology). [Note, even
if the $d_H(t)$ diverged, e.g., because $R \propto t^n$ ($n > 1$) for $t \lesssim t_{pl}$,
the physics horizon H^{-1} sets an absolute upper bound on ξ, which
is numerically identical.] On scales larger than ξ the Higgs
field must be uncorrelated, and thus we expect of order 1 monopole
per correlation volume ($\simeq \xi^3$) to be produced as a topological
defect when the Higgs field freezes out.

Let's focus on the case where the phase transition is either
second order or weakly-first order. Denote the critical
temperature for the transition by T_c ($\simeq 0$ (M)), and as before the
monopole mass by $m_M \simeq 0(M/\alpha)$. The age of the Universe when $T \simeq T_c$

is given in the standard cosmology by: $t_c \simeq 0.3 \ g_*^{-1/2} \ m_{pl}/T_c^2$, cf. Eqn. 15b. For SU(5): $T_c \simeq 10^{14}$ GeV, $m_M \simeq 10^{16}$ GeV and $t_c \simeq 10^{-34}$ s. Due to the fact that freezing of the Higgs <u>must</u> be uncorrelated on scales $\gtrsim \xi$, we expect an initial monopole abundance of $O(1)$ per correlation volume; using $d_H(t_c)$ as an <u>absolute</u> upper bound on ξ this leads to: $(n_M)_i \simeq O(1) \ t_c^{-3}$. Comparing this to our fiducials S_{HOR} and N_{B-HOR}, we find that the initial monopole-to-entropy and monopole-to-baryon number ratios are:

$$n_M/s \gtrsim 10^2 \ (T_c/m_{pl})^3, \tag{18a}$$

$$n_M/n_B \gtrsim 10^{12} \ (T_c/m_{pl})^3. \tag{18b}$$

Preskill[18] has shown that unless n_M/s is $> 10^{-10}$ monopole-antimonopole annihilations do not significantly reduce the initial monopole abundance. If $n_M/s > 10^{-10}$, he finds that n_M/s is reduced to $\simeq 10^{-10}$ by annihilations. For $T_c < 10^{15}$ GeV our estimate for n_M/s is $< 10^{-10}$, and we will find that in the standard cosmology T_c must be $\ll 10^{15}$ GeV to have an acceptable monopole abundance, so for our purposes we can ignore annihilations. <u>Assuming</u> that the expansion has been adiabatic since $T \simeq T_c$, this estimate for n_M/s translates into:

$$\langle F_M \rangle \simeq 10^{-3} \ (T_c/10^{14} \ \text{GeV})^3 \ \text{cm}^{-2} \ \text{sr}^{-1} \ \text{s}^{-1}, \tag{19a}$$

$$\langle \Omega_M \rangle \simeq 10^{11} \ (T_c/10^{14} \ \text{GeV})^3 (m_M/10^{16} \ \text{GeV}) \tag{19b}$$

--a flux that would make any monopole hunter/huntress ecstatic, and an Ω_M that is unacceptably large (except for $T_c \ll 10^{14}$ GeV). As was discussed previously, Ω can be at most $O(\text{few})$, so we have a <u>very</u> big problem with the simplest GUTs (in which $T_c \simeq 10^{14}$ GeV). This is the so-called 'Monopole Problem'. The statement that $\Omega_M \simeq 10^{11}$ for $T_c \simeq 10^{14}$ GeV is a bit imprecise; clearly if $k < 0$ (corresponding to $\Omega < 1$) monopole production cannot close the Universe (and in the process change the geometry from being infinite in extent and negatively-curved, to being finite in extent and positively-curved). More precisely, a large monopole abundance would result in the Universe becoming matter-dominated much earlier, at $T \simeq 10^3$ GeV $(T_c/10^{14}$ GeV$)^3$ $(m_M/10^{16}$ GeV$)$, and eventually reaching a temperature of 3 K at the young age of $t \simeq 10^4$ yrs $(T_c/10^{14}$ GeV$)^{-3/2}$ $(m_M/10^{16}$ GeV$^{-1/2}$. The requirement that $\Omega_M \lesssim O(\text{few})$ implies that

$$T_c \lesssim 10^{11} \ \text{GeV} \qquad (\Omega_M \lesssim \text{few})$$

where I have taken m_M to be $O(100 \ T_c)$. Note, given our generous estimate for ξ, even this is probably <u>not</u> safe; if one had a GUT in which $T_c \simeq 10^{11}$ GeV a more <u>careful</u> estimate for ξ would be called for.

The Parker bound[2-4] (see Fig. 1) on the average monopole flux in the galaxy, $\langle F_M \rangle \lesssim 10^{-15}$ cm^{-2} sr^{-1} s^{-1}, results in a slightly more stringent constraint:

$$T_c \lesssim 10^{10} \text{ GeV} \qquad \text{(Parker bound)}$$

The most restrictive constraints on T_c follow from the neutron star catalysis bounds on the monopole flux[6-11], and the most restrictive of those[8], $\langle F_M \rangle \lesssim 10^{-27}$ cm^{-2} sr^{-1} s^{-1}, implies that

$$T_c \lesssim 10^6 \text{ GeV} \qquad \text{(Neutron star catalysis bound)}$$

Note, to obtain these bounds I have compared my estimate for the average monopole flux in the Universe, Eqn. 19a, with the astrophysical bounds on the average flux of monopoles in our galaxy. If monopoles cluster in galaxies (which I will later argue is unlikely), then the average galactic flux of monopoles is greater than the average flux of monopoles in the Universe, making the above bounds on T_c more restrictive.

If the GUT transition is strongly first order (I am excluding inflationary Universe scenarios for the moment), then the transition will proceed by bubble nucleation at a temperature T_n ($\ll T_c$), when the nucleation rate becomes comparable to the expansion rate H. Within each bubble the Higgs field is correlated; however, the Higgs field in different bubbles should be uncorrelated. Thus one would expect O(1) monopole per bubble to be produced. When the Universe supercools to a temperature T_n, bubbles nucleate, expand, and rapidly fill all of space; if r_b is the typical size of a bubble when this occurs, then one expects n_M to be $\simeq r_b^{-3}$. After the bubbles coalesce, and the Universe reheats, the entropy density is once again $s \simeq g_* T_c^3$, so that the resulting monopole to entropy ratio is: $n_M/s \simeq (g_* r_b^3 T_c^3)^{-1}$. Guth and Weinberg[21] have calculated r_b and find that $r_b \simeq (m_{pl}/T_c^2)/\ln(m_{pl}^4/T_c^4)$, leading to a relatively accurate estimate for the monopole abundance:

$$n_M/s \simeq [\ln(m_{pl}^4/T_c^4)(T_c/m_{pl})]^3,$$

which is even more disasterous than the estimate for a second order phase transition [recall, however, estimate (18) was an absolute lower bound].

The bottom line is that we have a serious problem here--the standard cosmology extrapolated back to $T \simeq T_c$ and the simplest GUTs are incompatible (to say the least). One (or both) must be modified. This is a valuable piece of information.

A number of possible solutions have been suggested. To date the most attractive is the new inflationary Universe scenario (which will be discussed at length by P. Steinhardt). In this scenario, a small region (size $\lesssim$ the horizon) within which the Higgs field could be correlated, grows to a size which encompasses all of the presently observed Universe, due to the exponential expansion which occurs during the phase transition. This results in less than one monopole in the entire observable Universe (due to Kibble production).

Let me very briefly review some of the other attempts to solve the monopole problem. Several people have pointed out that if there is no complete unification [e.g., if G = H × U(1)], or if the full symmetry of the GUT is not restored in the very early Universe (e.g., if the maximum temperature the Universe reached was $< T_c$, or if a large lepton number[22], $n_L/n_\gamma > 1$, prevented symmetry restoration at high temperatures), then there would be no monopole problem. However, none of these possibilities seems particularly attractive.

Several authors[23-26] have studied the possibility that monopole-antimonopole annihilation could be enhanced over Preskill's estimate, due to 3-body annihilations or the gravitational clumping of monopoles (or both). Thus far, this approach too has been unsuccessful.

Bais and Rudaz[27] have suggested that large fluctuations in the Higgs field at temperatures near T_c could allow the monopole density to relax to an acceptably small value. They do not explain how this mechanism can produce the acausal correlations needed to do this.

Mechanisms have been proposed in which monopoles and anti-monopoles form bound pairs connected by flux tubes, leading to rapid monopole-antimonopole annihilation. For example, Linde[28] proposed that at high temperatures color magnetic charge is confined, and Lazarides and Shafi[29] proposed that monopoles and antimonopoles become connected by $Z°$ flux tubes after the SU(2) × U(1) SSB phase transition. In both cases, however, the proposed flux tubes are not topologically stable, nor has their existence even been demonstrated.

Langacker and Pi[30] have proposed a solution which does seem to work. It is based upon an unusual (and perhaps contrived) symmetry breaking pattern for SU(5):

$$SU(5) \rightarrow SU(3) \times SU(2) \times U(1) \rightarrow SU(3) \rightarrow SU(3) \times U(1)$$
$$T_c \simeq 10^{14} \text{ GeV} \qquad T_1 \qquad T_2$$

$$\overleftrightarrow{\text{superconducting phase}}$$

(note T_1 could be equal to T_c). The key feature of their scenario
is the existence of the epoch ($T \simeq T_1 \rightarrow T_2$) in which the U(1) of
electromagnetism is sponteneously broken (a superconducting
phase); during this epoch magnetic flux must be confined to flux
tubes, leading to the annihilation of the monopoles and antimonopoles which were produced earlier on, at the GUT transition.
Although somewhat contrived, their scenario appears to be viable
(however, I'll have more to say about it shortly).

Finally, one could invoke the Tooth Fairy (in the guise of a
perfect annihilation scheme).   E. Weinberg[31] has recently made a
very interesting point regarding 'perfect annihilation schemes',
which applies to the Langacker-Pi scenario[30], and even to a Tooth
Fairy which operates causally.   Although the Kibble mechanism
results in equal numbers of monopoles and antimonopoles being
produced, E. Weinberg points out that in a finite volume there can
be magnetic charge fluctuations. He shows that if the Higgs field
'freezes out' at $T \simeq T_c$ and is uncorrelated on scales larger than
the horizon at that time, then the expected net RMS magnetic
charge in a volume V which is much bigger than the horizon is

$$\Delta n_M \simeq (V/t_c^3)^{1/3}. \tag{20}$$

He then considers a perfect, causal annihilation mechanism which
operates from $T = T_1 \rightarrow T_2$ (e.g., formation of flux tubes between
monopoles and antimonopoles).   At best, this mechanism could
reduce the monopole abundance down to the net RMS magnetic charge
contained in the horizon at $T = T_2$, leaving a final monopole
abundance of

$$n_M/s \simeq 10^2 \ T_c \ T_2^2/m_{pl}^3, \tag{21}$$

resulting in

$$\Omega_M \gtrsim 0.1 \ (T_c/10^{14} \text{ GeV})(m_M/10^{16} \text{ GeV})(T_2/10^8 \text{ GeV})^2, \tag{22a}$$

$$\langle F_M \rangle \gtrsim 10^{-15} \ (T_c/10^{14} \text{ GeV})(T_2/10^8 \text{ GeV})^2 \ cm^{-2} \ sr^{-1} \ s^{-1}. \tag{22b}$$

It is difficult to imagine a perfect annihilation mechanism which
could operate at temperatures $\lesssim 10^3$ GeV, without having to modify
the standard SU(2) × U(1) electroweak theory; for $T_c \simeq 10^{14}$ GeV and
$T_2 \simeq 10^3$ GeV, E. Weinberg's argument[31] implies that $\langle F_M \rangle$ must be
$\gtrsim 10^{-25}$ cm^{-2} sr^{-1} s^{-1}, which would be in conflict with the most
stringent neutron star catalysis bound[8], $F_M < 10^{-27}$ $cm^{-2} sr^{-1}$ s^{-1}.

In a recent preprint A. Vilenkin[32] disputes E. Weinberg's
argument about the significance of magnetic charge fluctuations,
and provides several counterexamples.   In an even more recent
preprint Lee and E. Weinberg[33] refute Vilenkin's counterexamples,
and provide the results of numerical simulations which support E.

Weinberg's original conclusions[31]. Clearly, this important issue is not settled yet. It does, however, seem clear that a perfect annihilation mechanism which operates down to a temperature T_2 can do no better than to reduce the monopole abundance to 1 per horizon volume at $T = T_2$, or $n_M/s \simeq 10^2 (T_2/m_{pl})^3$--to do better would require Higgs field correlations on scales larger than the horizon. Therefore, at the very best, the T_c's in Eqns. 22a,b could be replaced by T_2's.

Finally, I should emphasize that the estimate of n_M/s based upon $\xi \lesssim d_H(t)$ is an absolute (and very _generous_) lower bound to n_M/s. Should a model be found which succeeds in suppressing the monopole abundance to an acceptable level (e.g., by having $T_c \ll 10^{14}$ GeV or by a perfect annihilation epoch), then the estimate for ξ _must_ be refined and scrutinized.

If the glut of monopoles produced as topological defects in the standard cosmology can be avoided, then the only production mechanism is pair production in very energetic particle collisions, e.g., particle(s) + antiparticles(s) $\rightarrow$ monopole + antimonopole. [Of course, the 'Kibble production' of monopoles might be consistent with the standard cosmology (and other limits to the monopole flux) if the SSB transition occurred at a low enough temperature, say $\ll O(10^{10}$ GeV).] The numbers produced are intrinsically small because monopole configurations do not exist in the theory until SSB occurs ($T_c \simeq M$ = scale of SSB), and have a mass $O(M/\alpha) \simeq 100 M \simeq 100 T_c$. For this reason they are never present in equilibrium numbers; however, some are produced due to the rare collisions of particles with sufficient energy. This results in a present monopole abundance of[34-36]

$$n_M/s \simeq 10^2 (m_M/T_{max})^3 \exp(-2m_M/T_{max}), \qquad (23a)$$

$$\Omega_M \simeq 10^{26} (m_M/10^{16} \text{ GeV})(m_M/T_{max})^3 \exp(-2m_M/T_{max}), \qquad (23b)$$

$$\langle F_M \rangle \simeq 10^{12} \text{ cm}^{-2} \text{ sr}^{-1} \text{ s}^{-1}(m_M/T_{max})^3 \exp(-2m_M/T_{max}), \qquad (23c)$$

where T_{max} is the highest temperature reached after SSB.

In general, $m_M/T_{max} \simeq O(100)$ so that $\Omega_M \simeq O(10^{-40})$ and $\langle F_M \rangle \simeq O(10^{-32} \text{ cm}^{-2} \text{ sr}^{-1} \text{ s}^{-1})$--a negligible number of monopoles. However, the number produced is _exponentially_ sensitive to m_M/T_{max}, so that a factor of 3-5 uncertainty in m_M/T_{max} introduces an enormous uncertainty in the predicted production. For example, in the new inflationary Universe, the monopole mass can be $\propto$ the Higgs field responsible for SSB, and as that field oscillates about the SSB minimum during the reheating process m_M also oscillates, leading to enhanced monopole production [m_M/T_{max} in Eqns. 23a, b, c is replaced by fm_M/T_{max}, where $f < 1$ depends upon the details of reheating; see Refs. 37, 38].

Cosmology seems to leave the poor monopole hunter/huntress with two firm predictions: that there should be equal numbers of north and south poles; and that either far too few to detect, or far too many to be consistent with the standard cosmology should have been produced. The detection of any superheavy monopoles would necessarily send theorists back to their chalkboards!

FROM BIRTH THROUGH ADOLESCENCE ($t \simeq 10^{-34}$ s to $t \simeq 3 \times 10^{17}$ s)

As mentioned in the previous section, monopoles and anti-monopoles do not annihilate in significant numbers; however, they do interact with the ambient charged particles (e.g., monopole + $e^- \overset{\rightarrow}{\leftarrow}$ monopole + e^-) and thereby stay in <u>kinetic equilibrium</u> (KE $\simeq 3T/2$) until the epoch of $e^{\pm}$ annihilations ($T \simeq 1/2$ MeV, $t \simeq 10$ s). At the time of $e^{\pm}$ annihilations mono-poles and antimonopoles should have internal velocity dispersions of:

$$\langle v_M{}^2 \rangle^{1/2} \simeq 30 \text{ cm s}^{-1} \ (10^{16} \text{ GeV}/m_M)^{1/2}.$$

After this monopoles are effectively collisionless, and their velocity dispersion decays $\propto R(t)^{-1}$, so that if we neglect gravitational and magnetic effects, today they should have an internal velocity dispersion of

$$\langle v_M{}^2 \rangle^{1/2} \simeq 10^{-8} \text{ cm s}^{-1} \ (10^{16} \text{ GeV}/m_M)^{1/2}.$$

Since they are collisionless, only their velocity dispersion can support them against gravitational collapse. With such a small velocity dispersion to support them they are gravitationally un-stable on all scales of astrophysical interest ($\lambda_{Jeans} \simeq 10^{-8}$ LY).

After decoupling ($T \simeq 1/3$ eV, $t \simeq 10^{13}$ s) [or the epoch of matter domination in scenarios where the mass of the Universe is dominated by a nonbaryonic component], matter can begin to clump, and structure can start to form. Monopoles, too, should clump and participate in the formation of structure. However, since they cannot dissipate their gravitational energy, they cannot collapse into the more condensed objects (such as stars, planets, the disk of the galaxy, etc.) whose formation clearly must have involved the dissipation of gravitational energy. Thus, one would only expect to find monopoles in structures whose formation did not require dissipation (such as clusters of galaxies, and galactic haloes). However, galactic haloes are not likely to be a safe haven for monopoles in galaxies with magnetic fields; monopoles less massive than about 10^{20} GeV will, in less than 10^{10} yrs, gain sufficient KE from a magnetic field of strength a few × 10^{-6} G to reach escape velocity[4]. So we are led to the conclusion that initially monopoles should either be uniformly distributed through the cosmos, or clumped in clusters of galaxies or in the haloes of galaxies with weak or non-existent magnetic fields. Since our own galaxy has a magnetic field of strength $\simeq$ few $\times$ 10^{-6} G, and is not a member of a cluster of galaxies, we would expect the local flux of monopoles to be not too different from the average monopole flux in the Universe.

Although monopoles initially have a very small internal velocity dispersion, there are many mechanisms for increasing their velocities. First, typical peculiar velocities (i.e., velocities relative to the Hubble flux) are $O(10^{-3}$ c), leading to a typical monopole-galaxy velocity of 10^{-3} c. Monopoles will be accelerated by the gravitational fields of galaxies (to $\simeq 10^{-3}$ c $\simeq$ orbital velocity in the galaxy), and if they encounter them, clusters of galaxies (to $\simeq 3 \times 10^{-3}$ c). A typical monopole, however, will never encounter a galaxy or a cluster of galaxies, the respective mean free paths being: $L_{gal} \simeq 10^{26}$ cm ($\simeq 10^{-2}$ c $\times$ age of the Universe) and $L_{cluster} \simeq 3 \times 10^{28}$ cm.

Monopoles will also be accelerated by magnetic fields. The intragalactic magnetic field strength is $\lesssim 3 \times 10^{-11}$ G (Ref.39), and results in a monopole velocity of

$$v_M \simeq 3 \times 10^{-4} \text{ c } (B/10^{-11} \text{ G})(10^{16} \text{ GeV}/m_M).$$

The galactic magnetic field will accelerate monopoles in our galaxy to velocities of[4]

$$v_M \simeq 3 \times 10^{-3} \text{ c } (10^{16} \text{ GeV}/m_M)^{1/2}.$$

Taking all of these 'sources of velocity' into account, we can make an educated estimate of the typical monopole-detector relative velocity.

From Table 1 below it should be clear that the typical monopole should be moving with a velocity of <u>at least</u> a few x 10^{-3} c with respect to an earth-based detector. It goes without saying that 'this fact' is an important consideration for detector design.

Although planets, stars, etc. should be monopole-free at the time of their formation, they will accumulate monopoles during their lifetimes. The number captured by an object is where M, R

$$N_M = (4\pi R^2)(\pi - sr)(1 + 2GM/Rv_M^2)\langle F_M \rangle \varepsilon \tau, \qquad (24)$$

and τ are the mass, radius and age of the object, v_M is the monopole velocity, and ε is the efficiency with which the object stops monopoles which strike its surface. The efficiency of capture ε depends upon the mass and velocity of the monopole, and

Table 1. Typical Monopole-Detector Relative Velocities

DETECTOR VELOCITY	MONOPOLE VELOCITY
orbit in $2/3 \times 10^{-3}$ c galaxy	galactic 3×10^{-3} c $(10^{16}$ GeV/$m_M)^{1/2}$ B-field
orbit in 10^{-4} c solar system	grav. acceleration 10^{-3} c by galaxy
	grav. acceleration 10^{-4} c by sun
	monopole-galaxy 10^{-3} c relative velocity

---

its rate of energy loss in the object. The quantity $(1 + 2GM/R\ v_M^2)$ is just the ratio of the capture cross section to the geometric cross section. Main sequence stars of mass $(0.6 - 30)M_\odot$ will capture monopoles less massive than about 10^{18} GeV with velocities $\lesssim 10^{-3}$ c with good efficiency ($\varepsilon \simeq 1$); in its main sequence lifetime a star will capture approximately 10^{24} F_{-16} monopoles[40] (essentially independent of its mass). Here $\langle F_M \rangle = F_{-16}\ 10^{-16}$ cm^{-2} sr^{-1} s^{-1}. Neutron stars will capture monopoles less massive than about 10^{20} GeV with velocities $\lesssim 10^{-3}$ c with unit efficiency, capturing about 10^{21} F_{-16} monopoles in 10^{10} yrs. Planets like Jupiter can stop monopoles less massive than about 10^{16} GeV with velocities $\lesssim 10^{-3}$ c, accumulating about 10^{22} F_{-16} monopoles in 10^{10} yrs.[41]. A planet like the earth can only stop light or slowly-moving monopoles (for $m_M = 10^{16}$ GeV, v_M must be $\lesssim 3 \times 10^{-5}$ c). Once inside, monopoles can do interesting things, like catalyze nucleon decay, which keeps the object hot (leading to a potentially observable photon flux), and eventually depletes the object of all its nucleons. A monopole flux of $F_{-21}\ 10^{-21}$ cm^{-2} sr^{-1} s^{-1} will cause a neutron star to evaporate in 10^{11} $F_{-21}^{-1/2}$ yrs, a Jupiter-like planet to evaporate in 5×10^{15} $F_{-21}^{-1/2}$ yrs, and an Earth-like planet to evaporate in 10^{18} $F_{-21}^{-1/2}$ yrs[42]. Accretion of monopoles by astrophysical objects, however, does not significantly reduce the monopole flux; the mean free path of a monopole in the galaxy is $\simeq 10^{42}$ cm.

CONCLUDING REMARKS

What have we learned about GUT monopoles? (1) They are exceedingly interesting objects, which, if they exist, must be relics of the earliest moments of the Universe. (2) They are one of the very few predictions of GUTs that we can attempt to verify and study in our low energy environment. (3) Becuase of the glut of monopoles that should have been produced as topological defects in the very early Universe, the simplest GUTs and the standard cosmology (extrapolated back to times as early as $\simeq 10^{-34}$ s) are not compatible. This is a very important piece of information about physics at very high energies and/or the earliest moments of the universe. (4) There is no believeable prediction for the flux of relic, superheavy magnetic monopoles. (5) Based upon astrophysical considerations, we can be reasonably certain that the flux of relic monopoles is small. Since it is not obligatory that monopoles catalyze nucleon decay at a prodigious rate, a firm upper limit to the flux is provided by the Parker bound[43], $\langle F_M \rangle \lesssim 10^{-15}$ cm^{-2} sr^{-1} s^{-1}. Note, this is not a predicted flux, it is only a firm upper bound to the flux. It is very likely that flux has to be even smaller, say $\lesssim 10^{-18}$ cm^{-2} sr^{-1} s^{-1} or even 10^{-22} cm^{-2} sr^{-1} s^{-1}. (6) There is every reason to believe that typical monopoles are moving with velocities (relative to us) of at least a few $\times 10^{-3}$ c. [Although it is possible that the largest contribution to the local monopole flux is due to a cloud of monopoles orbiting the sun with velocities $\simeq (1 - 2) \times 10^{-4}$ c, I think that it is very unlikely.[44-45]]

Based upon the above (not unbiased) 'list of facts', I believe that when designing a monopole detector, the monopole hunter/huntress must give highest consideration to building a detector which is sensitive to a monopole flux at least as small as 10^{-15} cm^{-2} sr^{-1} s^{-1}, for monopole velocities $\simeq 10^{-3\pm0.5}$ c. The risks involved in monopole hunting are very great, but the potential payoff is even greater!

I apologize for any omissions I may be guilty of in this mini-review of monopoles and the standard cosmology. More complete reviews can be found in Refs. 36 and 46. This work was supported by the DoE at Fermilab and Chicago (AC0280ER10773A004), by NASA at Fermilab, and by an Alfred P. Sloan Fellowship.

REFERENCES

1. E.W. Kolb and M.S. Turner, Ann. Rev. Nucl. Part. Sci. 33, 645 (1983).
2. E.N. Parker, Astrophys. J. 160, 383 (1970).
3. G. Lazarides, Q. Shafi, and T. Walsh, Phys. Lett. 100B, 21 (1981).
4. M.S. Turner, E.N. Parker, and T.J. Bogdan, Phys. Rev. D26, 1296 (1982).
5. Y. Rephaeli and M.S. Turner, Phys. Lett. 121B, 115 (1983).
6. E.W. Kolb, S. Colgate, and J. Harvey, Phys. Rev. Lett. 49, 1373 (1982).

7. S. Dimopoulos, J. Preskill, and F. Wilczek, Phys. Lett. 119B, 320 (1982).

8. K. Freese, M.S. Turner, and D.N. Schramm, Phys. Rev. Lett. 51, 1625 (1983).

9. F. Bais, J. Ellis, D.V. Nanopoulos, and K.A. Olive, Nucl. Phys. B219, 189 (1983).

10. T. Walsh, in Proceedings of the XXIIth Intl. Conference on High Energy Physics (Paris, 1982).

11. E.W. Kolb and M.S. Turner, Astrophys. J., in press (1984).

12. For a more detailed description of the standard cosmology, see, e.g., S. Weinberg, Gravtiation and Cosmology (Wiley: NY, 1972), chapter 15.

13. J. Yang, M.S. Turner, G. Steigman, D.N. Schramm, and K.A. Olive, Astrophys. J., in press (1984).

14. P.A.M. Dirac, Proc. Roy. Soc. (London) A133, 60 (1931).

15. G. 't Hooft, Nucl. Phys. B79, 276 (1974).

16. A.M. Polyakov, JETP Lett. 20, 194 (1974).

17. T.W.B. Kibble, J. Phys. A9, 1387 (1976).

18. J. Preskill, Phys. Rev. Lett. 43, 1365 (1979).

19. Ya.B. Zel'dovich and M. Yu Khlopov, Phys. Lett. 79B, 239 (1978).

20. M.B. Einhorn, D.L. Stein, and D. Toussaint, Phys. Rev. D21, 3295 (1980).

21. A.H. Guth and E. Weinberg, Nucl. Phys. B212, 321 (1983).

22. J.A. Harvey and E.W. Kolb, Phys. Rev. D24, 2090 (1981).

23. T. Goldman, E.W. Kolb, and D. Toussaint, Phys. Rev. D23, 867 (1981).

24. J. Fry, Astrophys. J. 246, L93 (1981).

25. J. Fry and G. Fuller, Univ. of Chicago preprint (1983).

26. D.A. Dicus, D.N. Page, and V.L. Teplitz, Phys. Rev. D26, 1306 (1982).

27. F. Bais and S. Rudaz, Nucl. Phys. B170, 507 (1980).

28. A. Linde, Phys. Lett. 96B, 293 (1980).

29. G. Lazarides and Q. Shafi, Phys. Lett. 94B, 149 (1980).

30. P. Langacker and S.-Y. Pi, Phys. Rev. Lett. 45, 1 (1980).

31. E. Weinberg, Phys. Lett. 126B, 441 (1983).

32. A. Vilenkin, Tufts Univ. preprint (1983).

33. K. Lee and E. Weinberg, Columbia Univ. preprint (1984).

34. M.S. Turner, Phys. Lett. 115B, 95 (1982).

35. G. Lazarides, Q. Shafi, and W.P. Trower, Phys. Rev. Lett. 49, 1756 (1982).

36. J. Preskill, in The Very Early Universe, eds. G.W. Gibbons, S.W. Hawking, S. Siklos (Cambridge Univ. Press, Cambridge 1983).

37. W. Collins and M.S. Turner, Phys. Rev. D, in press (1984).

38. A. Goldhaber and A. Guth, in preparation (1984).

39. J.P. Vallee, Astrophys. Lett. 23, 85 (1983).

40. K. Freese, J. Frieman, and M.S. Turner, Univ. of Chicago preprint (1984).

41. M.S. Turner, Nature 302, 804 (1983).

42. M.S. Turner, Nature 306, 161 (1983).

43. The Parker bound can be evaded if monopoles play a role in the generation of the galactic magnetic field (e.g. by magnetic plasma oscillations); see Ref. 4; J. Arons and R. Blandford, Phys. Rev. Lett. 50, 544 (1983); or E. Salpeter, S. Shapiro, and I. Wasserman, Phys. Rev. 49, 1114 (1982). In order for this mechanism to work, the phase velocity of the plasma oscillations must be greater than the gravitational velocity dispersion of the monopoles $\simeq 10^{-3}$ c (otherwise Landau damping will rapidly damp the oscillations). This results in a lower bound to the monopole flux:
$\langle F_M \rangle \gtrsim 1/4\ m_M v_{grav}{}^3\ (gl)^{-2} \approx 10^{-12}\ (m_M/10^{16}\ GeV)cm^{-2}\ sr^{-1}\ s^{-1}$;
here g $\simeq$ 69e is the monopole's magnetic charge, l $\simeq$ 300 pc is the characteristic length scale of the galactic B-field, and $v_{grav} \simeq 10^{-3}$ c. For $m_M \gtrsim 10^{18}$ GeV the monopole density required is inconsistent with the mass density density of the galaxy (see Fig. 1). I believe that this scenario is very unlikely; in addition, it is becoming observationally untenable because of the large monopole flux that it predicts.

44. S. Dimopoulos, S. Glashow , E. Purcell, and F. Wilczek, Nature 298, 824 (1982).

45. K. Freese and M.S. Turner, Phys. Lett 123B, 293 (1983).

46. P. Langacker, Phys. Rep. 72, 185 (1981).

PROGRESS AND PROSPECTS FOR THE INFLATIONARY UNIVERSE

Paul Joseph Steinhardt

Department of Physics
University of Pennsylvania
Philadelphia, Pennsylvania   19104

ABSTRACT

The present status of the inflationary universe scenario is briefly reviewed.   In particular, the experimentally testable "predictions" of the model are reassessed.   Recent progress in obtaining a fully workable inflationary universe model based on supersymmetric field theories is discussed.

THE INFLATIONARY UNIVERSE SCENARIO

The inflationary universe[1] is a radically new model of the very evolution of our universe.   The model is designed to resolve many problems of the standard model of cosmology --- the hot big bang picture --- while maintaining all the successes.   In fact, the inflationary universe agrees precisely with the hot big bang model in its description of the expansion of our universe for all times greater than one second or so after the big bang.   The two cosmologies differ only in their descriptions of the first instants after the birth of our universe.

The key new feature of the inflationary universe is the assumption that the universe underwent a special "slow rollover" type[5] phase transition during its early history.[1-4] The phase tran-sition may be associated with the spontaneous symmetry breaking in grand unified theories, as was suggested in the earliest renditions of the scenario,[2-4] or the transition may be due to some other physics, such as supersymmetry.   For the sake of this general discussion, we need only suppose that the phase transition involves a change from some sort of "symmetric state" to some sort of "symmetry breaking state." We will call the order parameter for the transition ϕ, where $\phi = 0$ in the symmetric state and ϕ is non-zero in the symmetry breaking state. In most cases, ϕ represents the expectation value of some scalar field (or fermion condensate) responsible for the symmetry breaking, e.g. a Higgs field.

The assumption that the universe underwent a slow rollover phase transition means, first of all, that at the high temperatures found in the very early universe it was energetically favorable for the universe to be in the symmetric state (as would be the case in a grand unified theory, for example). As the temperature decreased, a symmetry breaking state became more favored energetically, but the universe remained trapped in the symmetric state because a large energy barrier separated it from the symmetry breaking state. The energy barrier was caused by thermal effects that are well understood and calculable.[6] The process of being trapped in a "metastable" phase because of such a barrier is known as supercooling and it is characteristic of any first order phase transition. In a slow rollover transition, the universe supercooled to zero temperature, and precisely as the temperature approached zero the energy barrier (which was caused by thermal effects) disappeared. The previously metastable phase became altogether unstable. The energy density as a function of ϕ was as in Fig. 1 and the universe was balanced at $\phi = 0$.

Quantum fluctuations could then drive a region of the universe a small distance away from $\phi = 0$, and then the potential would carry ϕ in that region towards a symmetry breaking state. Different regions of the universe could be driven towards different spontaneous symmetry breaking minima (all of which are equivalent physically). The regions were initially of order of a causal horizon volume in size or smaller, which, if the transition is associated with the spontaneous breaking of the grand unified symmetry, had a radius of roughly 10^{-25} when the transition began (at a temperature $T \sim 10^{15}$ GeV). As ϕ first began to grow inside each region, it grew very slowly because the potential is very flat near $\phi = 0$ for the slow rollover transition. Also during this period the energy density was nearly constant in the universe, equal roughly to the vacuum energy density of the symmetric phase. [The potential energy density curve has been adjusted so that the symmetry breaking state corresponds to zero energy density, which is equivalent to setting the symmetry breaking state to have zero cosmological constant. In the inflationary universe model, the observation that our universe today (which lies in the symmetry breaking state near zero temperature) has zero cosmological constant is taken as a boundary condition on the model without explanation.] According to Einstein's equations, a region with constant energy density expands exponentially with time (like a de Sitter universe).[2] If the interior of one of the expanding regions

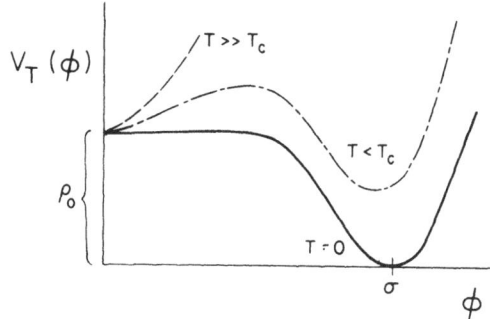

Fig. 1. The free energy density, $V_T(\phi)$, versus order parameter (ϕ)
at various temperatures for a slow rollover phase transi-
tion.   A state with $\phi = 0$ corresponds to the symmetric
phase and $\phi = \sigma$ corresponds to the   symmetry breaking
phase. (V_T has been shifted so that $V_T(0)$ is the same for
all the curves.)   At high temperatures, $\phi = 0$ is the
energetically favored state.   At low temperatures the
symmetry breaking phase becomes the more energetically
favored state.   However, an energy barrier keeps the
symmetric phase metastable. $V_0(\sigma)$ is set to zero because
today we live in the symmetry breaking phase at T near
zero and we find that the cosmological constant, or
equivalently, V_0 is zero.   The energy barrier separating
the symmetric from the symmetry breaking phase disappears
just as the temperature approaches zero.   Then the slow
roll to $\phi = \sigma$ begins.

is described by a Robertson-Walker metric in which R(t) is the
scale factor, $R(t) \sim \exp(Ht)$ during this expansion; $H^2 = 8\pi V(0)/3M_p^2$
where M_p is the Planck mass and V(0) is the energy density of the
symmetric state. Thus, during the early growth of ϕ each region of
symmetry breaking phase -- henceforth called a   "domain" --
continued to expand exponentially until ϕ grew to the steep part
of the potential.   This period of exponential expansion can easily
have lasted hundreds or thousands of Hubble times $[O(H^{-1})]$, so R(t)
would have grown by a factor of $e^{100-1000}$ or more in each domain.
This means that a single domain of radius 10^{-25} cm grew to a size
of 10^{20} cm or possibly much greater.   Our entire observable
universe (which today is over 10 billion light years in radius) was
only a few centimeters in radius by the time the transition is
supposed to have been completed.   Thus, our entire observable
universe, according to this picture, lies inside just a single
domain; and we have no information as to how many other domains
there might be.

Once ϕ reached the steep part of the potential, it oscillated rapidly about the symmetry breaking minimum. The rapid oscillation of the Higgs field corresponds to a rapid time variation of some field or fields and, as a result, induces radiation from the "vacuum" of the particles to which the field couples. The oscillations were then damped by two competing processes. First, the radiation itself converts energy from the ϕ field to ordinary matter and radiation.[7,8] Secondly, the continued expansion of the universe leads to a red-shifting of the kinetic energy associated with the oscillations. Which effect wins out depends upon the strength of the couplings of the field to ordinary matter.[8] In grand unified models the coupling is typically so strong that nearly all the original vacuum energy density is converted into ordinary matter and radiation; the universe reheats to a temperature nearly equal to the temperature before the transition initiated.[7] In some other models, such as supersymmetric models to be discussed later in this paper, the coupling is so weak that only a very small percentage of the vacuum energy is converted into ordinary matter and radiation.

After the oscillations were damped, the universe was in a state with zero cosmological constant with all of its energy density in the form of ordinary matter and radiation. The subsequent evolution of the universe was therefore just as in a hot big bang model beginning from that temperature. Provided the reheating temperature was greater than 10^{10} GeV or so after the transition, all the successful predictions of the hot big bang model are automatically reaccrued in the inflationary model. All the most successful predictions of the hot big bang picture occur when the universe is at temperatures of 1 MeV or less. Many regard the baryon asymmetry generated through combining the hot big bang model with grand unified theories as being an additional success of the hot big bang cosmology,[9] since the asymmetry generated is roughly in accord with what is observed astrophysically. The baryon asymmetry is generated at a temperature of 10^{10} GeV or more. Provided that the universe reheated after inflation to a temperature of 10^{10} GeV or more, a baryon asymmetry could be expected to be generated that is roughly equal to whatever would have been generated in the absence of inflation.[7,8,10] The baryon asymmetry represents by far the most stringent constraint on the reheating.

The period of tremendous exponential expansion or "inflation" induced by the slow rollover phase transition can play a crucial role in resolving many of the problems associated with the hot big bang picture:

Horizon Problem[11] - Our entire observable universe was totally contained within a single domain in the inflationary universe scenario. We have argued that a single domain was initially a causal horizon length or smaller, so our observable universe before the phase transition was necessarily totally causally connected. Thus, the observable universe could before the phase transition become homogeneous and isotropic through causal processes. Through exponential expansion, a tiny causally connected region could grow large enough to account for the present size of our observable universe. Thus the inflationary cosmology does not suffer from the horizon problem of the hot big bang model. According to the hot big bang model, our observable universe was always much larger than a single causal horizon length and there is no natural explanation as to how our universe became homogeneous and isotropic.

Flatness Problem[12] - According to Einstein's equations for a Friedman-Robertson-Walker universe,

$$\frac{\rho}{\rho_c} = 1 + \frac{k}{R^2 H^2} \qquad (1)$$

where ρ is the energy density, ρ_c is the critical energy density and k is the space curvature constant. If k is positive and the energy density is greater than the critical energy density, the universe is said to be closed. If k is negative and the energy density is less than the critical energy density, the universe is said to be open. If k is precisely equal to zero and the energy density precisely equals the critical energy density, the universe is in the flat condition. The second term in Eq.(1) represents, then, the deviation from flatness. During Friedmann-Robertson-Walker expansion, R(t) grows like t^α where α is less than one. The Hubble parameter, H, equals $\dot{R}/R$ where $\dot{R}$ is the time derivative of R. Thus, the second term in Eq.(1) grows with time. From astrophysical measurements we know that ρ/ρ_c is somewhere between 0.1 and 10 today. Extrapolating backwards to times before the first second after the big bang, we find that ρ/ρ_c had to be exponentially close to unity. In an inflationary universe, ρ/ρ_c is driven exponentially close to unity by the very nature of the inflationary epoch. R(t) grows exponentially with time and H [which is roughly proportional to $V(\phi)$] is essentially constant. Thus, the second term on the right hand side of Eq.(1) is driven exponentially close to zero and ρ is driven to ρ_c. No natural explanation of why ρ should have been so close to ρ_c in the early universe exists in the hot big bang picture.

Monopole and Domain Wall Problems[13] - Monopoles (and domain walls in some cases) are predicted to be produced when the universe underwent the phase transition associated with the spontaneous symmetry breaking of grand unified theories. In the hot big bang picture, so many monopoles and domain walls are expected to have been produced that their mass should have dominated the energy density of the universe soon after the phase transition. This would mean that the subsequent evolution of the universe would be as in a matter dominated universe, speeding up the expansion of the universe compared to the hot big bang picture. Such speeding up totally destroys all the successful predictions of the hot big bang picture. If the inflationary phase transition is the one associated with grand unified theories (GUTs) or one that occurs after the GUT phase transition, the density of monopoles or domain walls produced during the GUT transition are diluted by a huge exponential factor; thus monopoles and domain walls would have no (deleterious) effect on the subsequent evolution of the universe. As of this writing, inflation seems the only practical scheme for solving the cosmological problems associated with such topological defects. (To first approximation, inflation dilutes the density of monopoles to an exponentially small (unobservable) value; the validity of this conclusion will be examined below.)

Inhomogeneity (Galaxy Formation) Problem - In the hot big bang picture, the universe is assumed to be perfectly homogeneous and isotropic, but some deviation from perfect homogeneity is required in order for the inhomogeneities we observe in our universe (galaxies, clusters of galaxies, and the like) to evolve. For many years cosmologists have considered how fluctuations must vary with distance scale -- the fluctuation spectrum -- to account for the formation and organization of galaxies. Observations of the cosmic microwave background indicate that the universe is amazingly uniform on large scales; yet, sufficient inhomogeneities are required on smaller scales in order to form galaxies, galaxy clusters, etc. The inflationary universe scenario leads to an unambiguous prediction as to the nature of the fluctuations in the universe after the inflationary phase transition.[14] Inflation has two effects on fluctuations in the universe. First, inflation erases any fluctuations that exist in the universe before the phase transition by spreading them to such large scales that they have no effect on our observable universe. Second, inflation provides a natural mechanism to produce a new spectrum of inhomogeneities from quantum fluctuations. Quantum fluctuations produce inhomogeneities on very small scales, but then inflation expands them to large scales. The process continues over time until fluctuations are produced over a spectrum of scales ranging over all those scales we observe in our universe and beyond. During the inflationary phase, the universe is de Sitter-like and, to first approximation, time translation invariant. The quantum effects therefore produce equal amplitude fluctuations over time. One can further show that the amplitude of the fluctuations are not changed during the inflation (basically, because they are expanded to such large scales that no causal influences can change their amplitude).[14] The fluctuation on a given scale cannot be changed, then, until the Hubble horizon grows to encompass that scale during the Friedman-Robertson-Walker phase that follows the phase transition. By the argument, then, each fluctuation "re-enters the Hubble horizon" with the same amplitude as it had when it was produced, which is in turn the same for each scale. Such a spectrum of inhomogeneities is referred to as a "scale invariant" spectrum, and it is just the qualitative spectrum that many cosmologists claim is necessary to explain the formation and organization of galaxies in our universe.[15] The amplitude of the fluctuation when it re-enters the Hubble horizon is dependent upon the detailed shape of the potential energy density shown in Fig. 1. The fractional energy density perturbation is given by:

$$\frac{\delta\rho}{\rho}(x) = 0(1) \frac{H^2}{\dot{\phi}} \qquad (2)$$

where $\dot{\phi}$ is the expectation value of the time derivative of ϕ evaluated at the time during the inflationary epoch when the fluctuation on comoving scale, x, was produced (or, more precisely, was expanded beyond the Hubble horizon in the inflationary epoch). The fluctuations on the scales we observe in our universe were produced in the last 60 e-foldings or so of the inflationary period,[14] so it is the expectation value of $\dot{\phi}$ near the end of inflation (which typically involves a total of hundreds or thousands of e-foldings) that is relevant to determining the amplitude of the fluctuations in our own universe. The bad news is that the simplest GUT models yield fluctuations so large $(0(1))$ that they are inconsistent with the observed isotropy of the cosmic microwave background.[14] The good news is that inflation seems capable of resolving yet another fundamental problem of cosmology, provided that the right $V(\phi)$ can be found. How the right $V(\phi)$ might be found in a supersymmetric field theory will be discussed below.

The inflationary universe, then, should be viewed as a powerful and self-consistent cosmology that avoids many problems that have plagued other cosmological pictures. So far as cosmology is concerned, I believe the picture is complete. The missing ingredient now is determining what is the microphysics -- presumably particle physics -- that is responsible for the slow rollover phase transition required by inflation, a subject discussed in some detail below. It is also important to understand what predictions, if any, are made by the inflationary universe scenario so we can evaluate how the model might be tested. This will be the subject of the next section.

WHAT DOES THE INFLATIONARY UNIVERSE SCENARIO PREDICT?

The inflationary universe is a beautiful theory, but does it provide us with any experimentally testable predictions that allow us to positively determine whether our universe is inflationary or not. I know of no feasible test that proves that our universe is necessarily inflationary, but there are two issues that might be considered for proving that our universe is not inflationary. One issue concerns $\Omega = \rho/\rho_c$; during inflation, the universe is driven exponentially close to the flat condition, or $\Omega = 1$. This, it will be recalled, is crucial since only if Ω is exponentially close to one after the transition can it possibly be in the range 0.1 to 10 today, in accordance with astrophysical observation (see first section. However, as has been commonly observed in the past,[2] unless the amount of inflation is just the minimal amount of inflation necessary to solve the horizon problem (roughly 60 e-foldings), Ω can be driven so close to unity just after the phase transition that even today it is exponentially close to unity.

Since astrophysicists are making more and more accurate measurements of Ω, it is important to determine what to conclude if they find Ω is not equal to unity (recent measurements of the dynamical mass of the Virgo supercluster suggest that Ω is less than 50% of ρ_c).[16] First, one might suppose that there is inflation, but only the minimal amount. That possibility is incredibly ugly, but would not be impossible in the original inflationary scenario; however, with the new inflationary scenario and the need for small density fluctuations, even this remote possibility is not acceptable. To obtain small enough density perturbations on large scales to be consistent with the observed isotropy of the microwave background, the flat portion of $V(\phi)$ must extend over a range of ϕ which is large compared with H; this guarantees that near the end of the inflationary slow rollover transition $\dot{\phi}$ is large compared with H and $\dot{\phi}$ is large compared to H^2 (see Eq.(2)). For the scale invariant spectrum required by cosmologists to explain galaxy formation, $\dot{\phi}$ has to be of order $10^4 H^2$. Although it is not a rigorous theorem, in all potentials I know this leads to more than ten thousand e-foldings of inflation, much more than the minimal number of e-foldings necessary to resolve the horizon problem. Thus, the requirement of small density fluctuations means that the universe is flattened to such a degree after the transition that even today it should be exponentially close to flat.

What then, if astrophysicists inform us that Ω is definitely less than unity?[16] All is not necessarily lost. It is possible that the universe is very flat, but the cosmological constant is not zero! The cosmological constant is known to be extraordinarily small compared to the grand unification or weak interaction scale, but there is no evidence that the cosmological constant is not comparable to the present critical density. In this case inflation would drive the universe very close to the flat condition, which constrains the sum of the ordinary energy density and the vacuum energy density (equivalent to the cosmological constant) to be equal to ρ_c; but the ordinary energy density, which is what is measured by astrophysicists, may not equal ρ_c. There still remains some predictive power to inflation in this case, since there is a constraint on the difference between the ordinary energy density and the vacuum energy density, which is proportional to the

deceleration of the universe. The experimental bound on the deceleration is not very stringent nor are measurements of the age of
the universe very precise. Nevertheless, putting these constraints
together with the condition that our universe is (nearly) perfectly
flat due to inflation, Frampton and Lipton have recently concluded
that Ω must be less than 2.3 in inflationary scenarios.[17] Thus, a
loop-hole does exist if astrophysics can prove that Ω is not unity,
but one is then left with the mystery of explaining why the
cosmological constant should be so small, yet non-zero.

A second issue which may be used to test the inflationary
universe scenario is the monopole abundance. In spite of present
searches for monopoles, no convincing evidence exists that even one
monopole has been detected.[18] In addition, we have numerous
theoretical bounds that make the prospects of observing monopoles
in the near future highly unlikely.[19] Inflation at this point
provides the only workable mechanism for getting rid of the
monopoles produced by the grand unified phase transition. However,
suppose that an observable number of GUT monopoles are observed,
albeit with densities consistent with the theoretical bounds. Such
a small but observable density cannot be explained by any other
theory, but can it be made consistent with inflation?

If the inflationary phase transition is the GUT phase transition or one that occurs after it, any monopoles produced before the
inflation are separated by distances larger than the radius of our
observable universe by the inflationary process, even if the amount
of inflation were the minimal inflation necessary to resolve the
horizon problem (60 e-foldings). One might wonder about monopoles
produced through thermal processes once the universe has fully
reheated, but the maximum reheating temperature is of order the GUT
scale, and the mass of a monopole is at least a factor of 100
greater. Thus, there is a devastating Boltzman suppression factor
(the exponent is proportional to the monopole mass over the
reheating temperature) which makes the number of monopole pairs
produced after the phase transition unobservably small. Some,
however, have considered the possibilities that an "interesting"
number of monopoles might be produced at the end of the inflationary process or during reheating. During inflation the expectation
value of ϕ is much smaller than in the symmetry breaking state and
so the monopole mass, which is proportional to ϕ, is much lighter.
Furthermore, during inflation there always remains a minimal
"Hawking temperature" associated with the quantum fluctuations in
de Sitter space. Perhaps, it might be argued,[20] some interesting
number of monopoles might be produced while the monopole mass is
less than or comparable to the Hawking temperature (ϕ is less than
of $O(H)$). Others have proposed that during reheating when ϕ is
oscillating about the symmetry breaking minimum there is some
enhancement of the number of monopoles produced.[21] During the
oscillations ϕ spends a certain amount of time with an

expectation value much smaller than the symmetry breaking minimum, so the monopole mass is smaller; during that point in the oscillation the Boltzmann suppression factor might be less severe. Before the issue of the density perturbations was raised both of these suggestions seemed plausible; in Coleman-Weinberg SU(5) models,[3-4] $\phi \sim H$ (the Hawking temperature) and the oscillations could carry ϕ to values comparable to the reheating temperature.

Both of these approaches to enhancing monopole production are made impossible by the constraints imposed by the density fluctuations. In order to have small enough density fluctuations, a potential must be arranged where the potential is flat for a range of ϕ much larger than H so that $\dot{\phi}$ is much larger than H^2 (see Eq.(2)) by a factor of 10^4 or more. What is required is a "long, squat" potential in which ϕ in the symmetry breaking state is much larger than the scale which determines the height of the potential. During the last 60 e-foldings of inflation in such a model, ϕ is much greater than H (the Hawking temperature) and the monopole mass is much greater than $0(10^4)$ or more times the Hawking temperature; thus, only an insignificant number of monopoles can be produced. Similarly, in spite of the oscillations during reheating, the scale which determines the height of the potential and, consequently, the maximum possible reheating temperature remains much smaller than ϕ. Thus, achieving small enough density fluctuations automatically suppresses all means of producing an "interesting" monopole number.

Is there any hope at all of achieving an "interesting" (GUT) monopole number with inflation? All schemes of suppressing monopoles, even aside from inflation, have the characteristic that they tend to suppress them totally or not at all; but I can conceive of one way of achieving an interesting monopole number. Suppose that the transition responsible for inflation is not due to a GUT model, but to some other physics (e.g. supersymmetry). Suppose, furthermore, that the transition has a critical temperature greater than the GUT phase transition temperature. Then the universe can reheat after the transition to a temperature greater than the GUT transition temperature; if the reheating temperature is also less than the monopole mass (a factor of 100 or so times the GUT transition temperature) there will be some Boltzmann suppression of monopole pair production after the phase transition, but by controlling the reheating temperature one can control the number of monopoles produced. The scenario would proceed as follows: The universe would first supercool in the inflationary transition. The supercooling would drive the temperature well below the GUT phase transition temperature. The GUT transition can easily be arranged so that it is completed well before the inflation is completed. Any monopoles produced at the time are separated by inflation to scales large compared to our observable universe. Next, the universe reheats to a temperature between the GUT transition temperature and the monopole mass.   Monopoles are pair produced in numbers, if we choose the reheating temperature properly, that can be "interesting".   Then the universe proceeds to cool as in the ordinary hot big bang model.   One may be concerned that, because the universe has reheated to a temperature greater than the GUT transition temperature, the universe will undergo the GUT transition again and produce too many monopoles.   However, this can be avoided if the reheating temperature is only moderately (an order of magnitude or so) above the GUT transition temperature.   The GUT transition can easily be arranged so that it superheats; that is, when the universe reheats to above the GUT transiton temperature, the GUT symmetric phase becomes more energetically favored, but an energy barrier keeps the universe trapped in the symmetry breaking phase.   After reheating, the universe just remains in the symmetry breaking phase.

In a sense, the conclusions of this section are somewhat unfortunate; for they show that there are not hard and fast predictions of the inflationary universe scenario.   It is still fair to say that the most natural expectations for an inflationary scenario is that $\Omega = 1$ and there are essentially no (GUT) monopoles in our observable universe.   However, there is enough flexibility to soften both of these predictions.   Only time will tell if this is an unfortunate feature or a saving grace.

WHAT PHYSICS PROVIDES THE SLOW ROLLOVER PHASE TRANSITION?

As the inflationary universe scenario was originally conceived, the phase transition responsible for the exponential expansion was the phase transition associated with the spontaneous breaking of the grand unified symmetry.[2]   Grand unified theories naturally require there to be a phase transition associated with the spontaneous breaking of the grand unified symmetry, and there seemed to be enough freedom to adjust the parameters of a GUT model so that the transition can be appropriate for inflation.   Furthermore, the inflation has to take place after (or around the same time of) monopole and domain wall production yet before baryon asymmetry is generated (since inflation dilutes any baryon asymmetry before the phase transition to essentially nothing) to solve the monopole and domain wall problems while maintaining the successful resolution of the baryon asymmetry problem.   The GUT transition is the obvious candidate for inflation.   In fact, the initial candidate for a slow rollover phase transition was an SU(5) grand unified theory of the Coleman-Weinberg type.[3,4]   The Coleman-Weinberg type potential is the flattest possible near $\phi = 0$ and so is a natural candidate for the new inflationary universe scenario. However, with the study of density perturbations in inflationary universe models, it became clear that even the Coleman-Weinberg GUT models were not flat enough near $\phi = 0$.[14]   Radiative corrections to the effective potential always spoil attempts to make the potential flat enough over a long enough range of ϕ. Only by fine-tuning couplings of the Higgs particle to be extraordinarily small compared to unity can one hope to reduce the radiative corrections, but such small couplings are not desirable for phenomenological reasons.

For these reasons, several groups have begun to examine locally supersymmetric (or supergravity) models as candidates for providing the slow rollover transition required for a successful inflationary universe model.[22-24] The radiative corrections are controllable in supersymmetric models and "flat" potentials can be maintained with all radiative corrections taken into account. Although there have been several proposals for supersymmetric inflationary models over the past year,[22-23] in this section I will be discussing a systematic investigation of supersymmetric inflationary models that has just recently been completed by Burt Ovrut and myself.[24] Our studies indicate that all previous attempts at supersymmetric inflationary models have serious flaws; at the same time, we propose some new and exciting possibilities.

To set the context for discussing the results of the new study, it is worthwhile briefly reviewing the previous approaches to supersymmetric inflation. The previous efforts can be divided into two categories. In the first category,[22] only a special subclass of supersymmetric potentials was studied -- the O'Raifeartaigh models.[25] This class of supersymmetric models was studied for several reasons. First, O'Raifeartaigh models appeared, at the time, to be attractive models for phenomenological purposes; examples of such models are geometric (or "inverse") hierarchy models[26] and Polonyi potentials.[27] Second, the models automatically provide phase transitions in which the effective potential is rather flat without any fine-tuning of parameters. Let ϕ be the expectation value of the field which serves as the order parameter for the phase transition and let M be the fundamental mass scale in the theory. Then for ϕ greater than M the globally supersymmetric effective potential decreases logarithmically with increasing ϕ for an O'Raifeartaigh potential without any fine-tuning of parameters. Thus, the potential becomes flatter and flatter with increasing ϕ, ideal for inflation. In the investigation of inflationary O'Raifeartaigh models it was presumed that, for large ϕ (near the Planck scale), gravitational effects would cause the potential to steepen, forming a minimum near $\phi = M_p$ which could be adjusted to have zero cosmological constant. As ϕ evolves along the logarithmic portion of the potential the universe inflates and then ϕ oscillates around the minimum of the potential.

As originally advertised, the model was described as inflating but not reheating.[22] The problem is that the ϕ responsible for the inflation is necessarily highly decoupled from ordinary matter and

radiation in the O'Raifeartaigh models. (The coupling is suppressed by many powers of $1/M_p$). Describing the model as not reheating
is a bit of an exaggeration, though, since as long as there is any
non-zero coupling there will eventually be some time when the
energy density once again becomes dominated by ordinary matter and
radiation.[28],[8] For $M = 10^{12}$ GeV as appears in the Dimopoulos-Raby
geometric hierarchy model,[26] the reheating temperature is of order
10 MeV or so. This is much too low to generate a baryon asymmetry
after the transition; but even in this very depressing case the
reheating temperature is sufficient to recover all the other
successful predictions of the hot big bang model. Furthermore, the
reheating temperature is extremely sensitive to M; by raising M to
10^{16} GeV or so the reheating temperature can be raised to 10^{10} GeV,
marginally sufficient for baryon asymmetry generation.[28] At the
time this approach to supersymmetric inflation was proposed,
though, raising M to this large value was not phenomenologically
attractive; therefore, the O'Raifeartaigh approach to inflation lay
on shaky foundations when Ovrut and I began our investigation.

A second approach to supersymmetric inflation has been to
introduce a new superfield into the theory called an "inflaton"
whose sole purpose is to insure that the conditions for a
successful inflationary universe are met. The scalar component of
the superfield acts as the order parameter for the inflationary
phase transition. In the superpotential the inflaton couples only
to itself so that, in the absence of gravity, the effective
potential for the inflaton depends only upon its self couplings.
The inflaton plays no role in ordinary phenomenology. The self
couplings in the superpotential can be adjusted term by term so
that the effective potential has the correct form for a successful
inflationary model (or so it was thought). Reheating is possible
after inflation because gravity induces some couplings between the
inflaton field and the superpotential to ordinary particles (see
discussion of Eq.(3) below). This approach is not "pretty" since
an extra field must be introduced with specially adjusted couplings, but it could at least serve as an example to show that all
the constraints for a successful inflationary model can be met.

Surprisingly, the investigation by Burt Ovrut and myself
indicates that neither of these previous approaches to supersymmetric inflation is workable. The fact that the first approach does
not work may not seem surprising since it already seemed that
reheating is problematic in the model. However, it will be shown
that the problem with inflationary O'Raifeartaigh models has
nothing to do with reheating. The fact that the second approach is
unworkable is very surprising since it appears that one can
introduce through the superpotential even an infinite number of
adjustable parameters, if necessary to obtain a successful inflationary model.

The key feature of the investigation of the new investigation of locally supersymmetric potentials is that all the constraints necessary for a successful inflationary universe model have been systematically and simultaneously applied to a wide class of supersymmetric models. In the past, a few of the necessary constraints were considered and it was assumed that the others could be satisfied by adjusting extra parameters. However, super-symmetry turns out to be both subtle and malicious. Constraints that appear to have nothing to do with one another are subtly and non-linearly related. In many cases where it was known that two constraints could be satisfied individually we discovered that they could not be satisfied simultaneously by the same set of para-meters.

We begin by considering a completely general superpotential, W, which, for simplicity, we will take to be a function of a single superfield, ϕ. (The field can be thought of as analogous to the inflaton, or one can introduce couplings to other fields.)

$$W(\phi) = \mu^4 \sum_{n=0}^{\infty} a_n (\phi/M)^n, \qquad (3)$$

where $M^2 = 8\pi M_p^2$ and μ is the mass scale associated with this inflation sector of the theory. In terms of W the effective potential for ϕ (at zero temperature) is given by:

$$V(\phi) = \exp(|\phi|^2/M^2) \{|D_\phi W|^2 - (3/M^2)|W|^2\}, \qquad (4a)$$

$$= \sum_{n=0}^{\infty} c_n (\phi/M)^n \qquad (4b)$$

where

$$D_\phi W = \frac{\partial W}{\partial \phi} + \frac{\phi^\dagger}{M_p} W \qquad (5)$$

is the Kahler derivative of the superfield ϕ. $D_\phi W$ is a very significant physical quantity since it is zero for a state that is supersymmetric and non-zero for a state that is supersymmetry breaking (it can be thought of as the order parameter for super-symmetry.) We consider W's which include even an infinite number of terms (and adjustable parameters) in our investigation. If the theory were a globally supersymmetric model, a W with terms more than cubic in the fields would generate non-renormalizable terms in the effective potential and would be considered undesirable. However, in a locally-supersymmetric potential one always obtains an infinite number of non-renormalizable terms for any non-trivial W because of the exponential pre-factor in Eq.(4). I, for one,

believe that there is no reason not to consider terms of higher
order than cubic provided that they disappear when gravity is
turned off (M_p approaches infinity); this is the case for the
superpotentials we consider in Eq.(3).

The constraints for a successful inflationary universe scenario that have been considered by Ovrut and myself fall into three
categories. The first set of constraints can be called "slow
rollover" constraints; these are the constraints on the effective
potential V that demand: (1) that the potential be flat enough
near $\phi = 0$ to obtain sufficient inflation to solve the horizon,
flatness, monopole and domain wall problems; and (2) that the
potential be flat over the correct range so that perturbation in
the energy density is $\delta\rho/\rho \sim 10^{-4}$ on all observable scales, just what
cosmologists desire to explain the formation of galaxies. These
constraints have recently been codified into a precise prescription
for a completely general $V(\phi)$.[8] The slow rollover constraints are
most important for the linear and quadratic terms in $V(\phi)$; only by
tuning c_1 and c_2 to be much less than all the other c_n's can the
constraints be met. Roughly speaking, c_1 must be fine-tuned by
four orders of magnitude and c_2 by two orders of magnitude to
obtain a successful inflationary model. Perhaps this tuning can be
understood sometime in the future in terms of physical principles,
but at worst this is not horrific tuning. All other terms in W can
automatically be made to satisfy the slow rollover constraints
simply by adjusting μ to $O(10^{-(3-4)})$ times M. Notice that higher
and higher terms in ϕ are suppressed by factors of μ/M; thus,
higher order couplings can be made to be very small by this one
adjustment of μ in a supersymmetric model without any further
tuning of parameters.

The second constraint is the "cosmological constant" constraint which demands that $V(\phi)$ in the state to which we roll
during inflation is equal to zero. (The constraint is necessary
because the cosmological constant is measured to be zero in the
present phase of our universe.) In grand unified theories this is
a trivial constraint that can be satisfied by simply adding an
overall constant to $V(\phi)$ and adjusting it so that $V(\phi)$ is equal to
zero for the state corresponding to our present universe. In
supersymmetric models the constraint is much more subtle because
one cannot adjust an overall constant to $V(\phi)$ without explicitly
breaking supersymmetry (which destroys the whole purpose of considering supersymmetric models in the first place). A means of
setting the cosmological constant that preserves supersymmetry is
to add an overall constant to W and use it to adjust the
cosmological constant. However, in local supersymmetry, adjusting
the overall constant in W can affect an infinite number of terms in
the expansion of $V(\phi)$ because of the exponential prefactor in
Eq.(4). Setting the cosmological constant equal to zero can
interfere with attempts to make a flat potential. Thus, supersym-

metry connects two apparently unrelated sets of constraints: one a constraint on the potential for ϕ near zero (the high temperature phase) and the other a constraint on the potential near large ϕ (our present phase).

A third constraint that has been considered is a new one that we have called the "thermal constraint."[24] From the constraints that have already applied, it has been insured that the effective potential has the right shape between $\phi = 0$ and $\phi = \sigma$ (let's call $\phi = \sigma$, σ positive, the state to which ϕ rolls during inflation). However, one must also check that at high temperatures the favored state corresponds to ϕ equal to (or less than) zero so that a transition will occur in which ϕ rolls across the flat portion of the potential that has been designed according to the slow rollover constraints. If ϕ near σ is favored at high temperatures, which is possible in many cases, then as the temperature decreases ϕ will never have to roll over the flat region of the potential; there will be no inflation even though the slow rollover constraints have been satisfied. In Ref. 24 we describe how this "thermal constraint" is implemented technically. Basically, it represents a constraint on the relative signs of the first few terms in the superpotential, W (no further fine-tuning required).

At this point it may seem that we have applied an awesome set of constraints and that the whole picture is becoming unattractive. In my opinion this view is unwarranted. Although the array of constraints for a perfectly general (non-supersymmetric) potential may seem large, the number of parameters we have had to adjust in our supersymmetric potentials is just two, plus we have had to adjust the relative signs of a pair of other terms (I have not included in this count the additional fine-tuning that is required to make the cosmological constant zero since such an adjustment is required in any field theoretic model.) I do not believe that the supersymmetric models are significantly less attractive than the Coleman-Weinberg GUT models that served as previous candidates for inflation. Furthermore, in more recent schemes that Burt Ovrut and I have been considering, even less adjustment of parameters may be necessary.

The simplest non-trivial superpotential that can be studied is one which contains a constant term and a term linear in ϕ. Such a superpotential corresponds to a Polonyi potential,[27] the type of superpotential that is employed in so-called "hidden sector" theories.[29] The superpotential spontaneously breaks supersymmetry; for our purposes, the model is completely analogous to the more general O'Raifeartaigh potentials in which the field responsible for the inflation appears at most linearly in the superpotential. For even the most general potential of this type, though, the slow rollover constraints and the cosmological constant constraint cannot be satisfied simultaneously. In Fig. 2 two curves are shown.

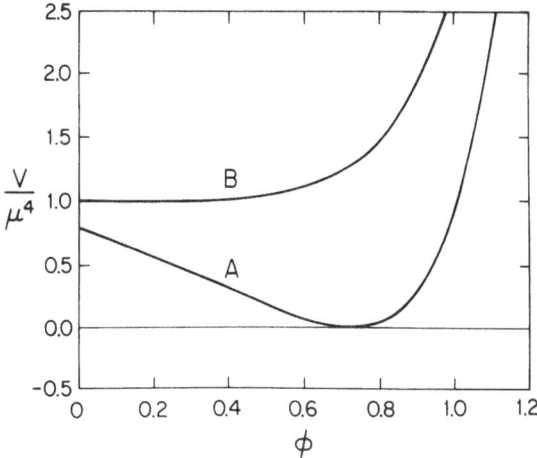

Fig. 2.   The energy density versus ϕ for the Polonyi model, plotted for the real ϕ direction.   Curve A is the result of satisfying the slow rollover constraints.   Curve B is the result of satisfying the cosmological constant constraint. The two constraints cannot be satisfied simultaneously.

In the upper curve the slow rollover constraints have been satisfied and the potential is flat near $\phi = 0$, but the energy density is always positive.   In the second curve the cosmological constant constraint has been satisfied, but it is obvious that the potential has a large linear term that spoils that flatness near $\phi = 0$.   Radiative corrections can add terms logarithmic in ϕ to the effective potential but always the large linear term induced by supergravity dominates.   Thus for the Polonyi and O'Raifeartaigh potentials a successful inflationary model is impossible, not because of problems with reheating as was originally supposed, but because gravitational corrections prevent the model from inflating in the first place.   (This presumes that the gravitational corrections are of the form demanded by local supersymmetry.   If one imbeds an O'Raifeartaigh model in a non-locally supersymmetric setting, in principle one could make a workable model; but there would be very poor motivation for considering such a model.)   Here is an example where the systematic application of all constraints can be a powerful tool in eliminating possibilities.

Superpotentials with ϕ terms higher order than just linear can then be considered.   However, even if one includes terms through fourth order in ϕ one finds that all the constraints for a successful inflationary model cannot be satisfied.   (Here the thermal constraint plays an important role in eliminating some possibilities.   Since there are ways of avoiding the thermal constraint with the addition of other fields, superpotentials which include terms at most cubic in the fields may also be workable.)

Once terms of fifth order or higher in ϕ are added to the superpotential, very interesting things begin to happen. (Again, some may regard having to consider such complicated superpotentials as unattractive, but aesthetic judgements should be put aside for the moment.) First, one can search for a superpotential which yields a transition from $\phi = 0$ to a state which is supersymmetric. This approach just corresponds to working out the details of the approach of Ref. 23. However, the big surprise is that no matter how many terms are added to the superpotential, no effective potential can be found which satisfies the inflationary constraints. The authors of Ref. 23 presumed that with enough free parameters it is always possible to satisfy all the constraints and roll to a supersymmetric state (with cosmological constant equal to zero), but here is where supersymmetry displays its maliciousness. Suppose one demands that the effective potential be flat near $\ell = 0$ and that there be a state at some $\phi = \sigma$ which is supersymmetric and has cosmological constant equal to zero. Fig. 3 displays the typical behavior. The constraints near $\phi = 0$ and $\phi = \sigma$ are met, but in between is horrible junk -- in particular there is a state with negative energy density which is the real state to which the universe would evolve in a slow rollover transition. One can prove that for a completely general superpotential there is always such a state with negative energy density separating $\phi = 0$ from the supersymmetric minimum $\phi = \sigma$ towards which the universe was supposed to roll.

Is there any hope left for supersymmetric inflation? The answer is a very strong yes! However, what should be considered is the case where the minimum towards which the slow rollover drives the universe is supersymmetry breaking. To a phenomenologist it may seem silly that this possibility was not suggested in the first place since, after all, our present phase of our universe is supersymmetry breaking. However, so far as I can tell, it was never noticed or, at least, never appreciated that the kind of superpotentials we have considered permit supersymmetry breaking minima. In fact, if only global supersymmetry had been considered, one can show that no such minimum is possible. Spontaneously breaking supersymmetry has always been difficult and there have previously been only two methods known to break supersymmetry -- the O'Raifeartaigh approach[25] and the Fayet-Iliopoulos approach,[30] the latter of which is not relevant because our inflationary sector does not involve gauge fields. A superpotential which includes terms up to fifth order in ϕ does not come under either of those categories and in global supersymmetry would only have (local) minima which are supersymmetric. In local supersymmetry one finds that the same superpotential can generate local minima which spontaneously break supersymmetry. (This difference is related to the fact that in global supersymmetry the sum over boson masses equals the sum over fermion masses and, if supersymmetry is spontaneously broken, the fermions are massless. In local

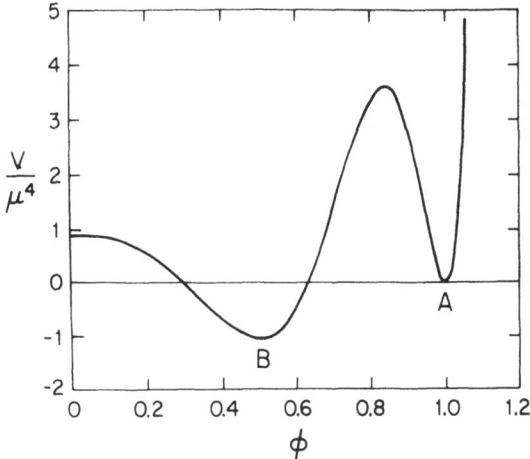

Fig. 3.   The energy density versus ϕ for a superpotential which
          satisfies the slow rollover constraints near $\phi = 0$ and
          which has a supersymmetric minimum with zero cosmological
          constant (point A on the curve).   The figure represents
          what typically results from trying to enact the program
          proposed in Ref. 23.   Between $\phi = 0$ and ϕ corresponding to
          point A there is always a state with negative energy
          density (point B), which is the real state to which the
          universe would roll in a slow rollover transition.   State
          B corresponds to a supersymmetric anti-de Sitter state.
          This approach is a disaster cosmologically.

supersymmetry this relation is modified and the sum over boson
masses is the sum of the fermion masses plus some factor times the
gravitino mass.   Thus, in global supersymmetry at least one boson
has negative mass in a supersymmetry breaking state but this need
not be the case in a locally supersymmetric model.)

        In fact, one can try to "push" the minimum with negative
energy density in Fig. 3 up to zero energy density provided one
relaxes the constraint that the minimum be supersymmetric.   One can
then obtain the effective potential shown in Fig. 4.   This
potential meets all the constraints for a successful inflationary
universe model.   The universe inflates by more than 10^4 e-foldings
(only 60 are necessary to satisfy the horizon, flatness, monopole
and domain wall problems) and the density fluctuations in this
model are $\delta\rho/\rho = 10^{-4}$.   We have satisfied the new thermal constraint and we have checked that the transition proceeds directly
from $\phi = 0$ to the supersymmetry breaking minimum without being
"side-tracked" towards other minima.   The reheating temperature is
computed to be of order 10^{10} GeV, marginally sufficient for baryon
asymmetry generation.

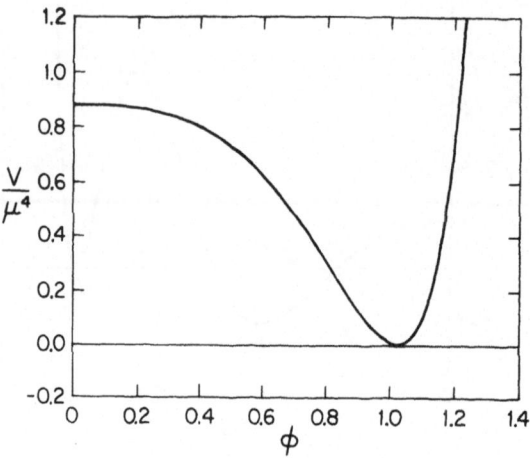

Fig. 4.   The energy density versus ϕ for a superpotential which
satisfies all the conditions for a successful inflationary
universe scenario.   This example corresponds to a superpotential:

$$W(\phi) = \mu^2 \left(\beta + \phi + \beta\phi^2 + \frac{\beta}{2} a_4\phi^4 + \frac{2\beta}{5} a_5\phi^5\right) \quad ,$$

where $\beta = .2$, $a_4 = -2.191$, and $a_5 = 1$.   There is a second
minimum for ϕ much greater than unity (not shown) which
preserves supersymmetry but which has negative energy
density.

At the same time we have been forced to stumble across a new
way to spontaneously break supersymmetry in locally supersymmetric
models.[31]   This discovery could potentially have important impact
in the phenomenology since it offers a new approach to making
realistic models. At the very least, this kind of superpotential
can be used to replace the Polonyi potentials that were introduced
in hidden sector theories to break supersymmetry.[29]   With this
superpotential in the place of Polonyi potentials, one breaks
supersymmetry and attains inflation; in fact, since the only
successful supersymmetric inflationary models (so far) require the
spontaneous breaking of supersymmetry this way, we have tied
together the notions of inflation and supersymmetry breaking in a
way that few would have thought possible previously.

One other curious feature of the effective potentials we have
generated is that, in all the cases found so far, the supersymmetry
breaking minimum which should correspond to our present universe is
metastable.   For much greater ϕ one finds a supersymmetric minimum
with negative energy density.   The two minima are separated by an

enormous barrier so the decay of our universe is exponentially suppressed.   Further, because the negative energy density minimum corresponds to ϕ beyond the Planck mass, maybe it should not be taken seriously.   Nevertheless, an automatic result of this ap- proach to supersymmetric inflation might be that our universe is necessarily metastable.   Thus, supersymmetric inflation has the potential of tying together three notions -- inflation, supersym- metry breaking, and the metastability of the universe -- that were totally unrelated previously.

Much more work is being done to integrate this kind of inflationary model into phenomenologically acceptable settings. We are still trying many possibilities that can lead to even simpler superpotentials that can satisfy all the inflationary constraints with even less adjustment of parameters.   I believe that there is every reason at this point to be optimistic.   At the very least though, we have demonstrated how the cosmological constraints associated with inflation can play an important role, even if only accidentally, in probing physics at very high energies.   I hope that this bodes well for the future development of the relation between particle physics and cosmology.

ACKNOWLEDGEMENTS

The author wishes to thank his colleagues Burt Ovrut and Michael Turner who were both major contributors to the work reviewed in this paper.   This work was supported in part by DoE under contract No. EY-76-C-02-3071 and the Alfred P. Sloan Foundation.

REFERENCES

1. For an excellent review of the inflationary universe scenario, see G. Gibbons, S. Hawking and S. Siklos The Very Early Universe, (Cambridge University Press, 1983).
2. A. Guth, Phys. Rev. D23, 347 (1981).
3. A. Linde, Phys. Lett. 108B, 389 (1982).
4. A. Albrecht and P. Steinhardt, Phys. Rev. Lett. 48, 1220 (1982).
5. P. Steinhardt, in The Birth of the Universe edited by J. Audouze and J. Tran Thanh Van (Editions Frontieres, France, 1982).
6. D. Kirshnitz and A. Linde, Phys. Lett. 42B, 471 (1972); L. Dolan and R. Jackiw, Phys. Rev. D9, 3320 (1974); S. Weinberg, Phys. Rev. D9, 3357 (1974).
7. A. Albrecht, P. Steinhardt, M. Turner and F. Wilczek, Phys. Rev. Lett. 48, 1437 (1982).
8. P. Steinhardt and M. Turner, "Prescription for Successful New

Inflation," University of Chicago Preprint (1983).

9. See, for example, E. Kolb and M. Turner, Ann. Rev. Nucl. Part. Sci. 335, (1983).

10. L. Abbott, E. Farhi, and M. Wise, Phys. Lett. 117B, 29 (1982); A. Dolgov and A. Linde, Phys. Lett. 116B, 329 (1982).

11. W. Rindler, Mon. Not. R. Astron. Soc. 116, 663 (1956).

12. R. Dicke and P.J.E. Peebles, in General Relativity: An Einstein Centenary Survey; edited by S. Hawking and W. Israel (Cambridge University Press, 1979).

13. J. Preskill, Phys. Rev. Lett. 43, 1365 (1979); Ya. Zeldovich and M. Khlopov, Phys. Lett. 79B, 239 (1978).

14. A. Guth and S.-Y. Pi, Phys. Rev. Lett. 49, 1110 (1982); S.W. Hawking, Phys. Lett. 115B, 295 (1982); A.A. Starobinskii, Phys. Lett. 117B, 175 (1982); J. Bardeen, P. Steinhardt and M. Turner, Phys. Rev. D28, 679 (1983).

15. Ya. Zeldovich, Mon. Not. Roy. Astron. Soc. 160, 1P (1972); E. Harrison, Phys. Rev. D1, 2726 (1970).

16. For a discussion of astrophysical limits on Ω, see J. Primack in Fourth Workshop on Grand Unification, edited by A. Weldon, P. Langacker, G. Segre and P. Steinhardt (Birkhauser Press 83)

17. P. Frampton and G. Lipton, "Age and Flatness of Inflationary Universe with Cosmological Constant," IFP-203-UNC (1983).

18. With the possible exception of B. Cabrera, Phys. Rev. lett. 48, 1378 (1982). See other papers in this proceedings.

19. See for example, J. Preskill, in The Very Early Universe, edited by G. Gibbons, S. Hawking and S. Siklos, (Cambridge University Press, 1983).

20. A. Guth, in Magnetic Monopoles edited by R. Carrigan and W.P. Trower (Plenum Press, New York, 1983).

21. M. Turner, private communication.

22. A. Albrecht, S. Dimopoulos, W. Fischler, E. Kolb, S. Raby and P. Steinhardt, Nucl. Phys. B in press (1983).

23. J. Ellis, D.V. Nanopoulos, K.A. Olive and K. Tamvakis, Nucl. Phys. B in press (1983); Phys. Lett. 120B, 331 (1983); D.V. Nanopoulos, K.A. Olive, M. Srednicki and K. Tamvakis, Phys. Lett. 124B, 171 (1983); Phys. Lett. 123B, 41 (1983); D.V. Nanopoulos, K.A. Olive, and M. Srednicki, Phys. Lett. 127B, 30 (1983).

24. B. Ovrut and P. Steinhardt, Phys. Lett. in press (1983).

25. L. O'Raifeartaigh, Nucl. Phys. B96 331 (1975).

26. S. Dimopoulos and S. Raby, Nucl. Phys. B219, 479 (1983).

27. J. Polonyi, Budapest Preprint KFK1-1977-93, unpublished. See also E. Cremmer, B. Julia, J. Scherk, P. van Nieuwenhuizen, S. Ferrara, L. Girardello, Phys. Lett. 79B, 231 (1978); Nucl. Phys. B147, 105 (1979).

28. A. Albrecht and P. Steinhardt, Phys. Lett. in press (1983).

29. See, for example, J. Polchinski, in Fourth Workshop on Grand Unification edited by A. Weldon, P. Langacker, G. Segre and P. Steinhardt (Birkhauser Press, 1983).

30. P. Fayet and J. Iliopoulos, Phys. Lett. 31B, 461 (1982).

31. B. Ovrut and S. Raby, Phys. Lett. in press (1983).

PRIMORDIAL INFLATION AND THE MONOPOLE PROBLEM

Keith A. Olive and David Seckel

Theoretical Astrophysics Group
Fermi National Accelerator Laboratory
Batavia, IL

We discuss the cosmological abundance of magnetic monopoles in locally supersymmetric GUTs and primordial inflation. Depending on the temperature scale of the GUT phase transition (Λ_5 in this model) monopoles may or may not be suppressed sufficiently to satisfy cosmological and astrophysical limits. For example, if the GUT transition occurs after inflation $\left(\Lambda_5 < T_H \sim 0 \ (10^{10}\text{-}10^{11} \ \text{GeV}), \text{ where } T_H \text{ is the temperature at which inflation occurs}\right)$ too many monopoles will be produced unless $\Lambda_5 < 10^9$ GeV. Even then, although the cosmological density limits are satisfied, neutron star limits on the monopole abundance may rule this situation out. If on the other hand $\Lambda_5 > T_H$, SU(5) breaking may occur during inflation and hence the monopole abundance is greatly suppressed as it was in non-primordial inflation. We show that both scenarios are possible with the latter ($\Lambda_5 > T_H$) being preferred for monopole suppression.

Phase transitions in grand unified theories (GUTs) have drastically changed the scenario for the very early Universe by providing mechanisms for baryon generation[1] and inflation[2] with all its benefits. However, many of these phase transitions also produce magnetic monopoles,[3] and their history has been the cause of a number of worries to cosmologists. One of the benefits of new inflation[4,5] was to solve the magnetic monopole problem.[6] However, as models of inflation have become more complicated the GUT transition and the inflation transition have become separate phenomena in many models.[7] As a result the magnetic monopole problem has reappeared. In this paper we discuss how the magnetic monopole problem may be solved (or not solved) in variants of broken N = 1 supergravity primordial inflation.[8,9] We begin by reviewing the monopole problem and its solution in inflationary

models.   We then discuss difficulties with a recent suggestion by
Linde[10] to solve the problem in primordial inflation. Finally, we
suggest some modifications to Linde's arguments that may be more
successful.

The monopole problem[5] arises whenever a simple group breaks
to   one   containing   an   explicit U(1) factor,[3] such   as
SU(5)→SU(3)×SU(2)×U(1).   In general, it is difficult to calculate
the precise number of monopoles produced by a GUT transition.
However, for a first order transition one can estimate that there
should be roughly one monopole per bubble of the new phase.   It
has been shown[6] that if the GUT transition takes place at
$T_c \sim 10^{15}$ GeV, the number of monopoles will greatly exceed their
cosmologically acceptable abundance.

There have been a number of approaches to calculating the
monopole abundance[6,11,12] but here we will outline only the
simplest (however naive) approach to the problem.   In an adiabatically expanding Friedmann-Robertson-Walker Universe, the expansion timescale can be related to the temperature by

$$t = (90/32\ \pi^3\ g_T)^{1/2}\ M_p/T^2 \tag{1}$$

where $M_p \simeq 1.2\times10^{19}$ GeV is the Planck mass and g_T is the total
number of relativistic degrees of freedom at the temperature T.
Monopoles are produced during a GUT phase transition due to a lack
of correlation among the Higgs fields driving the transition.[3]
These scalars could be correlated over distances of at most the
horizon scale, 2t.   If we then assume that at least one monopole
is produced per horizon volume at temperature T_c, then the number
density of monopoles is

$$n_m \sim (2t)^{-3} \sim 4\times10^3\ T_c^6/M_p^3 \tag{2}$$

when compared to the number of photons present at T_c

$$n_\gamma \sim (g_T/4)T_c^3 \tag{3}$$

the monopole to photon ratio is

$$r \equiv n_m/n_\gamma \sim 10^4\ (T_c/M_p)^3/g_T. \tag{4}$$

This is actually an underestimate,[6,11,12] because the Higgs
correlation length is probably much smaller than the horizon.   A
more realistic estimate using one monopole per bubble in a first
order transition might be[13]

$$r \sim 10^6\ (T_c/M_p)^3. \tag{5}$$

Even if a large number of monopoles are produced their number will be reduced by monopole-antimonopole annihilation until[6,11]

$$r \sim \left(1/g_T^{1/2}\right)\alpha_5^3 \, m/M_p \simeq 10^{-10} \qquad (6)$$

where α_5 is the SU(5) fine structure constant and $m \approx 10^{16}$ GeV is the mass of the monopole. In some non-standard scenarios[14] the ratio, r, could be further reduced.

To see that there is a problem, one must compare the predicted abundance with the allowed cosmological and astrophysical bounds. The surest limit comes from the overall mass density of the Universe.[6] The energy density stored in monopoles of mass m (in GeV) and density n_m is constrained by

$$\rho_m = m \, n_m < \Omega \rho_c \qquad (7)$$

where Ω is the ratio of the total mass density to the critical density $\rho_c = 1.88 \times 10^{-29} \, h_o^2 \, g \, cm^{-3}$, $h_o = (H_o/100 \, km \, M_{pc}^{-1} \, s^{-1})$ and H_o is the present value of the Hubble parameter. The most conservative bounds on the mass density and the Hubble parameter require $\Omega h_o^2 < 4$, however consistency with the age of the Universe requires a tighter limit on $\Omega h_o^2 \lesssim 1/4$ (see ref. 15). The monopole to photon ratio is then constrained by

$$r \lesssim 10^{-8}/m \qquad (8)$$

comparing this result with (6) one has a monopole problem.

There are other limits on the monopole abundance which are comparable[16] and stronger.[17] However, these limits are more model dependent and we only quote their results. Limits coming from the survival of galactic magnetic fields[16] can be expressed as

$$r \lesssim 10^{-38} \, m\beta \qquad (9a)$$

where β is the monopole velocity (typically $\beta \sim 10^{-3}$). For m and β this limit is comparable to the cosmological limit (8). A much stronger limit comes with the inclusion of baryon number violating interactions around a monopole.[18] Limits coming from neutron stars[17] are roughly

$$r \lesssim 10^{-29} \, \beta/\sigma_o \qquad (9b)$$

where σ_o is the magnitude of the baryon number violating cross-section normalized to a typical strong cross-section ($\sigma \sim 4 \times 10^{-28} \, cm^2$).

If taken seriously, both limits put strong constraints on the primordial abundance of monopoles.

A possible solution to the monopole problem[6] lies in the inflationary Universe scenario.[5] The general idea is that bubbles of the broken symmetric phase get expanded by the inflation. After the phase transition the Higgs coherence length is then much longer than the horizon size. For example, the "new" inflationary scenario[4,5] offers the possibility that there is only a single monopole inside the visible portion of the Universe. This is because in these models the whole Universe originates from a single bubble during the GUT phase transition. Unfortunately, these models cannot produce the perturbation spectrum necessary to explain the large scale structure of the Universe.[19]

A class of supersymmetric models known as primordial inflation[7,8,9] offers the possibility of explaining the large scale structure of the Universe as well as having the benefits of inflation. We now wish to discuss the monopole problem in the context of primordial inflation. In these models, inflation is no longer associated with the breakdown of SU(5), but rather, it is due to a phase transition involving an SU(5) singlet, the inflation, ϕ. In such models, it is possible to write down a single superpotential, f, which will describe all scalar inter-actions and the evolution of the Universe from the Planck time till the present. However, because inflation has been separated from SU(5) breaking, one must take special care that the number of monopoles produced during the SU(5) transition still satisfies the cosmological and astrophysical limits. For example, in the models[8,9] utilizing $N = 1$ supergravity, inflation occurs at $T_H \sim 10^{11}$ GeV, where $T_H = H/2\pi$ and H is the Hubble parameter. The model can be arranged so that SU(5) is broken at $T \sim 10^9$ GeV and, hence, after inflation has occurred. In this case, the number of monopoles will not be inflated away. Although $r \, \eta \, 10^6$ $(T_c/M)^3 \sim 10^{-25}$ might satisfy the cosmological density limit (8), it does not come close to the neutron star limit (9b). Hence there may still be a monopole problem. To illustrate these remarks we would like to work with the following model.[8,9,20] The superpotential is divided into three main parts

$$f = f_I + f_5 + f_B \tag{10}$$

where f_I solely accounts for inflation[6]

$$f_I = m_I^3 \left[\sum_{n=0}^{\infty} \left(\lambda_n/(n+1) \right)(\phi/M)^{n+1} + \lambda' \right] \tag{11}$$

where ϕ is the inflation and m_I is an overall scale for the super-potential and $M = M_p/\sqrt{8\pi}$. To get sufficient inflation, couplings

λ_i are all $O(10^{-3})-O(1)$ with $m_I \sim O(10^{-2})M$. The f_5 part accounts for the breaking of SU(5)[20] and local supersymmetry[21]

$$f_5 = (a_1/M)X^4 + (a_2/M^2)X^2 Tr(\Sigma^3) + h(z) \tag{12}$$

where Σ is the adjoint, X and z are SU(5) singlets and the coupling $a_i \sim O(1)$. The function $h(z)$ is normally taken to be

$$h(z) = \mu^2(z+\Delta) \tag{13}$$

and is the hidden sector used to break local supersymmetry. This simple form for h is known[22] to have problems. After inflation the z field is not exactly at its global minimum and it is difficult to dissipate the energy stored in the z field. More complicated forms for h may[23] resolve this problem. Finally, f_B is used to account for baryon generation.[9,20] The details of f_B are not relevant for this discussion.

To analyze the behavior of the model, one must write down the scalar potential in terms of f. For N = 1 supergravity,[24] the minimal expression is

$$V(y_i) = e^{\Sigma |y_i|^2/M^2} \left[|f_{y_i}|^2 - 3|f|^2/M^2 \right] \tag{14}$$

where

$$f_{y_i} = \partial f/\partial y_i + y^* f/M^2 \tag{15}$$

for each chiral supermultiplet y_i. In this model (Eqns. 10-12) one can separate out, to a large extent, the inflation self-interactions from the rest of V. One then has a scalar potential $V(\phi,\phi^*)$ in terms of the λ_i which will be required to meet the conditions of sufficient inflation.[7] The goal is to have a potential with a barrier near the origin while still being very flat to allow a long roll over time scale to a local minimum at $\langle\phi\rangle \sim M$. Recently it has been pointed out[5,25] that the effects of finite temperature[8] render this scheme inconsistent by proving the existence of a local minimum at some $\phi_o < M$ with $V(\phi_o) < 0$. Thus the inflation would produce a negative[25] cosmological constant. However, the finite temperature corrections used were computed neglecting the effects of other chiral supermultiplets in the model. It turns out that these are dominant and a suitable change of signs of the couplings λ_i (among other possible variations) will negate the existence of the troublesome minimum. It should also be noted that trying to use this minimum as a broken supersymmetric minimum with zero cosmological constant as in Ref. 25 would be disastrous. Supersymmetry would then be broken at a scale $M_S \sim 10^{16}$ GeV producing a gravitino of mass $m_{3/2} \sim 10^{13}$ GeV and thus destroying the gauge hierarchy and one of the original motivations for supersymmetry.

Using the procedure of Eqs. 14 and 15, we can write down the scalar potential for X and Σ. Before looking at this, it is important to realize that the breaking of SU(5) in supersymmetric theories is complicated by the existence of several degenerate vacua.[26] At zero temperature in globally supersymmetric models there are at least three degenerate minima, SU(5), SU(3)×SU(2)×U(1) and SU(4)×U(1). At finite temperature, the degeneracy is broken by the differing number of particle degrees of freedom in each phase, with SU(5) being the lowest minimum. This presents a problem as to how one ever gets out of SU(5). To accomplish this it has been observed[26,27] that when the temperature drops below Λ_5 (defined to be the scale at which the GUT fine structure constant, α_5, becomes O(1)) the above picture breaks down. At T ~ Λ_5 strong coupling phenomena decreases the effective number of degrees of freedom in the SU(5) phase. This effect may push the SU(5) minimum above the others. The tunnelling rate out of SU(5) will be vanishingly small unless the barrier separating these phases is kept small.[28] This can be done by either introducing a small coupling[27] λ ~ 10^{-14} or by looking at N = 1 supergravity[26] with couplings of order unity.

In the superpotential (10), f_5 (12) will accomplish the above without small couplings. The scalar potential for (14) can be simplified to

$$V/M^4 = |\chi^3 + \chi\sigma^3|^2 + |\chi^2\sigma^2|^2$$
$$+ \epsilon(\chi^4 + \chi\sigma^3 + \text{h.c.}) + \epsilon^2(|\chi|^2 + |\sigma|^2) \qquad (16)$$

where χ = X/M; σ = Σ/M and ϵ = $(\mu/M)^2 \approx 10^{-16}$ for a supersymmetry breaking scale μ ~ 10^{10} GeV. In the analysis of this potential (16), it was found[20] that the global minimum is an SU(3)×SU(2)×U(1) symmetric minimum with

$$\langle\Sigma\rangle = \epsilon^{1/4}M = (\mu M)^{1/2} \sim 10^{15} \text{ GeV} \qquad (17a)$$

and

$$\langle X\rangle = \epsilon^{3/8}M = 10^{-6}M \qquad (17b)$$

and hence also predicts the GUT scale from the planck scale and the supersymmetry breaking scale μ. (The weak scale is governed by the gravitino mass scale which is ϵM ~ μ^2/M ~ 10^2 GeV). Furthermore, the barrier height is only $\epsilon^{5/2}M^4$ ~ $(10^8 \text{ GeV})^4$, without small couplings. It is possible, therefore, that if Λ_5 > 10^8 GeV, strong coupling phenomena will be able to drive the transition.

In the past, we have discussed values of $\Lambda_5 \sim 0(10^9 \text{ GeV})$. The goal was to push Λ_5 (and hence T_c) as low as possible to produce few monopoles. For example, with this value of Λ_5, we might expect a monopole abundance $r \sim 0(10^{-26})$. This is sufficient for the cosmological (8) and galactic magnetic field (9a) limits, but falls six orders of magnitude short of the neutron star limits (9b). In fact, to make r consistent with (9b) one would need $T_c \sim \Lambda_5 \sim 10^7$ GeV. However, this value could no longer drive a phase transition over a barrier of 10^8 GeV and would also begin to make baryon generation very difficult.

Before proceeding, we note that in this scenario the breaking of SU(5) occurs after the exponentially expanding phase. Although $\langle\phi\rangle \approx M$, the Hubble parameter at the time of inflation is

$$H^2 \approx 1/3 \, m_I^6 \, \lambda_o^2 \sim 5\times10^{-14} \, M^2 \tag{18}$$

$$H \sim 5 \times 10^{11} \text{ GeV}$$

Thus inflation takes plase at $T_H \sim 10^{11}$ GeV, i.e., before the breaking of SU(5) at $T_c \sim \Lambda_5$.

Recently Linde[10] has argued that SU(5) will break during primordial inflation (even if $T_H > \Lambda_5$) and as a result the monopole abundance will dilute away exponentially. We believe that although the general idea is a good one the mechanism is not correct. Linde's argument depends on scalar field fluctuations in de Sitter space.[29] Define the total mass[2]

$$D = M_o^2 + cT^2 + 12\lambda^2\langle\Sigma^2\rangle \tag{19}$$

where M_o^2 is a bare mass and cT^2 is the Σ^2 coefficient of the temperature correction to the effective potential

$$V_T =(T^2/8)\partial^2V/\partial\Sigma\partial\Sigma* \tag{20}$$

and λ is the Σ^4 coupling. In the limit $D \ll H^2$, one finds that the fluctuations in the adjoint scalar field are[29]

$$\langle\Sigma^2\rangle \sim H^4/D \tag{21}$$

which must then be solved self consistently. Linde argues that D is small, so that $\langle\Sigma^2\rangle$ is large enough to populate the different symmetry breaking minima. The probability of finding our universe to be in the 3-2-1 phase is then a fraction of $0(1)$.

We believe that D is not small. In (19) the $\lambda^2\langle\Sigma^2\rangle$ term is negligible compared to the other terms because λ is so small in the effective potential derived from (12). (The Σ^4 term is in fact negligible to the terms included in (16) and for that reason left out.) The M_o^2 term is also very small with a value given by

$$M_o^2 \sim \mu^4/M^2 \sim (100 \text{ GeV})^2. \tag{22}$$

The temperature correction term comes from two sources. The first term arises from the F term in the potential. It is small because it is proportional to λ. However, there is a second contribution coming from the D-term and it is proportional to α_5. Therefore, the fluctuations are of order

$$\langle \Sigma^2 \rangle \sim H^4/\alpha_5 \, T_H^2 \sim 4\pi^2 H^2/\alpha_5. \tag{23}$$

These are not large enough to drive the transition.

The solution to the monopole problem can still be realized if we can raise the value of Λ_5 and hence T_c to a value $> T_H$. In fact, by modifying the Higgs sector and by choosing slightly different initial parameters one can find a value $\Lambda_5 \sim 10^{12}$ GeV. With this value of Λ_5, the SU(5) transition will take place during the epoch of primordial inflation[7-10] and the monopoles are inflated away. The transition may take place in two ways; first, as previously described,[20] via strongly coupling effects. In that case strong coupling effects are of order Λ_5^4 near the origin. We[30] warn the reader that the potential (16) is now altered by the fact that ϕ is not at its global minimum. The effect is to raise the barrier from $\varepsilon^{5/2}M^4$ to $\left(f_I(0)/M^3 \right)^{5/2} M^4$. In this model $f_I(0) \sim M_I^3 \lambda$ and must hence be constrained so that

$$\left(f_I(0)/M^3 \right)^{5/8} < \Lambda_5/M. \tag{24}$$

Note that the barrier is not affected by the D-term finite temperature correction, as the barrier is at $\Sigma \gg T_H$ or Λ_5. The correction is cut off by $\exp(-\Sigma/T)$. Another possibility is present for the SU(5) transition in the case of primordial inflation. Suppose strong coupling phenomena do not drive the transition, but instead Σ finds itself in a new minimum near the origin. Depending on the parameters, if the tunnelling rate through this barrier (the tunnelling rate is still very low) is greater than the rate for forming a bubble due to the inflation transition, one could simply wait inside an SU(3)×SU(2)×U(1) bubble for inflation to occur. So long as the probability of tunnelling to the 3-2-1 phase is comparable to the 4-1 phase one does not have to worry about which one is closest.[31]

To summarize, we have seen that contrary to past fears, primordial inflation coupled to supersymmetric SU(5) breaking can cure the monopole problem. Independent of whether the SU(5) transition proceeds via strong coupling phenomena or tunnelling the monopole abundance is exponentially reduced if $\Phi_5 > T_H$. For 10^8 GeV $< \Lambda_5 < 10^9$ GeV the monopole abundance is low enough to pass the cosmological density limits but probably not the neutron star limits. Other versions of supersymmetric SU(5) breaking in which the SU(5) transition takes place before primordial inflation offer the same types of solution.

ACKNOWLEDGEMENTS

K.A.O. would like to thank J. Ellis, G. Gelmini, D.V. Nanopoulos, M. Srednicki and K. Tamvakis for a series of fruitful collaborations.

REFERENCES

1. A.D. Sakharov, ZhETF Pis'ma 5, 32 (1967).
2. A.H. Guth, Phys. Rev. D 23, 347 (1981).
3. G. 't Hooft, Nucl. Phys. B 79, 276 (1974); A.M. Polyakov, Zh. Eksp. Teor. Fiz. 20, 430 (1974); JETP Lett. 20, 194 (1974); E.J. Weinberg, these proceedings.
4. A.D. Linde, Phys. Lett 108B, 389 (1982); A. Albrecht and P.J. Steinhardt, Phys. Rev. Lett 48, 1220 (1982); for a review see: A.D. Linde, P.N. Lebedev, Physical Institute preprint Nos. 30 and 50, to be published in The Very Early Universe, ed. by S.W. Hawking, G.W. Gibbons, and S. Siklos (1983).
5. P.J. Steinhardt, these proceedings.
6. Ya.B. Zel'dovich and M.Y. Khlopov, Phys. Lett 79B, 239 (1978); J.P. Preskill, Phys. Rev. Lett 43, 1365 (1979); D.A. Dicus, D.N. Page, and V.L. Teplitz, Phys. Rev. D 26, 1306 (1982).
7. J. Ellis, D.V. Nanopoulos, K.A. Olive, and K. Tamvakis, Phys. Lett. 118B, 335 (1982); Nucl. Phys. B 221, 524 (1983); Phys. Lett. 120B, 331 (1983).
8. D.V. Nanopoulos, K.A. Olive, M. Srednicki, and K. Tamvakis, Phys. Lett. 123B, 41 (1983); G.B. Gelmini, D.V. Nanopoulos, and K.A. Olive, CERN preprint TH. 3629, Phys. Lett. B (in press) (1983).
9. D.V. Nanopoulos, K.A. Olive, and M. Srednicki, Phys. Lett 127B, 30 (1983); see also, K.A. Olive, CERN preprint TH. 3587 to be published in the Proc. of the 3rd Moriond Astrophysics Meeting, ed. by J. Audouze and J. Tran Thanh Van (1983).
10. A.D. Linde, P.N. Lebedev, Physical Institute preprints No. 151 and "Inflation Can Break Symmetry in SUSY" (1983).
11. T. Goldman, E.W. Kolb, and D. Toussaint, Phys. Rev. D 23, 867 (1981); D.A. Dicus, D.N. Page, and V.L. Teplitz, Phys. Rev. D 26, 1306 (1982).

12. A.H. Guth and Si H.H. Tye, Phys. Rev. Lett. $\underline{44}$, 631, 963
    (1980); M.B. Einhorn, D.L. Stein and D. Toussaint, Phys. Rev.
    D $\underline{21}$, 3295 (1980); F.A. Bais and S. Rudaz, Nucl. Phys. B $\underline{170}$
    [F51], 507 (1980); A.H. Guth and E.J. Weinberg, Phys. Rev. D
    $\underline{23}$, 876 (1981); M.B. Einhorn and K. Sato, Nucl. Phys. B $\underline{180}$
    [F52], 385 (1981); G. Lazarides, Q. Shafi and T.F. Walsh,
    Phys. Lett. $\underline{100B}$, 21 (1981).
13. E.J. Weinberg, private communication (1983).
14. P. Langacker and S.-Y. Pi, Phys. Rev. Lett. $\underline{45}$, 1 (1980); E.J.
    Weinberg, Phys. Lett. $\underline{126B}$, 441 (1983).
15. K. Freese and D.N. Schramm, submitted to Nucl. Phys. B (1983),
    EFI preprint 83-22.
16. E.N. Parker, these proceedings.
17. E.W. Kolb, these proceedings.
18. V.A. Rubakov, Zh. Eksp. Teor. Fiz Pis'ma Red. $\underline{33}$, 6658 (1981);
    JETP Lett. $\underline{33}$, 644 (1981); Nucl. Phys. B $\underline{203}$, 311 (1982); C.G.
    Callan, Phys. Rev. D $\underline{25}$, 2141 (1982); Nucl. Phys. B $\underline{212}$, 391
    (1983), and these proceedings.
19. S.W. Hawking, Phys. Lett. $\underline{115B}$, 295 (1982); A.A. Starobinskii,
    Phys. Lett. $\underline{117B}$, 175 (1982); A.H. Guth and S.-Y. Pi, Phys.
    Rev. Lett $\underline{49}$, 1110 (1982); J. Bardeen, P.J. Steinhardt, M.S.
    Turner, Phys. Rev. D $\underline{28}$, 679 (1983).
20. D.V. Nanopoulos, K.A. Olive, M. Srednicki, and K. Tamvakis,
    Phys. Lett. $\underline{124B}$, 171 (1983).
21. J. Polonyi, Budapest preprint KFKI-1977-93 (1977).
22. G.D. Coughlan, W. Fischler, E. Kolb, S. Raby and G.G. Ross,
    Los Alamos preprint LA-UR 83-1423 (1983).
23. M. Dine, W. Fischler, and D. Nemeschansky, preprint (1983).
24. E. Cremmer, B. Julia, J. Scherk, S. Ferrara, L. Girardello and
    P. Van Nieuwenhuizen, Nucl. Phys. B $\underline{147}$, 105 (1979); E.
    Cremmer, S. Ferrara, L. Girardello and A. Van Proeyen, Phys.
    Lett. $\underline{116B}$, 231 (1982); Nucl. Phys. B $\underline{212}$, 413 (1983).
25. B.A. Ovrut and P.J. Steinhardt, Rockefeller University pre-
    print RU 83/B/65.
26. M. Srednicki, Nucl. Phys. B $\underline{202}$, 327 (1982); D.V. Nanopoulos,
    and K. Tamvakis, Phys. Lett. $\underline{110B}$, 449 (1982).
27. D.V. Nanopoulos, K.A. Olive, and K. Tamvakis, Phys. Lett.
    $\underline{115B}$, 15 (1982).
28. M. Srednicki, Nucl. Phys. B $\underline{206}$, 132 (1982).
29. T.S. Burch and P.C.W. Davies, Proc. R. Soc. London A $\underline{360}$, 117
    (1978); A. Vilenkin, Phys. Lett. $\underline{115B}$, 91 (1982); A. Vilenkin
    and L.H. Ford, Phys. Rev. D $\underline{26}$, 1231 (1982); A.D. Linde, Phys.
    Lett. $\underline{116B}$, 335 (1982); A.A. Starobinskii, Phys. Lett. $\underline{117B}$,
    175 (1982).
30. We thank M. Srednicki for bringing this to our attention.
31. C. Kounnas, J. Leon, and M. Quiros, CERN preprint TH. 3554
    (1983); C. Kounnas, D.V. Nanopoulos, and M. Quiros, CERN
    preprint TH. 3573 (1983).

KALUZA-KLEIN COSMOLOGY

C. Wetterich

Institut für theoretische Physik
Universität Bern
Bern, Switzerland

As we have learned from previous talks of this conference[1] by Perry and Sorkin, monopoles occur naturally in Kaluza-Klein theories. Kaluza-Klein theories are unified theories of gravitation and gauge interactions. In these theories, monopoles can be understood as geometrical objects. We want to know if such monopoles could be observed and therefore have to investigate how many of them are produced in the very early universe, and how many should be left in our present universe. This necessitates a study of the cosmology derived from Kaluza-Klein theories. The work on Kaluza-Klein cosmology reported here was done in collaboration with Q. Shafi.

Production of monopoles is of course not the only reason for studying Kaluza-Klein cosmology. Another motivation comes from the difficulties to relate scenarios of an inflationary universe to a phase transition during which the unified gauge theory of weak, electromagnetic and strong interactions is spontaneously broken. This symmetry breaking has an associated energy scale, the unification mass M_x, which is typically of the order 10^{15} GeV. The Hubble "constant" $\overset{.}{H}$ during such a transition is typically of the order M_x^2/M_p, several orders of magnitude smaller than the characteristic scale M_x. (M_p denotes the Planck mass.) Many of the problems to construct a realistic inflationary scenario in this context are related to this fact. These problems can be avoided if the characteristic length scale during inflation is itself roughly of the order M_p, inducing a Hubble constant in the same order of magnitude. The scalar field driving the inflation must be a singlet with respect to the gauge group in this case. Kaluza-Klein theories provide a natural origin for such a scalar singlet: it corresponds to the characteristic length scale of the

117

internal space.   I will show how inflation may have its origin in the development of the large asymmetry between the curvature of our known four dimensional universe and the curvature of the internal space.

Let me briefly recall some of the main features of Kaluza-Klein theories.   Consider a theory of pure gravity in 3+D spacelike and one timelike dimensions. The action of this theory, invariant under 4+D dimensional general coordinate transformations, should lead to a ground state characterized by a direct product of the usual four dimensional Minkowski space M^4 and a compact "internal" space K^D.   The characteristic length scale of K^D is typically of the order of the Planck length.   For simplicity, assume that the internal space is a D dimensional sphere S^D.   Due to the very small characteristic length scale (in our case the radius of S^D), the D internal dimensions cannot be observed directly.   Integrating out the internal coordinates, one ends with an effective four dimensional theory.   In our case, the isometries of the internal space consist of the rotations leaving S^D invariant.   They form the group $SO(D+1)$.   The ground state is invariant under $SO(D+1)$ transformations which can be made independently at every space-time point in M^4.   The effective four dimensional theory exhibits therefore local $SO(D+1)$ gauge invariance.   The gauge coupling is of the order $g^2 = L^{-2}M_p^{-2}$ with L the radius of S^D.

For all times sufficiently large compared to the Planck time $t_p = M_p^{-1}$, the radius L is constant to a good approximation. Cosmology is then described by the usual four dimensional Friedmann universe of hot big bang cosmology.   At very early times, however, the curvature of four dimensional space, characterized by the Robertson-Walker scale factor $R_3(t)$ $(H = \dot{R}_3/R_3)$ becomes comparable to the curvature of internal space.   At these times, one has to study the time evolution of L(t) as well as $R_3(t)$.   A time dependent L(t) corresponds to a time dependent scalar singlet field in the effective four dimensional theory.

The higher dimensional gravitational theory which determines this very early cosmology must certainly be a quantum theory. Gauge interactions are described by a quantum theory, and they represent a subtheory of this D+4 dimensional theory of gravity. The quantity relevant for a discussion of the ground state and the field equations determining cosmology is the effective action at the scale characteristic for the internal space.   This effective action includes quantum fluctuations.   Unfortunately, no consistent quantum theory of gravity is known so far, and we are unable to calculate the effective action from a more basic short distance theory.

I will give here a toy model to show how a simple effective action can lead to very interesting cosmology. I will assume that the effective action can be approximated by

$$S = -\frac{1}{V_0} \int d^{D+4} x \; |\det \hat{g}_{\hat{\mu}\hat{\nu}}|^{1/2} (\alpha\hat{R}^2 + \beta\hat{R}_{\hat{\mu}\hat{\nu}}\hat{R}^{\hat{\mu}\hat{\nu}}$$

$$+ \gamma\hat{R}_{\hat{\mu}\hat{\nu}\hat{\sigma}\hat{\lambda}}\hat{R}^{\hat{\mu}\hat{\nu}\hat{\sigma}\hat{\lambda}} + \delta\hat{R} + \epsilon) \qquad (1)$$

Here $\hat{R}_{\hat{\mu}\hat{\nu}\hat{\sigma}\hat{\lambda}}$ is the D+4 dimensional curvature tensor and $\hat{R}_{\hat{\mu}\hat{\nu}}$ and $\hat{R}$ the corresponding Ricci tensor and curvature scalar. The normalization V_0 is chosen to be the volume of the internal space at the ground state. This is the most general form of the effective action involving up to four derivatives of the metric $\hat{g}_{\hat{\mu}\hat{\nu}}$. For

$$D(D-1)\alpha + (D-1)\beta + 2\gamma \equiv \phi > 0$$

$$(D-1)\beta + 2\gamma > 0$$

$$\gamma > 0, \quad \delta > 0 \qquad (2)$$

$$\epsilon = \frac{1}{4} \delta^2 D(D-1)/\phi \qquad (3)$$

the effective action leads to a ground state $M^4 \times S^D$ corresponding to the absolute minimum of the effective four dimensional scalar potential.[2] (Equation (3) is a fine tuning of the cosmological constant.) For the ground state, the radius of S^D is given by

$$L_0^2 = 2\phi/\delta \qquad (4)$$

and one has

$$M_p^2 = 16\pi[(D-1)\beta+2\gamma]\delta/\phi \qquad (5)$$

Let me now discuss the cosmology derived from this effective action. I will make the assumption, that for the time scales discussed here the D+4 dimensional universe at a given time t is well approximated by a direct product of S^D and a homogeneous and isotropic three dimensional space. The degrees of freedom of this simplified system are described by $L(t)$ and $R_3(t)$. At time scales much larger than the Planck time the D+4 dimensional universe is well approximated by $L(t) = L_0 = $ const and $R_3(t)$ describing a Friedmann universe. Around the Planck time, other solutions of the field equations are relevant. We are interested in an inflationary phase during which $H = \dot{R_3}/R_3$ is almost constant. During this phase, $L(t)$ is almost constant too, $L(t) = L_H$ but L_H is different from L_0 and induces an effective cosmological constant for the four dimensional universe. Quantitatively, these de Sitter solutions[3] of the D+4 dimensional field equations derived from Eq. (1) are given by:

$$H^2 = \frac{1}{24} \delta \frac{D(D-1)}{\phi} (1-z)^2 (1+\sigma z)^{-1}, \tag{6}$$

where

$$z = L_o^2/L_H^2 \tag{7}$$

is determined from

$$a_1 z^3 + a_2 z^2 + a_3 z + a_4 = 0$$

$$a_1 = D(D-1) + D(D-1)\rho + [D(D-1) - 12]\sigma - 12\sigma^2$$

$$a_2 = -3D(D-1) - 3D(D-1)\rho - \frac{3}{D} [D^2(D-1) - 4D + 16]\sigma$$

$$+ 12 \frac{(D-4)}{D} \sigma^2$$

$$a_3 = \frac{3}{D} [D^2(D-1) - 4D - 16] + 3D(D-1)\rho$$

$$+ \frac{3}{D} [D^2(D-1) - 4D - 16]\sigma$$

$$a_4 = -D(D-1) + 12 - D(D-1)\rho - [D(D-1) - 12]\sigma \tag{8}$$

and

$$\sigma = -D(D-1)\alpha/\phi$$

$$\rho = -(D-4)\beta/\phi \tag{9}$$

The Hubble constant H is the characteristic scale for the exponential expansion of R_3 during the inflationary phase. Compared with the Planck mass, it is given by

$$\frac{H^2}{M_p^2} = \frac{1}{384\pi} \frac{D(D-1)}{(D-1)\beta+2\gamma} (1-z)^2 (1+\sigma z)^{-1} \tag{10}$$

Note that for $(D-1)\beta+2\gamma$ not too small, this ratio may be considerably smaller than 1.

I do not assume that the universe was ever described exactly by the above solution (or a corresponding exact de Sitter solution of a more elaborate model). However, at some time the universe

may have been near such a solution.   In this case one has to ask:
How long did it stay in the vicinity of a de Sitter solution? The
characteristic time t_c during which the universe could have been
in an inflationary phase can be calculated by making an approximation linear in the deviations from the de Sitter solutions.
Typically, a deviation s grows exponentially

$$s \approx s_0 \exp \omega t, \quad \omega = t_c^{-1} \tag{11}$$

Inserting all this in the field equations, one finds

$$\frac{\omega}{H} = \frac{4z(1+\sigma)}{D} \frac{(1+\sigma z)^2}{(1-z)^2} \frac{\tilde{A}}{\tilde{B}} \tag{12}$$

with

$$\tilde{A} = (1+\sigma z)(1-z) - \frac{2}{D} [1 + 2(1+\sigma)z + \sigma z^2] \tag{13}$$

and $\tilde{B}$ a lengthy function of σ and z typically of order one.   For
ω/H smaller than about 1/60 the inflationary period would last
long enough to solve some of the outstanding puzzles of present
cosmology--and to dilute to an unobservably small amount all
monopoles produced previous to this inflation. This can be rather
easily achieved, especially for small values of z (compare Eq.(8))
and large values of D.   For small values of z and sufficiently
large values of $(D-1)\beta+2\gamma$, the second characteristic scale associated with the de Sitter phase (the other scale is H)

$$L_H^{-2} = \frac{z}{32\pi} \frac{1}{(D-1)\beta+2\gamma} M_p^2 \tag{14}$$

is also smaller than the Planck mass.

     Is this inflation obtained from Kaluza-Klein theories an
accident with a rather low probability of the D+4 dimensional
universe coming near such a de Sitter solution?   Or is it generic
in these theories?   What about reheating after the inflationary
phase?   What density fluctuations are produced, and how many
monopoles are produced and left over?   And what about the
stability of the solution described above?   Many questions are
still open, and I do not expect that all of them can be answered
satisfactorily in this toy model.   However, some of them can be
investigated in this simple model.   The first step in this
direction is the reformulation of this theory in an equivalent
effective four dimensional theory.   This will allow an easier
contact with previous work on the inflationary universe.   The
degrees of freedom $R_3(t)$ and $L(t)$ are contained in the four
dimensional metric and in a scalar field.   This corresponds to an

ansatz for the D=4 dimensional metric $\hat{g}_{\mu\nu}(y,x)$:

$$\hat{g}_{\mu\nu} = g_{\mu\nu}(x), \quad \hat{g}_{\alpha\beta} = \frac{L(x)^2}{L_0^2} \dot{g}_{\alpha\beta}(y), \quad \hat{g}_{\mu\alpha} = 0 \tag{15}$$

where $\dot{g}_{\alpha\beta}$ is the metric of S^D with radius L_0.

I will not describe here the details of deriving the effective four dimensional theory, but only indicate the resulting equations of motion for the scalar singlet field. After performing the appropriate Weyl scalings, one finds in a good approximation for the vicinity of a de Sitter phase

$$\ddot{\phi} + 3H \dot{\phi} = - \frac{\partial}{\partial\phi} \overline{W}(\phi) \tag{16}$$

The derivative of the de Sitter potential $\overline{W}(\phi)$ is given by

$$\frac{\partial\overline{W}}{\partial\phi} = \frac{\partial s}{\partial\phi} \left\{ D\mu \ \exp\text{-}Ds(1\text{-}\exp\text{-}2s)(1+\sigma \ \exp\text{-}2s)^{-4} \ \frac{D(D\text{-}1)}{4\phi} \ \left(\frac{M_p^2}{16\pi}\right)^2 \right\}$$

$$\times \left\{ D_\mu \ (1\text{-}\exp\text{-}2s)^3 - D(1\text{-}\exp\text{-}2s)(1+\sigma \ \exp\text{-}2s)^2 \right.$$

$$+ 4 \ \exp\text{-}2s(1+\sigma \ \exp\text{-}2s)^2 \tag{17}$$

$$\left. + 4\sigma \ \exp\text{-}2s(1\text{-}\exp\text{-}2s)(1+\sigma \ \exp\text{-}2s) \right\}$$

with

$$\mu = \frac{1}{12} \left[D(D\text{-}1) + [D(D\text{-}1) - 12] \ \sigma + D(D\text{-}1)\rho \right] \tag{18}$$

and $\partial s/\partial\phi$ a scaling factor roughly of the order M_p^{-1} coming from the correct normalization of the scalar kinetic term. The Hubble constant is given by

$$H^2 = \frac{M_p^2}{384\pi} \frac{D(D\text{-}1)}{\phi} \ \exp\text{-}Ds(1\text{-}\exp\text{-}2s)^2(1+\sigma\exp\text{-}2s)^{-2} \quad . \tag{19}$$

(Note that due to the Weyl scaling, the Hubble constant H is now different from its unscaled value in Eq. (10).) The de Sitter potential $\overline{W}(\phi)$ has stationary points at s_0 and one can identify de Sitter solutions found previously from the D+4 dimensional field equations with these stationary points for $z = \exp\text{-}2s_0$. The factor A in Eq. (12) has now a simple meaning: It is proportional to the second derivative of the de Sitter potential. (The factor $\tilde{B}$ comes from the rescaling $\partial s/\partial\phi$.) The way is now open to study the inflationary phase in the usual four dimensional language.

At this point, let me add another interesting observation. For $D > 4$, the de Sitter potential has not always an extremum for $s > 0$. However, for large positive s the de Sitter potential becomes exponentially flat. If the universe ever comes into this region (which means that the internal radius must have expanded beyond its ground state value L_0) it must undergo an inflationary phase before developing into its ground state. This inflation seems to be more generic than the inflation associated with a local maximum of the de Sitter potential. This possibility deserves further study. It is an intrinsic feature of Kaluza-Klein theories related to the gravitational origin of the scalar field. Whereas the potential is typically a polynomial in $1/L$, the typical gravitational kinetic term involves always $\dot{L}/L$ and the resulting four dimensional scalar field ϕ is therefore proportional to $\ln L$.

Kaluza-Klein cosmology is still at the beginning. I did not try to give an overview about the work already done by several authors. However, I hope I could convince you that Kaluza-Klein cosmology opens a lot of interesting possibilities and questions.

ACKNOWLEDGEMENTS

I would like to thank the organizers of the Monopole '83 Conference for their hospitality at Ann Arbor. Work supported by Schweizerischer Nationalfonds.

REFERENCES

1. M. Perry, talk given at this conference; R.D. Sorkin, talk given at this conference.
2. C. Wetterich, Phys. Lett B 113, 377 (1982).
3. Q. Shafi and C. Wetterich, Phys. Lett. B 129, 387 (1983).

GALACTIC MAGNETIC FIELDS AND MAGNETIC MONOPOLES

Eugene N. Parker

Departments of Physics, Astronomy and Astrophysics
Enrico Fermi Institute
University of Chicago
Chicago, Illinois

INTRODUCTION

Magnetic fields are associated with astronomical objects, from planets and stars to galaxies and clusters of galaxies. In contrast to magnetic fields, strong electric fields are absent throughout most of space. The free electrons reduce electric fields essentially to zero in the frame of reference of the local matter. Exceptions are cold planetary atmospheres and the thin isolated shear planes produced by the dynamical nonequilibrium of strained magnetic fields. The general absence of free electrons in the terrestrial atmosphere is attested by such phenomena as lightning and the free motion of the air across the geo-magnetic field. The occasional shear planes in planetary and stellar magnetic fields induce such high electric current densities as to accelerate the local electrons to large velocities, producing the aurora, the stellar flare, and the X-ray corona. Apart from these fascinating exceptions, however, electrical potential differences are limited to the thermal energy of the local electrons.

One should not overlook the fact that the existence of a general magnetic field (carried bodily with the fluid, i.e. moving in the local frame of reference in which $\vec{E} = 0$) means that there are electric fields in other frames of reference. Thus, if the conducting fluid has a velocity $\vec{v}$ relative to some other (perhaps inertial) frame, then a Lorentz transformation from the fluid to the chosen frame yields an electric field $\vec{E} \cong -\vec{v} \times \vec{B}/c$ (assuming that $v \ll c$). There has been some confusion on this point in the literature, with some authors pointing out the potential difference of 10^8 volts across the orbit of Earth (as a consequence of the 10^{-5} gauss carried in the solar wind) and the 10^{15} volts across a galactic arm (as a consequence of the field of 3×10^{-6} gauss carried with the 250 km s^{-1} rotation of the galaxy). These electric fields cause electrically charged particles to drift (with a velocity $< \vec{E} \times \vec{B}/B^2$) in the frame of reference of the conducting fluid, in which frame $\vec{E} = 0$. So the fields do not have the effect of accelerating cosmic rays to kinetic energies comparable to the potential differences in the "fixed" frame of reference (or in any other frame of reference in which we might compute $\vec{E}$).

Now as everyone is aware, magnetic charges are rare if indeed they exist at all. Hence magnetic fields are not neutralized by flows of magnetic charge. They occur everywhere, evidently produced by the dynamo effect of the nonuniform rotation and internal convection of the fluids in the cores of planets, the convective zones of stars, and perhaps even in the gaseous disk of the galaxy (the interstellar medium)[1]. A magnetic field in a body of dimension L and electrical conductivity σ has a resistive decay time of the order of L/η (where η is the resistive diffusion coefficient $c^2/4\pi\sigma$) which is typically 10^4 years for the liquid metal core of Earth, 10^9 years for a star, and 10^{24} years for the gaseous disk of the galaxy. Turbulent mixing greatly enhances the escape of the magnetic field, producing an effective turbulent diffusion coefficient $\eta_T \cong 0.1\,w\ell$ for eddies of scale ℓ and velocity w. The effect is not large in the core of Earth, where $\eta_T = O(\eta)$, but it reduces the diffusion time for magnetic fields in the convective zone of the sun to something of the order of 10 years, and the diffusion time in the gaseous disk of the galaxy to $10^8 - 10^9$ years. The general dynamical instability of the magnetic field (of $2-3 \times 10^{-6}$ gauss) embedded in the gaseous disk of the galaxy further enhances the escape of magnetic field, so that 10^8 years seems to be a reasonable estimate of the decay time.

The age of the galaxy is estimated at 10^{10} years, from which one concludes that the galactic field must be regenerated on a time scale of the order of 10^8 years, presumably by the dynamo effect of the nonuniform rotation and turbulence of the gaseous disk. The formal theory of the galactic dynamo indicates that the observed motion of the interstellar gas regenerates the magnetic field in about this length of time, so that the existing galactic field is readily understood if there are no magnetic monopoles to put an additional load on the dynamo.

The galactic field has a strength of $2-3 \times 10^{-6}$ gauss in the nearby (Orion) galactic arm. The field is aligned more or less along the Perseus arm and the local Orion arm,[2] over a width of the order of 3 kpc, and has the opposite sense in the Sagittarius arm.[3] The large-scale pattern of the galactic field is best observed in other galaxies, where the observer can see the whole picture (cf.

M51[4],[5], NGC6946[6], M31[7], NGC253[8]).   The basic facts are fields of a
few microgauss over widths of 1-4 kpc extending along spiral arms
with lengths of 5-20 kpc. Observations of the local galactic field
indicate ripples of large amplitude with scales of 0.1 kpc in the
large-scale pattern of the field.[9],[10]

ACCELERATION OF MAGNETIC MONOPOLES

     It is obvious that a free magnetic monopole of charge g and
mass M would be freely accelerated by the galactic (interstellar)
magnetic field.   The magnetic   potential difference Φ = Bℓ over a
length ℓ = 1 kpc (3 $\times$ 10^{21} cm) along a field B = 3 $\times$ 10^{-6} gauss is
3 $\times$ 10^{18} volts. A magnetic   charge (g = 137e/2 = 3.3 $\times$ 10^{-8} cgs)
moving down such a field would gain an energy gBℓ = 2 $\times$ 10^{20}eV =
3 $\times$ 10^8 ergs at the expense of the magnetic field.   A flux F
monopoles cm^{-2} s^{-1} extracts energy at a rate FgB ergs cm^{-3} s^{-1}.

     The energy density B^2/8π of the galactic field (3 $\times$10^{-6} gauss)
is 4 $\times$ 10^{-13} ergs cm^{-3}. If we assume that the field can be
replenished on a time-scale of T = 10^8 years by the motions of the
interstellar gas, the energy input is of the order of

$$\epsilon = B^2/8\pi T = 1.3 \times 10^{-28} \text{ ergs cm}^{-3} \text{ s}^{-1} \tag{1}$$

which, as already noted, is supplied by the calculated dynamo
effect of the interstellar gas. The existence of the field implies
that the field does not transfer energy to magnetic monopoles at a
rate in excess of ϵ, for in that case the drain would be too much
for the dynamo. Thus, for instance, a directed monopole flux F_D
along the magnetic field must not exceed ϵ/gB, which is 1.3 $\times$ 10^{-15}
monopoles cm^{-2} s^{-1} for the above numbers.   This is a low flux,
equivalent to one monopole per year across an area of 2500 m^2
(roughly a football field).

     Now the current belief is that the magnetic monopole, if it
exists, has a mass M of the order of 10^{16} GeV (1.7 $\times$ 10^{-8} gm).   In
that case, the monopoles are strongly accelerated along the
galactic magnetic field and their flux F is limited by the above
figure F_D $\cong$ 1.3 $\times$ 10^{-15} cm^{-2} s^{-1}. Suppose, however, that monopoles
are more massive, with M $\cong$ 10^{19} GeV.   In that case a monopole is
deflected but little in crossing a galactic arm; the potential
difference gBℓ = 3 $\times$ 10^8 ergs is only a fraction of the kinetic
energy 2 $\times$ 10^{10} ergs of a monopole falling into the galaxy.   We
would expect a more or less random isotropic velocity distribution
for such massive monopoles, with only a small change $\Delta\vec{v}$ in velocity
$\vec{v}$ as each monopole crosses a region of scale ℓ containing a mean
coherent field B.   The transit time across the region would be

$\Delta t \cong \ell/v$ so that $\Delta v \cong gB\Delta t$. If the monopole enters the region with velocity $\vec{v}$ and departs with $\vec{v} + \Delta\vec{v}$, the change in kinetic energy is

$$\Delta E = M \left(\vec{v} \cdot \Delta\vec{v} + \frac{1}{2} (\Delta v)^2 \right) \quad .$$

Averaging over an isotropic distribution of $\vec{v}$ leads to the result that $\langle\vec{v}\cdot\Delta\vec{v}\rangle$ is of the same order as $\langle(\Delta v)^2\rangle$, so that $\Delta E = O[M(\Delta v)^2]$. A two-dimensional isotropic monopole flux incident on a circular region of radius R containing a uniform magnetic field with random orientation yields the precise result

$$\Delta E = \frac{2}{3} M v^2 \left(\frac{gBR}{Mv^2}\right)^2 \tag{2}$$

neglecting terms third order in gBR/Mv^2. If a monopole undergoes this experience every ℓ cm along its trajectory, then a monopole flux F implies a mean energy gain $F\Delta E/\ell$ which must be less than ε. Hence,

$$F < F_D \frac{3\ell}{2R} \frac{Mv^2}{gBR} \tag{3}$$

where F_D $(= B^2/8\pi TgB)$ is the limiting flux of monopoles directed along the field, equal to 1.3×10^{-15} cm^{-2} s^{-1} for the numbers quoted above.

It is evident that fast, massive monopoles are the least affected by the galactic field, and extract the least energy from the field as they pass through it. However, there is a limit imposed on the monopole mass and monopole flux by the mean density of the universe, which evidently does not exceed about 4×10^{-29} gm/cm^3. We can ease this restriction somewhat by supposing that the monopoles are trapped by the gravitational field of the galaxy, so that they occupy only a fraction 3×10^{-5} of space, but we are restricted, then to monopole velocities $v \lesssim 10^{-3}$ c. The optimum mass for minimizing the damage to the galactic field appears to be $M \cong 10^{19}$ GeV (about a factor of 10^3 larger than the canonical 10^{16} GeV). We do not expect v to be less than 10^{-3} c = 300 km s^{-1}, because: (a) this is approximately the orbital velocity of a monopole in the galaxy at the position of the sun and because; (b) for $M = 10^{19}$ GeV $= 1.7 \times 10^{-5}$ gm the monopole kinetic energy is $1/2$ $Mv^2 = 7 \times 10^9$ ergs, comparable to the magnetic potential difference over 10 kpc along a galactic arm. Thus, slow monopoles are quickly accelerated to $v \cong 10^{-3}$ c. Indeed, if the mass M is much less than 10^{19} GeV the monopoles are accelerated enough to escape the galaxy so that they do not remain confined. It follows, therefore, that $M = 10^{19}$ GeV with $v = 10^{-3}$ c are the most optimistic values from the point of view of permitting the existence of monopoles in the face of the observed magnetic field.[16] With $\ell = 2R = 1$ kpc we obtain

$$F \lesssim 140 \; F_D \cong 2 \times 10^{-3} \; cm^{-2} \; s^{-1} \quad (4)$$

from Eq. (3). A further factor of ten can be gained by supposing
that the galactic field is coherent over a length ℓ of only 0.1 kpc
for which the limit is $F \lesssim 2 \times 10^{-12}$ cm^{-2} s^{-1}.[11] But so short a
hypothetical coherence length ℓ is becoming increasingly difficult
to maintain in the face of the recent observational studies[3,7],
which make it clear that ℓ is more like 1 or 2 kpc, or more.   In
the same direction, it should be noted that there is no apriori
reason to think that M is larger than 10^{16} GeV, in which case the
potential energy difference $gB\ell$ is comparable to 1/2 Mv^2 (for
$v = 10^{-3}$ c), the monopoles are quickly boiled out of the galaxy,
spreading them uniformly over the universe. The mass limit is then
imposed, which brings us to $F \lesssim 2 \times 10^{-15}$ cm^{-2} s^{-1}. Coincidentally, this is about the same as $F_D \cong 1.3 \times 10^{-15}$ cm^{-2} s^{-1} which must
also be imposed.

These results are summarized in Figs. 1, 2, and 3 for assumed
monopole velocities $v = 10^{-3}$ c, 10^{-2} c, and   c, based on the
optimistic assumption that the coherence length is only 0.1 kpc.[11]
The more realistic value of 1 kpc has the effect of lowering the
sloping part of the curve labelled "survival of the galactic
magnetic field" by a factor of ten.

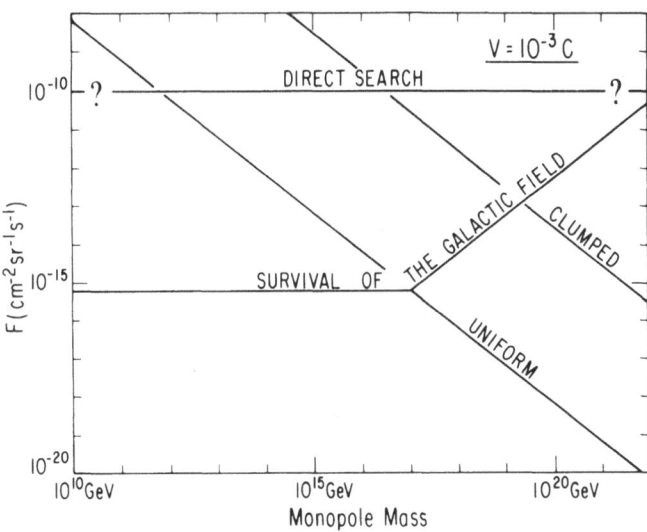

Fig. 1.   Summary of monopole flux limits as a function of monopole
          mass for a monopole velocity $v = 10^{-3}$ c. The lines marked
          "uniform" and "clumped" are based upon the mass density of
          the universe and galaxy, respectively. The limits imposed
          by direct search for monopoles are obtained from Refs.
          12 and 13, applicable for 2×10^{-2} c $\gtrsim$ v $\gtrsim 3 \times 10^{-4}$ c.

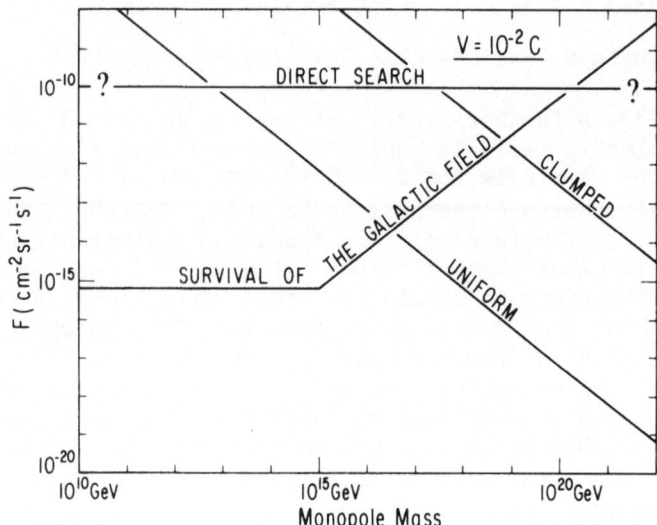

Fig. 2.   Same as Fig. 1 except for monopole velocity v = 10^{-2} c.
It is unlikely that monopoles with velocities as large as
10^{-2} c would remain clumped about the galaxy, but the
"clumped" curve has been included for completeness.

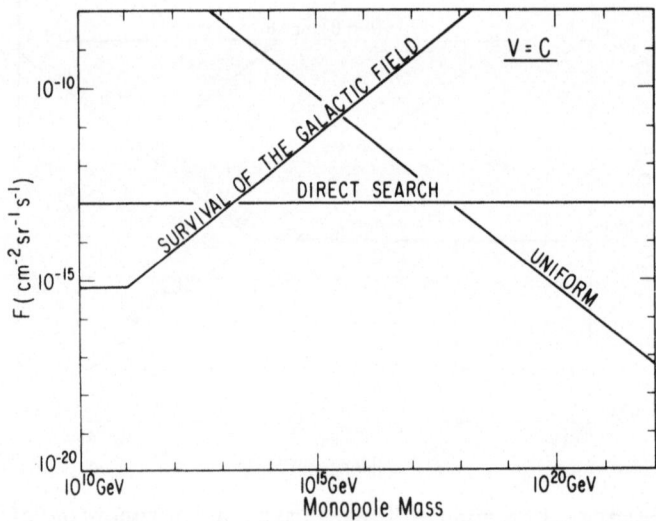

Fig. 3.   Same as Fig. 1 except for monopole velocity v ~ c.   The
direct search limit is valid for v > 2 × 10^{-2} c   and
follows from Refs. 13, 14, and 15.

Finally, some authors have argued that the galactic magnetic field is not regenerated by the motions of the interstellar gas.[17] They assert that the field is primordial, with an age $T = 10^{10}$ years. If that were the case, the curves labelled "survival of the galactic magnetic field" would be lowered an additional factor of 10^2 over their entire length.

We suggest that 10^{-15} monopoles cm^{-2} s^{-1} is a reasonable upper limit on the flux of free monopoles imposed by the existence of the galactic magnetic field. Unless the mass of the monopole proves to be much larger than 10^{16} GeV, a larger flux of monopoles would confront us with a serious dilemma.

MAGNETIC MONOPOLE PLASMA OSCILLATIONS

One may ask at this point whether there might not be some way in which magnetic monopoles avoid being accelerated by the galactic magnetic field, thereby avoiding the flux limit just imposed. Several individuals have turned their thoughts to this basic question, producing several clever schemes. Unfortunately the ideas involve special symmetries, such as exist in the minds of mathematicians but not in the real world.

It is appropriate to mention some of the more ingeneous ideas here to illustrate the lengths to which we must go to avoid the foregoing restrictions. Thus, for instance, Blandford[18] pointed out that the monopoles in the magnetic field $\hat{e}_y B_0 \sin(kx)$ have velocities $[v_x, v_y -(gB/kM) \cos(x_0 + v_x t), v_z]$ at time t, where v_x, v_y, v_z are constants. There is no net gain of energy for any choice of (v_x, v_y, v_z). However, averaging over an initial isotropic distribution gives a <u>continual average</u> energy gain $\langle dE/dt \rangle = 1/4 \pi g^2 B^2 / Mkv$, which is essentially the same as (2) for random fields. The mean energy gain arises from those particles with initial velocities nearly in the negative direction which have not yet moved across the first band of field to be decelerated again in the next band. This condition exists for all t. It can be avoided by breaking the perfect symmetry of the field, of course, but in that case the whole scheme collapses and something similar to (2) applies again.

Ostriker[19] pointed out the interesting fact that a magnetic monopole in a sufficiently anisotropic harmonic gravitational potential may be prevented from gaining energy in an azimuthal magnetic field. Consider a monopole moving in the xy - plane in an azimuthal field $B(\phi) = B_0 W/h$, where $W = (x^2 + y^2)^{1/2}$ and B_0 and h are constants, and a gravitational potential $V = 1/2(\omega_1^2 x^2 + \omega_2^2 y^2)$ with the basic anisotropy that one might encounter in the disk of an open spiral.

The equations of motion in the xy-plane are

$$\ddot{x} + \omega_1^2 x = -\Omega^2 y, \quad \ddot{y} + \omega_2^2 y = +\Omega^2 x$$

where Ω is the characteristic frequency $(gB_0/Mh)^{1/2}$ for accelera-tion of the monopole in the field B_0 at a radius h. Solutions of the form exp(iωt) yield the dispersion relation

$$\omega^2 = 1/2 \, (\omega_1^2 + \omega_2^2) \pm \left[(\omega_1^2 - \omega_2^2)^2 - 4\Omega^4\right]^{1/2} \quad . \tag{5}$$

The frequency ω is real, so that there are no growing solutions, when the anisotropy $|\omega_1^2 - \omega_2^2|$ exceeds $2\Omega^2$. In that case the individual modes consist of oscillations along straight lines through the origin. Combinations of modes give Lissajous figures that circle successively clockwise and counterclockwise so that there is no net energy gain. Unfortunately the azimuthal field in our galaxy seems too strong to permit this interesting circum-stance, even if we suppose the system to be sufficiently symmetric as to be described by these simple equations. The harmonic oscillation of a particle in the gravitational field at the position of the sun is approximately 2.5×10^8 years, so that the mean value of ω_1 and ω_2 is of the order of 0.8×10^{-15} s^{-1}. Their difference is less. On the other hand, with an azimuthal field $B_0 = 3 \times 10^{-6}$ gauss at a radial distance h=10 kpc, we find that $\sqrt{2}\Omega = 2 \times 10^{-14}$ s^{-1} for monopoles of mass = 10^{16} GeV, so that $2\Omega^2$ is considerably larger than $|\omega_1^2 - \omega_2^2|$. If we suppose that M is as large as 10^{19} GeV, then $\sqrt{2}\Omega = 0.7 \times 10^{-15}$ s^{-1}, and the idea might be applied if the galaxy were as rotationally anisotropic as some pure barred spirals appear to be. However, in spite of the possibility of a barred structure in the inner galaxy, there is no evidence for $\omega_1^2 - \omega_2^2$ as large as $2\Omega^2$ at the sun.

Now there is a different approach to the galactic field-monopole relationship if we abandon the conventional view that the magnetic field is rooted in the gaseous disk of the galaxy by the electric currents that it induces. Suppose instead that the observed magnetic field is a consequence of oscillations in a plasma composed of monopoles of opposite sign.[11] In the simplest case, N monopoles/cm^3 with charge +g and N/cm^3 with -g (both signs with mass M) oscillate relative to each other with the monopole plasma frequency

$$\omega_p = (8\pi Ng/M)^{1/2} \quad . \tag{6}$$

The magnetic field follows from Maxwell's equations $\partial \vec{B}/\partial t = -4\pi \vec{J} = -4\pi Ng(\vec{v}_+ - \vec{v}_-)$ where $\vec{v}^{\pm}$ represents the velocity of the oscillation of the monopoles with charge ±g, respectively. As a specific example, suppose that the magnetic monopole mass is 10^{16} GeV (M = 1.7×10^{-8} gm), the flux, F, has the limiting value 10^{-15} cm^{-2} s^{-1}, and the random thermal velocity of the monopoles is

10^{-3} c. Then $N \cong 2.5 \times 10^{-23}$ cm^{-3} and $\omega_p = 6 \times 10^{-15}$ s^{-1}. The period of the oscillation is 3×10^7 years. Who is to say that the presently observed galactic field is not an oscillatory field, with a period of tens of millions of years?

There are a number of difficulties with the idea, unfortunately. First of all, the monopole plasma oscillations are subject to strong Landau damping if their characteristic phase velocity ω_p/k is not large compared to the random velocity of the monopoles. A coherence length $\ell = 1$ kpc yields a wave number $k = \pi/\ell = 10^{-21}$ cm^{-1}, for which the above value of ω_p yields a phase velocity of 60 km s^{-1}. This is small compared to the random velocities of 300 km s^{-1} in the gravitational field of the galaxy so the oscillations would not survive one period.[11]

Arons and Blandford[20] pointed out, however, that there are no obvious limits on the monopole density if the galactic field is principally the product of monopole plasma oscillations. Then, if F is as large as 10^{-13} cm^{-2} s^{-1}, so that $N = 2.5 \times 10^{-21}$, we have $\omega_p = 3 \times 10^{-6}$ years and $\omega_p/k = 600$ km s^{-1}, which may be enough larger than the assumed random velocity of 300 km s^{-1} to avoid too much Landau damping.

But there is another difficulty, that the monopole plasma frequency depends upon $N^{1/2}$, which varies across the galaxy, so that any initial pattern of oscillation is soon corrupted by phase mixing, i.e. the fields at neighboring points soon get out of phase, producing steep gradients in the field patterns. One can avoid this to some degree by a judicious balance of Landau damping with phase mixing, and a hypothetical source of excitation, but in any case it is difficult to understand how to produce plasma oscillations that follow the visible galactic arms with alternating signs in adjacent arms.

This brings us to the final point. The galaxy is filled with a moving conducting gas, in which electric currents may be induced, whatever may be the role of magnetic monopole oscillations. The combined effects of monopoles and electrical conductivity give some interesting results for the dynamics of the galactic magnetic field. To derive the appropriate field equation we adopt the elementary view that there is a uniform number density N of cold positive and negative monopoles, (i.e. negligible random velocities) so that the monopoles may be treated as separate fluids, with velocities $\vec{v}_+$ and $\vec{v}_-$, respectively. The magnetic current density is, then, $\vec{J} = Ng (\vec{v}_+ - \vec{v}_-)$, as already noted. Using Ohms law in the frame of reference moving with the conducting gas, $\vec{j} = \sigma\vec{E}'$, with $\vec{E}'$ related by the Lorentz transformation $\vec{E}' \cong (\vec{E} + \vec{v} \times \vec{B}/c)$ to the fields $\vec{E}$ and $\vec{B}$ in the fixed frame of the magnetic monopoles relative to which the conducting gas has a velocity $\vec{v}$, we have $\vec{j} = \sigma(\vec{E} + \vec{v} \times \vec{B}/c)$, neglecting terms $O(v^2/c^2)$

compared to unity. Then, excluding electron plasma oscillations, so that the displacement current $\partial \vec{E}/\partial t$ may be neglected, Maxwell's equations

$$4\pi \vec{j} = c\nabla \times \vec{B}, \qquad 4\pi \vec{J} + \partial \vec{B}/\partial t = -c\nabla \times \vec{E},\tag{7}$$

together with the linearized equations of motion for the monopole fluids,

$$\partial \vec{v}_\pm/\partial t = \pm g\,\vec{B}/M,$$

yield

$$\vec{E} = -\vec{v} \times \vec{B}/c + (c/4\pi\sigma)\,\nabla \times \vec{B} \quad ,\tag{8}$$

$$\partial \vec{J}/\partial t = Ng^2\vec{B}/M\tag{9}$$

for monopole oscillations of small amplitude. Using these two expressions to eliminate $\vec{E}$ and $\vec{J}$, we arrive at the relation

$$\partial^2 \vec{B}/\partial t^2 + \omega_p^2 \vec{B} = (\partial/\partial t)\left[\nabla \times (\vec{v} \times \vec{B}) - \nabla \times (\eta\nabla \times \vec{B})\right]\tag{10}$$

where $\eta = c^2/4\pi\sigma$ is the resistive diffusion coefficient. Consider solutions of the form of monopole oscillations with time dependent form $F(\vec{r},t)$ so that

$$\vec{B}(\vec{r},t) = \vec{F}(\vec{r},t)\,\exp(i\omega_p t)\,.\tag{11}$$

Since ω_p is taken to be independent of position (i.e. $\nabla N = 0$), the equation reduces to

$$\left(1 + \frac{1}{2i\omega_p}\frac{\partial}{\partial t}\right)\frac{\partial \vec{F}}{\partial t} = \frac{1}{2}\left(1 + \frac{1}{i\omega_p}\frac{\partial}{\partial t}\right)\left[\nabla \times (\vec{v} \times \vec{F}) - \nabla \times (\eta\nabla \times \vec{F})\right].$$

We have already noted that ω_p must be large to avoid Landau damping suggesting that the operators can be expanded in descending powers of $(2i\omega_p)^{-1}\,\partial/\partial t$. The result can be written

$$\frac{\partial \vec{F}}{\partial t} = \frac{1}{2}\left[1 - \sum_{n=1}^{\infty}\left(\frac{-1}{2i\omega_p}\right)^n\left(\frac{\partial}{\partial t}\right)^n\right]\left[\nabla \times (\vec{v} \times \vec{F}) - \nabla \times (\eta\nabla \times \vec{F})\right]\,.$$

The lowest order terms in the expansion are

$$\partial \vec{F}/\partial t = \nabla \times \left(\tfrac{1}{2}\vec{v} \times \vec{F}\right) - \nabla \times \left(\tfrac{1}{2}\eta\nabla \times \vec{F}\right),$$

which is to be compared to the conventional hydromagnetic equation

$$\partial \vec{F}/\partial t = \nabla \times (\vec{v} \times \vec{F}) - \nabla \times (\eta \nabla \times \vec{F})$$

when $\omega_p = 0$. It is evident that, if the magnetic field is directly associated with monopole oscillations, so that it is described by (10), then the large-scale field is carried bodily with <u>half</u> the velocity of the gas relative to the monopole background. The resistive decay proceeds at just <u>half</u> the conventional rate too. The important point is that the field is not frozen into the conducting gas, but lags behind the gas, with a velocity $-1/2\vec{v}$ relative to the gas. Small-scale, rapid variations produced by local velocities such as stellar winds, supernovae, etc. (for which $\partial/\partial t \gg \omega_p$) satisfy the conventional hydromagnetic equation (11), of course. But, if there are enough monopoles to avoid Landau damping of the monopole oscillations, then $\partial/\partial t \ll \omega_p$ on a galactic scale, and the galactic magnetic field moves only half as fast as the interstellar gas. It is an interesting but doubtful possibility to contemplate.

In conclusion, then, we suggest that the existence of the galactic field implies that the flux of magnetic monopoles is not in excess of 10^{-15} monopoles cm^{-2} s^{-1} unless one is willing to suppose that the monopole mass is much larger than 10^{16} GeV or that the field of the galaxy, and other galaxies as well, is a manifestation of magnetic monopole plasma oscillations, with all the necessary caveats to avoid excessive Landau damping and phase mixing noted above. So special a situation as a general rule in spiral galaxies seems unlikely and we have placed our bets accordingly. But, of course, bets are not collected until the race is finished. It remains to be seen whether the monopole flux is even so large as 10^{-15}, of course.

ACKNOWLEDGEMENTS

We gratefully acknowledge discussion with many colleagues, too numerous to mention, which has contributed so much to the ideas presented in this brief review. This work was supported by NASA through grant NGL-14-001-001.

REFERENCES

1. E.N. Parker, Astrophys. J. <u>122</u>, 293 (1955); Astrophys. J. <u>160</u>, 383 (1970); <u>Cosmical Magnetic Fields</u>, (Oxford, Clarendon Press) 1979.
2. W.A. Hiltner, Astrophys. J. <u>109</u>, 471 (1949); Astrophys. J. <u>114</u> 241 (1951); Astrophys. J. Suppl. <u>2</u>, 389 (1956).

3. J.P. Vallee, Astrophys. J. 124, 147 (1983).
4. A. Segalovitz, W.W. Shane, and A.G. DeBruyn, Nature 264, 222 (1976).
5. M. Tosa and M. Fujimoto, Publ. Astron. Soc. Japan 30, 315 (1978).
6. U. Klein, R. Beck, U.R. Buczilowski and R. Wielebinski, Astron. Astrophys. 108, 176 (1982).
7. R. Beck, Astron. Astrophys. 106, 121 (1982).
8. U. Klein, M. Urbanik, R. Beck, and R. Wielebinski, Astron. Astrophys., in press (1983).
9. J.R. Jokipii and I. Lerche, Astrophys. J. 157, 1137 (1969).
10. J.R. Jokipii, I. Lerche, and R.A. Schommer, Astrophys. J. Letters 157, L119 (1969).
11. M.S. Turner, E.N. Parker, T.J. Bogdan, Phys. Rev. D. 26, 1296 (1982).
12. J.D. Ullman, Phys. Rev. Lett. 47, 289 (1981).
13. M. Longo, Phys. Rev. D. 25, 2399 (1982).
14. P.B. Price, E.K. Shirk, W.Z. Osborne, and L.S. Pinsky, Phys. Rev. Lett. 35, 487 (1975).
15. K. Kinoshita and P.B. Price, Phys. Rev. D. 24, 1707 (1981).
16. For R=1kpc, $B=3\times10^{-6}$ gauss, $M=10^{19}$ GeV=1.7×10^{-5} gm, $v=10^{-3}$ c the small parameter gBR/Mv^2 has a value 0.02.
17. J.H. Piddington, Cosmic. Electrodyn. 3, 129 (1972); Astrophys. Space Sci. 37, 183 (1975); Astrophys. Space Sci. 59, 237 (1978).
18. R.D. Blandford, private communication (1982).
19. J.P. Ostriker, private communication.
20. J. Arons and R.D. Blandford, Phys. Rev. Letters 50, 544 (1983).

NEUTRON STAR PHYSICS AND MONOPOLE FLUX LIMITS

Jeffrey Harvey

Joseph Henry Laboratories
Princeton University
Princeton, New Jersey 08544

INTRODUCTION

Stringent bounds on the galactic monopole flux have been derived using monopole catalysis of nucleon decay inside neutron stars.[1] Currently the most stringent limit comes from the pulsar PSR 1929+10 with the inclusion of monopole capture by the precursor main sequence star.[2] This limit is $F_M < 10^{-28}$ cm^{-2} s^{-1} sr^{-1} assuming a strong interaction catalysis cross section. If this bound is correct then such monopoles are very rare indeed and detecting them will be virtually impossible. It is therefore of utmost importance to examine these catalysis bounds for possible loopholes. One possibility is that there are uncertainties in the observations of neutron star X-ray luminosities[1] or in the calculations of monopole catalysis of nucleon decay. Another possibility is that something happens to the monopoles once they enter the neutron star that invalidates the catalysis bounds. I will focus on this latter possibility in my talk. Since neutron stars are complicated objects which are imperfectly understood, I will rely mainly on order of magnitude estimates and qualitative arguments and attempt to indicate where these may go wrong. A more detailed analysis will be presented in Ref. 3. I will give a brief review of the relevant properties of neutron stars.

NEUTRON STARS

A typical neutron star[4] has a mass of approximately one solar mass $(2 \times 10^{33}$ g) and a radius of 10 km and is therefore at roughly nuclear density. It consists of at least three distinct regions; an outer crust with 10^6 g $cm^{-3} < \rho < 4 \times 10^{11}$ g cm^{-3}

consisting of a lattice of heavy nuclei in β equilibirum with a degenerate electron gas, an inner crust with 4×10^{11} g cm^{-3} < ρ < 2.4×10^{14} g cm^{-3} consisting of degenerate electrons, a lattice of neutron-rich nuclei, and superfluid neutrons paired in a $^1S_0$ state, and a core region with ρ > 2.4×10^{14} g cm^{-3} which contains degenerate electrons, superfluid neutrons paired in a $^3P_2$ state and superconducting protons paired in a $^1S_0$ state. The core may also contain a condensate of pions and perhaps other more exotic possibilities such as quark matter or a neutron solid. Detailed calculations of the structure of neutron stars tend to agree on the existence of the various superfluids but may differ in the exact values of the critical density and temperature. The theoretical interpretation of pulsar glitches in terms of depin-ning of vortices in the neutron superfluid also provides indirect observational evidence for the existence of superfluidity in neutron stars.

The magnetic field structure of neutron stars will be particularly important in our considerations. Typical pulsars have surface magnetic fields of ~ 10^{12} G as determined by fitting measured pulsar spin-down rates to energy loss through dipole radiation and by the observation of what are probably cyclotron lines in the spectrum of certain binary X-ray pulsars. The spin-down rates of the two recently discovered millisecond pulsars indicate that they probably have surface magnetic fields in the range 10^8-10^9 G. We have no direct information concerning the interior magnetic fields of neutron stars but it is reasonable to assume that they are at least comparable to the surface fields. The behavior of magnetic monopoles in neutron stars is determined primarily by the structure of the magnetic field in and sur-rounding the superconducting core region.

The superconducting core is characterized by two lengths, the proton coherence length $\xi_p \simeq (2/\pi k_F)(E_p/\Delta_p)$ where $E_p = h^2 k_F^2/2m_p$, k_F is the proton Fermi wave number and Δ_p is the energy gap, and the penetration length λ_p which is given by the London formula $\lambda_p^2 \simeq m_p c^2/4\pi n_p e^2$ with n_p the number density of protons. For typical neutron star parameters $\lambda_p \sim 50$ fm and $\xi_p \sim 10$ fm so that $\xi_p/\lambda_p < \sqrt{2}$. This means that the core is a type II superconductor and hence the magnetic field will permeate the region as a series of flux tubes,[5] each carrying one flux quantum $\phi_0 = hc/2e \simeq 2 \times 10^{-7}$ G cm^2. The area density of flux tubes is $B/\phi_0 \sim 5 \times 10^{18}$ (B/10^{12} G) cm^{-2}. A neutron star will thus have ~πR_{NS}^2 (B/ϕ_0) ~ 10^{31} (B/10^{12} G) flux tubes. To understand how monopoles behave we will need to know the energy per unit length of the flux tubes and the drag force on a moving flux tube. The energy per unit length is just of order the magnetic field energy per unit length or more precisely

$$\varepsilon = (\phi_0/4\pi \lambda_p)^2 \ln (\lambda_p/\xi_p) \sim 2 \times 10^7 \text{ erg cm}^{-1} \quad .$$

The drag force is more uncertain since even in laboratory superconductors there are questions concerning the proper calculation of the drag force on moving flux tubes. In neutron stars the problem is actually somewhat simpler since the superconductor there is very pure and is at a temperature far below the critical temperature. An order of magnitude estimate of the energy loss due to scattering of electrons off of the core field gives a viscous drag force per unit length of $f_v = \eta v$ where v is the velocity of the flux tube and

$$\eta \sim \frac{\phi_0^2 \, n_e \, e^2}{\pi \lambda_p \, m^*_e \, c^3} \sim 10^{10} \text{ dyne s cm}^{-2}$$

where n_e is number density of electrons and m_e^* is the effective electron mass.

MONOPOLES AND NEUTRON STARS

I would now like to describe what happens when a monopole encounters a neutron star. As we will see the crucial quantity is the ratio of the gravitational force to the magnetic force on the monopole. At the surface of the neutron star the gravitational force is $F_{gr} \sim GM_{ns} M_m/R^2_{ns} \sim 10^6 M_{16}$ dynes where M_{16} is the monopole mass in units of 10^{16} GeV/c^2 while the magnetic force is $F_m = gB \sim 3 \times 10^4 B_{12}$ dynes where B_{12} is the magnetic field in units of 10^{12} G so that $F_{gr}/F_m \sim 30 B^{-1}_{12} M_{16}$. Therefore the magnetic field is negligible for monopoles heavier than 10^{16} GeV/c^2. If $B_{12}M_{16} \lesssim 10^{-2}$ then the magnetic force will dominate and will accelerate the monopoles to escape velocity. Therefore, bounds on the monopole flux from neutron stars apply only to monopoles with mass $>10^{14}$ GeV/c^2 for most neutron stars or $>10^{10}$ GeV/c^2 for the two millisecond pulsars. It is also interesting to note that the ratio F_{gr}/F_m is constant for fixed stellar mass and magnetic flux, and is therefore roughly the same for the sun as it is for a neutron star.

Monopoles which are heavy enough to remain trapped will fall through the outer layers of the neutron star until they encounter the superconducting core. The only way a monopole can penetrate the core is by forming a flux tube, but the magnetic field in a flux tube is $B_f \sim \phi_0/\pi \lambda_p^2 \sim 3 \times 10^{15}$ G so that $F_{gr}/F_m \sim 10^{-2} M_{16}$ in a flux tube. Therefore, the Meissner force will suspend the monopole near the surface of the superconducting region. I would like to leave the monopole hanging for a moment while I digress to explain what seems at first sight to be a remarkable coincidence. The coincidence is the fact that the ratio F_{gr}/F_m in a flux tube is so close to one (note that for a Planck mass monopole it is 10) yet F_m and F_{gr} involve wildly different physics. I want to show you that this is in fact not a coincidence at all. First consider the magnetic force. Setting $h=c=2=\pi=1$ we have

$$F_m = gB \sim (1/e)\,(\phi_0/\lambda_p^2) \sim n_p/m_p \quad.$$

Note that this is independent of e and depends only on the proton mass and number density. The gravitational force is

$$F_g = G\,M_{ns}\,M_m/R_{ns}^2 = (M_{ns}\,M_m/R_{ns}^2\,M_{p\ell}^2)$$

where $M_{p\ell}$ is the Planck mass. Now the mass of a typical star is determined by equating the radiation pressure and gas pressure which gives the result $M_{star} \sim M_{p\ell}^3/m_p^2$. Substituting this for M_{ns} and using $(M_{ns}/R_{ns}^3) \sim m_p\,n_n$ where n_n is the nucleon number density gives

$$F_g = (M_m/M_{p\ell})\,n_n^{2/3} \quad.$$

If $n_n \sim n_p \sim m_p^3$ then the two forces would be equal for $M_m = M_{p\ell}$. In fact n_p is only a few percent of n_n and n_n is closer to m_π^3 than m_p^3, this plus the numerical factors that were ignored accounts for the values found in the previous paragraph.   This ends the digression.

The eventual fate of the monopole we left hanging near the normal-superconducting boundary depends on the mobility of the monopole. If the monopole is very near the surface it can move easily under the influence of the tangential magnetic field until it encounters a flux tube or perhaps an antimonopole, but if the monopole is embedded deeper in the superconductor then it will move very slowly if at all.   In order to decide between these two possibilities we need to know something about the transition from the superconducting to the normal phase.   Detailed calculations suggest that at $\rho < 2.4 \times 10^{14}$ g cm^{-3} the protons are normal so that λ_p is effectively infinite while at $\rho > 2.8 \times 10^{14}$ g cm^{-3} the protons superfluid is well established with $\lambda_p \sim 50$ fm.   The distance between these two density regimes is of order 100 m.   On the basis of existing calculations it is difficult to decide whether the transition from $\lambda_p = 50$ fm to $\lambda_p = \infty$ occurs gradually or suddenly.   If the transition is gradual then the monopole will sink into the superconductor dragging behind it a flux tube until the energy per unit length of the flux tube equals the gravita-tional force at $\lambda_p \sim 150\,M_{16}^{-1/2}$ fm. In this case the monopole is suspended from the boundary by a flux tube that is   ~100 m long. If the transition is sudden then at first one would think the monopole would sit right on the surface except for a small dimple of order the penetration length. This would be correct for a solid but the superconductor under consideration is a liquid.   In a liquid the pressure inside and outside the flux tube must be equal.   The pressure inside has an additional contribution from the magnetic field pressure $\Delta p \sim B^2/8\pi$ which must be compensated for by a difference in the fluid pressure inside and out. This difference will be of order $\rho a z$ where ρ is the density, a is the acceleration due to gravity, and z is the distance down from the normal-superconducting boundary. Thus, we have $\rho a z \sim B^2/8\pi$ or substituting $B \sim \phi_0/\pi \lambda_p^2$,

$$z \sim \frac{\phi_0^2}{8\pi^3 a\rho \lambda_p^4} .$$

As before the monopole will be in equilibrium when $\lambda_p \sim 150$ fm which corresponds to $z \sim 0.01 - 0.1$ cm. Again the monopole is suspended by a flux tube but by a much shorter one. The monopole-flux tube configuration will move under the influence of the magnetic field tangential to the normal-superconducting boundary. Equating the magnetic force and the drag force gives a velocity of $v \sim 10^{-3}$ $(10^{-2}$ cm/z$)$ cm s^{-1}. The time required for the monopole to move from one pole of the neutron star to the other is $\sim 10^2$ years for $z = 10^{-2}$ cm (sudden transition) or $\sim 10^8$ years for $z \sim 10^4$ cm (gradual transition).

MONOPOLES AND FLUX TUBES

So far I have ignored the effects of the flux tubes that carry the magnetic field in the core. There are $\sim 10^{31}$ of these flux tubes with a spacing of $\sim 10^{-9}$ cm. The monopole will interact with these flux tubes either by being attracted to them as it first enters or by encounters between the monopole and its flux tube and pre-existing flux tubes. The flux emanating from a magnetic monopole is $4\pi g = 2\phi_0$ so the flux tube connected to the monopole actually carries two flux quanta. Since this is a type II superconductor it will probably be energetically favorable for this flux tube to split into two. To see what happens when a monopole encounters a pre-existing flux tube I will consider two scenarios that give the same end result.

First suppose that the monopole is approaching the normal-superconducting boundary from the normal side. To the monopole, the mouths of the flux tubes look like half monopoles or half antimonopoles since they carry flux ϕ_0. Assume that the monopole is at the pole of the neutron star such that the flux tubes look like half antimonopoles (to a first approximation we can take the core field to be a dipole field with all the flux tubes parallel). The monopole will then be attracted to the mouths of the flux tubes and will drop down a flux tube. This switches the orientation of the magnetic field above the monopole so that the mouth of this flux tube now looks like half a monopole. This is illustrated in Fig. 1. On the other hand suppose that the monopole penetrates the superconductor without finding a flux tube and is suspended by its own double flux tube as described earlier.

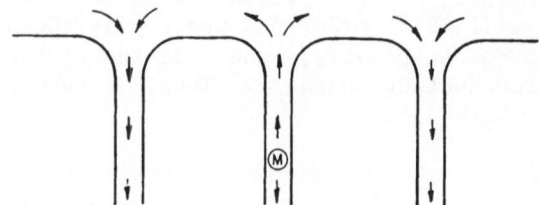

Fig. 1. Flux   tube   configuration   after   swallowing   monopole.

If the monopole is at the same pole of the neutron star as in the
previous scenario then the magnetic field in the monopole flux
tube is oriented oppositely to the field in the adjacent flux
tubes and there will be an attractive force between them. Because
of the spreading of the flux tubes at the surface this will
rapidly draw the monopole-flux tube configuration into an adjacent
flux tube and the result will be the same as shown in Fig. 1.

    Once the monopole has been swallowed by a flux tube it will
drop down the flux tube towards the center of the neutron star.
The flux tube containing the monopole will be attracted to an
adjacent flux tube since the magnetic fields are oppositely
oriented.   The spreading of the flux tubes at the surface will
result in the pinching off of the flux tubes with the formation of
a loop of flux tube as shown in Fig. 2.   The tension in the flux
tube will pull this loop through the superconducting core at a
rate determined by equating the energy per unit length in the flux
tube and the viscous drag force.   This gives a velocity $V = \varepsilon/\eta d_f$
where d_f is the width of the loop which is roughly the original
flux tube spacing $\sim 10^{-9}$ cm.   Therefore the flux tubes moves
through the core with velocity $V \sim 10^6$ cm/s and will cross the core
in a time of ~ 1 s.   When it encounters the monopole near the
center of the core it will simply drag it along as well.   As the
monopole plus flux loop approaches the opposite side of the core
it will slow down due to the spreading of the flux tubes.
Depending on how fast the tubes spread out the monopole will

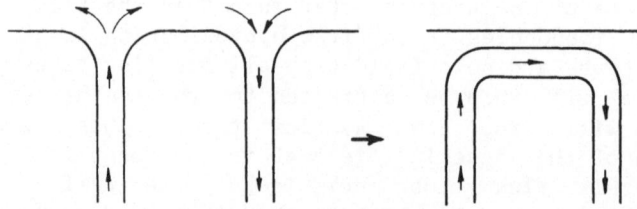

Fig. 2 Pinching off of flux tubes at normal-superconducting
        interface.

either remain suspended beneath the surface or will be ejected some small distance above the surface and then fall back and sink into the superconductor and be suspended by a flux tube. Now the flux tubes will not pinch off since they have aligned magnetic fields and the monopole will move very slowly if at all in the direction of the magnetic field tangential to the surface. Eventually this motion may return the monopole to the opposite side where it can repeat the above scenario.

MONOPOLE BOUNDS

Each orbit of a monopole through the core via a flux tube and then back due to the surface magnetic field destroys two flux tubes. Eventually this process will destroy the core field but it would take $\sim 10^{24}$ monopoles to destroy the core field in $\sim 10^{10}$ years and since 10^{21} monopoles would be captured over 10^{10} years if a Parker flux of monopoles was incident on the neutron star this will not provide a very stringent bound on the monopole flux. In addition the destruction of the core field will be much slower if the monopole ranges a distance of 100 m into the core rather than 10^{-2} cm as I have assumed above.

We are now in a position to answer the question of whether any of this affects the catalysis bounds on the monopole flux. Since I don't see any way of keeping the monopoles from catalyzing nucleon decay the only way out is to get rid of the monopoles. One way of getting rid of monopoles is to annihilate them with antimonopoles. The only efficient way of bringing monopoles and antimonopoles together is in a flux tube, but there are 10^{31} flux tubes and the catalysis bounds correspond to only 10^{14} monopoles and antimonopoles so it is unlikely that a monopole and anti-monopole will find themselves in the same flux tube. If the protons are not superconducting then previous analyses[6,7] show that again monopole-antimonopole annihilation is negligible except perhaps for very old neutron stars with small interior magnetic fields. Monopole-antimonopole annihilation in the presence of superconductivity has also been analyzed in Ref. 7. I believe their scenario is incorrect both because they assumed proton-neutron pairing rather than proton-proton pairing which is certainly incorrect since the difference in proton and neutron Fermi energies is greater than the energy gap and because they neglected the spreading of flux tubes near the normal-superconducting boundary.

THE 10^{17} GeV MONOPOLE ACCELERATOR

If we can't annihilate monopoles with antimonopoles maybe we can eject them from the neutron star. In our previous analysis the monopole-flux configuration crosses the star at v $\sim 10^6$ cm s^{-1}

which is small compared to the escape velocity of ~ 0.5 c. In addition the monopole loses energy due to encounters with electrons with $dE/dx \sim 10^{11} \beta_M$ GeV/cm so that it quickly falls back down to the core. However if the energy per unit length of the flux tube were a few orders of magnitude larger than the monopole would be accelerated to $v \sim c$ and might be able to escape. This may in fact happen if the core contains a pion condensate.

At a few times nuclear density the difference in the neutron and proton chemical potentials may become of order the effective pion mass so that the process $n \to p^+\pi^-$ becomes energetically favorable. The pions would then Bose condense to form a new type of superconductor which could fill a fair protion of the core of the neutron star. In terms of the linear sigma model this condensate corresponds to a chiral rotation of the vacuum induced by the background nucleon fields[8] so that

$$\langle \sigma \rangle = f_\pi \cos\theta$$

$$\langle \pi^- \rangle = f_\pi \sin\theta$$

with $\theta = 0$ corresponding to the normal vacuum. There is no convincing evidence for or against the existence of pion condensates in neutron stars so I will take θ to be a free parameter. The photon mass is given by $m_\gamma = ef_\pi \sin\theta \sim 32$ MeV $\sin\theta$ which is less than the effective pion mass so the π^- condensate will be a type II superconductor. The main differences between the pion superconductor and the proton superconductor are that the flux quantum is now $2\phi_0 = hc/e$ and the penetration length $\lambda_p \sim 1/m_\gamma \sim 6$ fm $\sin^{-1}\theta$ is much smaller unless θ is very small. As a result the energy per unit length of a flux tube in the pion superconductor is $\varepsilon_\pi \sim 10^{12} \sin^2\theta$ GeV/cm.

Monopoles will eventually end up in flux tubes in the pion condensate by the mechanism described earlier. Since the flux quantum of the flux tubes now matches that of the monopole, the monopole will be attached to a single flux tube and there won't be a loop of flux tube that has to be dragged along. The monopole velocity depends on the balance between the energy gain from the flux tube of $\sim 10^{12} \sin^2\theta$ GeV/cm and the energy loss to electrons of $\sim(10^{11}-10^{12}) \beta_M$ GeV/cm. Which of these is larger depends on the exact values of θ and the electron number density. It is certainly possible however that the energy gain is a few times larger so that the monopole reaches the surface of the neutron star with an energy of $\sim 10^{17}$ GeV, or $v \sim c$ for 10^{16} GeV mass monopoles. The neutron star would absorb monopoles with $v \sim 10^{-3}$ c and shoot them back out with $v \approx c$! How this affects the catalysis bounds depends on the time it takes for monopoles to enter the core. If the incident flux is isotropic then this time is roughly the time it takes the monopole-flux tube to move from one pole to the other which is 10^2–10^8 years depending on the nature of the normal-superconducting transition. If it takes 10^2 years then the catalysis bounds are only comparable to the Parker limit. If it takes 10^8 years then the bounds from old neutron stars are 10^2 times larger and the bounds from young pulsars are essentially unchanged. I should point out that only $\sim 10^{-18}$ of the incident galactic flux of monopoles gets processed by neutron stars so one should not start looking for relativistic GUT monopoles.

CONCLUSIONS

I have tried to convince you that it is possible to gain a qualitative understanding of the behavior of monopoles in neutron stars. Monopole-antimonopole annihilation appears to be unimportant which leaves monopole ejection by a pion condensate as the possible loophole in the catalysis bounds. Until we gain more information about the existence of pion condensates the most reliable catalysis boudns will probably come white dwarfs.[9] There is a general feeling which I may have reinforced that astrophysical bounds always contain loopholes and therefore should not be taken seriously. In the case of magnetic monopoles I don't think this is true, but I would be delighted to be proved wrong.

ACKNOWLEDGEMENTS

I have benefited greatly from conversations with my collaborators M. Ruderman and J. Shaham. I would also like to acknowledge the hospitality of the Aspen Center for Physics where this collaboration began. This work was supported in part by NSF Grant PHY80-19754.

REFERENCES

1. E. Kolb, talk at this workshop.
2. K. Freese, M.S. Turner, and D.N. Schramm, Phys. Rev. Lett.
3. J. Harvey, M. Ruderman, and J. Shaham, in preparation.
4. S.L. Shapiro and S.A. Teukolsky, "Black Holes, White Dwarfs, and Neutron Stars," (John Wiley and Sons, 1983).
5. R.P. Heubener, "Magnetic Flux Structures in Superconductors," (Springer-Verlag, 1979).
6. J.A. Harvey, Nucl. Phys. B, in press.
7. V.A. Kuzmin and V.A. Rubakov, Phys. Lett. 125B, 372 (1983).
8. D.K. Cambell, R.F. Dashen, J.T. Manassah, Phys. Rev. D12, 974 (1975).
9. K. Freese and R. Kron, "Do Monopoles Keep White Dwarfs Hot?", Univ. of Chicago preprint (1983).

MONOPOLES IN PULSAR PSR 1929+10

Katherine Freese

Astronomy and Astrophysics Center
University of Chicago
Chicago, Illinois 60637

Using x-ray observations of old radio pulsars, especially PSR 1929+10, one can derive a very stringent bound on the flux of monopoles times cross-section for catalysis, viz., $(F/cm^{-2}s^{-1}sr^{-1})$ $\times$ $(\sigma v/10^{-28} cm^2) \leqslant 6.7 \times 10^{-22}$. If the monopoles captured by the main sequence progenitor are also included, the limit improves by about 6 orders-of-magnitude.

The monopoles predicted to exist in grand unified theories[1], such as the monopoles of mass $\approx 10^{16}$ GeV in SU(5), can give rise to the catalysis of nucleon decay with a cross-section characteristic of strong interactions[2], $\sigma v \approx 10^{-28}$ cm^2. The astrophysical consequences of this catalysis process can in turn be used to place bounds on the monopole flux. I am going to report on using results from the Einstein Observatory to find a stringent limit. M.S. Turner, D.N. Schramm and I used the x-ray emission from old nearby pulsars; the best limit was obtained from consideration of PSR 1929+10.

For our neutron star model, we use $M = M_{1.4}M_\odot$, $R = R_{15}15$ km, $\bar{\rho} = M/(4\pi R^3/3) = 2.0 \times 10^{14}$ gm cm^{-3} $M_{1.4}R_{15}^{-3}$, and central density $\rho_c = 3 f \bar{\rho}$. Following Rubakov[2], we take the cross-section for catalyzed nucleon decay times relative velocity of monopole and nucleon to be constant, $\sigma v = (\sigma v)_{-28}10^{-28}$ cm^2.

Monopoles traveling through objects in the galaxy lose energy by electronic interactions[4]. Neutron stars can easily stop monopoles moving with the virial velocity of the galaxy, $v \approx 10^{-3}c$, as long as the monopole mass is less than 10^{22} GeV. Once the monopoles are trapped inside the neutron star, because of their large masses they quickly sink towards the star's center where they can catalyse nucleon decay. The energy released by this catalysis process then gives rise to a flux of photons from the star.

A neutron star exposed to a monopole flux $F = F_{-16}10^{-16}$ cm^{-2}s^{-1}sr^{-1} for a time $\tau = \tau_6\ 10^6$ years will capture a number of monopoles

$$N_M = F\ (\pi - sr)\ A\ \tau = 7.8 \times 10^{16}\ F_{-16}\tau_6 a_1 \tag{1}$$

where the effective capture area $A = 4\pi R^2\left[1 + (v_{esc}/v_M)^2\right]$, $a_1 = R_{15}$ $M_{1.4}\beta_{-3}^{-2}$, far from the star the monopole's velocity is $v_M = \beta_{-3}10^{-3}$ c, and v_{esc} is the escape velocity from the surface of the neutron star. To find the rate of energy released due to catalysis by one monopole, we multiply the rate for catalysis times the energy released per decay, about one nucleon rest-mass:

$$L_1 = \rho_c\ \sigma v = 1.6 \times 10^{18}\ (\sigma v)_{-28}\ a_2\ erg/s, \tag{2}$$

where $a_2 = R_{15}^{-3}\ M_{1.4}\ f$. The total luminosity of a neutron star due to catalysis is then

$$L_{mon} = N_M L_1 = 1.3 \times 10^{35}\ F_{-16}\ (\sigma v)_{-28}\ \tau_6\ a_3\ erg/s, \tag{3}$$

where $a_3 = a_2 a_1$.

This photon flux can then be used to limit the monopole abundance. Because of the catalysis process, neutron stars are x-ray sources, with the energy of the photons ≈ 100 eV. Adding all the photons due to catalysis from all the neutron stars, Dimopoulos et al.[6] found the contribution to the overall diffuse x-ray background. Requiring the monopole-induced luminosity not to exceed observed backgrounds one finds a flux limit[12],

$$F\ (\sigma v)_{-28} < 10^{-19}\ cm^{-2}s^{-1}sr^{-1}.$$

Alternatively, one can look at individual discrete sources, such as those expected in the serendipitous surveys of the Einstein satellite.[5] Both of these techniques, the diffuse x-ray background and the serendipitous surveys, have several inherent uncertainties, in the assumed number density of old neutron stars and in consideration of interstellar absorption.

Perhaps a better way to find flux limits is to look at individual known objects. In work with M.S. Turner and D.N. Schramm[3], we compared predicted photon fluxes with the x-ray emission of old nearby radio pulsars[8] (including PSRs 0031-07, 0355+54, 0655+64, 0809+74, 0950+08, 1055-52, 1133+16, 1508+55, 1642-03, 1929+10, and 1959+29). The most favorable of these old pulsars for our purposes is PSR 1929+10. It is very nearby (distance $\approx$ 60 pc) and its spindown age $\tau \approx 3 \times 10^6$ yr. Of these pulsars it has the lowest detected surface temperature $T_2 \approx 2 \times 10^5$ K) corresponding to photon luminosity $L_\gamma \approx 2.6 \times 10^{30}$ R_{15} erg/s. One might worry that the energy released due to catalysis might come out as neutrinos and go undetected. However, van Riper and Lamb[9] have shown that at these low luminosities, the contribution from neutrinos is negligible, so that L_γ may be taken as the total luminosity of PSR 1929+10. By requiring the energy due to catalysis not to exceed the observed luminosity, and by using Eq. 3 we find the flux limit

$$F \ (\sigma v)_{-28} \ < \ 6.7 \times 10^{-22} \ a_4 \ cm^{-2} s^{-1} sr^{-1}, \tag{4}$$

corresponding to a number of monopoles inside the star,

$$N_M \ (\sigma v)_{-28} \ < \ 1.6 \times 10^{12} \ f^{-1} R_{15}^{5} \ M_{1.4}^{-1} \quad . \tag{5}$$

$\left[\text{Here } a_4 = (\tau_6/3)^{-1} R_{15}^{4} \ \beta_{-3}^{2} \ M_{1.4}^{-2} \ f^{-1}. \right]$

To get a stronger limit one can also include those monopoles captured by the progenitor of the neutron star while it was on the main sequence (MS). Integrating the energy loss[4] of a monopole as it passes through a main sequence star, we find[10] that for a large range of stellar masses (3 – 30 $M_\odot$), the number of monopoles captured is approximately $N_M \approx 10^{24} \ F_{-16}$. Comparing this number to the limit on N_M obtained from PSR 1929+10 (Eq. 5), we obtain the more stringent but less secure bound,

$$F \ (\sigma v)_{-28} \ < \ 2 \times 10^{-28} \ a_5 \ cm^{-2} s^{-1} sr^{-1}, \tag{6}$$

where $a_5 = f^{-1} \ R_{15}^{5} \ M_{1.4}^{-1} \quad .$

One must also consider the possible effect of monopole-antimonopole annihilations on the limits discussed above. It is generally believed that the interiors of neutron stars are super-conducting.[11] In this case, monopoles will be confined to flux tubes and as long as the number of flux tubes ($\approx 10^{31} \ B_{12}$; magnetic field strength $B \approx B_{12} 10^{12} G$) exceeds the number of monopoles, annihilations will not be important;[12] for the limits discussed here this requirement is clearly satisfied. In the unlikely case that neutron star interiors are not superconducting, and the fields are too weak to separate monopoles and antimonopoles ($\ll 10^8$ G, see Ref. 12), monopole-antimonopole annihilations may be significant. The equilibrium abundance is determined by balancing the incoming monopole flux with the rate of annihilations. Harvey[12] finds that the equilibrium abundance $N_{eq} \approx 1.6 \times 10^{15} \ (F_{-16} M_{1.4} R_{15} \beta_{-3}^{-2})^{1/2}$. For PSR 1929+10 the constraint on N_M from x-ray observations (Eq. 4) implies that the number of monopoles captured since the pulsar's birth could not have attained the equilibrium abundance, and thus annihilations do not affect our bound (Eq. 4). If, in addition, we include the monopoles captured by its MS progenitor, then the number of monopoles can easily reach N_{eq}, and our bound (6) becomes

$$F < 1.0 \times 10^{-22} \, a_6 \, cm^{-2}s^{-1}sr^{-1}, \tag{7}$$

when annihilations are taken into account, where

$$a_6 = (\sigma v)_{-28}^{-2} \, f^{-2} \, R_{15}^{9} \, M_{1.4}^{-3} \, \beta_{-3}^{2} \qquad .$$

We have used x-ray observations of old radio pulsars to derive a very stringent bound on the average galactic flux of monopoles over the past 10^6 yrs: $F < 6.7 \times 10^{-22} \, a_4 \, cm^{-2}s^{-1}sr^{-1}$. When the monopoles captured by the MS progenitor are also taken into account, our bound improves by about 6 orders-of-magnitude.

Finally, R. Canal, R.G. Kron, and I are considering the catalysis process in white dwarfs.[13] The heat released in monopole-catalyzed nucleon decay may explain the paucity of cool white dwarfs. The largest abundance of white dwarfs is observed with a luminosity of $10^{-4}L_\odot$; monopole heat may explain why no cooler white dwarfs are observed. By requiring the monopole luminosity not to exceed this value of $10^{-4}L_\odot$, we find a flux limit

$$F (\sigma v)_{-28} < 1.8 \times 10^{-18} \, cm^{-2}s^{-1}sr^{-1},$$

or if we again include monopoles captured by MS progenitors,

$$F (\sigma v)_{-28} < 1.4 \times 10^{-20} \, cm^{-2}s^{-1}sr^{-1}.$$

REFERENCES

1. G. 't Hooft, Nucl. Phys. B79, 276 (1974); A. Polyakov, Pis'ma Zh. Eksp. Teor. Fiz. 20, 430 (1974) [JETP Lett. 20, 194 (1974)].
2. V.A. Rubakov, Pis'ma Zh. Eksp. Teor. Fiz. 33, 658 (1981) [JETP Lett. 33, 644 (1981)]; Nucl. Phys. B203, 311 (1982).
3. K. Freese, M.S. Turner, and D.N. Schramm, Phys. Rev. Lett. 51, 1625 (1983).
4. G. Tarle and S.P. Ahlen, in preparation (1983).
5. E.W. Kolb, S.A. Colgate, and J.A. Harvey, Phys. Rev. Lett. 49, 1373 (1982).
6. S. Dimopoulos, J. Preskill, and F. Wilczek, Phys. Lett. 119B, 320 (1982).
7. E.W. Kolb and M.S. Turner, Fermilab Astrophysics Preprint-'83.

8. D.J. Helfand, "X-rays from Radio Pulsars: the Portable
    Supernova Remants," to be published in the proceedings of IAU
    Symposium 101 (Supernova remnants and their x-ray emission),
    held Aug. 30 - Sept. 9, 1982 (Venice, Italy); R. Novick, G.
    Chanan, and D.J. Helfand, Bull. Am. Astron. Soc. 11, 779
    (1979); D.J. Jelfand, G. Chanan, and R. Novick, Nature 283,
    337 (1980); D.J. Helfand, private communication (1983).
9. See, eg., K.A. van Riper and D.Q. Lamb. Ap. J. 244, L13
    (1981); or S. Tsuruta, Phys. Rep. 56, 237 (1979).
10. K. Freese, J. Frieman, and M.S. Turner, in preparation
    (1983).
11. See, eg., G. Baym and C. Pethick, Ann. Rev. Astron. Astrophys.
    17, 415 (1979).
12. J. Harvey, Nucl. Phys. B, to be published (1983).
13. K. Freese, R. Canal, and R.G. Kron, in preparation.

MONOPOLE CONTAMINATION OF NORMAL SEQUENCE STARS

A.K. Drukier and G.G. Raffelt

Max-Planck-Institut für Physik und Astrophysik
Werner Heisenberg Institut für Physik
München, Federal Republic of Germany

We consider the monopole contamination of the normal sequence, 1 M companion star of the fast binary pulsar PSR 1953+29. We find that the photosphere of this star is practically monopole free, i.e. the number ratio r of photospheric monopoles to nucleons is $r < 3 \times 10^{-29}$. If the Callan-Rubakov effect exists at the predicted level, the much stronger limit $r < 1.2 \times 10^{-42}$ is obtained and the galactic isotropic flux is bounded by $F < 1.3 \times 10^{-20}$ cm^{-2}sr^{-1}s^{-1}.

Magnetic monopoles[1,2] are a general feature[3,4] of gauge theories such as Grand Unified Theories. If one considers SU(5)-GUTs, monopoles have the following properties:

$$eg = \frac{n}{2}\, \hbar c, \quad n = 0, \pm1, \pm2, \dots \tag{1a}$$

$$m \gtrsim 10^{16} \text{ GeV} \tag{1b}$$

$$\sigma(\Delta B = 1) = \sigma_0 \frac{1}{\beta} 10^{-28} \text{ cm}^2, \tag{1c}$$

where e and g are the quantum of electric and magnetic charge. Eqn. (1c) refers to the GUM's (Grand Unified Monopole) alleged ability to catalyze nucleon decay[6-10], and σ_0 is a factor of order unity.

All of the above properties of monopoles have been used to set astrophysical limits on their possible presence in the universe.[11-33] The mass of an ambient monopole background would contribute to the overall slow-down of the Hubble expansion of the universe and thus their number density cannot exceed certain

limits.[11-33]   Monopoles   would   be   accelerated   by   galactic,[14-17]
intracluster,[18] and stellar[19-21] magnetic fields and thus consume
the energy of these fields.   The persistence of these fields sets
stringent limits   on the cosmic monopole background.   If GUMs were
gravitationally   confined   to   the   interior   of   stars   they   would
catalyze nucleon decay and thus contribute to the luminosity of
stars.[24-33]   This process   would be particularly efficient in the
interior of neutron stars.

A particularly fascinating star of this sort is the recently
discovered binary pulsar PSR 1953+29.[34]   Its pulse period is
$P = 6.133$ ms and $\dot{p} < 6 \times 10^{-16}$ ss^{-1}.   The binary orbit appears to
be roughly circular and the orbital period is 120 ± 4 days.   The
radius   of   the   pulsar's   orbit   is   calculated   to   be
$a_p = (0.99 + 0.01) \times 10^{12}$ cm/sin(i), where i is the unknown angle
of inclination of the orbital plane with the plane perpendicular to
the line of sight. For an assumed mass of the pulsar of about 1 $M_\odot$
the minimal mass for the companion is about 0.2 $M_\odot$, and for
$18^0 \lesssim i \lesssim 90^0$ it must be less than 1 $M_\odot$. The angle α between the
spin-axis and magnetic dipole moment is probably small[34] but
non-zero.   The surface field strength is B $< 4 \times 10^{10}$ G/sin α. The
distance to the system is about 3.5 kpc.   It is optically in-
visible, and from observational data of the Einstein Observatory a
bound to the X-ray luminosity of $L_x(0.25 - 3.5$ keV) $< 4 \times 10^{32}$
ergs^{-1} is inferred, which includes corrections for interstellar absorption.[39] For an assumed radius of 15 km, this translates into a
surface temperature of 1×10^6 K and into a limit for the total
photon luminosity of

$$L_{PSR\ 1953+29} < 1.6 \times 10^{33} \text{ erg s}^{-1}. \tag{2}$$

Two scenarios of evolution have been suggested to account for
the above properties of the system.[34-42] One begins with a neutron
star, the other with a white dwarf near its Chandrasekhar limit, in
orbit with a companion of about one solar mass.   In both scenarios
the companion evolves into a red giant.   When its radius reaches
its Roche-lobe, mass transfer begins.   For about 10^8 years mass
spills over to the primary, the total mass transfer being about
0.7 $M_\odot$. If the primary is a white dwarf it collapses under excess
accreted mass to a neutron star, and in both scenarios the primary
is spun up by mass accretion and the binary orbit is circularised.
The end state is an old, fast, low-magnetic field pulsar in wide
circular orbit with a helium white dwarf of mass 0.3 $M_\odot$.   The
magnetic field of the neutron star is expected to have been on the
order 10^9 G for the whole period of mass transfer and the time
thereafter.   We now discuss what can be learned about magnetic
monopoles if these scenarios are correct.

If monopoles were contained in the photosphere of the original
binary partner of the neutron star, they would be transferred in

the course of mass accretion and thus be subject to acceleration by the pulsar's magnetic field. Therefore they would gain an additional amount of kinetic energy beyond that gained from gravity. This energy would be dissipated when entering the star, although very heavy monopoles may swing several times through the star before settling to its center.[24] The pulsar's stock of magnetic energy would be consumed, and the persistence of the magnetic field allows us to set limits on the number of monopoles accreted.

A magnetic field which is an exact dipole field exterior to the star with polar field strength B contains an amount W_{mag} of energy in the region exterior to the star (in Gaussian units):

$$W_{mag} = \frac{1}{12} B^2 R^3.$$ (3)

For our example with $B = 10^9$ G and $R = 15$ km this yields 2.8×10^{35} ergs. The magnetic potential on the surface of the pulsar can be expressed as $\phi = 1/2 \, BR \cos \Theta$, where Θ is the polar angle to the magnetic dipole axis.

If monopoles were so light that their trajectories were dominated by magnetic forces, north- and south-poles would be essentially funneled to their respective stellar poles and thus gain energy

$$W_1 = \frac{1}{2} gBR$$ (4)

beyond gravitational acceleration. For our standard parameter values $B = 10^9$ G, $R = 15$ km this yields $W_1 = 2.5 \times 10^7$ ergs. We tacitly assume that monopoles carry one Dirac quantum g of magnetic charge. Equation (4) can be interpreted as a differential energy loss $W_1 = dW_{mag}/dN$ of the star. By equating $W_1 dN$ from Eq. (4) with the differential $dW_{mag} = 1/6 \, R^3 BdB$ from Eq. (3) and integrating we find that

$$N_{destr} = \frac{1}{3} R^2 Bg^{-1}$$ (5)

monopoles are required to completely destroy the field. We note that the quantity R^2B is proportional to the magnetic flux through the star. This flux would be roughly conserved when a white dwarf collapses to a neutron star. Thus, we find for both scenarios of formation of our system that the persistence of the magnetic field yields a bound to the number of accreted monopoles

$$N_{accr} < 2.3 \times 10^{28}.$$ (6)

Since $0.7 \, M_\odot = 1.4 \times 10^{33}$ g of photospheric material was accreted, we find that the number ratio of photospheric monopoles to nucleons is $r < 2.7 \times 10^{-29}$. These results are valid only if the motion of the monopoles is dominated by magnetic forces. A dimensionless figure of merit to characterize the importance of magnetic vs. gravitational forces is the ratio b of surface magnetic field at the poles to surface gravity, i.e.

$$b = \frac{gBR^2}{GM_m} \, , \tag{7}$$

where G is the gravitational constant and M and m are the masses of the star and monopole respectively. Thus for monopoles much heavier than 10^{11} GeV, such as GUMs, the motion is controlled by gravity.

Nevertheless, will the path of a heavy monopole in the presence of a weak magnetic dipole field (b $\ll$ 1) deviate from a pure gravitational trajectory? In a model calculation we assume that A monopole and an antimonopole start their motion at a distance y from the star. In the absence of magnetic forces they would fall on a radial path with polar angle Θ toward the stellar surface. In the presence of a weak dipole field their trajectories split and they hit the surface at angles $\Theta + \Delta\Theta$ and $\Theta - \Delta\Theta$ respectively. Therefore, they end at different magnetic potentials and gain a net amount of magnetic potential energy of $g\phi(\Theta - \Delta\Theta) - g\phi(\Theta + \Delta\Theta)$, where ϕ is the surface magnetic potential. Although one of them gains and one of them loses a large amount of magnetic energy, there is a small difference between gain and loss, and each of them can be said to gain half of the above net gain. From a nonrelativistic calculation we find for this average gain to first order in b

$$W_1 = \frac{1}{2}gBR \left[\frac{1}{3}(1 - \frac{R}{y} + \frac{1}{2} \log \frac{y}{R})\right] b \sin^2 \Theta. \tag{8}$$

Following the same line of reasoning as before, we find for the number, N_{red}, of monopoles necessary to reduce the field from the value B to a value B_{red}

$$N_{red} = \frac{1}{3} GMmg^{-2} \log \left(\frac{B}{B_{red}}\right) \left[\frac{1}{3}(1 - \frac{R}{y} + \frac{1}{2} \log \frac{y}{R})\right]^{-1} \sin^{-2} \Theta. \tag{9}$$

The pulsar's spin can be assumed to be essentially orthogonal to the accretion disc since it is spun up by mass accretion. If the angle, α, between spin axis and magnetic dipole axis is indeed small, the motion of the accreted monopoles essentially takes place in the magnetic equatorial plane and the $\sin^2 \Theta$ term can be approximated by unity. For y on the order of the orbital radius of the binary system, the expression in brackets is about 2.5. Assuming that the field was not reduced by more than a factor of 10 by monopoles and using our standard masses M = 1.4 $M_\odot$ and m=10^{16} GeV we find a limit $N_{accr} < 0.9 \times 10^{33}$ corresponding to r < 1 × 10^{-24}. Since Eq. 9 is essentially independent of B, our results are valid for both scenarios of formation of the pulsar system.

If monopoles were now present in the interior of the pulsar PSR 1953+29 they would catalyze neutron decay. Freese et al.[28] find that due to the Callan-Rubakov effect, the energy release from one monopole at the center of a neutron star is $L_1 = 1.6 \times 10^{18}$ erg s^{-1}. From the limit in Eq. 2 of the total photon luminosity, we find a limit to the number of monopoles now present in PSR 1953+29:

$$N_{present} < 1.0 \times 10^{15} \qquad . \tag{10}$$

Following Harvey[31,32], we assume that monopoles and antimonopoles do not annihilate because they are confined to separate magnetic flux tubes in the superconducting neutron star's interior. Then they would have been accumulated during the total lifetime of the star and no more than 10^{15} monopoles would have been accreted from the binary partner. This gives r < 1.2 × 10^{-42}.

Monopoles could also have been captured from a hypothetical cosmic background flux if they were lighter than about 10^{21} GeV. Freese et al.[28] find for the number N_{capt} of monopoles captured from an isotropic flux F (in cgs units) during a time interval τ

$$N_{capt} = 8\pi^2 \ (GMR/v^2) \ F \ \tau \quad , \tag{11}$$

where v is the velocity of the monopoles far away from the star. Assuming v = 10^{-3} c as the typical virial velocity of monopoles gravitationally confined to the galaxy and taking τ = 10^8 years for the age of the neutron star we find

$$F = 1.3 \times 10^{-35} \ N_{capt} \ cm^{-2} sr^{-1} s^{-1} \tag{12}$$

and this translates into a flux limit

$$F < 1.3 \times 10^{-20} \ cm^{-2} sr^{-1} s^{-1} \qquad . \tag{13}$$

This limit is about a factor 20 less stringent than the one obtained by Freese et al.[28] from the pulsar PSR 1929+10. They gain about three orders of magnitude from the lower surface temperature (2 × 10^5 K) due to the T^4-dependence of the Stefan Boltzmann-law, while we gain about two orders of magnitude from the age of the pulsar PSR 1953+29.

REFERENCES

1. P.A.M. Dirac, Proc. Roy. Soc. A133, 60 (1931).
2. B. Cabrera, Phys. Rev. Lett. 48, 1378 (1982).
3. G. 't Hooft, Nucl. Phys. B79, 276 (1974).
4. A.M. Polyakov, JETP Lett. 20, 194 (1974) and ZhETF Pis. Red. 20, 430 (1974).
5. R.D. Peccei, "Theoretical Review of Monopole Bounds," MPI-PAE/PTh 22/83.
6. V.A. Rubakov, JETP Lett. 33, 644 (1981) and Pis'ma Zh. Eksp. Teor. Fiz. 33, 658 (1981); Nucl. Phys. B203, 311 (1982).
7. C.G. Callan, Phys. Rev. D26, 2058 (1982).
8. F. Wilczek, Phys. Rev. Lett. 48, 1146 (1982).
9. F.A. Bais, J. Ellis, D.V. Nanopoulos, K.A. Olive, Nucl. Phys. B219, 189 (1983).
10. C.G. Callan, Monopole '83 Conference, Ann Arbor, Oct. 1983; W. Nahm, Workshop on Monopoles, GUTs and Early Universe, Copenhagen, Nov. 1983.
11. Ya.B. Zel'dovich, M.Yu. Khlopov, Phys. Lett. 79B, 239 (1978).
12. J.P. Preskill, Phys. Rev. Lett. 43, 1365 (1979).
13. G. Lazarides, Q. Shafi, T.F. Walsh, Phys. Lett. 100B, 21 (1981).
14. E.N. Parker, Ap. J. 160, 383 (1970).
15. S.A. Bludman, M.A. Ruderman, Phys. Rev. Lett. 36, 840 (1976).
16. M.S. Turner, E.N. Parker, T.J. Bogdan, Phys. Rev. D26, 1296 (1982).
17. E.E. Salpeter, S.L. Shapiro, I. Wasserman, Phys. Rev. Lett. 49 1114 (1982).
18. Y. Rephaeli, M.S. Turner, Phys. Lett. 121B, 115 (1983).
19. D.M. Ritson, "New Limits on Galactic Monopole Flux," SLAC-PUB-2977, Sept. 1982.
20. A.K. Drukier, "The Limits to the Cross-Section for Monopole Creation Obtained from the Persistance of the Magnetic Field of White Dwarfs," Acta Astronomica, to be published.
21. A.K. Drukier, Proc. Trieste Conference on Magnetic Monopoles, Dec. 1981.
22. S. Dimopoulos, S.L. Glashow, E.M. Purcell, F. Wilczek, Nature 298, 824 (1982).
23. K. Freese, M.S. Turner, Phys. Lett. 123B, 293 (1983).
24. S. Dimopoulos, J. Preskill, F. Wilczek, Phys. Lett. 119B, 320 (1982).
25. E.W. Kolb, S.A. Colgate, J.A. Harvey, Phys. Rev. Lett. 49, 1373 (1982).
26. J. Ellis, D.V. Nanopoulos, K.A. Olive, Phys. Lett. 116B, 127 (1982).
27. M.S. Turner, Nature 302, 804 (1983).
28. K. Freese, M.S. Turner, D.N. Schramm, "Monopole Catalysis of Nucleon Decay in Old Pulsars," Preprint Univ. of Chicago, May 1983.
29. K.A. Olive, D.N. Schramm, "One the Compatibility of Observable

Grand Unified Monopole Fluxes and Neutron Stars," Ref.TH.3595-CERN, May 1983.

30. K. Freese, R. Kron, "Do Monopoles Keep White Dwarfs Hot?," Univ. of Chicago Preprint, 1983.

31. J.A. Harvey, "Monopoles in Neutron Stars," Preprint Princeton University, 1982.

32. J.A. Harvey, Monopole '83 Conference, Ann Arbor, Oct. 1983; E.W. Kolb, ibid.

33. E.W. Kolb, M.S. Turner, Fermilab Astrophys. Group Preprint, 1983.

34. V. Doriakoff, R. Buccheri, F. Fauci, Nature 304, 417 (1983).

35. F. Pacini, Nature 216, 567 (1967).

36. P.C. Joss, S.A. Rappaport, Nature 304, 419 (1983).

37. B. Paczynski, Nature 304, 421 (1983).

38. G.J. Savonije, Nature 304, 422 (1983).

39. D.J. Helfand, M.A. Ruderman, J. Shaham, Nature 304, 423 (1983).

40. D.C. Backer, Sh.R. Kulkarni, C. Heiles, M.M. Davis, W.M. Goss, Nature 300, 615 (1982).

41. M.A. Ruderman, J. Shaham, Nature 304, 425 (1983).

42. M.A. Alpur, A.F. Cheng, M.A. Ruderman, J. Shaham, Nature 300, 728 (1982).

MAGNETIC MONOPOLES IN A MAGNETIC UNIVERSE

Daniele Fargion

Istituto di Fisica
Universita di Roma, "La Sapienza"
Rome, Italy

Large scale coherence of a cosmological magnetic field, independently of its strength, is incompatible with any detectable monopole flux. Consequently, the controversial evidence[1] of a coherent magnetic field up to a redshift $z \simeq 2$ infers the most severe cosmological bound on present number density (n_0) and fluxes (F) of magnetic monopoles: $n_0 < 2 \times 10^{-29} \, m_{16} \, H^2_{100} \, cm^{-3}$; $F < 2.77 \times 10^{-21} \, m^{-1}_{16} \, B_{-g} \, H^{-1}_{100} \, cm^{-2} \, s^{-1}$.

The cosmological expansion reduces the present velocities of the heavy GUT monopoles[2-3] to such a negligible level that it could easily allow slow magnetic charges to escape normal detection via ionization processes. Magnetic charges, however, may be accelerated by coherent magnetic fields up to high velocities. The consequent motion of the magnetic charges may erode the strength of the same fields responsible for their acceleration. As E. Parker has shown[4], the survival of our galactic magnetic field and the absence of coherent galactic electric fields probe the scarcity of free magnetic charges with respect to electric charges in our galaxy. This statement is known in a quantitative form as Parker's limit on the galactic monopole fluxes.

Naturally E. Parker assumed a dynamo mechanism as the source of the galactic field. However the present galactic field may be obtained from a primordial cosmological field. In this case the magnetic field is "frozen" in the galactic plasma during the galaxy formation. In this situation Parker's limit should be even more severe because of the absence of a regeneration process. A possible way to overcome this bound may be the inhomogeneous clustering of the monopole charges in stars with a net charge gradient from one extreme to the other of the protogalaxy. Then

the inhomogeneous monopole charge distribution in stars is the
source of present galactic fields. This scheme is only qualitative and needs further quantitative elaboration. An analogous
dynamical model is considered by other authors.[7]

The origin of the cosmological magnetic field may be postu-
lated ad hoc[5] or it may relate to some parity non-conservation
processes.[6] Stable and coherent magnetic fields, as large as
possible, are the best candidates for high speed GUT monopoles.
Therefore cosmological magnetic fields may not just be a source of
the galactic magnetic field, but also a source of fast (and
easily) detectable magnetic monopoles.[8] However, as we shall see,
the coherence of the field is not compatible with any relevant
monopole flux. On the contrary, a positive detection of monopoles
in the future may disprove the existence of a cosmological
magnetic field. Let use see how this limit is found from
cosmology. Because of the conservation law for a source free
magnetic field we are forced to consider only the anisotropic
homogeneous cosmological models; Bianchi type VII, IV, III, II and
I. For the sake of simplicity we chose the latter (Bianchi I) in
its axial symmetric form as a first approximation to the problem.

Present observations on the Faraday rotation measure of the
polarized radio emissions by distant radiosources support the
existence of a uniform intergalactic magnetic field[1] of a value
$B_0 \simeq 0.7 \times 10^{-9} \left(\rho_0/\rho_c\right)^{-1}$ Gauss, (where ρ_0 and ρ_c are the present
baryon density and the critical density for a present Hubble
constant of 100 km s^{-1} M$_{pc}$$^{-1}$). Its direction is in the galactic
direction $\ell = 100°$, $b = 15°$ in the space up to a redshift z = 2.

Even this magnetic field is small. Its value is not totally
compatible with other indirect upper bounds derived from different
arguments on the nucleosynthesis in the early universe.[8-9] Therefore in the following we shall assume B_0 as a free parameter in
units of 10^{-9} Gauss (B_{-9}).

The invariant space-time element axial symmetric Bianchi I
model reduces to

$$ds^2 = dt^2 - a^2(t)(dx^2 + dy^2) - b^2(t)dz^2 \quad .$$

Naturally the longitudinal magnetic field is assumed to point
along the "b" direction; in absence of the field we require a flat
isotropic Friedmann solution (a=b).

Then the Einstein field equations may be written as follows:

$$\left(\frac{\dot{a}}{a}\right)^2 + \frac{2\dot{a}\ddot{b}}{ab} = B^2 + 8\pi\rho; \tag{1a}$$

$$2\frac{\ddot{a}}{a} + \left(\frac{\dot{a}}{a}\right)^2 = B^2 - 8\pi p_r; \tag{1b}$$

$$\frac{\dot{\rho}}{\rho + p_r} = -(\ell na^2 b); \tag{1c}$$

$$\dot{B} = \frac{-2\dot{a}}{a} B - 4\pi \frac{ng}{m} p \tag{2a}$$

$$\dot{p} = \frac{-\dot{b}}{b}p + gB - F_\eta \tag{2b}$$

where the dot stands for time derivative, B^2 is the energy density of the magnetic field, p_r is the pressure density, ρ labels all the other energy densities, g is the Dirac magnetic charge, n is the magnetic charge number density, m is the monopole mass, and p is the monopole's longitudinal momentum along the B vector.

Opposite charges move in opposite directions[8] but both are ruled by the system.[1-2] Naturally, we neglect the fate of the transverse momentum of heavy monopoles because of the cosmological expansion. F_η is an undefined force due to the monopole baryon interactions. This force may be relevant at early universe, but after the recombination, F_η and the corresponding pressure p_r due to electromagnetic interaction of monopoles with neutral gas are negligible. In this framework (just after the recombination) system 1 decouples from system 2 and leads to a solution which is highly anisotropic in the early universe, but quasi-isotropic at present stages. This result[8] derives mainly from the small value of a typical adimensional parameter of the solution, which characterizes the present anisotropy: $B_*^2 = \left(B_0/B_u\right)^2$; $B_u \equiv CH_0/G^{1/2} = 3.94 \times 10^{-4} H_{100}$ Gauss, which is today, for $B_0 \lesssim 10^{-9}$ Gauss, much less than unity; namely:

$$a(\eta) \propto \eta^2 + B_*^2;$$

$$b(\eta) \propto a(\eta) + 2B_*^2 - 4B_*^4 a^{-1}(\eta);$$

$$(t - t_i) \propto \eta \left(\frac{\eta^2}{3} + B_*^2\right).$$

The quasi-isotropic solution behaves as $a \propto t^{2/3}$, $b \propto t^{2/3}$ and consequently the number density decreases as $\eta = n_0 t^{-2}$.

The results in Eq. (2) let us find a simple analytical solution[6]

$$B = B_0 t^{-3/2} (R_p t^\alpha + R_p^{-1} t^{-\alpha})$$

$$= \frac{\rho_0}{6 g t_0} \left(\frac{n_0}{n_c}\right)^{1/2} t^{-3/2} (N^{-1} \nu_p t^\alpha + N \nu_p^{-1} t^{-\alpha}); \qquad (3a)$$

$$\beta^* \simeq \frac{p}{mc} = \frac{6 g B_0 t_0}{mc} \left(\frac{n_c}{n_0}\right)^{1/2} (N R_p t^\alpha + N^{-1} R_p^{-1} t^{-\alpha}) = \frac{\rho_0}{mc} t^{-1/2}$$

$$(r_p t^\alpha + r_p^{-1} t^{-\alpha}) \qquad (3b)$$

where $n_c \equiv \dfrac{m}{144\pi g^2 t_0^2} = 3.76 \ 10^{-32} \ m_{16} \ H_{100}^2 \ cm^{-3}$; $\qquad (4)$

n_0 is the present monopole number density, and $M_{16} \equiv \dfrac{m}{10_{16} GeV}$,

$H_{100} \equiv \dfrac{H}{100 km \ s^{-1} \ Mpc^1}$

$$\alpha \equiv \frac{1}{6} \left(1 - \frac{n_0}{n_c}\right)^{1/2} \ ; \qquad N \equiv \left[\frac{1 - (1 - n_0/n_c)^{1/2}}{1 + (1 - n_0/n_c)^{1/2}}\right]^{1/2} \ .$$

(Naturally the equality in Eq. (3b) holds only at non-relativistic regime: for the range of values considered below this is the case.)

Let us remark that this critical density n_c is a very small number, almost 10 orders of magnitude less than the number density corresponding to the critical cosmological mass density

$$n_{cos} \equiv \frac{\rho cos}{m_{16}} = \frac{3 H_{100}^2}{8\pi G} \frac{1}{m_{16}} = 1 \times 2 \times 10^{-21} \ m^{-1}_{16} \ H_{100}^2 \ cm^{-3}.$$

Solutions in Eq. (3a) and (3b) behave as a monotonically decreasing function as long as $n_0 < n_c$, otherwise, for $n_0 > n_c$, they will behave as damped oscillating waves as considered below.

It is possible to show[8] that, for $n_0 \ll n_c$ the final (present) velocity is almost independent of any reasonable boundary conditions. Namely, either one assumes an initial vanishing momentum (at the recombination, at redshift $z_r = 1500$) or a stationary solution at relativistic speed (where $\dot{p} = 0$,

$F_\eta = 0$, b, a $\propto t^{1/2}$; $\dfrac{\beta_i}{(1 - \beta_i^2)^{1/2}} = 1.2 \ B_{-g} \ H_{100}$) the final

results are today respectively:

$$\beta_0 = \frac{3gB_0t_0}{mc} \left[1 - \frac{\left(\frac{1}{3}\right)x}{(1)x} (1 - Z_r)^{-1/2}\right] \tag{5}$$

$$= 5.38 \times 10^{-2} \times \left(\frac{99.1\%}{97.4\%}\right) B_{-g}m^{-1}16H^{-1}100;$$

i.e., the final results coincide within 1.7%.

Therefore, high speed magnetic monopoles are in principle possible. For $\eta_0 = \eta_c$ one obtains a maximum value for the β_0 modulo twice the value in Eq. (5), $\beta_0 \simeq 0.1 \times B_{-g}m^{-1}16H^{-1}100$. However, for larger values of, $\eta_0 > \eta_c$ we face a reduction of the velocity not only because of the damped oscillatory mode, but also because of a suppression in the modulo by a factor $\left(\eta_c/\eta_0\right)^{1/2}$. Indeed the solution in Eq. (3) for $\eta_0 > \eta_c$ may be better written as follows[8]:

$$\beta \simeq \frac{p}{mc} = \frac{6gB_0t_0}{mc} \left(\frac{\eta_c}{\eta_0}\right)^{1/2} t^{-1/2} \cos(\gamma + 1/2\phi + |\alpha|\ell nt)/\cos\gamma; \tag{6a}$$

$$B = B_0 t^{-3/2} \cos(\gamma + |\alpha| \ell nt)/\cos\gamma; \tag{6b}$$

where $\phi \equiv \arctan \left[\frac{2(\eta_0/\eta_c - 1)^{1/2}}{\eta_0/\eta_c - 2}\right]$,

and where γ is an arbitrary constant defining the present phase of the oscillation.

A remarkable feature of the solution in Eq. (6) is the appearance of a characteristic time scale corresponding to a complete oscillatory period for the cosmological magnetic field (and the monopole momentum). Its value is defined by the condition: $|\alpha| \ell n\frac{\Delta\tau}{t_0} = \pm 2\pi$ (7a)

i.e. $\Delta\tau = t_0 \left\{1 - \exp\left[\frac{-12\pi}{(\eta_0/\eta_c - 1)^{1/2}}\right]\right\}$ (7b)

This value ranges from $\Delta\tau = t_0$ (the present cosmological age) for $\eta_0 = \eta_c$ to a smaller fraction of t_0 for $\eta_0 > \eta_c$.

The existence[1] of a coherent metagalactic magnetic field up to a large redshift ($z \simeq 2$), infers a strong limit on the present number density of the GUT magnetic monopoles, independently on the value B_0 of the present cosmological magnetic field considered (i.e. this limit occurs also for $B_0 \ll B_{-g}$).

Indeed an interval time $\Delta\tau$ corresponding to a coherent distance, up to a redshift z implies $t_0 (1 - (1+z)^{-3/2}) < \Delta\tau$, therefore from Eq. (7)

$$n_o/n_c < \left[(8\pi)^2 \, \ell n^{-2} \, (1 + z) + 1 \right]; \tag{8a}$$

for $z = 2$ as the coherent redshift inferred by Japanese astronomers[1]

$$n_o < 524 \, n_c = 1.97 \times 10^{-29} \, m_{16} \, H_{100}^2 \, cm^{-3} \ . \tag{8b}$$

This limit corresponds to an upper bound to the present monopole flux:

$$F = \frac{6gB_o t_o}{m} \left(\frac{n_o}{n_c} \right)^{1/2} < 2.77 \times 10^{-21} \, m_{16}^{-1} \, B_{-g} \, H_{100}^{-1} \, cm^{-2} \, s^{-1} \ , \tag{9}$$

which is almost six orders of magnitude below Parker's limit (few tens of monopoles/day on the whole Earth!) The corresponding speed modulo is $|\beta_o| < 4.7 \times 10^{-3} \, B_{-g}$.

The consequences of limits in Eqs. (8) and (9) are quite important because they imply that:

i) Either there is some definitive evidence for a coherent cosmological magnetic field, and therefore there are no hopes of observing magnetic monopoles in the foreseeable future; or,

ii) a serendipitous detection of monopoles will disprove at once the controversial presence of large scale coherent metagalactic fields. Let us remark that these considerations do not depend on the value of the magnetic field but only on the coherence scale ($z = 2$).

The limit in Eq. (9) is probably at present the most stringent cosmological and astrophysical bound on the monopole fluxes, independent of the Callan-Rubakov effect derived from the luminosity of neutron stars.

Are there loopholes in the above arguments? For instance, let us consider briefly a different scenario that apparently may relax or invalidate the bound in Eq. (9):

i) The primordial cosmological field first "polarizes" the motion of the free magnetic charges as described by the solution of Eq. (6);

ii) then the monopoles cluster gravitationally into galaxies together with normal baryonic matter;

iii) while the clustering occurs, the monopoles "freeze" their peculiar monopole-antimonopole asymmetric motion into an aligned "galactic" magnetic dipole,i.e. into a galactic charge gradient frozen into a net stellar charge,

    iv) which is itself responsible for the galactic and the present overall cosmological magnetic field.

In this "dielectric" picture the bound in Eq. (9) becomes even more stringent.

    Indeed, before galaxy formation, suppose at redshift $Z_G \stackrel{>}{\sim} 5/20$ monopoles are still free to move and oscillate as prescribed by Eq. (6). The characteristic time of the "field reversal", (for $\eta_0 > \eta_c$), and the corresponding coherent length should be "frozen" at an epoch with a typical smaller time scale $t_G = t_0(1+Z_G)^{-3/2}$ derived by condition, Eq. (7). The consequent cosmological expansion will increase the space coherence of the field, but it is not enough to compensate the smaller time scale of the oscillations. Therefore the limit, Eq. (9), becomes even more severe:

$$\frac{\eta_0}{\eta_c} < (12\pi)^2 \times \ell n^{-2} \left[1 - (1 + Z_G)^{1/2} + (1 + Z)^{-3/2} (1 + Z_G)^{1/2}\right]$$

$$+ 1 < \left[(8\pi)^2 \ell n^{-2} (1 + Z) + 1\right] . \qquad (10a)$$

This limit exists only if:

$$Z_G < \frac{2(1 + Z)^{3/2} - 1}{\left[(1 + Z)^{3/2} - 1\right]^2} , \qquad (10b)$$

where Z_G is the galaxy formation redshift and Z is the coherent field redshift. For $Z = 2$, $Z_G \leqslant 0.53$, for $Z = 1$, $Z_G \leqslant 1.4$ (and $\eta_0 = 0$); therefore for even smaller coherent fields $Z \stackrel{>}{\sim} 1$, this scenario is not plausible. Consequently, limits Eqs. (8) and (9) are still loose bounds.

CONCLUSION

    Either a non-vanishing cosmological magnetic B does exist and exhibits large scale coherence (up to a redshift $Z = 2$) in which case it would be difficult to find monopole fluxes above 3×10^{-21} cm^{-1} sr^{-1} s^{-1}; or GUT monopoles are found at Parker's level and those who advocate cosmological magnetic fields will be in serious trouble.

REFERENCES

1. Y. Sofue, M. Fujimoto and K. Kawabata, Publ. Astron. Soc.

Japan, $\underline{31}$, 125-142 (1979); $\underline{21}$, 293 (1969); $\underline{20}$, 388 (1968).

2. G. 't Hooft, Nucl. Phys. $\underline{B79}$, 276 (1974).
3. A. Polyakov, JEPT Lett. $\underline{20}$, 194 (1976).
4. E.N. Parker, Ap. J. $\underline{163}$, 255 (1971); Cosmological Magnetic Fields (Oxford Clarendon) Chap. 22 (1979); M.S. Turner, E.N. Parker and T.J. Bogdan, Phys. Rev. $\underline{D26}$, 1296 (1982).
5. B.Ya. Zel'dovich, (1964), Soviet Phys. JEPT $\underline{21}$, 656 (1964). M. Fujimoto, Oki, Z. Hitotuyangi, Prog. Theor. Phys. Suppl. $\underline{31}$, 77 (1964).
6. H. Vilenkin and D.A. Leahy, Ap. J. $\underline{254}$, 77 (1982).
7. E.E. Saltpeter, S.L. Shapiro and I. Wasserman, Phys. Rev. Lett. $\underline{49}$, 1114-1117 (1982).
8. D. Fargion, Phys. Lett. $\underline{127B}$, 35-39 (1983); Lett. Nuovo Cimento $\underline{36}$, 449 (1983); Ph.D. Thesis Rome (1977).
9. S.L. Shapiro, I. Wasserman, Nature $\underline{289}$, 657 (1981).

CATALYSIS WITHOUT TEARS

Curtis G. Callan, Jr.

Joseph Henry Laboratories
Princeton University
Princeton, New Jersey   08544

ABSTRACT

A semiclassical model of baryon-monopole interactions is presented. The model, which could serve as the basis for practical calculations of catalysis cross sections, interpolates between a soliton model of baryons at large distances and a radial field theory near the monopole core.

INTRODUCTION

Recently it has been found that in the presence of a magnetic monopole, reactions that are ordinarily very unlikely can have a greatly increased cross section.[1,2] For instance, in the SU(5) grand unified theory, the cross section for monopole induced proton decay is determined entirely by the strong interactions and is free of any suppression associated with the tiny size of the magnetic monopole core. At present, we understand these processes reasonably well at the quark and lepton level. To make realistic estimates of the cross sections, however, it is necessary to take account of the binding of quarks into color singlet hadrons. We wish to study not a quark-monopole collision but a nucleon-monopole collision, with a more or less realistic model of the initial state nucleon and the final state nucleons or mesons. There may be very large suppression or enhancement effects associated with the long distance aspects of strong interactions. Such enhancements or suppressions could be of great importance in planning or interpreting experimental searches for monopole catalysis since the low velocities expected for cosmic monopoles could make the effects of strong interactions particularly important.

In this talk we will describe a model which we hope can be used to make fairly realistic estimates of nucleon-monopole cross sections. A model also helps clarify the basic physics of monopole catalysis in that it shows how the localized nucleons of ordinary physics can evolve in a monopole field into the spherically symmetric, s-wave objects of previous theoretical analyses of the monopole-fermion problem. The discussion is based on work done by the author with E. Witten and reported in greater detail in a paper to be published in Nuclear Physics B.

Low energy hadron physics can be well described by simple quark models, bag models, or QCD sum rules. For our purposes, however, these models have a drawback: they cannot be conveniently coupled to magnetic monopoles. We will consider, therefore, a more exotic model that is more suitable for our purposes.

It has long been known that the non-linear sigma model summarizes all the soft-pion theorems of current algebra. Many years ago, Skyrme pointed out that this model has a conserved topological charge and suggested that it should be identified with baryon number.[3],[4] It has recently become apparent that this long neglected idea is perfectly correct. The identification of topological charges with baryon number is one of the manifestations of the anomalies in QCD current algebra[5] and the solitons of the non-linear sigma model correspond to the nucleon and delta (or to the baryon octet and decuplet in the three flavor case).

The perturbative expansion about a classical soliton solution corresponds to the 1/N expansion in QCD. (Properly,[6] the baryons should be found as soliton solutions of the as-yet-unknown large N effective action of mesons, not the simple non-linear sigma model.) Despite the absence of explicit valence quarks in the non-linear sigma model, the soliton picture gives a fair account of the static properties of nucleons (such as magnetic moments and charge radii).[7] Although this approach will probably never be competitive in most respects with more standard phenomenological models (unless the true large N meson action is somehow discovered) it has for us a great virtue: it can be readily coupled to a magnetic monopole.

In this talk, we will show that the baryon decay catalysis phenomenon occurs very naturally within the non-linear sigma model when a background monopole gauge field is included in that model. Roughly speaking, the monopole provides a defect on which, in the absence of special boundary conditions to forbid it, the topological charge (which carries the baryon number) can unwind. At the same time, the defect provides a natural opportunity for coupling the sigma field to a lepton field in such a way that topological charge, when it disappears, is converted to lepton number. Taken together, these two effects allow us, in a rough but essentially parameter-free way, to describe all the physics of monopole catalysis in the non-linear sigma model.

From the purely theoretical point of view, this exercise throws fascinating new light on the properties of topological charge and solitons. Besides identifying an explicit finite-action method of unwinding a topological charge, we also encounter new objects which share fermion number between a conventional fermi field and a topological structure. There is a more practical side to all of this as well. The tools we will develop may make possible semiquantitative estimates of catalysis cross-sections and branching ratios. Such estimates could be of considerable experimental interest, as we have indicated previously. Whatever reservations one may have about the detailed accuracy of a non-linear sigma model calculation, it surely gives a much closer approximation to the truth about catalysis than any other currently practical method.

KINEMATICS

Although grand unification monopoles can have very complicated structure on short distance scales, on scales greater than the confinement scale they look exactly like Dirac's abelian magnetic monopole. Since the non-linear sigma model only purports to describe physics on scales greater than the confinement scale, it suffices for our purposes to study the effect of an ordinary U(1) monopole on the sigma model. The microscopic physics in the monopole core will be simulated by a boundary condition at r = 0, where the U(1) monopole field is singular.

Let us first give a nutshell description of the phenomenon we will discover. In the absence of electromagnetism, the baryon current of the sigma model is

$$B_\mu = \frac{1}{24\pi^2} \, \epsilon_{\mu\nu\alpha\beta} \, \text{Tr} \, U^{-1} \partial_\nu U \, U^{-1} \partial_\alpha U \, U^{-1} \partial_\beta U \tag{1}$$

where in (say) the two flavor case, $U = \exp(2i/F_\pi)\vec{T}\cdot\vec{\pi}$. This vanishes unless all three meson fields, π°, π^+, and π^- are excited. In a monopole field, the charged pions $\pi^\pm$ have an anomalous angular momentum, egx, and an associated centrifugal barrier which keeps them away from the monopole core. These facts appear to mean that a baryon cannot be influenced by the microscopic physics inside the core. We will find, however, that in the presence of electromagnetism, Eq. (1) must be modified. It turns out that near a monopole, a chiral field with only π° excited can carry baryon number. Such a field experiences no centrifugal barrier in approaching the monopole core. In the sigma model, the catalysis process will be described by a sequence in which an incident Skyrme soliton unwraps itslef into a radial π° field which then falls into the monopole core and is turned into a lepton.

We will restrict ourselves to the problem of two quark flavors (u and d) to avoid extraneous complications, such as the need to consider the Wess-Zumino interaction. The simplest form of the sigma model with stable solitons is the Skyrme model,

$$\mathcal{L} = \frac{F_\pi^2}{16} \, \text{Tr} \, \partial_\mu U \, \partial_\mu U^{-1} + \frac{1}{32a^2} \, \text{Tr} \left[U^{-1} \, \partial_\mu U, \, U^{-1} \, \partial_\nu U \right]^2 \tag{2}$$

where $F_\pi \simeq 190$ MeV. The quartic term is needed to prevent the soliton from shrinking to zero radius and energy; a^2 is a dimensionless constant which can be adjusted to fit the mass of the proton. A typical vacuum is described by the configuration $U(\vec{x}) = 1$. The field U is related to the conventional pion field, $\vec{\pi}$, by

$$U(x) = \exp \frac{2i}{F_\pi} \, \vec{T} \cdot \vec{\pi} \tag{3}$$

The nucleon and its excitations are described by soliton configurations of the form

$$U_N(\vec{x}) = -\exp i \, F(r) \, \hat{x} \cdot \vec{T} \tag{4}$$

where F(r) runs from 0 at the origin to π at infinity. The detailed shape of F(r) is determined by minimizing the total energy. One inevitably finds that the radius of the region in which $U(\vec{x})$ differs significantly from the vacuum value U = 1 is of order $(aF_\pi)^{-1}$. Actually, this soliton is degenerate with rotations of it of the form

$$U_{N,A}(\vec{x}) = A \, U_N(\vec{x}) \, A^{-1} \tag{5}$$

where A is an arbitrary SU(2) matrix. A must be treated as a collective coordinate, and quantization of its motion yields a sequence of states whose lowest members are the N and Δ.

We wish to generalize the Lagrangian, Eq. (2), to be gauge invariant under $U(x) \rightarrow U(x) + ie\alpha(x)[Q,U]$, $A_\mu \rightarrow A_\mu + \partial_\mu \alpha$, where $Q = \begin{pmatrix} 2/3 & \\ & -1/3 \end{pmatrix}$ is the charge matrix of quarks, e is the charge of the proton, and A_μ is the photon field. The minimal way to achieve this is to replace all derivatives $\partial_\mu U$ by covariant derivatives $D_\mu U = \partial_\mu U - ieA_\mu[Q,U]$. This minimal prescription yields a gauge invariant Lagrangian but not the correct one; it fails to include the effects of QCD anomalies. The terms that reflect QCD anomalies have been worked out previously.[*] The full Lagrangian, including the minimal terms and the terms associated with anomalies is

$$\mathcal{L} = \frac{F_\pi^2}{16} \text{Tr } D_\mu U \ D^\mu U^{-1} + \frac{1}{32a^2} \text{Tr}[U^{-1} D_\mu U, U^{-1} D_\nu U]^2 + \frac{e}{16\pi^2} \ \varepsilon^{\mu\nu\alpha\beta} \ A_\mu$$

$$\times \ \text{Tr } Q\big(\partial_\nu U \ U^{-1}\partial_\alpha U \ U^{-1}\partial_\beta U \ U^{-1} + U^{-1}\partial_\nu U \ U^{-1} \ \partial_\alpha U \ U^{-1}\partial_\beta U\big)$$

$$+ \frac{ie^2}{8\pi^2} \ \varepsilon^{\mu\nu\alpha\beta} \ (\partial_\mu A_\nu)A_\alpha \ \text{Tr } \big(Q^2\partial_\beta U\cdot U^{-1} + Q^2 U^{-1}\partial_\beta U + \tfrac{1}{2} \ Q\partial_\beta UQU^{-1}$$

$$- \tfrac{1}{2}QUQ\partial_\beta U^{-1}\big) \quad . \tag{6}$$

It is not difficult to see that this Lagrangian correctly describes anomalous processes such as $\pi^0 \to \gamma\gamma$ or the $\gamma\pi^+\pi^-\pi^0$ vertex.

For our applications, we also need the proper form of the baryon number and electromagnetic currents, including all anomalous contributions. As we have noted previously, the baryon number current in the absence of electromagnetism is

$$B_\mu = \frac{1}{24\pi^2} \ \varepsilon_{\mu\nu\alpha\beta} \ \text{Tr } U^{-1}\partial^\nu UU^{-1}\partial^\alpha UU^{-1}\partial^\beta U \tag{7}$$

As is usual for topological currents, it is conserved whether or not the equations of motion are obeyed. In the presence of electromagnetism Eq. (7) is unsatisfactory as it is not gauge invariant. Of course, Eq. (7) can be made gauge invariant by replacing derivatives with covariant derivatives, but the resulting formula is not conserved. A little experimentation leads to the gauge invariant, conserved generalization of Eq. (7):

[#]See Eqs. (19) and (20) of Ref. 5; we set the number n of colors equal to 3. However, there is an error in Eq. (19) of Ref. (5), where parity conservation was not imposed properly.

$$B_\mu = \frac{\varepsilon_{\mu\nu\alpha\beta}}{24\pi^2} \left[\text{Tr } U^{-1}\partial_\nu UU^{-1}\partial_\alpha UU^{-1}\partial_\beta U + 3ieA_\nu \text{ Tr } Q\left(U^{-1}\partial_\alpha UU^{-1}\partial_\beta U \right.\right.$$

$$\left.\left. - \partial_\alpha UU^{-1}\partial_\beta UU^{-1}\right) + 3ie\partial_\nu A_\alpha \text{ Tr } Q\left(U^{-1}\partial_\beta U + \partial_\beta UU^{-1}\right)\right] \quad . \tag{8}$$

This can be written alternatively as

$$B_\mu = \frac{\varepsilon_{\mu\nu\alpha\beta}}{24\pi^2} \text{ Tr } U^{-1}\partial^\nu UU^{-1}\partial^\alpha UU^{-1}\partial^\beta U$$

$$+ \frac{\varepsilon_{\mu\nu\alpha\beta}}{24\pi^2} \partial_\nu\left[3ieA_\alpha \text{ Tr} Q\left(U^{-1}\partial_\beta U + \partial_\beta UU^{-1}\right)\right] \tag{9}$$

The latter form demonstrates that the addition to the current is a total divergence, so that the baryon number of any state is still an integer, as long as there are no singularities and all surface terms vanish.

As for the electromagnetic current J_μ^Q, it may be determined in a similar way. Alternatively, it may be obtained as

$$eJ_\mu^Q = \frac{\delta \mathscr{L}}{\delta A_\mu}$$ where $\mathscr{L}$ is the Lagrangian of Eq. (6). We find

$$J_\mu^Q = J_\mu^3 + \frac{1}{16\pi^2} \varepsilon_{\mu\nu\alpha\beta} \text{ Tr } Q\left(\partial^\nu UU^{-1}\partial^\alpha UU^{-1}\partial^\beta UU^{-1} + U^{-1}\partial^\nu UU^{-1}\partial^\alpha\right.$$

$$\left. UU^{-1}\partial^\beta U\right) + \frac{ie}{4\pi^2} \varepsilon_{\mu\nu\alpha\beta} \partial^\nu A^\alpha \text{ Tr}\left[Q^2\partial^\beta UU^{-1} + Q^2 U^{-1}\partial^\beta U + \right.$$

$$\left. \frac{1}{2}Q\partial^\beta UQU^{-1} - \frac{1}{2} QUQ\partial^\beta U^{-1}\right] \tag{10}$$

Here J_μ^3 is the third component of the isospin current, obtained from the conventional part of the Lagrangian; we have explicitly written only the terms that reflect anomalies.

SPECIAL FEATURES OF THE MONOPOLE VECTOR POTENTIAL

If the background gauge field, A_μ, is taken to be the potential of the standard Abelian Dirac magnetic monopole,

$$A\phi = \frac{g(1 + \cos\theta)}{r \sin\theta}$$

$$F_{\theta\phi} = \frac{g}{r^2} \quad , \tag{11}$$

some very surprising things happen. Because of the singular nature of the monopole field, the chiral field can not take on arbitrary values at the monopole location. If we solve the Klein Gordon equation for spin-zero particles moving in a monopole background field, we readily find that the wave function of particles of charge $\pm e$ must <u>vanish</u> at the origin at least as fast as $r^{(\sqrt{3}-1)/2}$, while the wave functions of neutral particles are <u>not</u> required to vanish. The chiral field is neutral if it commutes with $Q = 1/6 + T_3/2$, and we take the above result to mean that we must set $U = \exp i\phi T_3$ at the monopole location. Of course, ϕ is related to the standard π^0 field by $\phi = \dfrac{2\pi^0}{F_\pi}$.

The question then arises whether ϕ is fixed, or whether it can vary with time. The internal dynamics of the chiral model does not tell us anything one way or the other about this question. To see what a time-varying ϕ would mean, let us consider the net flux of baryon number through a small sphere of radius r about the monopole. In the vicinity of the monopole we may take $U(\vec{x},t) = e^{i\phi(t)T_3}$, and immediately find that if eg = 1/2 then

$$J_r^B = -\frac{ie}{8\pi^2}\, \epsilon^{r\theta\phi t}\left(\partial_\theta A_\phi\right) \operatorname{tr}\left(-QU\partial_t U^{-1} + QU^{-1}\partial_t U\right)$$

$$= \frac{\dot{\phi}(t)}{8\pi^2\, r^2} \tag{12}$$

The net flux of baryon number <u>into</u> the monopole is $\dot{\phi}/2\pi$! A similar calculation using Eq. (10) shows that if Q is the usual electric charge matrix of light quarks, then $\dot{\phi}/2\pi$ is likewise the electric flux into the magnetic monopole. In other words, a time variation of the chiral field at the monopole location means that baryon number (topological charge) and electric charge are disappearing from the outside world into the monopole!

The correct "boundary condition" on U at the monopole location clearly is determined by short-distance physics not directly incorporated in the sigma model. If the underlying gauge theory responsible for the existence of the monopole is strictly baryon-number-conserving, then a proper microscopic treatment must lead to the condition ϕ = constant to ensure this conservation. If the underlying theory is like SU_5, and does <u>not</u> conserve baryon number, then the microscopic dynamics may lead to an effective theory in which ϕ obeys a free boundary condition, $\left(\dfrac{\partial\phi}{\partial r}\right)_{r=0} = 0$. With this boundary condition, the classical field equations of our sigma model will permit baryons to disappear at r = 0.

This is not quite the end of the story, however, since the SU(5) theory also does not conserve lepton number and satisfies the relation $\Delta B = \Delta L$. Thus, if one unit of baryon number disappears, a positron must appear. This has the side benefit of restoring charge conservation -- as the chiral field system loses baryon number, it also loses charge in the proportion $\Delta B = \Delta Q$. If leptons are coupled in according to $\Delta B = \Delta L$, then all the charge lost by the chiral field is transferred to the leptons, leaving the monopole uncharged. This is quite essential, because if the monopole is treated as a point object, it would cost an infinite amount of energy to put a charge on it.

This sigma model, of course, makes no reference to leptons, and we have to find a sensible and, if possible, correct way of coupling leptons to $U(\vec{x})$. From the previous discussions, it is apparent that any such coupling must happen at the monopole location and would most conveniently present itself as some sort of boundary condition. In fact the field theory of an electron moving in the background field of a monopole is known to need a specially imposed boundary condition at $r = 0$. In order for the Dirac Hamiltonian to be Hermitean, it is necessary to impose on the electron field the boundary condition

$$\psi_R(0) = ie^{i\theta}\psi_L(0) \tag{13}$$

where θ is an angle which may be chosen arbitrarily.

The remarkable thing is that the physics of the lepton Dirac sea depends on θ.[8] In particular, if the Dirac sea is normalized so that the charge (and lepton number) is zero at $\theta = 0$, then for general θ, it turns out that $Q = -L = \theta/2\pi$. If we imagine varying θ, then the net charge of the lepton world varies continuously and for every 2π increase in θ one "whole" electron is expelled to infinity. The obvious choice for our problem then is to couple the chiral field to the lepton system by setting $\theta = \phi(r = 0)$. Then as ϕ changes by 2π the system gains one unit of lepton number. The net effect of such a process is monopole catalysis of baryon decay.

A different argument for this boundary condition, based on our understanding of the physics of monopole catalysis at the quark level, can also be given. Consider SU(5) grand unification with, for simplicity, one generation of quarks and leptons. Consider the interaction of these fermions with the basic SU(5) monopole. The u_1, u_2, $\bar{d}_3$, and e^+ fields have a state of total angular momentum zero with respect to the monopole. This distinguished partial wave has strong interaction with the monopole core and requires a boundary condition, similar to the boundary condition on the electron in the field of an abelian monopole discussed in the previous paragraph.

The simplest way to approach this problem is to use the bosonization trick, replacing the $J = 0$ fermi field by an equivalent boson field. The essential boundary condition on the boson fields at the monopole core reads

$$\phi_e+(0) + \phi_{\bar{d}_3}(0) + \phi_{u_1}(0) + \phi_{u_2}(0) = 0 \tag{14}$$

To connect this to the sigma field, let us imagine a modified bag model of the strong interaction structure around the monopole: For $r > R$ ($\sim$ confinement scale) we have only the usual chiral field $U(\vec{x})$. For $r < R$ we have only the u and d quark fields. The two are connected at $r = R$ by the chirally invariant "bag" boundary condition

$$q_R = U(r) \, q_L \tag{15}$$

where q is the flavor doublet field built out of the u and d quarks. If we set $U = e^{i\phi\tau_3}$ and replace u and d by their corresponding bosonization fields, the bag boundary condition becomes

$$\phi_u(R) = - \phi_{\bar{d}}(R) = \frac{\phi}{2\sqrt{\pi}} \tag{16}$$

The kinetic energy carried by the quark fields inside the bag is minimized if $\phi_u(r)$ and $\phi_d(r)$ are actually r-independent.

In an obvious adiabatic approximation, then, we may insert the values of the quark fields at the bag boundary ($r = R$) into the monopole boundary condition ($r = 0$) to obtain

$$\phi_e+(0) = - \frac{\phi}{2\sqrt{\pi}} \tag{17}$$

When translated back into a statement about the lepton fermi field, this becomes

$$\psi_{eR}(0) = i e^{-i\phi} \, \psi_{eL}(0) \tag{18}$$

This is exactly what we guessed earlier.

To summarize, the sigma model in a monopole background requires boundary conditions on the sigma field which are similar in structure to the boundary conditions needed to define the Hamiltonian of an electron in a monopole field. The precise form of these conditions must reflect unification-scale physics not otherwise accessible to the sigma model. A small number of parameters (in simple cases, none at all!) will suffice to summarize the (possibly baryon-number-violating) effect of very short distances on low-energy pion-nucleon physics.

MONOPOLES WITH NON ZERO BARYON NUMBER

The presence of the monopole permits us to construct remarkable configurations which have baryon number but no "three-dimensional" topology -- despite the fact that baryon number in the sigma model originally appeared to be entirely topological. Since these configurations present a particularly clear and explicit scenario for the monopole to eat baryon number, we will explain their properties in some detail.

Let the monopole be located at r = 0. Consider the chiral field configuration

$$U(\vec{x}) = e^{if(r)T_3} \tag{19}$$

which is constructed entirely out of neutral (π^0) chiral fields. Since we always impose the vacuum condition $U(\infty) = 1$, we set $f(\infty) = 2\pi$. In accord with the previous discussion, $f(0)$ is not necessarily constrained. If we refer back to the expressions for baryon number and charge current in the field of a monopole, we can easily see that for this configuration

$$J_0^B = J_0^Q = \frac{1}{4\pi r^2} \frac{f'(r)}{2\pi} \quad . \tag{20}$$

Consequently the baryon number and charge of this pure π^0 configuration satsify

$$B = Q = \frac{1}{2\pi} \left[f(\infty) - f(0) \right] \quad . \tag{21}$$

Although it is built out of neutral fields and has no obvious topological properties, this configuration has both charge and baryon number! It has the quantum numbers of the proton if $f(0) = 0$ and $f(\infty) = 2\pi$.

It turns out that this configuration is one which can smoothly be reached by bringing in a Skyrme soliton from infinity and wrapping it around the monopole without ever changing the value of U at the monopole center from its initial value of 1. In that sense, our configuration represents a situation in which a monopole has captured a proton. What happens next depends on whether $f(0)$ is tied down or not. Before we discuss this, note that what we have here is clearly invariant to rotations and must be a <u>spin zero</u> object. This is just right since the proton has intrinsic spin 1/2 but has an extra spin 1/2 field angular momentum in the presence of a monopole. The neutron on the other hand can't be captured into an s-wave, and this is why the charge of this state is definite.

Suppose $U(0)$ is fixed at 1. Then the kink $f(r)$ $\bigl(f(0) = 0,\ f(\infty) = 2\pi\bigr)$ is trapped on the monopole and there will be some lowest energy "kink" configuration which represents the stable monopole plus trapped proton. A bit of analysis shows that the quartic term in the Lagrangian, which was introduced to stabilize the topological solitons, makes no contribution to the energy of a neutral radial kink $\bigl(U(\vec{x}) = \exp if(r)T_3\bigr)$. Since this kink carries electric charge, it will eventually be stabilized by its Coulomb self energy. The energy function, including Coulomb energy, is

$$E = \int 4\pi r^2 dr\ \{\frac{F_\pi{}^2}{8}\ f'(r)^2 + \frac{e^2}{128\pi^4 r^4}\ f^2(r)\} \tag{22}$$

The minimum-energy solution satisfying the boundary conditions $F(0) = 0$ and $f(\infty) = 2\pi$ is

$$f(r) = 2\pi\ e^{-e/4\pi^2 F_\pi r} \tag{23}$$

and the energy of that solution is $\frac{\pi}{2}\ eF_\pi$. If, as is equally possible, the boundary condition is chosen so that the kink has baryon number $N\ \bigl(f(0) = 0,\ f(\infty) = 2\pi N\bigr)$, the minimum energy is $E_N = N^2 \pi eF_\pi/2$. Since $\pi eF_\pi/2$ is much smaller than the nucleon mass, it appears that, even if baryon number is strictly conserved and $U(0) = 1$, the monopole can bind a number of protons with a large energy release which might be phenomenologically confused with monopole catalyzed proton decay!

It is perfectly possible for $U(0)$ to be frozen at some value other than 1. A particularly interesting possibility is $U(0) = -1$. The lowest energy state will always be a non-trivial kink and by arguments similar to those of the preceding paragraph, we readily see that it has half-integral charge and baryon number! By the arguments of the last section, one could simultaneously choose a "half integer" boundary condition for the lepton field, thereby constructing a system which is electrically neutral but has half integer baryon number! This is quite reminiscent of the structure found by Jackiw and Rebbi[9] in their study of 't Hooft-Polyakov monopoles coupled to fermions.

Suppose that $U(0)$ is <u>not</u> fixed, but is connected to the boundary condition of the lepton field in the way discussed in the last section. What is the fate of an initial soliton

U = $e^{if(r)T_3}$, $0 < f(r) < 2\pi$? Clearly, it is energetically favored for the soliton to be drawn into the origin with simultaneous production of an outgoing lepton, whose energy can be as small as the electron rest mass. This is most easily described using the bosonized version of the lepton theory. Then we have a scalar field $\phi(r)$ describing the lepton and the kink field f describing the sigma field. The effective Lagrangian for this system consists of kinetic energy terms for ϕ and f separately and a joint Coulomb energy term since both fields carry charge. The f kinetic energy term has already been constructed earlier in this section as has a Coulomb energy term which can readily be generalized to this case. The kinetic energy term for the bosonized lepton has been derived elsewhere.[1,2] The full effective action is

$$= \int_0^\infty 4\pi r^2 dr \; \frac{F_\pi^2}{8} \left(\dot{f}^2 - f'^2\right) + \int_0^\infty dr \; \frac{1}{2} \left(\dot{\phi}^2 - \phi'^2\right)$$

$$+ \int_0^\infty dr \; \frac{e^2}{32\pi^2 r^2} \; (f - \phi)^2 \; . \tag{24}$$

In theories in which baryon number is not conserved, the microscopic physics can be expected to give "free" boundary conditions for f and ϕ at r = 0: $\dot{f}(0)$ and $\dot{\phi}(0)$ are allowed to be non-zero, and lepton and baryon number are independently allowed to flow in and out of the monopole. However, by a mechanism completely analogous to that which has been explained in the context of the fundamental field theory version of monopole catalysis, the Coulomb energy dynamically enforces the condition $\phi(0) = f(0)$. This is precisely the coupling between sigma field and lepton that we described in a previous section and which enforces the SU(5) conditions $\Delta B = \Delta L$. We could of course solve the equations of motion which follow from the above Lagrangian to see in detail how an initial B = 1 soliton decays into an outgoing s-wave electron.

UNWRAPPING OF TOPOLOGICAL CHARGE

In the previous sections, we have explored the peculiar behavior of baryon number in the sigma model when monopoles are present. We would now like to study in more detail what happens when a Skyrme soliton collides with a monopole. If baryon number is conserved, it should be possible to capture the nucleon on the monopole in one of the radial kink states just described. If baryon number is not conserved, it should be possible for the monopole to convert the nucleon into a collection of pions and a lepton. In either case, it must be possible for something unusual to happen to the topological charge of the incoming soliton when it encounters the monopole.

Although we will not explain the details, a nucleon can be captured by the monopole in the baryon-number conserving case. We suppose that the vacuum state at infinity is $U = 1$ and that there is a fixed boundary condition, say $U = 1$, at the monopole location. It turns out to be possible to smoothly turn a neutral radial kink centered on the monopole, into a topological soliton centered at some distance from the monopole, without ever changing the value of U at the monopole core.

The starting configuration is a radial kink, say $U(x,y,z) = \exp iT_3 f(r)$, where $f(r) = 0$ for $r < r_1$ and $f(r) = 2\pi$ for $r > r_2$. It is sketched in Fig. 1. The Dirac string is explicitly indicated in this figure and will play a crucial role. The final configuration is a Skyrmion centered at some distance from the monopole much as in Fig. 2. Thus a state in the topological class of the Skyrmion can turn smoothly into the spherically symmetric state consisting of a baryon captured by a proton, without ever changing the value of U at the monopole or at infinity.

This conclusion should not be too surprising. We have seen previously that in the presence of the monopole the radial kink has the same baryon number as the Skyrme soliton. Since baryon number is the only topological invariant in the problem $\left(\text{as } \pi_3(SU(2)) = Z\right)$, the radial kink must be deformable into the Skyrme soliton.

Now let us study the kinematics of monopole catalyzed proton decay in the event that baryon number is not conserved and the sigma field at the monopole is free to vary. Consider a soliton moving along the z-axis toward a monopole at the origin, with its configuration oriented so that $U \sim e^{i\alpha T_3}$ at all points on the z-axis. This is a very special initial configuration, but instructive enough for our purposes.

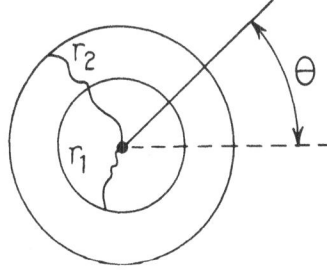

Fig. 1 A radial kink configuration of baryon number 1. It has $U = 1$ except for $r_1 < r < r_2$. The Dirac string is indicated by the dashed line.

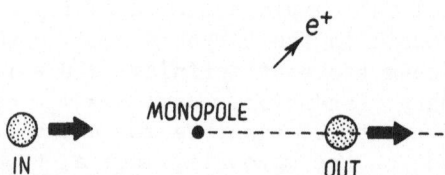

Fig. 2   A high energy soliton-monopole encounter with zero impact
         parameter.   The incident   Skyrmion (left) passes through
         the monopole essentially undisturbed (right) but the
         outgoing configuration is topologically trivial and can
         disperse into pions.   A positron is emitted at the same
         time.

     Let us imagine simply translating the soliton along the
z-axis so that the monopole passes through the soliton, with no
change in the soliton configuration.   This is a "sudden" approximation to the scattering problem and is probably a good
description of what happens in a relativistic collision with this
choice of impact parameter. In the course of the translation, the
field at the monopole makes one complete circuit around the $e^{i\alpha T_3}$
subgroup of SU(2).   If we adopt SU(5)-style boundary conditions,
this causes one positron to be emitted from the monopole.   In
general, there is some back-reaction on the soliton due to the
coupling to the lepton field, but in the spirit of the sudden
approximation we will neglect this effect.   The final state
consists of a positron, plus an unmodified soliton moving away
from the monopole along the Dirac string.   This situation is
described in Fig. 2.

     At first sight this may seem surprising. A positron has been
emitted yet the Skyrme soliton is still there, in the right-hand
side of Fig. 2.   This seems to conflict with conservation of B-L.
The resolution of this paradox is simply that the Skyrmion on the
right of Fig. 2 is "drilled" by the Dirac string; the baryon
number of the drilled Skyrmion is actually zero.   Indeed, suppose
the drilled Skyrmion has the prototypical form

$$U(\vec{x}) = - \text{expi}F(r)T \cdot x$$

$$= - \text{expi}\left[F(r)(\cos\theta T_3 + \sin\theta \cos\phi\, T_1 + \sin\theta \sin\phi\, T_2\right] \quad (30)$$

after being translated to the origin.   Normally this object has
baryon number one.   If, however, it is drilled by a Dirac string
which·can be removed by the gauge transformation $\Lambda = e^{-iT_3\phi/2}$,
then in the new gauge U becomes

$$\tilde{U} = \Lambda\, U\, \Lambda^{-1}$$

$$= - \text{expi}F(r)(T_3 \cos\theta + T_1 \sin\theta) \tag{31}$$

Since $\tilde{U}$ is excited only in the T_1 and T_3 directions, it has winding number zero. Thus, the drilled Skyrmion or the right-hand side of Fig. 2 is not a baryon but a mere collection of pions, which can disperse. We are describing the reaction $p + M \rightarrow e^+ + M + $ pions.

This is not quite the whole story. Suppose we replace the monopole by an antimonopole. Then the gauge transformation needed to "flip" the string from the positive z-axis to the negative z-axis is $\Lambda = \exp + i\phi T_3/2$ rather than $\Lambda = \exp - i\phi T_3/2$. The effect of this gauge transformation on our standard soliton is to convert Eq. (30) into

$$\text{expi}F(r)(\cos\theta T_3 + \sin\theta (\cos2\phi T_1 + \sin2\phi T_2)) \quad . \tag{32}$$

Once the string is removed, the soliton is seen to have <u>increased</u> its topological charge to two units! The overall effect of this scattering process, due account being taken of the lepton boundary conditions, is

$$p + \overline{M} \rightarrow \overline{M} + p + p + e^- \quad .$$

Because we are using the sudden approximation, we do not trouble ourselves about energy conservation. In fact, this process has an energy threshold comparable to the proton mass. How the reaction actually goes below that threshold is not clear.

SUMMARY

The above remarks are meant to give an overview of the peculiar kinematics of the capture, destruction (or even augmentation!) of topological charge by a magnetic monopole. One should, of course, study these processes dynamically, solving the equations of motion of the model to determine whether and in what detailed manner the monopole modifies the nucleon's topological charge. It seems to us that it would make sense to attempt to estimate the total monopole catalysis cross-section by solving the sigma model field equations for a variety of impact parameters of the incident soliton on the monopole. The previous discussion suggests that for zero impact parameter the topological charge will be destroyed and a lepton produced. It is equally obvious that for large impact parameter, the nucleon will retain its identity and no lepton will be produced. At some intermediate impact parameter, which can only be determined by detailed numerical calculation, there will be a crossover between the two regimes. A calculation of this critical impact parameter would be tantamount to a calculation of the monopole catalysis cross-section.   Despite the difficulties and uncertainties of such a calculation, it would   be very interesting to attempt,   given that there currently exists no other tractable method for computing the catalysis cross-section.

ACKNOWLEDGEMENTS

This research is supported in part by the National Science Foundation under grant No. PHY80-19754.

REFERENCES

1. V. Rubakov, Nucl. Phys.   203, 311   (1982), JETP Lett.   33, 644 (1981).
2. C.G. Callan, Jr., Phys. Rev. D25, 2141 (1982), Phys. Rev. D26, 2058 (1982), Nucl. Phys.   B212, 391 (1983),   "Baryon Decay Catalysis S-Matrix", in "Grand Unification and Supergravity", eds. G. Farrar and F. Henyey, AIP Conference Series.
3. T.H.R. Skyrme, Proc. Roy. Soc. A260, 127 (1961), Nucl. Phys. 31, 556 (1962).
4. D. Finkelstein and J. Rubinstein, J. Math. Phys. 9, 1762 (1968); J.G. Williams, J. Math. Phys. (N.Y.) 11, 2611 (1970); L.D. Fadeev, Lett. Math. Phys. 1, 289 (1976); N.K. Pak and H. Ch. Tze, Ann. Phys. (N.Y.) 117, 164 (1979); J.M. Gipson and H.Ch. Tze, Nucl. Phys. B183, 524 (1981); A.P. Balachandran, V.P. Nair, S.G. Rajeev, and A. Stern, Phys. Rev. Lett. 49, 1124 (1982), Phys. Rev. D27, 1153 (1983); J. Boguta, Phys. Rev. Lett. 50, 148 (1983).
5. E. Witten, Nucl. Phys. B223, 433 (1983); J. Goldstone and F. Wilczek, Phys. Rev. Lett. 47, 986 (1981); J. Goldstone and R.L. Jaffe, Phys. Rev. Lett. 51, 1518 (1983).
6. E. Witten, Nucl. Phys. B160, 57 (1979).
7. G. Adkins, C. Nappi, and E. Witten, "Static Properties of Nucleons in the Skyrme Model, " to appear in Nuclear Physics B; A.D. Jackson and M. Rho, Phys. Rev. Lett. 51, 751 (1983); M. Rho, A.S. Goldhaber, and G.E. Brown, Phys. Rev. Lett. 51, 743 (1983).
8. B. Grossman, Phys. Rev. Lett. 50, 464 (1983); H. Yamagishi, Phys. Rev. D28, 977 (1983).
9. R. Jackiw and C. Rebbi, Phys. Rev. D13, 3398 (1976).

CONSERVATION LAWS IN THE MONOPOLE-FERMION SYSTEM

Ashoke Sen

Fermi National Accelerator Laboratory
P.O. Box 500
Batavia, Illinois 60510

It is shown that the monopole induced baryon number non-conservation is a necessary consequence of the exact conservation laws of the full four dimensional fermion-gauge field-Higgs system and properties of the J = 0 partial wave fermions.

It is also shown that the charge associated with the unbroken gauge symmetry is exactly conserved in the monopole-fermion interaction.

The subject of monopole induced baryon number violation, first introduced by Rubakov[1] and subsequently by Callan[2], has been of great interest in the recent past.[3-11] In this lecture I shall try to give a simple account of the subject, based on the conservation laws in the monopole fermion system. I shall show that the conservation laws of the full four dimensional gauge theory uniquely forces us to baryon number violating processes in the monopole fermion system. These conservation laws also help us determine the origin of the monopole induced baryon number violation, and, in particular, the role of anomaly in such processes. Finally, I shall discuss the conservation of electric charge in such systems, since it has also been a subject of great controversy in the last year.[6-9]

I shall start the discussion with an SU(2) gauge theory with the massless Dirac doublet of fermions:

$$\psi_{1\uparrow} \qquad \text{and} \qquad \psi_{2\uparrow}$$
$$\psi_{1\downarrow} \qquad \qquad \qquad \psi_{2\downarrow}$$

which I shall identify with

$$U_1 \qquad\qquad e^+$$
$$\qquad\qquad \text{and}$$
$$U_2^c \qquad\qquad d_3$$

respectively, keeping in mind the lowest charge monopole in SU(5). We also need a triplet of Higgs ϕ whose vacuum expectation value (VEV) breaks SU(2) to U(1). The Lagrangian for such a system is,

$$\mathcal{L} = -1/4 \, \mathrm{Tr} \, (F_{\mu\nu} \, F^{\mu\nu}) + \sum_{i=1}^{2} \bar{\psi}_i iD\psi_i + (D_\mu\phi)^\dagger (D_\mu\phi) - V(\phi) \qquad (1)$$

In terms of the four Dirac fields $\psi_{1\uparrow}$, $\psi_{1\downarrow}$, $\psi_{2\uparrow}$, $\psi_{2\downarrow}$, we may define eight charges:

$$Q_{i\uparrow} = \int \bar{\psi}_{i\uparrow} \gamma^0 \psi_{i\uparrow} d^3x \qquad Q_{i\downarrow} = \int \bar{\psi}_{i\downarrow} \gamma^0 \psi_{i\downarrow} d^3x$$

$$\qquad\qquad\qquad\qquad\qquad\qquad\qquad\qquad (2)$$

$$Q_{i\uparrow}^5 = \int \bar{\psi}_{i\uparrow} \gamma^0 \gamma^5 \psi_{i\uparrow} d^3x \qquad Q_{i\downarrow}^5 = \int \bar{\psi}_{i\downarrow} \gamma^0 \gamma^5 \psi_{i\downarrow} d^3x$$

These eight charges completely determine the fermion content of the system, as well as the helicities of the fermions, up to fermion–antifermion pairs. The SU(2) Lagrangian, given in Eq. (1), has three exact global symmetries, giving rise to three conserved charges:

Symmetry	Conserved Charge
$\begin{pmatrix} \psi_{1\uparrow} \\ \\ \psi_{1\downarrow} \end{pmatrix} \rightarrow e^{i\Theta_1} \begin{pmatrix} \psi_{1\uparrow} \\ \\ \psi_{1\downarrow} \end{pmatrix}$	$S_1 = Q_{1\uparrow} + Q_{1\downarrow}$
$\begin{pmatrix} \psi_{2\uparrow} \\ \\ \psi_{2\downarrow} \end{pmatrix} \rightarrow e^{i\Theta_2} \begin{pmatrix} \psi_{2\uparrow} \\ \\ \psi_{2\downarrow} \end{pmatrix}$	$S_2 = Q_{2\uparrow} + Q_{2\downarrow}$

Symmetry   Conserved Charge

$$\begin{pmatrix} \psi_{1\uparrow} \\ \\ \psi_{1\downarrow} \end{pmatrix} \rightarrow e^{i\Theta_3\gamma^5} \begin{pmatrix} \psi_{1\uparrow} \\ \\ \psi_{1\downarrow} \end{pmatrix}$$

$$\left. \begin{matrix} \\ \\ \end{matrix} \right\} \quad S_3 = Q_{1\uparrow}^5 + Q_{1\downarrow}^5 - Q_{2\uparrow}^5 - Q_{2\downarrow}^5$$

$$\begin{pmatrix} \psi_{2\uparrow} \\ \\ \psi_{2\downarrow} \end{pmatrix} \rightarrow e^{-i\Theta_3\gamma_5} \begin{pmatrix} \psi_{2\uparrow} \\ \\ \psi_{2\downarrow} \end{pmatrix}$$

Besides these three symmetries, the total electric charge carried by the fermions:

$$S_4 = Q_{1\uparrow} - Q_{1\downarrow} + Q_{2\uparrow} - Q_{2\downarrow} \tag{3}$$

is conserved if we ignore all other charge degrees of freedom of the system (e.g. the dyonic excitation of the monopole).

[In writing down Eq. (3), I have assumed that the unbroken generator of the SU(2) group is T_3. When we are considering the interaction of fermions with the monopole, then, if we work in the spherically symmetric $A_0 = 0$ gauge, the total U(1) charge carried by the fermions is given by

$$\int \sum_{i=1}^{2} \bar{\psi}_i \gamma^0 \hat{r} \cdot \vec{\sigma} \psi_i d^3x \quad .$$

This can be cast in the form given in Eq. (2) by taking the fields $\psi_{i\uparrow}$ and $\psi_{i\downarrow}$ to be eigenstates of $\hat{r} \cdot \vec{\sigma}$ with eigenvalues +1 and -1 respectively. This does not change the expression for the conserved charges S_1, S_2 and S_3.]

Let us now concentrate on the monopole-fermion scattering. For a given initial state containing incoming fermions, we may compute all the charges Q_i and Q_i^5 given in Eq. (2). Of these, S_1, S_2, S_3 and S_4 are conserved in the scattering. Thus we need four more constraints on the Q's to completely determine the final state for a given initial state. These are obtained by restricting the fermions to the J = 0 partial wave. This is a plausible assumption, since only the J = 0 partial wave has a non-vanishing particle density at the core (the wave-function blows up as $1/r$ as $r \to 0$).

The restriction $J = 0$ implies that,

$$\hat{r} \cdot \vec{J} = 0 = \hat{r} \cdot (\vec{L} + \vec{S} + \vec{T}) \tag{4}$$

Since $\hat{r} \cdot \vec{L} = 0$, $\hat{r} \cdot \vec{S}$ measures the radial component of the spin of the particle (which may be identified with the helicity for outgoing particles) and $\hat{r} \cdot \vec{T}$ measures the unbroken $U(1)$ charge carried by the particle, we have, for outgoing particles,

$$\text{Helicity} = - \text{ the } U(1) \text{ charge} \quad . \tag{5}$$

This constraint says, for example, that only $\psi_{i \uparrow L}$ and $\psi_{i \downarrow R}$ can be in the outgoing state (where R and L refer to positive and negative helicity respectively in our convention), but not $\psi_{i \downarrow L}$ or $\psi_{i \uparrow R}$. Expressed in terms of the Q_i's, it reads as,

$$Q_{i\uparrow} = - \overset{5}{Q}_{i\uparrow} \;, \; Q_{i\downarrow} = \overset{5}{Q}_{i\downarrow} \; i = 1,2 \quad . \tag{6}$$

These four constraints, together with the conservation laws for S_1, S_2, S_3 and S_4, uniquely determine the final state for a given initial state. For example, for an initial state,

$$U_{1R} + d_{3L}$$

the reader can easily verify that the only possible final state consistent with all the conservation laws and the constraints given in Eq. (6) is $u_{2R}^c + e_L^+$. Similarly, the only possible final state for an initial state of $U_{1R} + U_{2R}$ is $d_{3L}^c + e_L^+$. Thus we see that the conservation laws imposed by the unbroken global and local symmetries of the theory, together with the restriction that all the fermions must belong to the $J = 0$ partial wave, uniquely leads to baryon number violating processes.

We may also use the conservation laws to find the origin of baryon number violation in the monopole-fermion scattering. To do this, we must look at the linear combination of the Q's other than the four conserved charges S_1, S_2, S_3 and S_4. Some of these linear combinations are conserved in the presence of the unbroken $U(1)$ field, but not in the presence of the full $SU(2)$ fields. These are associated with the approximate symmetries of the Lagrangian when we ignore the W^+ and the W^- fields. They are,

Approximate Symmetry   Associated Charge

$$\psi_{1\uparrow} \to e^{i\theta_1} \psi_{1\uparrow} \ , \ \psi_{1\downarrow} \to e^{-i\theta_1}\psi_{1\downarrow}$$

$$\psi_{2\uparrow} \to e^{-i\theta_1}\psi_{2\uparrow} \ , \ \psi_{2\downarrow} \to e^{i\theta_1}\psi_{2\downarrow}$$

$$N_1 = Q_{1\uparrow} - Q_{1\downarrow} - Q_{2\uparrow} + Q_{2\downarrow}$$

$$\psi_{1\uparrow} \to e^{i\theta_2\gamma^5}\psi_{1\uparrow} \ , \ \psi_{1\downarrow} \to e^{-i\theta_2\gamma^5}\psi_{1\downarrow} \qquad N_2 = Q^5_{1\uparrow} - Q^5_{1\downarrow}$$

$$\psi_{2\uparrow} \to e^{i\theta_3\gamma^5}\psi_{2\uparrow} \ , \ \psi_{2\downarrow} \to e^{i\theta_3\gamma^5}\psi_{2\downarrow} \qquad N_3 = Q^5_{2\uparrow} - Q^5_{2\downarrow}$$

Finally, there is an anomalous charge, associated with the anomalous symmetry transformation:

Anomalous Symmetry   Anomalous Charge

$$\begin{pmatrix} \psi_{i\uparrow} \\ \\ \psi_{i\downarrow} \end{pmatrix} \to e^{i\Theta\gamma^5} \begin{pmatrix} \psi_{i\uparrow} \\ \\ \psi_{i\downarrow} \end{pmatrix} \quad i = 1, 2 \qquad A = Q^5_{1\uparrow} + Q^5_{1\downarrow} + Q^5_{2\uparrow} + Q^5_{2\downarrow}$$

This charge has an anomaly in the presence of the U(1) gauge field and is non-conserved even in the absence of the $W^{\pm}$ fields.

Since the four exact conservation laws and the constraint imposed by $J = 0$ partial waves completely determine the final state for a given initial state, we cannot, in general, satisfy the conservation laws of N_1, N_2, N_3 and A, and some of them must necessarily be violated in the monopole-fermion interaction.[10] Of these, the violation of N_1, N_2, and N_3 takes place because of the presence of classical SU(2) gauge fields inside the monopole core, while the violation of A is due to the anomaly induced by the unbroken U(1) gauge field.

We may now try to investigate the origin of the baryon number violation by expressing it as,

$$B = \frac{1}{3}(\bar{u}_1\gamma^0 u_1 - \frac{1}{3}\bar{U}_2^c\gamma^0 U_2^c - \frac{1}{3}\bar{d}_3^c\gamma^0 d_3^c) = \frac{1}{3}(Q_{1\uparrow} - Q_{1\downarrow} + Q_{2\downarrow}) \quad (7)$$

and noting that it may be expressed as a linear combination of S_1, S_2, S_4 and N_1. This clearly shows that the baryon number violation in the monopole-fermion interaction is essentially due to the presence of the non-trivial SU(2) gauge fields inside the monopole core, and not due to anomaly. For processes like $U_{1R} + U_{2R} \rightarrow d_{3L}^c + e_L^+$, both the baryon number, and the anomalous charge A are violated. Hence, for this process both the anomaly and the non-trivial dynamics inside the monopole core play a vital role. This is apparent from the calculation of Rubakov[1] and Callan[2], where gauge field configurations with nonzero topological charge is needed to get a non-zero value of the condensate $\langle U_{1R} U_{2R} d_{3R} e_R^- \rangle$. However, for the process $U_{1R} + d_{3L} \rightarrow U_{2R}^c + e_L^+$, B is violated but A is conserved. Hence we expect the anomaly to play no role in these processes. This becomes apparent if we calculate the condensate $\langle U_{1R} d_{3L} U_{2L} e_R^- \rangle$ (i.e. $\langle \psi_{1\uparrow R}\psi_{2\downarrow L}\psi_{1\downarrow L}^c\psi_{2\uparrow R}^c \rangle$ in our notation) in the Rubakov-Callan model. Such condensates receive contributions from the gauge field configuration with zero topological charge[4], which shows that the anomaly does not play any role in these processes.

The conservation laws for the charges S_1, S_2, S_3 and S_4 may also be derived from the effective two dimensional bosonized Hamiltonian introduced by Callan.[2] The boundary conditions on the boson fields at the monopole core play a vital role in deriving these conservation laws.[5] This shows that these boundary conditions are necessary consequences of the conservation laws of the full four dimensional field theory, although they were originally derived by solving the one particle Dirac equation in the background of a classical monopole field. In the full SU(5) grand unified theory, the four conservation laws refer to the conservation of electric charge, color isospin, color hypercharge and weak isospin. In this case, however, the baryon number violation comes from two independent sources, the non-trivial gauge field configuration inside the monopole core, and the anomaly in the baryon number current due to the weak interaction gauge fields. A detailed analysis of the effect of the weak anomaly in a general grand unified theory has been made by A.N. Schellekens.[11]

Finally, I shall come to the question of charge conservation in the monopole-fermion system, since this has been a subject of great controversy in the last year. This work was done in collaboration with Y. Kazama.[7] As was shown by Callan[2], the monopole-fermion system in the J = 0 partial wave may be represented by a two dimensional boson field theory with the Hamiltonian:

$$H = \int_{r_0}^{\infty} dr \left[\sum_{i=1}^{2} (P_i^2 + \pi_i^2 + \Phi_i'^2 + Q_i'^2) + \frac{c}{r^2}(\Phi_1 + \Phi_2 + Q_1 + Q_2)^2 \right] \quad (8)$$

with the boundary conditions:

$$\Phi_i' + Q_i' = 0 \qquad \Phi_i = Q_i \qquad i = 1, 2 \tag{9}$$

where Φ_i, Q_i are two dimensional boson fields and π_i, P_i are their conjugate momenta. The charges Q given in Eq. (2) have simple expressions when expressed in terms of the bosonized fields Φ and Q:

$$Q_{i\uparrow} = \int_{r_0}^{\infty} \Phi_i' \, dr \qquad Q_{i\downarrow} = - \int_{r_0}^{\infty} Q_i' \, dr$$

$$\tag{10}$$

$$Q_{i\uparrow}^5 = \int_{r_0}^{\infty} \pi_i \, dr \qquad Q_{i\downarrow}^5 = \int_{r_0}^{\infty} P_i \, dr$$

The charges S_1, S_2 and S_3, constructed from the boson fields, may easily be shown to commute with H as a consequence of the boundary conditions in Eq. (9). The commutator of S_4 with H, on the other hand, is proportional to,

$$(\dot{Q}_1 + \dot{Q}_2 + \dot{\Phi}_1 + \dot{\Phi}_2)|_{r=r_0} \tag{11}$$

which is not zero as a consequence of the boundary conditions. However, as can be seen from Eq. (8), as $r_0 \rightarrow 0$, $\Phi_1 + \Phi_2 + Q_1 + Q_2$ must vanish at r_0, in order to satisfy the finiteness of the energy. Physically, this reflects the fact that for a finite r_0, it is possible for the fermions to dump charge into the monopole core, but as $r_0 \rightarrow 0$, it becomes more and more difficult, since the Coulomb energy associated with such dyonic excitations blows up.

The question then is, how do we see the conservation of the total charge in the monopole-fermion system for finite r_0? The problem is that in models of Rubakov and Callan, we summarize the full dynamics inside the monopole core by effective boundary conditions on the fermion fields at r_0, and then completely forget about the core. One way to see charge conservation in the monopole fermion system would be to introduce extra degrees of freedom which corresponds to the charge degree of freedom of the monopole, and then define a conserved charge as the sum of the charge carried by the fermions and the monopole core. Such treatments have been given by Balachandran et al.[8] and Yan.[9]

However, it is not really necessary to introduce new degrees of freedom to measure the charge inside the monopole core. The total charge inside the monopole core is measured by the radial electric field at r_0, which, in turn, may be expressed in terms of the Φ_i and Q_i fields at r_0. We may then define the total charge of the system as,

$$(r^2 E_r)_{r=r_0} + S_4 = (\Phi_1 + \Phi_2 + Q_1 + Q_2)|_{r_0} + S_4 \tag{12}$$

This can easily be shown to commute with H, hence the total charge is conserved.

The conservation of the total charge is reflected in the Green's functions as follows. In the present model, there are two ways to define a gauge invariant fermion field, e.g.

$$\psi_{\uparrow N}(r,t) = \exp\left(-ie \int_{r_0}^{r} A_r(r',t)\, dr'\right) \psi_\uparrow(r,t) \tag{13}$$

$$\tilde{\psi}_{\uparrow N}(r,t) = \exp\left(ie \int_{r}^{\infty} A_r(r',t)\, dr'\right) \psi_\uparrow(r,t) \tag{14}$$

and similarly for $\psi_\downarrow$.

$\psi_{\uparrow N}$ creates a gauge invariant state by creating a fermion at the point r, and an equal and opposite charge at the monopole core. On the other hand, $\tilde{\psi}_{\uparrow N}$ creates a fermion at a point r and an equal and opposite charge at infinity. When we calculate the Green's function involving the ψ_N fields, then, for a finite r_0, the Green's function is non-vanishing even if the total charge carried by all the fermion fields in the Green's function is non-zero. This is not surprising, since the operators ψ_N always create a charge neutral state by creating an equal and opposite charge at the monopole core together with the fermion. On the other hand, a Green's function involving the $\tilde{\psi}_N$ fields vanish identically unless the total charge carried by all the fermion fields in the Green's function is zero. This shows that the Green's functions in the monopole-fermion system are indeed charge conserving.

To summarize, I shall state the two main points of the talk:

1) Conservation laws derived from the full four dimensional Lagrangian of the fermion-gauge field-Higgs system, together with the restriction on the fermions to be in the J = 0 partial wave, uniquely lead to baryon number violating processes. The boundary condition on the boson fields at the monopole core, used by Callan in the bosonized version of the theory, then follow.

2) Total charge of the fermion-monopole system is exactly conserved, although the monopole ground state may make virtual transition to states containing fermion fields carrying a net charge, and an equal and opposite charge at the core.

ACKNOWLEDGEMENTS

I wish to thank Y. Kazama for discussions, who was a collaborator in part of this work. I also thank S. Das, A.S. Goldhaber and A.N. Schellekens for useful conservations during various stages of this work.

REFERENCES

1. V.A. Rubakov, JETP Lett. 33, 645 (1981), Inst. Nucl. Res. report No. P0211, Moscow (1981), Nucl. Phys. B203, 311 (1982), V.A. Rubakov and M.S. Serebryakov, Nucl. Phys. B218, 240 (1983).

2. C.G. Callan, Phys. Rev. D25, 2141 (1982), Phys. Rev. D26, 2058 (1982), Nucl. Phys. B212, 365 (1983).

3. S. Dawson and A.N. Schellekens, Phys. Rev. D27, 2119 (1983), Fermilab Rep. No. Fermilab-Pub-83/43-THY, to be published. B. Sathiapalan and T. Tomaras, California Institute of Technology Rep. No. Calt-68-987. F. Bais, J. Ellis, D.V. Nanopoulos and K. Olive, Nuc. Phys. B219, 189 (1983), A.S. Goldhaber, Stony Brook Rep. No. ITP-SB-83-2, ITP-SB-83-30.

4. K. Seo, Phys. Lett. 126B, 201 (1983).

5. A. Sen, Phys. Rev. D28, 876 (1983).

6. Y. Kazama, Kyoto Univ. Rep. No. KUNS 679 HE(TH) 83/06, to be published in Prog. Theor. Phys., T. Yoneya, Tokyo University Rep. No. UT-Komaba-83-3, Z.F. Izawa and A. Iwazaki, Tohoku University Rep. No. TU/83/260, T.M. Yan, Phys. Rev. D28, 1496 (1983).

7. Y. Kazama and A. Sen, Fermilab Rep. No. Fermilab-Pub-83/58-THY (KUNS 695 HE(TH) 83/12).

8. A.P. Balachandran and J. Schechter, Phys. Rev. Lett. 51, 1418 (1983), Syracuse University Report No. 260.

9. T.M. Yan, Lecture delivered at the Monopole '83 Conference held at University of Michigan, Ann Arbor.

10. This result is consistent with the analysis of C.G. Callan and S.R. Das, Phys. Rev. Lett. 51, 1155 (1983).

11. A.N. Schellekens, Stony Brook Report.

MONOPOLES WITH Z_n CHARGES

David London

Enrico Fermi Institute
University of Chicago
Chicago, Illinois 60637

We consider a simply connected gauge group G, and assemble the Higgs fields into a scalar multiplet S, in some representation of G. The Higgs' vacuum expectation value breaks G to H, some subgroup of G. If the second homotopy group $\pi_2(G/H) = \pi_1(H)$ is non-trivial, there is a conserved charge.[1]

The values of S which minimize the Higgs potential are related by gauge transformations:

$$S = gS_0 \tag{1}$$

where g is an element of G and S_0 is an arbitrary reference point. The Higgs field approaches its vacuum expectation value at spatial infinity:

$$S(\theta, \phi) = \lim_{r \to \infty} S(r, \theta, \phi) \quad . \tag{2}$$

We impose the requirement that $S(\theta = 0, \phi) = S_0$.

Two configurations, $S(\theta, \phi)$ and $S'(\theta, \phi)$ will be in the same homotopy class if one can be smoothly deformed to the other, keeping the value at $\theta = 0$ fixed, i.e. if there exists a continuous function $S(\theta, \phi, t)$, such that

$$S(\theta, \phi, 0) = S(\theta, \phi)$$

$$S(\theta, \phi, 1) = S'(\theta, \phi) \tag{3}$$

$$S(\theta = 0, \phi, t) = S_0 \quad .$$

We write:

$$S(\theta, \phi) = g(\theta, \phi)S_0, \tag{4}$$

with $g(\theta=0, \phi) = 1$. Although $S(\theta,\phi)$ must be single valued, $g(\theta,\phi)$ may be multiple-valued, as long as this corresponds to right-multiplication by an element of H. We can choose $g(\theta, \phi)$ such that the multiple-valuedness occurs at $\theta = \pi$:

$$g(\pi, \phi) = g(\pi, 0)h(\phi) \quad . \tag{5}$$

Therefore any two configurations $S(\theta,\phi)$ and $S'(\theta,\phi)$ will be homotopic if an only if $h(\phi)$ and $h'(\phi)$ belong to the same element of $\pi_1(H)$.

We now consider the case when $\pi_1(H) = Z_n$. In this case, the identity element of H corresponds to n elements in the center of $\tilde{H}$, the covering group of H. These elements may be written in the form

$$a^k = \exp(4\pi ikT_3), \quad k = 0, 1, \ldots n-1 \tag{6}$$

where T_3 is some generator of H.

Now suppose that we can find two other generators T_1 and T_2 such that

$$[T_i, T_j] = i\, G_{ijk}\, T_k \quad . \tag{7}$$

We now define

$$g_k(\theta,\phi) = e^{ik\phi T_3}\, e^{i\theta T_2}\, e^{-ik\phi T_3} \quad . \tag{8}$$

Therefore:

$$e^{2\pi ikT_3}\, e^{i\theta T_2}\, e^{-2\pi ikT_3} = e^{i\theta T_2}, \tag{9}$$

so $g_k(\theta,\phi)$ is single-valued for $\theta \neq \pi$. At $\theta = \pi$,

$$g_k(\pi,\phi)\, g_k^{-1}(\pi, 0) = e^{2ik\phi T_3}$$

$$\equiv h_k(\phi) \quad . \tag{10}$$

As ϕ runs from 0 to 2π, $h_k(\phi)$ goes from 1 to $e^{4\pi ikT_3}$. Viewed in $\tilde{H}$, this corresponds to a path from I to a^k; in H, this corresponds to a closed path from I to I. Hence this is a non-trivial element of $\pi_1(H)$. Therefore

$$S_k(\theta,\phi) = g_k(\theta,\phi)S_0 \tag{11}$$

is the asymptotic Higgs configuration with topological charge equal to k mod n.

When SU(2) is broken to U(1), the charges are Z-type. When we enlarge SU(2) to G, there is more deformation freedom, so that configurations which were topologically inequivalent are no longer so.

The magnetic field is:

$$B_i(k) = - k \, g_k \, T_3 \, g_k^{-1} \, \frac{\hat{r}_i}{r^2} \quad . \tag{12}$$

Note that the magnetic charge is not uniquely determined by topological charge, since two charges differing by n are topologically equivalent, but yield gauge-inequivalent magnetic fields. As usual, the only stable solutions correspond to the smallest k in each topological sector.[2,3]

Example (i): SU(3) → SO(3)

SO(3) is generated by[4]:

$$J_x = \begin{pmatrix} 0 & 0 & 0 \\ 0 & 0 & -i \\ 0 & i & 0 \end{pmatrix} = \lambda_7 \qquad J_y = \begin{pmatrix} 0 & 0 & i \\ 0 & 0 & 0 \\ -i & 0 & 0 \end{pmatrix} = -\lambda_5 \qquad J_z = \begin{pmatrix} 0 & -i & 0 \\ i & 0 & 0 \\ 0 & 0 & 0 \end{pmatrix} = \lambda_2 \quad . \tag{13}$$

SU(3) is broken to SO(3) by a sextet of Higgs, a symmetric tensor S^{ab}. In the string gauge, S^{ab} is:

$$S_o{}^{ab} = \sigma \delta^{ab} \quad . \tag{14}$$

Under a gauge transformation,

$$S^{ab} \to (gS)^{ab} = V^a{}_c V^b{}_d S^{cd} \tag{15}$$

$$= (VSV^T)^{ab} \quad .$$

To construct the monopole SU(2), let

$$T_1 = 1/2\lambda_3 = 1/2 \begin{pmatrix} 1 & 0 & 0 \\ 0 & -1 & 0 \\ 0 & 0 & 0 \end{pmatrix} \quad T_2 = 1/2\lambda_1 = 1/2 \begin{pmatrix} 0 & 1 & 0 \\ 1 & 0 & 0 \\ 0 & 0 & 0 \end{pmatrix} \quad T_3 = 1/2\lambda_2 = 1/2 \begin{pmatrix} 0 & -i & 0 \\ i & 0 & 0 \\ 0 & 0 & 0 \end{pmatrix} \tag{16}$$

The gauge transformation $g_k(\theta, \phi)$ of Eq. (8) is:

$$V_k(\theta, \phi) = e^{ik\lambda_2\phi/2} \, e^{i\lambda_3\theta/2} \, e^{-ik\lambda_2\phi/2} \quad . \tag{17}$$

In the hedgehog gauge,

$$S_k = \sigma \, V_k(\theta, \phi) \, V_k^T(\theta, \phi)$$

$$= \sigma \begin{bmatrix} \cos\theta + i\sin\theta\cos k\phi & -i\sin\theta\sin k\phi & 0 \\ -i\sin\theta\sin k\phi & \cos\theta - i\sin\theta\cos k\phi & 0 \\ 0 & 0 & 1 \end{bmatrix} . \tag{18}$$

We can see the Z_2 nature of this configuration in two ways. First, the $k = 1$ and $k = -1$ configurations are homotopic. Let

$$S(\theta, \phi, t) = e^{i\pi t\lambda_5} \times S_1(\theta, \phi) e^{-i\pi t\lambda_5} \quad . \tag{19}$$

The right-hand side is equal to $S_1(\theta, \phi)$ at $t = 0$, and is $S_{-1}(\theta, \phi)$ when $t = 1$. Also, $S(\theta = 0, \phi, t) = S_0$ for all t, so the conditions of Eq. (3) are met.

The second way of seeing the Z_2 nature is to show that, for any even k, the asymptotic Higgs field can be deformed to the trivial configuration, $S = S_0$. We must find a non-singular and single-valued gauge transformation $V_k(\theta, \phi)$ with $V_k(\theta = 0) = I$ and

$$V_k(\theta, \phi) S_k(\theta, \phi) V_k^T(\theta, \phi) = S_0 \qquad \text{(for all even } k\text{)} \quad . \tag{20}$$

The gauge transformation which does this is

$$V_k(\theta, \phi) = e^{i\lambda_5\theta} \, e^{i\lambda_2 k\phi/2} \, e^{-i\lambda_5\theta} \, e^{-i\lambda_3\theta/2} \, e^{-ik\lambda_2\phi/2} \quad . \tag{21}$$

$V_k(\theta, \phi)$ is non-singular for all k, and $V_k(\theta, \phi + 2\pi) = V_k(\theta, \phi)$ only if k is even.

In both cases, it was necessary to go outside the monopole $SU(2)$ to show the Z_2 nature of the monopole.

Example (ii): $SO(10) \to SO(6) \times SO(4)$

The symmetry breaking is accomplished by a 54 of Higgs, a symmetric, traceless, rank two tensor. In the string gauge,

$$S_0 = \sigma \, \text{diag}(2,2,2,2,2,2,-3,-3,-3,-3) \quad . \tag{22}$$

We choose the monopole $SU(2)$ to be

$$T_3 = 1/2(J_{56} - J_{78})$$
$$T_1 = 1/2(J_{67} - J_{58}) \tag{23}$$

$$T_2 = 1/2(J_{75} - J_{68})$$

where $(J_{ab})_{ij} = -i(\delta_{ai}\delta_{bj} - \delta_{aj}\delta_{bi})$.   (24)

The k = 1 and k = -1 are continuously connected by

$$S(\theta, \phi, t) = e^{i\pi tL} S_1(\theta, \phi)e^{-i\pi tL}$$ (25)

with $L = J_{45} + J_{79}$.   (26)

Now, suppose that the symmetry is broken in stages. Suppose that, at mass scale M, G is broken to H by a Higgs S, with $\pi_2(G/H) = Z_n$. Suppose that, at mass scale M' < M, the symmetry is broken further to H' by a Higgs S', with $\pi_2(G/H') = Z$. Two questions arise. First, what happens to the Z_n monopoles at the second stage of symmetry breaking? Second, does the breaking to H' give rise to new monopoles not related to Z_n monopoles?

We can get a lot of information from topology.[5] A closed path in H' is a closed path in H, so there is a mapping from $\pi_1(H')$ to $\pi_1(H)$. Therefore a set of H' topological charges (q'_{j1}, $q'_{j2},...$) gets mapped to each H topological charge q_j. So, when H is broken to H', each configuration with charge q_j is converted to a configuration with some charge q'_{jk}. If there are no H' charges corresponding to q_j, the H monopoles become bound by flux tubes.

If $\pi_1(H) = Z_n$ and $\pi_1(H') = Z$, there are three possibilities:

    (i)   all q' are mapped to q = 0
    (ii) q' = k (mod n) is mapped to q = k
    (iii) q' = k (mod n) is mapped to q = -k .

Topological considerations do not determine which of the q'_{jk} is selected.

Consider the Z_2 case [in which (ii) and (iii) are equivalent], and look again at the case of SU(3). Imagine SO(3) broken to U(1) in the 't Hooft-Polyakov[6] way, by a triplet of Higgs. We have

$$SU(3) \underset{S}{\rightarrow} SO(3) \underset{S'}{\rightarrow} U(1)$$ (28)

with $S_0{}^{ab} = \sigma \, \delta^{ab}$ (29)

$$S_0'{}^a = \sigma' \, \delta^{a3} \; .$$ (30)

The Z_2 monopole was constructed using $T_3 = 1/2\lambda_2 = 1/2J_2$. However, we could have chosen $T_3 = 1/2n\cdot\vec{J}$. The direction of $\hat{n}$ has no physical significance at this stage. But when we impose a uniform S' triplet field, the configuration has finite energy only if $\hat{n} = \pm\hat{Z}$. These correspond to a monopole and anti-monopole with Z-charge g' = ±1. Therefore Z_2 monopoles with arbitrary $\hat{n}$ must rotate in SO(3) space in order to maintain finite energy. Assuming it makes the minimum rotation, it becomes a Z monopole with the sign of the magnetic charge the same as the sign of $\hat{n}_3$. A similar mechanism operates for any G, H and H' with Z_2 monopoles at one stage of symmetry breaking and Z monopoles at the next.

Since there are two scales of symmetry breaking, there may be a monopole solution associated with each scale. If the lower mass monopole has the higher charge, both will be stable. The mass of the monopole is determined by which Higgs fields are 'twisted' in the monopole solution.

Suppose M' << M. A configuration with both Higgs fields will be assigned both a Z charge (q') and a Z_n charge (q). The relation between the two charges must be in accord with the homotopy groups. If q' is not a multiple of n, then q is non-zero. Hence S is twisted, so this is a heavy monopole. If q' is a multiple of n, then q = 0, which means S is not twisted. Therefore we have a low mass monopole with n units of magnetic charge.

In the case of SO(10), there is an intermediate stage containing two U(1) factors; SO(6) × SO(4) ≡ H, is broken to SU(3) × SU(2) × U(1) × U(1) ≡ H_2. This in turn is broken to SU(3) × SU(2) × U(1) ≡ H_3. At the H_2 stage, there will be two topological charges, one for each U(1) factor. The allowed charges are of the form (n_1, n_2), with $n_1 - n_2$ even. Therefore the Z_2 monopole will be converted to one with charges either (1,1) or (-1, -1). Also, there will be two light monopoles, with charges (2, 0) and (0, 2), which live in the SO(6) and SO(4), respectively. These do not catalyze baryon decay with strong cross-sections.

When this is broken further to H_3, we must look at the mapping from $\pi_1(H_2) = Z \times Z$ to $\pi_1(H_3) = Z$. No H_3 charges correspond to (n_1, n_2) unless $n_1 = n_2$, in which case the former Z_2 monopole becomes an ordinary monopole with unit charge. The lighter monopoles become bound by flux tubes, except for (2, 0) + (0, 2), which becomes a light monopole of charge 2.

REFERENCES

1. S. Coleman, "Classical Lumps and Their Quantum Descendants," in New Phenomena in Subnuclear Physics, ed. A. Zichichi, (New York, Plenum 1977), Part A, 297; Yu.S. Tyupkin, V.A. Fateev, and A.S. Shvarts, Pis'ma. Zh. Eksp. Teor. Fiz.

21, 91 (1975) [Sov. Phys. JETP Lett. 21, 42 (1975)]; M.I. Monastyrshii and A.M. Perelomov, ibid. 21, 94 (1975) [ibid., 21, 43 (1975)].

2. R.A. Brandt and F. Neri, Nucl. Phys. B161, 253 (1979).

3. S. Coleman, "The Magnetic Monopole Fifty Years Later," Harvard University report HUTP-82/A032, 1982, presented at 1981 International School of Subnuclear Physics, Erice (unpublished).

4. M. Gell-Mann and Y. Ne'eman, The Eightfold Way (New York, W.A. Benjamin, 1964).

5. F.A. Bais, Phys. Lett. 98B, 437 (1981).

6. G. 't Hooft, Nucl. Phys. B79, 276 (1974); A.M. Polyakov, Pis'ma. Zh. Eksp. Teor. Fiz. 20, 430 (1974) [Sov. Phys.-JETP Lett. 20, 194 (1974)].

INTERACTIONS OF FERMIONS WITH SPHERICALLY SYMMETRIC MONOPOLES

A.N. Schellekens

Institute for Theoretical Physics
State University of New York at Stony Brook
Stony Brook, New York 11794

The formalism of Rubakov and Callan is generalized to all spherically symmetric monopoles that satisfy the Wilkinson-Goldhaber conditions, coupled to fermions in any representation of the gauge group.

The observation[1,2] that SU(5) monopoles can catalyze proton decay has led to a revival of theoretical interest in monopole fermion interactions. The failure to observe spontaneous proton decay at the expected rate has cast some doubt on the correctness of the simplest unification model. Different models may be considered, and if these models have the desirable feature of explaining charge quantization, they are likely to have monopole solutions. Fermions will interact with these monopoles and one would like to understand in general what kind of processes are possible if the fermions pass through the core of such a monopole. We will answer that question for a large class of monopoles, although certainly not the most general class.

The monopoles we have studied are the most general spherically symmetric[3] ones, interacting with fermions in any representation of the gauge group G. The asymptotic behavior of these monopoles is characterized by SU(2) subgroups of G and the unbroken subgroup H, with generators $\vec{T}$ and $\vec{I}$. The magnetic field is proportional to some constant generator Q. The theorem of Wilkinson and Goldhaber[3] states that a necessary and sufficient condition for the existence of a spherically symmetric gauge is

$$Q = I_3 - T_3 \tag{1}$$

$$[Q,\vec{I}] = 0 \quad . \tag{2}$$

In that gauge, the vector potential $\vec{A}$ transforms like a vector with respect to $\vec{L} + \vec{T}$, and the Higgs field transforms like a scalar.

By solving radial equations of motion for the functions that parameterize the vector potential and the Higgs field, one can obtain the structure of the monopole core. These radial functions interpolate between the limiting behavior for $r \to \infty$, obtained from Ref.3, and for $r \to 0$, enforced by continuity and finiteness of the energy. We have not used the exact radial functions since they are in general not known, difficult to handle analytically, and complicate the discussion of higher fermion representations. Instead, we approximate the radial functions by step functions, so that we get the following monopole field

$$\vec{A} = \frac{1}{r} \left(\vec{I}(\hat{r}) - \vec{T} \right) \times \hat{r} \qquad\qquad r > a$$

$$\vec{A} = 0 \qquad\qquad\qquad\qquad r < a$$

(3)

where $\vec{I}(r)$ is the generator $\vec{I}$ in the spherically symmetric gauge and a is the size of the core. This has a straightforward generalization to higher representations.

We couple the field, Eq. (3), to massless lefthanded fermions and solve the Dirac equation

$$\sigma^\mu (\partial_\mu + A_\mu) \psi_L = 0 \quad .$$

(4)

This is reduced to a matrix problem by expanding ψ_L in terms of eigenstates of the operators J^2, J_z, T^2, $\vec{T} \cdot \hat{r}$ and $\vec{S} \cdot \hat{r}$, where $\vec{J} = \vec{L} + \vec{S} + \vec{T}$, and $\vec{S} = 1/2\vec{\sigma}$. Explicit expressions can be obtained for the matrix elements of the Dirac operator, in all partial waves. The details of this will be presented elsewhere.[4] By making a basis transformation, one can reduce the equation for $r > a$ and $r < a$ to a set of equations of the form

$$\{\frac{\partial}{\partial t} + \tau_3 [\frac{\partial}{\partial r} + \frac{1}{r}(1 - d_i \tau_1)]\} \chi_i = 0 \quad .$$

(5)

Apart from a trivial transformation, this is the same equation that was obtained by Kazama, Yang and Goldhaber,[5] for fermions in the field of a point monopole. The treatment of the singularity at $r = 0$ is, however, completely different in our case.

In the stepfunction approximation, the non-trivial structure of the core is represented by a matching condition for the fermion fields at $r = a$. This condition can be derived from the different basis transformations required to bring the Dirac operator in the form of Eq. (5) for $r > a$ and for $r < a$.

For future purposes, we need the expressions for some generators of H on this basis. Any generator that commutes with T_3 and $\vec{T}$ becomes an $\vec{L} + \vec{T}$-scalar on the spherically symmetric basis. On the basis on which the Dirac equation has the form of Eq. (5) for $r > a$, all these generators can be diagonalized.[4] Especially important is the generator Q.

A fermion interacts with the core only if $d_i = 0$. The appearance of doublets with $d_i = 0$ in a subspace with J^2, J_z eigenvalues (j, m_j) is determined by a theorem proven in Ref. 4. The rule works as follows: consider all fermions with T_3-eigenvalue $\pm(j + 1/2)$. Subtract all (i.e., the maximum number) of pairs with opposite T_3 eigenvalue and the same Q-eigenvalue. Then a doublet with $d_i = 0$ appears for every remaining pair. The physical interpretation of this result is that the fermions that interact with the core are precisely the ones that can only scatter to a final state with different charge.

To determine how the fermions scatter, we solve the Dirac equation for $r > a$ and $r < a$ and match the solutions at $r = a$. Then we take the limit $a \to 0$. In that limit, only the scattering of fermions with $d_i = 0$ is affected by the matching conditions. For these fermions, we obtain an "S-matrix" describing to which combination of states an incoming fermion scatters. In a given model, this matrix can be calculated by means of a series of matrix manipulations described in Ref. 4. In practice, this is hardly ever necessary. The equations usually block-diagonalize in such a way that, at most, one doublet with $d_i = 0$ appears per block, so that every incoming fermion can only go to one final state. For example, fermions from different representations of G can not be transformed into each other by the core.

This charge violating S-matrix for the $d_i = 0$ fermions is the generalization of the Rubakov-Callan boundary condition. To enforce charge conservation, we can include the effect of some of the electric fields produced by the fermions. If we consider a set of generators Q_ℓ forming an abelian subgroup of H, which commutes with T_3 and $\vec{T}$, we obtain the following two dimensional Lagrangian, analogous to the one of Refs. 1 and 2.

$$\mathcal{L} = -\frac{\pi r^2}{g^2} G^\ell_{\mu\nu} G^{\mu\nu\ell} + i \bar{\chi}_i \gamma^\mu D^i_\mu \chi_i + i \frac{d_i}{r} \bar{\chi}_i \gamma^5 \chi_i \tag{6}$$

where $G^\ell_{\mu\nu} = \partial_\mu B^\ell_\nu - \partial_\mu B^\ell_\nu$

$$D_\mu^i = \partial_\mu - i\, B_\mu^k\, (q_i^k + p_i^k\, \gamma_5).$$

The fields (B_o^ℓ, B_1^ℓ) are the temporal and radial components of the four dimensional gauge fields. The charges q_i^k and p_i^k are defined by writing the diagonal matrix Q_i^k (representing the charge Q^k on a doublet i) as follows

$$Q_i^k = q_i^k + \tau_3 p_i^k \quad . \tag{7}$$

The γ-matrices in Eq. (6) are $\gamma^\circ = \tau_1$, $\gamma^1 = -i\tau_2$, $\gamma^5 = \tau_3$. The boundary conditions on the fermions in Eq. (6) are

$$(1 + \gamma_5)\chi_i(0) = S_{ij}(1 + \gamma_5)\gamma^\circ\chi_j(0) \qquad (d_i = d_j = 0) \tag{8}$$

where S_{ij} is the aforementioned S-matrix of the fermions in the absence of electric fields. The fermion fields with $d_i \neq 0$ vanish at $r = 0$.

The simplest way to describe the possible processes is to determine the conservation laws[6] for Lagrangian Eq. (6). Due to anomalies and the "mass" term in Eq. (6), we get

$$\partial_\mu \bar{\chi}_i \gamma^\mu \chi_i = -\frac{1}{\pi} \sum_k p_i^k\, \epsilon_{\mu\nu}\, \partial^\mu\, B^{\nu,k} \tag{9}$$

$$\partial_\mu \bar{\chi}_i \gamma^\mu \gamma^5 \chi_i = -\frac{1}{\pi} \sum_k q_i^k\, \epsilon_{\mu\nu}\, \partial^\mu\, B^{\nu,k} - \frac{2d_i}{r} \bar{\chi}_i \chi_i \quad . \tag{10}$$

We will only consider the case $S_{ij} = \delta_{ij}$, which covers all cases of interest. Then we get the following conserved charges:

(i)   All gauge charges, corresponding to the generators Q_ℓ

(ii) All anomaly-free vector charges.

Point (i) has been addressed by other speakers at this conference.[7] The second set of charges is conserved because of the boundary condition. We can show[4] that for fermion doublets with $d_i \neq 0$, the axial charges p_i^k vanish, so that their vector current is anomaly free. This simply means that their net number is conserved; they scatter from the monopole without a change in any internal quantum number. Therefore, we can restrict ourselves to the fermions with $d_i = 0$. For these fermions, we get (at least) as many conservation laws as there are doublets, so that for every process, the final state is completely determined by conservation laws. This conclusion is strictly valid only in the massless limit.

Starting from Lagrangian Eq. (6), we can proceed in several other ways. For example, if we omit the doublets with $d_i \neq 0$, we can bosonize the model in such a way that all anomalies, Eqs. (9) and (10), are preserved. It is also possible to consider field configuration with non-zero winding number ν and construct the fermion zero modes as Rubakov has done.[1] This can be done for any charge $\tilde{Q}$ which commutes with T_3 and $\vec{I}$. We find that the number of zero modes is $2\nu \, \mathrm{Tr}\, Q\tilde{Q}$, where the trace is over all fermions in a representation. This result agrees with the index theorem. (This can easily be understood by the observation that $\vec{B} \sim Q$, $\vec{E} \sim \tilde{Q}$ and the index theorem involves the integral of $\vec{E}\,\vec{B}$).

Our results agree with those of Ref. 8 for fermions in higher representation coupled to a single strength monopole and with those of Ref. 9 for the purely electromagnetic monopole in SU(5), apart from some details. The structure of the multiplets and the breakdown into doublets is shown in Table I. For one generation, we obtain 12 doublets:

$$3 \begin{pmatrix} d_\alpha \\ e^+ \end{pmatrix}_L \; ; \; 3 \begin{pmatrix} e \\ d^c_\alpha \end{pmatrix}_L \; ; \; 4 \begin{pmatrix} u^c_\beta \\ u_\beta \end{pmatrix}_L \; ; \; 2 \begin{pmatrix} u^c_\gamma \\ u_\gamma \end{pmatrix}_L \; .$$

The multiplicities of the doublets are due to the m_j-degeneracy. The color indices indicate linear combinations of eigenstates of $\vec{T}\cdot r$, which are independent of m_j.

The only independent gauge charge that can be taken into account is the electromagnetic charge (weak charge is automatically conserved). By means of the conservation laws we find, for example, the following processes:

$$d_{\alpha L} + u_{\beta R} \rightarrow e^+_L + u^c_{\beta R}$$

$$(2u_\beta + 2u_\gamma + 2d^c_\alpha + 2s^c_\alpha + \mu^+)_L \rightarrow (2u_\beta + 3d^c_\alpha + 2s^c_\alpha + e^+ + \mu^+)_R \; .$$

Both processes violate baryon number, the second one also helicity. This second process appears also in Ref. 9, but with different color indices. We have added (μ, s) doublets to obtain this process.

Although states with different m_j are the same linear combination of $\vec{T}\cdot\hat{r}$ eigenstates, they do not have the same color. This is due to the fact that the generator T_3 in the spherically symmetric gauge is not $\vec{T}\cdot\hat{r}$, but $\vec{T}\cdot\hat{r} - \vec{I}(\hat{r})\cdot\hat{r} + I_3(\hat{r})$. Therefore, conservation of color is not obvious in these processes. This will be discussed in more detail in Ref. 4. The problem of color conservation is at

Table 1. Multiplets for the electromagnetic monopole in SU(5) (from Ref. 12) and the assignment of doublets interacting with the core.　Charge conjugate multiplets are omitted.　The states with $t = 2$, $m_t = 0$ are mixtures of Q-eigenstates.

Particle	t	T_3	I_3	$Q(= 3/2\ Q_{em})$	# of doublets	
d_1		3/2	1	−1/2		
d_2	3/2	1/2	0	−1/2	3 (j=1)	0
d_3		−1/2	−1	−1/2		
e^+		−3/2	0	3/2		
u_3^c		2	1	−1		
u_2^c		1	0	−1		
$\frac{1}{\sqrt{2}}\ (u_1^c + u_1)$	2	0	(−1,1)	(−1,1)	4 (j=3/2)	2 (j=1/2)
u_2		−1	0	1		
u_3		−2	−1	1		
$\frac{1}{\sqrt{2}}\ (u_1^c - u_1)$	0	0	(−1,1)	(−1,1)		

first sight more complicated for the purely electromagnetic monopole than for the single strength monopole, but is probably not
essentially different. Aspects of non-abelian interactions for the
single strength monopole have been discussed in Refs. 10 and 11.

The work reported here brings the discussion of fermion
interactions with general spherically symmetric monopoles to the
same level of approximation as was obtained for the simplest SU(5)
monopole. Understanding of the generalization provides new insight
in the special case. Whether our results have practical applications will depend on which unification model--if any--will turnout
to survive experimental tests.

ACKNOWLEDGEMENTS

Part of the results presented here was done at Fermilab and at
the Aspen Center for Theoretical Physics. This work was supported
in part by NSF Grant #PHY 81-09110 A-01.

REFERENCES

1. V.A. Rubakov, Nucl. Phys. B203, 311 (1982); Pis'ma Zh. Eksp.
   Teor. Fiz. 33, 658 (1981). [JETP Lett. 33, 644 (1981)].
2. C.G. Callan Jr., Phys. Rev. D25, 2141 (1982) Phys. Rev. D26,
   2058 (1982); Nucl. Phys. B212, 391 (1983).
3. D. Wilkinson and A.S. Goldhaber, Phys. Rev. D16, 1221 (1977).
4. A.N. Schellekens, in preparation.
5. Y. Kazama, C.N. Yang and A.S. Goldhaber, Phys. Rev. D15, 2287
   (1977).
6. For a discussion of the conservation laws of the "standard"
   model see A. Sen, Phys. Rev. D28, 876 (1983); A. Sen, in
   this volume; A.N. Schellekens, Stony Brook preprint ITP-
   SB-83-53.
7. T.M. Yan, in this volume; A. Sen, in this volume; Y. Kazama
   and A. Sen, Fermilab Pub. 83/58-THY.
8. B. Sathiapalan and T. Tomaras, CALT-68-987 (1982).
9. V.A. Rubakov and M.S. Serebryakov, Nucl. Phys. B218, 240
   (1983).
10. N. Craigie, in this volume.
11. P. Nelson, Phys. Rev. Lett. 50, 939 (1983). P. Nelson and A.
    Manohar, ibid., 943 (1983); A. Abouelsaoud, Harvard preprint
    HUTP-82/A058; A.P. Balachandran, G. Marmo, N. Mukunda, J.S.
    Nilsson, E.B.G. Sudarshan and F. Zaccharia, Austin preprint
    DOE-ER-03992-502; A.P. Balachandran, in this volume.
12. S. Dawson, A.N. Schellekens, Phys. Rev. D27, 2119 (1983).

ON THE QFT OF MONOPOLE INDUCED BARYON NUMBER VIOLATING PROCESSES

INCLUDING NON-ABELIAN FORCES

Neil S. Craigie

International Centre for Theoretical Physics
Trieste, Italy

Istituto Nazionale de Fisica Nucleare
Sezione di Treiste, Trieste, Italy

In this talk I would like to briefly outline an attempt by Werner Nahm, Valerie Rubakov and myself, to take into account all the long-range quantum fields outside a 't Hooft-Polyakov monopole in grand unified theories such as SU(5). This means we must also deal with non-Abelian color strong interactions and the effect they have on s-wave quarks, the latter being the relevant quark degrees of freedom regarding the interaction with the monopole core.

One remarkable feature of the monopole-fermion system is that it reduces to a two-dimensional QFT problem and in the case of massless fermions and Abelian gauge fields this is completely solvable. A study of the effect of an SU(2) monopole's U(1) electromagnetic quantum field on s-wave fermions led Rubakov[1] and Callan[2] to predict that fermion number violating processes can occur with probability of order unity for s waves, in particular, by showing the existence of the corresponding condensates, which are not suppressed by any small parameter. In the full SU(5) system these authors conjectured the existence of corresponding baryon number violating processes such as proton decay catalyzed by a monopole. The complete treatment of the effect of all the quantum fields on s-wave fermions so far has relied on bosonization (see Callan in Ref. 2 and Rubakov and Serebryakov in Ref. 3), however it is far from clear as to the validity or usefulness of the latter, when non-Abelian fields are present. Thus in Ref. 4 we offer a new approach which exploits special properties of two-dimensional non-Abelian field theories, first noted by 't Hooft in the case of QCD_2. Specifically, we show that one can separate out all the Abelian factors so that one is left with a radically reduced QCD system of s-wave quarks, which shares many of the features of 't Hooft's QCD_2, in particular, it is solvable in planar graph (or large N_c) approximation. We are thus able to show that Rubakov's original cluster argument for the existence of fermion number violating condensates in fact gives the correct answer, despite his neglect of QCD strong interactions. The reason is reminiscent of current algebra properties of QCD, where the computation of certain Green's functions using free quarks gives the correct answer. This was because, for certain questions, the summation over the bound state spectrum is totally "dual" to free quarks. In the monopole problem, a zero mass bound state appears in the spectrum near the monopole core and as far as the cluster argument is concerned, this state is totally dual to free quarks.

Let me start by writing down schematically the relevant functional integral and action for the full SU(5) monopole system and give its radial reduction to a two-dimensional field theory [details of what follows are available in Ref. 4, where all relevant further references can be found]:

$$Z = [\Pi dA_\mu][\Pi d\psi][\Pi d\bar{\psi}] \ e^{-S[A,\psi,\bar{\psi}]}, \tag{1}$$

where

$$S = \int dt \int_{r>r_o} dr \ 4\pi r^2 \ [-1/2\Sigma \ F^2 + \Sigma \ \bar{\psi}_L D\psi_L] \quad . \tag{2}$$

The monopole is taken to be infinitely heavy and the core radius r_o is essentially zero. The fermion integration is restricted to s wave, since only this sector feels the boundary conditions at the core and the monopole's axial anomaly, which is associated with rearrangements of the fermionic vacuum [higher partial waves can be treated perturbatively]. Let us recall that $J = 0$ fermions occur because the spin contribution to the total angular momentum is exactly cancelled by a half unit of angular momentum generated by the fermion charge field interaction with the monopole. Further, we recall up to a color rotation that the monopole fits in an SU(2) subgroup of SU(5), and this breaks down to $U(1)_Q$ - the monopole electromagnetic group. The generators of this Abelian "electromagnetic" field outside the core are made up of the ordinary electromagnetic charge and the color hypercharge, i.e. $Q = Q_{em} + Y_c$. If we choose an $SU(3)_{col}$ gauge so that the monopole sits in (d_3, e^-) space, then the relevant $J = 0$, i.e. s-wave fermion $SU(2)_{Mon}$ doublets, are

$$\begin{bmatrix} \bar{d}_3 \\ e^- \end{bmatrix}_L \quad \begin{bmatrix} e^+ \\ \bar{d}_3 \end{bmatrix}_L \quad \begin{bmatrix} u_1 \\ \bar{u}_2 \end{bmatrix}_L \quad \begin{bmatrix} u_2 \\ -\bar{u}_1 \end{bmatrix}_L \quad . \tag{3}$$

In addition to the $U(1)_Q$ Abelian interaction, the fermion fields have $SU(2)_{col} \times U(1)_{Y_c}$ gauge interactions. Note if we keep only $J = 0$ quarks for consistency, we should do likewise for the gauge boson (i.e. gluons). [Due to the charge field interaction only part of the gluon octet is in the $J = 0$ sector]. Thus the monopole breaks the $SU(3)_{col}$ group to $SU(2)_{col} \times U(1)_{Y_c}$. We will return later to what might be a reasonable approximation, in which the color group is dynamically restored to $SU(3)_{col}$ by quantum fields.

Having described the system, let us now note that s-wave fermions in the $SU(2)$ monopole system can be decomposed according to $\psi_{L,\alpha,j}^{(i)} = (8\pi)^{-1/2} r^{-1} [g^{(i)}\sigma_2 + (\hat{n}\sigma) \sigma_2 h^{(i)}]_{\alpha,j}$, where α refers to spin and j to isospin. Using this decomposition one has

$$4\pi r^2 \ \bar{\psi}_L^{(i)} \rlap{/}{D} \psi_L^{(i)} = \bar{f}^{(i)} \hat{\rlap{/}{D}} f^{(i)} \ , \tag{4}$$

where

$$f = \begin{bmatrix} g \\ h \end{bmatrix} \quad , \quad \hat{\rlap{/}{D}} = \sigma_3 D_0 + \sigma_1 D_r \quad , \quad \hat{\gamma}_5 = i\sigma_2 \ .$$

Here we have only exhibited the $SU(2)_{Mon}$ indices. The different flavors (i) will, in general, form different representations of the other gauge subgroups. All the s-wave fermions are divided into incoming and outgoing fermions, according to their charge and helicity and are linked only by the b.c. at $r = 0$. The upper line in Eq. (3) corresponds to incoming and the lower line to outgoing fermions. Apart from the b.c. at $r = 0$, in the absence of quantum fields, the two-dimensional radial fields satisfy the free Dirac equation. However when we include quantum fields, we arrive at a two-dimensional $SU(2)_{col} \times U(1)_{Y_c} \times U(1)_Q$ gauge theory on a half space $0 < r < \infty$ with a special boundary condition at $r = 0$. The relevant functional integral and action is:

$$Z = \int [\Pi da_\mu][\Pi df][\Pi d\bar{f}] \ e^{-S[a,f,\bar{f}]}$$

$$S = \int_{-\infty}^{\infty} dt \int_0^{\infty} dr \ [-\sum \frac{1}{4}(\frac{4\pi r^2}{g^2})F_{\mu\nu}F^{\mu\nu} + \sum i\bar{f} \ \rlap{/}{D} \ f] \ . \tag{5}$$

$$\underset{\substack{\text{gauge} \\ \text{fields}}}{} \qquad \underset{\substack{\text{fermion} \\ \text{copies}}}{}$$

[The approximations involved in arriving at Eq. (5) are explained in Ref.4.] The interaction of the fermions with the Abelian fields can be factored out as Schwinger pointed out many years ago. One simply makes a chiral and $U(1)$ transformation $f \rightarrow \exp[i\alpha + \gamma_5\beta] f_0$ where β refers to an axial $U(1)$ gauge rotation and α is chosen to remove the $U(1)$ gauge interaction from $\bar{f}\hat{\gamma}\cdot\hat{D}f$, i.e. $a_\mu = \varepsilon_{\mu\nu}\partial_\nu\alpha + \partial_\mu\beta$ and $E = \partial_0 a_r - \partial_r a_0 = \Box\alpha$. The Jacobian of the transformation $f \to f_0$ of the fermion fields is simply related to the anomaly in the two-dimensional radial field theory by

$$\text{Log } J = \frac{N_D}{2\pi} \int \alpha(x) E(x) \, d^2 x \tag{6}$$

which is actually nothing but the four-dimensional anomaly projected by the monopole's specially symmetric magnetic field $B = \hat{n}/e \, r^2$ into the radial field theory. An arbitrary correlation function can thus be calculated using

$$\overline{f(1) f'(2) \ldots} = \int [d\alpha] \, e^{-S[\alpha]} \prod_j e^{i\alpha(j)} \, \overline{f_0(1) f_0'(2) \ldots} \tag{7}$$

where

$$\overline{f_0(1) f_0'(2) \ldots} = \int [da_\mu^{\text{non Ab.}}][df_0][d\bar{f}_0] f_0(1) f_0'(2) \ldots$$

$$\exp[-\int dt dr[-(\frac{4\pi r^2}{g^2})^{1/4} F^2{}_{\substack{\text{non} \\ \text{Ab.}}} + f_0 i \not{D} f_0] \,] \quad . \tag{8}$$

Let us note that the two U(1) factors in the full $SU(2)_{\text{col}} \times U(1)_{Y_c} \times U(1)_Q$ radial gauge theory outside the core are mixed by the boundary condition, it is therefore appropriate to define $U(1)_{\bar{Q}} \times U_Q(1)$, where $\bar{Q} = Q_{\text{em}} - Y_c$. In this case $U(1)_{\bar{Q}}$ is vector, while $U(1)_Q$ as stated above is axial. The respective α fields have the distinct boundary conditions (i) $\alpha = 0$ and (ii) $d\alpha/dr = 0$ at $r = 0$, respectively. Note that (ii) admits twisted boundary conditions or instanton-like configurations, which is intimately tied to the presence of the axial U(1) anomaly and fermion number violating processes. Indeed if we examine the two-dimensional fermion number current $J_\mu^F = \sum f^{(i)} \gamma_\mu f^{(i)}$, then it has the two-dimensional anomaly $\partial^\mu J_\mu^F = e/2\pi \, \epsilon_{\mu\nu} F^{\mu\nu}$. Further, if we consider a gauge transformation $a_\mu \to a_\mu + 1/e \, \partial_\mu \beta$, in which the gauge function β has the twisted boundary condition condition $\beta(t = +\infty, r_0) - \beta(t = -\infty, r_0) = 2\pi$, then the corresponding fermion number

$$n_f = \int_{-\infty}^{\infty} dt (dQ_f/dt) = \frac{1}{2\pi} \int_{-\infty}^{\infty} dt \int_{r_0}^{\infty} dr \partial_t \partial_r \beta$$

changes by one unit. The vacuum state is in fact made up of a superposition of different gauge or instanton sectors, which correspond to vacua with different fermion number, in accordance with properties of the ordinary Schwinger model.

The existence of fermion number violating condensates was established by considering the pairing parameter $F = f^{(1)}(r) \, f^{(2)}(r)$ and taking the product $\langle F(x_1), \, F^+(x_2) \rangle$ at different points. At large time separations $T = |t_1 - t_2|$, one can show factorization with $\langle F \rangle \sim 1/r$ (in the two doublet case). In terms of four-dimensional spinors this corresponds to $\langle \overline{\psi}^{(1)} \, \psi^{(2)} \rangle \sim 1/r^3$. In the original works[1,2], after separating out the $U_Q(1)$ gauge field interaction f_0 was replaced by free fields and one was left to estimate the free propagator factor shown in Fig. 1:

(9)

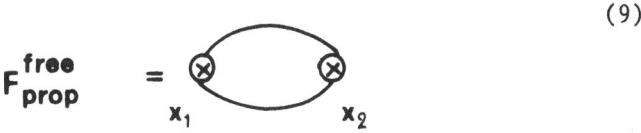

$$F^{\text{free}}_{\text{prop}} \quad = \quad \text{(diagram)}$$

Fig. 1 Free propagator factor.

the time dependence of which cancelled with that of the $U(1)_Q$ factor $\langle e^{i2\alpha(x_1)} \, e^{-i2\alpha(x_2)} \rangle_\alpha$. The two-dimensional non-Abelian field theory described by Eq. (8) can however be treated by analyzing the basic Dyson equation in the temporal gauge and perturbing the whole system about planar diagrams. Initially it is convenient to consider arbitrary $SU(N_c)$ color group. In the temporal gauge, i.e. $a_0 = 0$, the gauge action becomes quadratic,

$$\left(\frac{4\pi r^2}{g^2}\right) F^a_{\mu\nu} \, F^{a\mu\nu} = 2 \sum_{c=1}^{N_c^2-1} \left(\frac{4\pi r^2}{g^2}\right) (\partial_t \, a^c_r)^2 \tag{10}$$

and the corresponding gluon propagator, i.e. $\Delta = - \left[\frac{4\pi r^2}{g^2} \, \partial_t^2 \right]^{-1}$ is ultralocal in the radial co-ordinate r.

The Dyson equation for the fermion propagator, namely

$$\int d^2x [\hat{\gamma} \cdot \partial \cdot \delta^{(2)}(x - x') + \Sigma[x, \, x']] S_F(x', \, y) = \delta^{(2)}(x - y) \tag{11}$$

with

$$\Sigma[x, \, y] = (\Delta S_F)_{x,y}$$

$$= C_F \, \frac{\alpha_s}{2r^2} \, |t-t'| \, \delta(r - r')\sigma_1 \, S_F(x, \, y) \, \sigma_1$$

reduces in energy space to the first order differential equation:

$$[\sigma_1 \, \partial_r - i\omega\sigma_3 + \tilde{\Sigma}(r,\omega)]\tilde{S}_F(r, \ r', \ \omega) = \delta(r - r') \tag{12}$$

with

$$\tilde{\Sigma}(r, \ \omega) = iC_F \frac{\alpha_s}{2\pi r^2} \, \sigma_3 \left(\frac{1}{\omega} - \frac{\varepsilon(\omega)}{\lambda}\right) \ ,$$

where λ is an infr-red cut-off. The exact solution of Eq. (12) is given by

$$S_F(r, \ r', \ \omega) = \exp[-C_F \frac{\alpha_s}{2\pi} \ |\frac{1}{r} - \frac{1}{r'}| \ (-\frac{1}{|\omega|} + \frac{1}{\lambda})] \ . \tag{13}$$

In the limit $\lambda \rightarrow 0$ the full propagator vanishes. This means that s-wave radial quarks do not propagate in the presence of their non-Abelian color interactions. Instead color singlet bound states form in complete analogy to 't Hooft's QCD_2.

The corresponding Bethe-Salpeter equation is given by

$$\frac{2\pi r^2}{C_F\alpha_s}E \ \Phi(r, \ E, \ \omega) = [\frac{1}{\omega} + \frac{1}{E-\omega} + i \ \frac{2\pi r^2}{C_F\alpha_s} \frac{d}{dr}]\Phi(r, \ E, \ \omega) \tag{14}$$

$$+ P \int \frac{d\omega'}{(\omega-\omega')^2} \ \Phi(r, \ E, \ \omega'),$$

where the Bethe-Salpeter wave function at equal radial distance is defined by

$$\Phi(r, \ t, \ t') = \langle 0 \ |f(t, \ r) \ f^+(t', \ r)| \ n\rangle \ . \tag{15}$$

Equation (14) admits a zero energy solution, which does not depend on the radial co-ordinate r and consequently is translational invariant. In the case of $N_c = 2$ the above system admits a special symmetry originally pointed out by Pauli and Gursey and recently in connection with gauge theories analyzed by Jan Stern, namely since f and its charge conjugate f_c have the same gauge transformation the system is invariant under $f \rightarrow a \ f + b\gamma_5 f_c$ with $|a|^2 + |b|^2 = 1$. One consequence of this symmetry is that the corresponding SU(2) color singlet diquark channel has the same Bethe-Salpeter equation and zero energy pole.

These zero energy bound states are in fact totally dual to free quarks in the cluster argument. An explicit computation of the contribution of diagrams in the cluster argument (see Fig. 2) gives

$$F_{ladder} = F^{free}_{prop} \tag{16}$$

where F_{prop}^{free} was defined in Eq. (9). One can in fact use a generalized local current algebra in this radial QCD together with a study of the two-dimensional anomaly to argue that the results are valid beyond the planar graph approximation.

$$\mathbf{F_{ladder}} \quad = \quad$$

Fig. 2 Contribution of diagrams in the cluster argument.

Thus in summary we see that there are three aspects of monopole-fermion physics, each playing an essential role, namely,

  i) the charge exchange boundary condition at the core i.e. $\bar{d}_{3L}^{in} \rightarrow e_L^{-\,out}$, $e_L^{+\,in} \rightarrow d_{3L}^{out}$, $u_{1L}^{in} \rightarrow u_{2L}^{out}$ and $u_{2L}^{in} \rightarrow u_{1L}^{out}$ for the first SU(5) generation.

  ii) instanton-like configurations in the monopole radial axial U(1) field, which is intimately tied to the fact that the monopole is a superposition of vacua with different fermion number;

  iii) non-Abelian two-dimensional confining color interactions, which only allow s-wave quarks to propagate in color singlet bound states.

The question is can we simultaneously deal with these and arrive at a computational framework for calculating branching ratios and cross-sections? We have gone a little way in answering this question and we end this talk by briefly outlining our thoughts so far on this.

Let us first make some remarks on 4-fermion condensates in the case of 4 $SU(2)_{Mon}$ doublets, i.e. one light SU(5) generation. The condensates can be deduced directly from the functional integral using Schwinger-model-instanton-calculus, i.e. we can calculate $\langle f^{(1)}(r)\ f^{(2)}(r)\ f^{(3)}(r)\ f^{(4)}(r)\rangle$ directly and obtain the same result as that obtained from the cluster argument. Here four different fermions disappear into the Fermi vacuum near the monopole as represented in Fig. 3(a).

In the case in which we only have the monopole's $U(1)_Q$ gauge field there is another four fermion condensate, which is due solely to the charge exchange boundary condition at $r = 0$. This mechanism is shown in Fig. 3(b), where we notice two helicities are reversed, so the $U(1)_Q$ factors cancel and one is left to calculate free quarks. However when we include the $SU(2)_{col}$ gauge forces, Fig. 3(b) is replaced by Fig. 3(c), where an incoming $u_{1L} - u_{2R}$ diquark bound state must replace the incoming u_{1L} outgoing u_{2L} free quark system. However this bound state does not occur and thus this four-fermion condensate is suppressed by the color strong interactions. In the light of these remarks a re-appraisal of the effect of color interactions on the formation of condensates is required. There will of course also be non-trivial consequences on the kind of scattering processes that can occur.

The dynamics can be considerably simplified if we make an approximation in which we assume color symmetry is restored to $SU(3)_{col}$. This amounts to supposing that the color magnetic charge of the monopole is effectively screened in the s-wave quark sector until they actually reach the core. This approximation is actually self consistent, since only $SU(3)$ color singlet bound states can propagate, in which color is confined ultra-locally in space. This means the quarks will not see the color magnetic charge until they reach the core. In the case of $SU(3)$ color, we can completely rewrite the original non-Abelian action, Eq. (8), in terms of an equivalent string or bilocal theory, in which the fermion fields have been integrated out, leaving the action expressed in terms of the auxilliary string variable $\sum(x, y)$. The latter describes the theory in terms of color singlet composite fields $f(x) \overline{f}(y)$. The corresponding form of the action is given by (initially neglecting the interaction with the monopole core or the $U(1)_Q$ field):

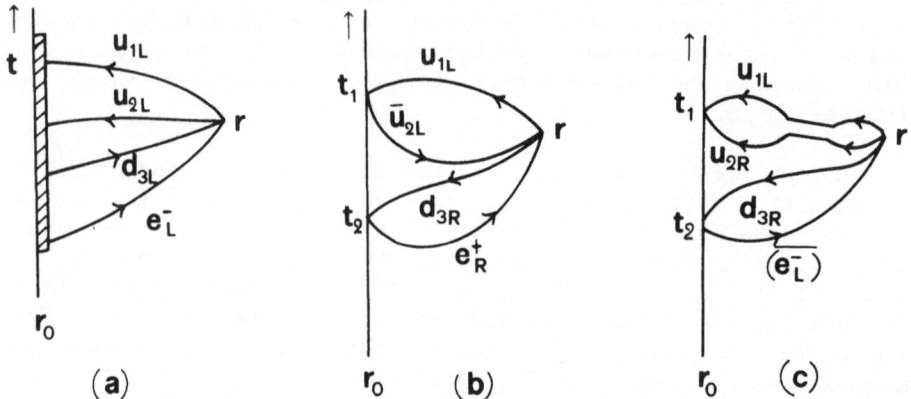

Fig. 3 (a) Four different fermions disappear into the Fermi vacuum near the monopole; (b) Monopole's $U(1)_Q$ gauge field; (c) $SU(2)_{color}$ gauge forces included.

$$Z = \int [d\Sigma] \ e^{-S_{eff}[\Sigma]}$$

$$S_{eff} = \frac{1}{2} (\Sigma, K^{-1} \Sigma) + \text{Tr Log}(S_F^{0-1} + \Sigma) \ . \tag{17}$$

The minimum of this effective action corresponds to the Dyson Eq. (11) i.e. $\delta S_{eff}/\delta\Sigma = 0$ is simply $(S_F^{0-1} + \Sigma_0)^{-1} = K^{-1}\Sigma_0$. The quadratic part is given by

$$S_{eff}^{quad} = (\sigma, D_\sigma \ ^\sigma) \ \ ; \ D_\sigma = \frac{\delta S_{eff}}{\delta\Sigma\delta\Sigma}\Big|_{\Sigma = \Sigma_0} \ , \tag{18}$$

where $\Sigma = \Sigma_0 + \sigma$ and Σ_0 is a solution of the Dyson equation. The inverse D_σ^{-1} satisfies the 't Hooft equation described earlier and is made up of a superposition of the bound states. Finally, the interaction term is given by

$$S_{eff}^{int} = \sum_n \frac{1}{n} \text{Tr}[(S_F)^n] \ . \tag{19}$$

In order to include the effect of the b.c. at $r = 0$ which is due to the interaction with the heavy SU(5) bosons condensed in monopole core, as well as the fermion number violating axial $U(1)_Q$ instanton-like effects, we make use of the ultra localization in space of these effects near the core. This means we can assume that for $r > r_0$ the dynamics are uneffected and at $r = r_0$ we have an effective interaction vertex. In order to see what this is, we simply note that the b.c. at $r = r_0$ corresponds to the following additional piece in the fermion action:

$$S^{b.c.} = \int dt\Sigma_1 \Big[\ \overline{f}_L^{(i)}(t, \ r_0)f_R^{(i)}(t, \ r_0) + h.c. \Big] \ . \tag{20}$$

Similarly the one-instanton effects correspond to the additional piece of the action (for the $N_D = 2$ case)

$$S^{inst} = \int d^2z_1 \ d^2z_2 \ K(z_1, \ z_2|a_\mu) \ \overline{f}^{(1)}(z_1)\overline{f}^{(2)}(z_2) \tag{21}$$

where $K(z_1, \ z_2|a_\mu) = e^{\frac{1}{\pi} \int \alpha \ \Box \ \alpha d^2z} \ \partial_{z_1} \ f^\circ(z_1) \ \partial_{z_2} \ f^\circ(z_1)$ and f° is the zero mode function in background field $a_\mu = \epsilon_{\mu\nu}\partial_\nu\alpha + \partial_\mu\beta$, β having the twisted b.c. we mentioned earlier. The inclusion of these b.c. and instanton effects leads to a modification of Eq. (19) in the form

$$S_{eff}^{int} (\sigma) = \sum_n \frac{1}{n} \text{Tr}[S_F(\sigma + M^{b.c.} + M^{inst}(a_\mu)] \ , \tag{22}$$

where $M^{b.c.}$ and M^{inst} are localized effectively at $r = 0$. In general, both $M^{b.c.}$ and M^{inst} will involve bilinears in the lepton fields, so Eq. (22) will correspond to generalized hadron-lepton vertices. We thus arrive at an action principle, which could be used to derive generalized Feynman rules for computing monopole induced lepton-hadron transitions.

The approach we have outlined in the last part of this talk can also be linked directly to the Skyrmion model discussed by Callan at this meeting. The essential point is that the exact duality between zero mass bound states in the radial QCD and free quarks has its origin in the existence of an underlying current algebra. If we switch off the monopole U(1) electromagnetism, then this $SU_L(2) \times SU_R(2)$ current algebra has a realization in a two-dimensional non-linear sigma model in radial space. On including the electromagnetism we are left with only the $U_{I3}(1)$ and an effective Lagrangian

$$\mathcal{L} = \int_0^\infty dr \ [\partial^\mu \pi_0 \partial_\mu \pi_0 \ -V(\pi_0)] \quad . \tag{23}$$

If this system has kink solutions, then the kink number is given by

$$Q = \frac{1}{2\pi} \int_0^\infty dr \ \frac{d}{dr} \ \pi_0(t, \ r) \quad . \tag{24}$$

This is a topological invariant corresponding to $\Pi_1\big(U(1)\big) = Z$. However the latter is precisely the Skyrmion topological baryon charge in the presence of a monopole [see talk by Callan]. In the presence of the monopole electromagnetism, the original SU(2) flavor symmetry is broken down to $U_{I3}(1)$ and for a Skyrmion localized near the core of a monopole, the topological charge becomes

$$Q_B = \frac{1}{8\pi^2} \ \int \ dr \ 4\pi r^2 \ e\vec{B}\cdot\vec{V}\pi_0(t, \ r) \tag{25}$$

which is nothing but Eq. (24), since $\vec{B} = \frac{\hat{n}}{er^2}$ for a monopole. Away from the core Q_B corresponds to $\Pi_3\big(SU(2)\big) = Z$, which is the topological invariant in the four-dimensional non-linear sigma model.

REFERENCES

1. V.A. Rubakov, Zh ETF Pis'ma 23, 658 (1980); Nucl. Phys. B203, 211 (1982).

2. C.G. Callan, Phys. Rev. D25, 2142 (1982); D26, 2058 (1982).
3. V.A. Rubakov and M.S. Serebryakov, INR, Moscow, preprint (1983) (submitted to Nucl. Phys. B.).
4. N.S. Craigie, W. Nahm and V.A. Rubakov, ICTP, Trieste, preprint IC83/180 (submitted to Nucl. Phys. B.).

A FRIEDEL SUM RULE FOR THE DYON CHARGE

Bernard Grossman

Department of Physics
The Rockefeller University
New York, New York   10021

In understanding the ground state of the magnetic monopole, insight can be gained by regarding the magnetic monopole as an impurity in the Dirac sea, much as solid-state theorists investigate magnetic impurities in the Fermi sea. The angle θ that measures the amount of CP violation in the theory can be regarded as a probe of the ground, much as solid-state physicists use an external magnetic field to probe the Fermi sea. In a CP violating theory, the magnetic monopole acquires a non-zero electric charge that does not violate any quantization condition.[1,2,3] The value of this charge

$$Q = - \frac{e\theta}{2\pi}$$

can be thought of as arising from a polarization of the ground state of the magnetic monopole by the CP violating term $\theta \vec{E} \cdot \vec{B}$. A connection can be made with the Friedel Sum Rule that solid state physicists use. This determines the polarization charge, Z, in terms of the scattering phase shift, δ, of condition band electrons at the Fermi level off the impurity

$$Z = \frac{2}{\pi} \Sigma_\ell \ (2\ell + 1) \ \delta_\ell \ (E_F) \quad .$$

We will restrict the discussion to a point abelian magnetic monopole. This can be described in a non-singular way by regarding the vector potential as a connection on a U(1) bundle over S^2 with the fermion considered as a section.[4] We cover S^2 by two patches, one for $0 \leqslant \theta < \pi/2 + \delta$ with

$$A{\binom{1}{\phi}} = g (1 - \cos \theta)$$

and one for $\pi/2 - \delta < \theta \leqslant \pi$ with

$$A{\binom{2}{\phi}} = -g (1 + \cos \theta)$$

such that

$$A{\binom{2}{\phi}} - A{\binom{1}{\phi}} = -2g \qquad \pi/2 - \delta < \theta < \pi/2 + \delta$$

and so that the two connections, $A{\binom{1}{\phi}}$ and $A{\binom{2}{\phi}}$, are related by a gauge transformation U on the overlap of the two patches with

$$U = e^{2ieg\phi}.$$

We note that

$$F_{\theta\phi} = \partial_\theta A_\phi - \partial_\phi A_\theta$$

$$= g \sin \theta$$

so that

$$\int \vec{B} \cdot d\vec{s} = \int F_{\theta\phi} \; d\theta \; d\phi = 4\pi g$$

In order that U be single-valued, we must have the Dirac quantization condition

$$eg = \frac{n}{2}$$

so that

$$\frac{e}{2\pi} \int F_{\theta\phi} \; d\theta \; d\phi = n$$

This last formula is the important two-dimensional version of the Atiyah-Singer index theorem.[5] The left hand side is the topological invariant determined by the curvature (magnetic field) on the two-sphere S^2. This is an integer, n, which must equal the number of positive chirality zero modes of the Dirac operator on S^2 minus the number of negative chirality zero modes.

We make use of the above Atiyah-Singer index theorem by dimensional reduction from four-dimensions. We can choose a representation of the γ-matrices so that

$$\overline{\gamma} = \begin{pmatrix} 0 & -\vec{\sigma} \\ \vec{\sigma} & 0 \end{pmatrix} \qquad\qquad \gamma_0 = \begin{pmatrix} 1 & 0 \\ 0 & -1 \end{pmatrix}$$

$$\gamma^5 = \begin{pmatrix} 0 & 1 \\ 1 & 0 \end{pmatrix}$$

The Dirac equation

$$(D - m) \ \psi = 0$$

becomes

$$H \ \tilde{\psi} = E\tilde{\psi}$$

with $\tilde{\psi}$ a 2-component spinor after dimensional reduction. This loss of degrees of freedom occurs because

$$\vec{\sigma}\cdot\hat{r} \ \psi = \psi$$
$$\vec{\sigma} \times \hat{r}\cdot\vec{D}\psi = 0 \quad .$$

With n = 1, there is one zero mode with positive chirality, as the above two equations state. We recognize this in a more familiar form by looking at the conserved angluar momentum

$$\vec{J} = \vec{L} + 1/2 \ \vec{\sigma} -eg\hat{r}.$$

With eg = 1/2 (n = 1), there exists an S-wave, J = 0 state with

$$1/2 \ \vec{\sigma}\cdot\hat{r} = 1/2$$
$$\vec{\sigma}\cdot\vec{L} = 0$$

The Hamiltonian, H, for the S-wave has an interesting property; it is not self-adjoint.[6]

$$H = -i\gamma_5 \frac{d}{dr} + \beta M$$
$$\gamma_5 = \sigma, \quad \beta = \sigma_3$$

In order to make this Hamiltonian well-defined, we must impose a boundary condition on the wave function, $\tilde{\psi}$, at the origin. For example,

$$(H)^2 = -\frac{d^2}{dr^2} + M^2 \quad .$$

Solutions to the eigenvalue problem

$$(H)^2 \, \phi = E^2 \phi$$

consist of plane waves $e^{\pm ikr}$ with $E^2 = k^2 + M^2$. However, proper scattering solution is obtained by a linear superposition,

$$e^{-ikr} + e^{i\theta} \, e^{ikr}$$

where the phase $e^{i\theta}$ is determined by a boundary condition at the origin. In the case of H, we also obtain a one-parameter family of self-adjoint extensions by the boundary conditions

$$\frac{F(0)}{G(0)} = i \, \tan \, (\frac{\theta}{2} - \frac{\pi}{4})$$

with $\tilde{\psi}(r) = \binom{F(r)}{G(r)}$.

The angle θ that appears here can be identified with the angle θ that measures CP violation. There exists a self-adjoint Hamiltonian

$$H' = -i\gamma_5 \frac{d}{dr} + \beta M e^{i\theta\gamma_5} + \beta\frac{\kappa eg}{2Mr^2} \, e^{i\theta\gamma_5}$$

that yields the same boundary condition. However, because of the singularity of the anomalous moment term,

$$F(r) \sim e^{-\kappa eg/2Mr} \quad \text{as } r \to 0.$$

This Hamiltonian is easily recognized as a chiral rotation by angle θ of H plus an anomalous magnetic moment. θ cannot be rotated away because the charge and the chiral charge do not commute in the presence of a magnetic monopole.

The charge of the ground state can be computed by computing the contribution of all the negative energy states in the Dirac sea.[2,3]

$$\tilde{\psi}(r) = N \left(\begin{array}{c} ke^{-ikr} + s(k) \, e^{ikr} \\ \frac{k}{E+M} \, (e^{ikr} - s(k) \, e^{ikr}) \end{array} \right) \quad S(k) = e^{2i\delta(k)}$$

$$\tilde{\psi}_B(r) = (-2M \, \cos\theta)^{1/2} \exp[i(\frac{\theta}{2} - \frac{\pi}{4})\gamma_5](\begin{smallmatrix}1\\0\end{smallmatrix}) \, \exp(M \, \cos\theta), \, \cos\theta < 0$$

$$\langle Q \rangle = e \int_0^R dr \; \{ \int_{-\infty}^{-M} [\tilde{\psi}^+ \tilde{\psi} - \psi_0^+ \psi_0] \; dE + \tilde{\psi}_B^+ \tilde{\psi}_B \}$$

$$= \int_0^R dr \int_{-\infty}^{\infty} dk \; e^{2ikr} \; \frac{e}{\pi} \; \frac{M}{\sqrt{K^2 + M^2}} \; S(k)$$

$$+ \text{ Bound State Contribution}$$

$$\langle Q \rangle = - \frac{e}{\pi} \sin \theta \int_0^{\infty} \frac{d\chi \; (1 - e^{-MR \cosh \chi})}{\cosh \chi + \cos \theta}$$

$$= \begin{cases} \dfrac{-e\theta}{2\pi} & \text{M finite, } R \to \infty \\ \\ 0 & \text{R finite, } M \to 0 \end{cases} .$$

Finally, a Friedel sum rule relation between the scattering phase shifts $\delta(\omega)$ can be obtained

$$\int_M^{\infty} \delta' \Delta(\omega) \; d\omega + \int_{-\infty}^{-M} \delta'(\omega) \; d\omega = -\theta$$
$$+ \pi(\# \text{ Bound states})$$

with $\delta'(\omega) = $ time delay .

This can be derived from the definition of the charge

$$\langle Q \rangle = - \frac{ie}{2} \lim_{t \to 0^+} \int_{-\infty}^{\infty} \frac{d\omega}{2\pi} \; e^{-i\omega t} \; TR_p \; [G(p, \omega) - G^{\circ}(p, \omega)]$$
$$= - \frac{e}{2} \eta(0)$$

where $\eta(0)$ is what mathematicians call the η-invariant?

REFERENCES

1. E. Witten, Phys. Lett. 86B, 283 (1979).
2. B. Grossman, Phys. Rev. Lett. 50, 464 (1983); 51, 959 (1983).
3. H. Yamagishi, Phys. Rev. D27, 2383 (1983); D28, 977 (1983).
4. T.T. Wu and C.N. Yang, Nucl. Phys. B107, 365 (1976).
5. M.F. Atiyah and I.M. Singer, Bull. Amer. Math. Soc. 69, 422 (1963).
6. A.S. Goldhaber, Phys. Rev. D16, 1815 (1977); C.J. Callias, Phys. Rev. D16 3068 (1977).
7. M.F. Atiyah, V.K. Patodi, and I.M. Singer, Bull. London Math. Soc. 5, 229 (1973).

MONOPOLE CATALYSIS: AN OVERVIEW

Sally Dawson

Lawrence Berkeley Laboratory
University of California
Berkeley, California 94720

ABSTRACT

A summary of the talks presented in the topical workshop on monopole catalysis at this conference is given. We place special emphasis on the conservation laws which determine the allowed monopole-fermion interactions and on catalysis as a probe of the structure of a grand unified theory.

INTRODUCTION

At this conference, there were nine talks presented dealing with the general topic of monopole catalysis of baryon number violating processes. We attempt--with no pretense at completeness--to summarize these talks here. The talks fall into two general categories. The first is a discussion of the allowed fermion-monopole scattering processes. In the presence of a finite size monopole, it is not a priori clear that charge is conserved in all interactions. Charge conservation in the fermion-monopole system and its relationship to gauge invariance is the focus of the second section of this paper. The second topic discussed in this workshop was the connection between catalysis and the underlying structure of a grand unified theory. In the third section, we discuss the monopole catalysis resulting from monopoles with charges larger than a Dirac charge and from monopoles with a different topology than those of the SU(5) grand unified theory.

One of the most interesting developments in the study of monopole catalysis was not discussed in this workshop. This is the hope, as given by Callan at this conference, of performing a reliable calculation of the total cross section and branching ratios for monopole catalysis of baryon number violating processes. The interested reader is referred to Callan's contribution to these proceedings.

CONSERVATION LAWS AND CHARGE CONSERVATION

The use of conservation laws to study monopole-fermion scattering has been discussed by Sen.[1] We summarize his results here. For simplicity, consider the Georgi-Glashow SU(2) model. In this model, monopoles are produced when the gauge symmetry is broken by the vacuum expectation value of a triplet of Higgs, ϕ, to a residual U(1) symmetry. The model contains two Dirac doublets of fermions,

$$\psi_i = \begin{pmatrix} a_i{}^+ \\ b_{i-} \end{pmatrix}_L \qquad i = 1, 2 \tag{1a}$$

with charges,

$$Q_{em} = \begin{pmatrix} -1/2 \\ 1/2 \end{pmatrix} \qquad , \tag{1b}$$

and is described by the Lagrangian,

$$L = -\frac{1}{4} F_{\mu\nu}F^{\mu\nu} + \sum_{i=1}^{2} \bar{\psi}_i \not{D} \psi_i + \frac{1}{2}|D_\mu \phi|^2 + V(\phi) \quad . \tag{2}$$

(In the SU(5) grand unified model, we make the identifications, $a_1 = e^-$, $b_1 = \bar{d}_3$, $a_2 = \bar{u}_2$, $b_2 = \bar{u}_1$).

The theory can be described by eight charges,

$$Q_{a_i(b_i)} = \int d^3x \, \bar{\psi}_{a_i(b_i)} \gamma^0 \, \psi_{a_i(b_i)},$$

$$Q^5{}_{a_i(b_i)} = \int d^3x \, \bar{\psi}_{a_i(b_i)} \gamma^0 \, \gamma_5 \, \psi_{a_i(b_i)} \quad . \tag{3}$$

It is possible to construct linear combinations of these charges which correspond to global symmetries of the Lagrangian of Eq. (2) and hence are conserved in all scattering processes. For example, L is invariant under the transformation $\psi_1 \to e^{i\alpha_1}\psi_1$ which corresponds to the conserved charge,

$$S_1 = Q_{a_1} + Q_{b_2} \quad . \tag{4}$$

The other conserved charges are

$$S_2 = Q_{a2} + Q_{b2},$$

$$S_3 = (Q^5_{a_1} + Q^5_{b_1}) - (Q^5_{a_2} + Q^5_{b_2}), \tag{5}$$

$$S_4 = (Q_{a_1} - Q_{b_1}) + (Q_{a_2} - Q_{b_2}) = -2Q_{em}.$$

Thus far we have assumed that the monopole is pointlike. In this case, the mass gap between a monopole and a dyon is infinite and so electric charge, $(-1/2S_4)$, must be conserved within the fermion system alone.

Monopole fermion scattering is further constrained because only the $\vec{J} = 0$ partial wave can penetrate to the monopole core. The conserved angular momentum is $\vec{J} = \vec{L} + \vec{S} + \vec{T}$ and so in the $\vec{J} = 0$ partial wave,

$$\vec{r} \cdot (\vec{S} + \vec{T}) = 0. \tag{6}$$

For an outgoing particle, $\vec{r} \cdot \vec{S}$ is the helicity and so there is a relationship between the helicity of a particle and its position in the doublets of Eq. (1):

Allowed incoming states: a^-_R, b^+_L, a^+_L, b^-_R

Allowed outgoing states: a^-_L, b^+_R, a^+_R, b^-_L $\qquad$ (7)

The helicity constraints and the conserved charges completely specify the allowed scattering process. Examples of allowed interactions are,

$$a^+_{1_L} + a^-_{2_R} \rightarrow b^-_{1_L} + b^+_{2_R}$$

$$a^+_{1_L} + b^+_{1_L} \rightarrow a^+_{2_R} + b^+_{2_R} \tag{8}$$

Note that both helicity violating and helicity conserving processes are allowed.

The role of the boundary conditions at the monopole core deserves special mention here. We must impose boundary conditions which conserve S_1, S_2, S_3, and S_4 since these correspond to symmetries of the full theory. However, the most general set of boundary conditions which conserves $S_1,...,S_4$ violates baryon number. It is therefore not possible to remove baryon number violating processes from the theory by changing the boundary conditions.

If the monopole has a finite size R_0, the picture changes.[2,3] The mass difference between the monopole and the dyon is no longer infinite, but is of order $1/R_0$. Since the boundary conditions on the fermion fields at the monopole core,

$$\psi_{a_i}(r = R_0) = \gamma^0 \psi_{b_i}(r = R_0), \tag{9}$$

violate charge, the conservation of charge in the monopole fermion system is not obvious at first glance.

Sen[1] has noted that charge conservation for a finite size monopole can be verified as follows. The charge deposited on the monopole core, Q, can be computed from the radial electric field, E_r,

$$\tilde{Q} = (r^2 E_r)_{r=R_0}. \tag{10}$$

The total charge Q_T of the monopole fermion system is $\tilde{Q}$ plus the fermion charge. Q_T commutes with the Hamiltonian and so is conserved in all interactions. An example of a process where charge is deposited on the monopole core is:

$$a^+_{1_L} + \text{Monopole}(Q_{em} = 0) \rightarrow a^-_{1_L} + \text{Dyon}(Q_{em} = -1). \tag{11}$$

Both Sen[1] and Yan[3] have noted that the amplitude for such processes is suppressed by powers of R_0,

$$\left(\bar{\psi}(r) \bar{\gamma}_5 \, \psi(r) \right) \sim \left[\frac{R_0 r}{(r + R_0)^2} \right]^{2/N}, \tag{12}$$

where N is the number of Dirac fermion doublets. In a realistic grand unified theory, R_0 is of order $1/M_x$, so the amplitude for creating a dyon is suppressed by powers of μ/M_x, where μ is an appropriate low energy scale and M_x is the unification scale.

In his talk at the conference, Yan made the interesting observation that it is possible to obtain a Lagrangian for fermion-monopole interactions which manifestly conserves charge. The usual decomposition of the monopole potential is not gauge invariant. When the potential is written in a gauge invariant fashion, charge conservation becomes an exact symmetry of the Lagrangian.

CATALYSIS AS A PROBE OF THE GUT STRUCTURE

It is an important phenomenological question to determine if different unified theories yield different selection rules for monopole catalysis. Unfortunately, however, this does not appear to be the case.

Most GUTs contain monopoles which catalyze $p \rightarrow e^+\pi^0$. Any GUT in which the $SU(3) \times SU(2) \times U(1)$ groups are unified in an $SU(N)$ group in which the fermions are embedded in the fundamental representation such that they decompose under $SU(5)$ as a [5] plus $N - 5$ singlets will have the same monopole catalysis of proton decay as the $SU(5)$ model, (i.e. $p \rightarrow e^+\pi^0$ will be the dominant decay mode of the proton).

Even introducing supersymmetry does not radically change the predictions for monopole catalysis. Since it is the gauge degree of freedom which is important for monopole catalysis, the supersymmetric $SU(5)$ GUT will catalyze the same proton decay events as the ordinary $SU(5)$ model. This model has the interesting feature that "ordinary" proton decay proceeds predominantly through Higgs exchange which yields $p \rightarrow \mu^+K^0$, while proton decay by monopole catalysis gives $p \rightarrow e^+\pi^0$.

We turn now to the catalysis induced by monopoles with a different topology from those of the $SU(5)$ model. At this meeting, London[4] spoke about Z_N monopoles--monopoles whose charges are additive modulo N. For example, a Z_2 monopole is its own anti-monopole. Such monopoles arise in a GUT theory where $SO(10)$ is broken to $SU(4) \times SU(2) \times SU(2)$ which is then broken to $SU(3) \times SU(2) \times U(1)$. At the first stage of symmetry breaking, Z_2 monopoles are produced with eg = 1/2. These monopoles catalyze the same proton decay as $SU(5)$ monopoles. At the second stage of the symmetry breaking "ordinary", Z, monopoles are produced with eg = 1. These monopoles do not catalyze proton decay.

We next consider monopoles with charges larger than the minimal Dirac charge. The interactions of fermions with these monopoles was discussed by Schellekens[5] at this meeting. He has solved the Dirac equation for fermions interacting with a monopole of arbitrary strength to find the allowed fermion monopole scattering processes. The trick is to find a basis in which the Dirac equation reduces to N non-interacting doublets, where N is an effective number of fermion doublets. The problem is then equivalent to that solved by Rubakov. It is no longer the $J = 0$ partial wave which interacts with the monopole to produce baryon number violating interactions, but rather it is the $J = T - 1/2$ partial wave. ($|eg| = T$).

The rules for constructing the effective doublets are easily found.   The Dirac   potential for the monopole can be written as

$$\vec{A}_D = Q_M(1 - \cos \theta)\frac{\hat{\phi}}{r \sin \theta},\tag{13}$$

where Q_M is in a representation of the unified gauge group. For a spherically symmetric monopole,

$$Q_M = I_3 - T_3,\tag{14}$$

where $\vec{I}$ is an SU(2) generator which commutes with Q_M and $\vec{T}$ is the generator of the SU(2) group which defines the monopole-fermion interactions.   The Dirac quantization condition, classical stability for a non-Abelian monopole, and the charge-triality relationship suffice to determine Q_M uniquely for a given eg.

The Dirac quantization condition requires that,

$$\exp(4\pi i Q_M) = 1,\tag{15}$$

while non-Abelian stability for SU(3) requires that if

$$Q_M = \alpha Q_{em} + Q_c,\tag{16}$$

where Q_{em} and Q_c are generators of U(1)$_{em}$ and SU(3)$_{color}$ and α is a constant, then

$$Q_c = (q_1, \ldots, q_n), \qquad \text{with } q_i - q_j = 0, \pm 1/2,\tag{17}$$

for every i, j.   Finally, if the only colored fields are color triplets,

$$Q_c = -\frac{1}{3} + m,\tag{18}$$

where m is an integer.   The monopole charge eg is then determined in terms of Q_{em} and $Y_c = 1/3, 1/3, -2/3$),

$$\begin{aligned}
eg &= \frac{1}{2}, & Q_M &= \frac{1}{2}(Q_{em} + Y_c) \\
eg &= 1, & Q_M &= Q_{em} + Y_c \quad . \\
eg &= \frac{3}{2}, & Q_M &= \frac{3}{2}Q_{em}
\end{aligned}\tag{19}$$

The rules for finding the effective doublets which interact with a monopole of arbitrary strength eg, are:

1. Find a basis in which T_3 and Q_M are diagonal,
2. The doublets are formed with fermions which have equal and opposite T_3 and different values of Q_M, and

3. The degeneracy of each doublet is $2|T_3|$.

For example the $eg = \frac{3}{2}$ monopole in SU(5) interacts with fermions in the following representations,[6]

$$
\begin{bmatrix} d_1 \\ d_2 \\ d_3 \\ e^+ \end{bmatrix}_L
\quad
\begin{bmatrix} e^- \\ \bar{d}_3 \\ \bar{d}_2 \\ \bar{d}_1 \end{bmatrix}_L
\quad
\begin{bmatrix} \bar{u}_3 \\ u_2 \\ \frac{1}{\sqrt{2}}(u_1 + \bar{u}_1) \\ u_2 \\ u_3 \end{bmatrix}_L
\quad .
\tag{20}
$$

Using the rules given above, there are twelve effective doublets,

$$
3 \times \begin{pmatrix} d_1 \\ e^+ \end{pmatrix}_L, \quad
3 \times \begin{pmatrix} e^- \\ \bar{d}^1 \end{pmatrix}_L, \quad
4 \times \begin{pmatrix} \bar{u}_3 \\ u^3 \end{pmatrix}_L, \quad
2 \times \begin{pmatrix} \bar{u}_2 \\ u^2 \end{pmatrix}_L.
\tag{21}
$$

The $eg = 3/2$ monopole is particulary interesting because its interactions are purely electromagnetic ($Q_M = 3/2\ Q_{em}$).

The allowed interactions can then be found by constructing the conserved charges. The simplest allowed process is

$$
d_{3L} + u_{1R} + \text{Monopole} \rightarrow e_L^+ + \bar{u}_{2R} + \text{Monopole}.
\tag{22}
$$

The next simplest allowed process involves nine particles in the incoming state.

Finally, we turn to a discussion of the connection between the zero energy states of the theory and catalysis. The presence of a zero energy bound state, (i.e. one for which $\partial_0 \psi = 0$), depends critically upon the structure of the mass terms in the theory. The SU(5) GUT with a [24] and a [5]-plet of Higgs does not possess such zero energy states, while the Georgi-Glashow SU(2) model with a triplet of Higgs bosons does. Hence if catalysis depends on the existence of these zero energy states, (as claimed by Walsh[7] at this meeting), the presence or absence of catalysis would be a sensitive probe of the Higgs structure of a grand unified theory. Walsh's argument is as follows: Rubakov's calculation[8] of the baryon number violating condensates which are formed in the presence of a monopole relies on the use of the cluster property to find the expectation value of two operators at infinite time separation. The claim by Walsh is that since the calculation is performed at an infinite time separation, it is sensitive to the zero energy states of the theory and catalysis will not occur without such states.

We disagree with Walsh's argument for a number of reasons. The first is that his argument is concerned with the use of the cluster property. It is possible, however, to calculate the expectation value of some condensates, (baryon number violating, but chirality conserving), without the use of the cluster property. Such condensates certainly lead to cross sections of strong interaction magnitude. The second reason for disagreement is more subtle. The physics of monopole catalysis is occurring through the boundary conditions and the anomaly at short distances near the monopole core. The effects of the fermion mass terms are important at distances of the order of $1/M_{fermion}$ and should not affect the physics near the monopole core. In sum, we do not believe that one particle zero energy bound states are a necessary prerequisite for catalysis.

CONCLUSION

Finally, we will mention briefly several other interesting talks which were given in this workshop. Craigie[9] has examined the effects of including the non-Abelian generators in the calculation of the baryon number violating condensates. His conclusion is that catalysis proceeds at strong interaction rates even when these non-Abelian interactions are included. There were also talks at this meeting by Grossman[10] and Fiorentini.[11] We have not discussed these talks since their results are well covered in the literature.

The conclusion to be drawn is that catalysis is in good shape--the selection rules and the conservation of charge in the monopole-fermion system are well understood. The baryon number violating processes which are catalyzed by monopoles have been examined in a variety of unified theories and the catalysis of $p \to e^+\pi^0$ seems to be a general effect. It remains only to calculate the magnitude of the cross section and the branching ratios!

ACKNOWLEDGEMENT

This work was supported by the Director, Office of Energy Research, Office of High Energy and Nuclear Physics, Division of High Energy Physics of the U.S. Department of Energy under Contract DE-AC03-76SF00098.

REFERENCES

1. A. Sen, talk given at this conference and Phys. Rev. $\underline{D28}$, 876 (1983).
2. Y. Kazama and A. Sen, "On the Conservation of Electric Charge Around a Monopole of Finite Size", Fermilab-Pub-83/58-THY, KUNS695 HE 83/12 (1983).
3. T.M. Yan, talk given at this conference and "Breaking of Conservation Laws Induced by Magnetic Monopole", CLNS 83/563 (1983).
4. D. London, talk given at this conference and D. London, J. Rosner, E. Weinberg, "Magnetic Monopoles with Z_N Charges", EFI 83-39-Chicago (1983).
5. A.N. Schellekens, talk given at this conference and preprint in preparation.
6. S. Dawson and A.N. Schellekens, Phys. Rev. $\underline{D27}$, 2119 (1983).
7. T. Walsh, talk given at this conference and T. Walsh, P. Weisz, and T.T. Wu, DESY 83-022 (1983).
8. V.A. Rubakov, Nucl. Phys. $\underline{B203}$, 311 (1982).
9. N. Craigie, talk given at this conference and "Status of Rubakov-Callan Effect", IC/83/131 (1983).
10. G. Grossman, Phys. Rev. Lett. $\underline{50}$, 464 (1983).
11. G. Fiorentini, Phys. Lett. $\underline{124B}$, 29 (1983).

MONOPOLE CATALYZED NUCLEON DECAY:

THE ASTROPHYSICAL CONNECTION

Edward W. Kolb

Theoretical Astrophysics Group
Fermi National Accelerator Laboratory
Batavia, Illinois 60510

If monopoles catalyze nucleon decay, limits on the product of the monopole flux and the catalysis cross section may be placed from "astrophysical" considerations. We review these limits and discuss their reliability.

One of the most remarkable developments in monopole physics over the past few years has been the observation by Callan[1] and Rubakov[2] that the monopole-fermion system opens a low-energy window to short-distance physics. In particular, they have demonstrated that for the massive magnetic monopoles expected in grand unified theories, $m_M \approx 10^{16}$ GeV, low-energy monopole fermion scattering could violate baryon number with a cross section on the order of a strong cross section.

The non-conservation of baryon number in the presence of a monopole results in monopole "catalysis" of nucleon decay. The nucleon lifetime in the presence of a monopole is very short compared to the usual GUT proton lifetime of $\sim 10^{31}$ years. In the presence of a monopole, nucleons decay at a rate[3]

$$\Gamma = n_N \sigma_{\Delta B} |v| \tag{1}$$

where n_N is the nucleon density, and $\sigma_{\Delta B}$ is the cross section for monopole nucleon scattering with concomitant nucleon decay. Since the rate is proportional to the nucleon density, the first guess for a place to look for an astrophysical effect is where the nucleon density is high. A good candidate for such a location is the interior of a neutron star.

Neutron stars are objects of roughly 1 solar mass $(2 \times 10^{33}$ g) at nuclear matter density $(\rho \sim 3 \times 10^{14}$ g cm^{-3}).[4] In the neutron star the monopole converts the rest mass of the nucleon into relativistic particles, releasing energy at a rate

$$L_M = m_N n_N \sigma_{\Delta B} |v| \qquad (2)$$

$$= 8.5 \times 10^{18} \sigma_0 |v| \text{ erg s}^{-1} \text{ monopole}^{-1} \quad ,$$

where we have defined σ_0 by

$$\sigma_{\Delta B} = \sigma_0 \, 10^{-27} \text{ cm}^2 \quad . \qquad (3)$$

If catalysis proceeds at a strong rate, $\sigma_0 \cong 1$.

The relative velocity appearing in Eq. (2) is of the order $0.1 - 0.3$, which is the Fermi velocity of the nucleons in the neutron star. One might naively expect for the exothermic decay reaction $\sigma_{\Delta B} \sim |v|^{-1}$, but it has been pointed out[5] that the situation, at least for low relative velocities, may be much more complicated. In this paper I will discuss limits on $\sigma_0 |v|$ for values of $|v|$ from $|v| \sim 1$ (in neutron stars) to $|v| \sim 10^{-5}$ (in the earth). When comparing limits, it should be remembered that the direct comparison of limits for very different values of $|v|$ may be dangerous. I should also note that catalysis of nucleon decay in neutron stars is closer to the "high-energy" regime where the calculations are more reliable than catalysis in detectors on earth where $|v| \lesssim 10^{-3}$.

The total energy released by the monopoles in the neutron star is proportional to the number of monopoles in the star. A terrestrial monopole hunter doesn't care about limits on the number of monopoles in neutron stars, but about the inferred limits on the galactic monopole flux. Unless there is some unexpected physics in the interior of neutron stars, the limit on the number in the star can be simply related to the monopole flux. Jeff Harvey discusses the possibility of unexpected physics in these proceedings.[6]

If we assume that all monopoles once captured by the neutron star are present (i.e. no M-$\overline{\text{M}}$ annihilation or ejection of monopoles[6]), the total number of monopoles in the neutron star is given by

$$N_M = \pi F_M A_c \tau f \qquad (4)$$

where F_M is the monopole flux, A_c is the capture area of the neutron star, τ is the age of the neutron star, and f is the fraction of monopoles incident on the neutron star which are captured. The capture area of the neutron star is larger than the geometrical area by a factor

$$\frac{A_c}{A_{NS}} = \frac{1 + 2M_{NS}G/v_M^2 R_{NS}}{1-R_S/R_{NS}} \quad , \tag{5}$$

$$\cong 4 \times 10^5 \ (v_M = 10^{-3})$$

where M_{NS} and R_{NS} are the mass and radius of the neutron star, v_M is the incident monopole velocity, and R_S is the Schwarzschild radius of the star. The neutron star has $M_{NS} \cong 1$ solar mass, $R_{NS} \cong 10$ km, and the initial velocity of the monopole should be a typical virial velocity of 10^{-3} c.

The initial monopole energy is $E_O = mv_M^2/2 = 5 \times 10^9$ $(m/10^{16}$ GeV$)(v_M/10^{-3}c)^2$ GeV. When the monopole hits the neutron star[7], it will have been accelerated to the escape velocity of the neutron star, $v_{esc} \sim 0.3$ c. If the monopole loses energy greater than E_O, it will eventually become trapped in the star. In scattering with electrons, the monopole suffers an energy loss of

$$dE/dx \cong 4\pi^2(n_e/p_e)|v| \quad , \tag{6}$$

$$\cong 10^{11} \ |v| \ \text{GeV cm}^{-1}$$

where n_e is the electron density ($\sim 10^{36}$ cm^{-3}), and p_e is the electron Fermi momentum (~ 50 MeV). Therefore for $|v|$ greater than escape velocity, the monopole will lose enough energy to become bound after traversing less than a centimeter. (It is not necessary for the monopole to be completely stopped in the first pass through the star, as long as it has lost enough energy to become gravitationally bound.) In addition to energy loss in electron collisions, there are additional energy loss mechanisms of less importance, but still sufficient to capture the monopole.[8] Therefore we will assume the fraction of incident monopoles captured by the neutron star f, is one. For f = 1, the number of monopoles in the neutron star is related to the galactic flux by

$$N_M = 5 \times 10^{20}F_{16} \quad , \tag{7}$$

where we have defined F_{16} as

$$F_{16} = F_M/10^{-16} \ \text{cm}^{-2} \ \text{s}^{-1} \ \text{sr}^{-1} \quad . \tag{8}$$

The luminosity of the neutron star due to monopole catalyzed nucleon decay will be

$$L^T = N_M L_M = 8.5 \times 10^{18} \ N_M \sigma_o|v| \ \text{erg s}^{-1} \ \text{monopole}^{-1} \tag{9}$$

$$= 4 \times 10^{39} \ F_{16} \ \sigma_o|v| \ \text{erg s}^{-1} \quad .$$

We will now discuss ways of limiting the luminosity of neutron stars, hence placing limits on $N_M \sigma_0 |v|$ and $F_{16} \sigma_0 |v|$.

Old neutron stars should be quite numerous in the galaxy. Published estimates of the density of old ($\tau \sim 10^{10}$ years) neutron stars range from $n_* = 10^{-1}$ to $n_* = 4 \times 10^{-3}$ pc^3.[9] The HEAO II "Einstein" observatory is capable of seeing neutron stars with a luminosity of 10^{31} erg s^{-1} at a distance of 100 pc. Beyond 100 pc, absorption by the interstellar medium prevents observations of a luminosity as low as 10^{31} erg s^{-1}. The surface density as low as old neutron stars within 100 pc is

$$\sigma \sim 0.4 \left(\frac{n_*}{4 \times 10^{-3} \text{ pc}^{-3}} \right) \text{ deg}^{-2} \quad , \tag{10}$$

or about 1 neutron star every 2.5 Einstein pictures. There have been surveys of "serendipitous" sources which cover ~ 50 deg^2 of sky and see no "blank field" x-ray sources.[10] We may interpret this lack of observation as evidence that the luminosity of old neutron stars is less than 10^{31} erg s^{-1}.[11] This limit is subject to two major uncertainties. If the number density of old neutron stars is an order of magnitude smaller than the lowest published limit, then an area of 50 deg^2 might only contain 1 or 2 sources. Therefore this limit is extremely sensitive to n_*. The second problem is that the surveys might have discriminated against sources with photon energy as low as expected from a 30 eV blackbody (the temperature of a neutron star with luminosity of 10^{31} erg s^{-1}). In order to answer these uncertainties a proposal has been made to perform a survey to cover a much larger area, and to look specifically for sources of low-energy photons.[12]

A second way to limit the luminosity of old neutron stars is to consider their effect upon the x-ray background. This was done in Ref. 11 by using the Silk limit, which is a limit on the total power radiated during the lifetime of the neutron star. The limit $P(E > 0.2 \text{ keV}) \lesssim 6 \times 10^{49}$ ergs over the lifetime of the neutron star results in a limit of $L^\gamma < 2 \times 10^{32}$ erg s^{-1} on the luminosity of old neutron stars. A similar bound may be found by calculating the present UV and x-ray flux from the old neutron stars in the galaxy and comparing it to the measured background. This method was used in Ref. 13, and resulted in a limit $L^\gamma \lesssim 3 \times 10^{30}$ erg s^{-1} on the luminosity of old neutron stars. However, this calculation did not consider the absorption of the photons by the interstellar medium. Recently there has been a better measurement of the soft x-ray background.[14] This observation is really four independent measurements of the flux in different energy bands, which in principle gives four independent limits on the luminosity of old neutron stars. This recent measurement has been used to limit the luminosity of old neutron stars after including corrections for absorption by the interstellar medium.[15] The limits found were

$L^Y \lesssim 1.4 \times 10^{32}$ erg s^{-1}, $L^Y \lesssim 9.7 \times 10^{31}$ erg s^{-1}, $L^Y \lesssim 1.8 \times 10^{32}$ erg s^{-1}, and $L^Y \lesssim 5.6 \times 10^{32}$ erg s^{-1} for the four energy bands measured. The above limits are limits on the total photon luminosities from the energy flux in different energy bands. Although the background limits are not as good as the serendipitous limit, they are not as sensitive to n_*. The flux from the background sources, hence the luminosity limit, is <u>linear</u> in n_*. If n_* is a factor of 10 smaller than expected, then the limit from the background is a factor of 10 higher, while the limit from the serendipitous sources may disappear altogether.

The above limits are on the <u>photon</u> luminosity, and to obtain limits on the total luminosity, it is necessary to correct for neutrino emission. The relationship between the observed limit on the photon luminosity, and the inferred limit on the total luminosity depends upon the equation of state and the structure of the neutron star. In Fig. 1, several examples of the conversion from photon to total luminosity are given. The curves π^a and π^b are equations of state with pion condensates, while the curves BPS, PS, I, II, IIB, III, A, B are equations of state without pion condensates.[16] For most models a lower photon luminosity results in a lower ratio of L^T/L^Y. Most corrections have been made using the π^a curve as the best model. Using the π^a curve the limit from

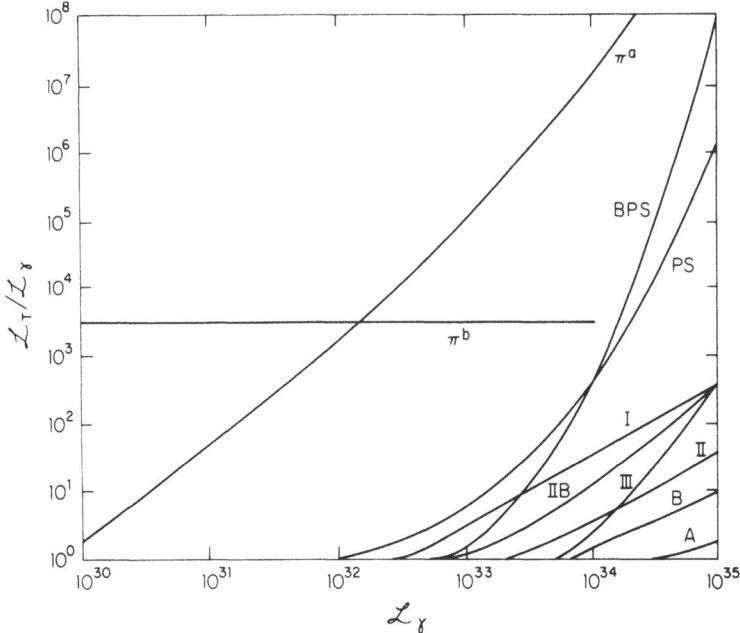

Fig. 1 The total (photon + neutrino) luminosity as a function of the photon luminosity. The different curves are for the different neutron star models discussed in ref. 16.

serendipitous   sources, $L^\gamma \lesssim 10^{31}$ erg s^{-1} translates into a limit
on the total luminosity of 10^{33} erg s^{-1}. Using Eq. 9 this results
in the limits $N_M \sigma_0 |v| \lesssim 10^{14}$ and $F_{16}\sigma_0|v| \lesssim 2.5 \times 10^{-7}$. The limit
from the background flux, $L^\gamma \lesssim 10^{32}$ erg s^{-1} translates   into a
limit   on L^T of 10^{36} erg s^{-1},   or $N_M\sigma_0|v| \lesssim 10^{17}$ and
$F_{16}\sigma_0|v| \lesssim 2.5 \times 10^{-4}$.   With the prejudice that π^a is the best
equation of state, an improvement of an order of magnitude in L^γ
resulted in a difference of 10^3 in the flux limit.

There is a window of vulnerability in the above arguments.
If the catalysis proceeds quickly enough, the neutron star might
be completely eaten in a time less than the age of the galaxy.[17]
The rate of nucleon decay is

$$R_M = 8.5 \times 10^{21}\sigma_0|v| \ s^{-1} \ \text{monopole}^{-1} \tag{11}$$

and the number of monopoles in the neutron star after a time t is

$$N_M = 2 \times 10^3 \ F_{16} \ t(sec) \quad . \tag{12}$$

Therefore the number of nucleons that have been eaten in a time t
is

$$N = \int_0^t R_M N_M \ dt$$

$$= 8.5 \times 10^{24} \ F_{16}\sigma_0|v|t^2, \tag{13}$$

where t is in seconds.   The monopole will eat the neutron star
$(N = 10^{57})$ in a time

$$t = \frac{4 \times 10^9}{(F_{16}\sigma_0|v|)^{1/2}} \ \text{years}, \tag{14}$$

Therefore if $F_{16}\sigma_0|v| \gtrsim 10^{-3}$, monopoles will eat the neutron star
in an age less than the age of the galaxy, and limits on the
luminosity of old neutron stars cannot be used to limit the
monopole flux.

There are several ways to close the window of vulnerability.
It is not at all clear that the neutron star will sit around
quietly as it is eaten.   There should be an instability in the
neutron star when its mass becomes less than about half its
initial mass.   This instability might result in a neutron star
explosion.   The details of this explosion have not been worked
out, but it might be suspected that such events are ruled out.
Another way to close the window of vulnerability is to look at
known pulsars.

There exist many nearby (d $\lesssim$ 100 pc) young (spin-down age $\sim 10^7$ years) pulsars that have limits on their luminosity. One such example is PSR 1929+10, which is a 3×10^6 yr old pulsar at a distance of $\sim$ 60 pc, and a maximum luminosity of 3×10^{30} erg s^{-1}. This pulsar has been used[18] to limit $N_M \sigma_o |v|$ to be less than 3.5×10^{11}. For a neutron star of age $3 \times 10^{-4}\tau$(galaxy), $N_M = 10^{17} F_{16}$. This results in the limit $F_{16}\sigma_o|v| \lesssim 2.5 \times 10^{-6}$. This limit is an order of magnitude less stringent than the serendipitous bound, but has several advantages. It is insensitive to the assumption of the number density of old neutron stars. For the π^a equation of state, there is negligible L^T/L^γ correction for this low luminosity. Finally it closes the window of vulnerability discussed above. It does depend on the assumed distance to the pulsar and the assumed age of the pulsar. In the analysis a distance of 60 pc was assumed, but a recent measurement has it as far away as 250 pc,[19] in which case the limit on L^γ would be 5×10^{31} erg s^{-1}, which due to neutrino emission (using π^a) results in a limit of 5×10^{34} erg s^{-1}. This results in the limits $N_M\sigma_o|v| \lesssim 6 \times 10^{15}$, and $F_{16}\sigma_o|v| \lesssim 4 \times 10^{-2}$. The true distance to PSR 1929+10 is probably somewhere in between the above extremes. At any rate, there are several other "nearby" pulsars with similar luminosity limits, and the limits from known pulsars are more reliable than the limits from (unseen) unknown pulsars, and it is worth paying the price of having τ/τ(galaxy) fewer monopoles than in the old neutron stars.

The observation of known pulsars is particularly interesting if we relax the assumption that the only monopoles in the neutron star have been accreted during the neutron star phase. It is believed that massive stars were the progenitors of neutron stars, and recent estimates of the number of monopoles captured in the pre neutron star stage is[20] $N_M \cong 10^{22}F_{16}$, independent of the age of the neutron star. In this case, the limit $N_M\sigma_o|v| \lesssim 3.5 \times 10^{11}$ from PSR 1929+10 results in $F_{16}\sigma_o|v| \lesssim 3.5 \times 10^{-11}$.

Neutron stars are not the only condensed objects where monopole catalyzed nucleon decay might be important. If there were 4×10^{19} monopoles in white dwarfs, the white dwarf luminosity would be about 10^{30} erg s^{-1}.[21] Remarkably, there are no observations of white dwarfs with luminosity below 10^{30} erg s^{-1}.[22] This results in the limit $F_{16}\sigma_o|v| \lesssim 2 \times 10^{-4}$.[21] There are several uncertainties in the above arguments. It is yet to be demonstrated that M-$\overline{\text{M}}$ annihilations do not reduce the number of monopoles in the white dwarfs. Possible nuclear suppression factors have not been considered, and since the relative velocity of the monopole - nucleon systems may be small, and the interior of the white dwarfs contain spinless nuclei, there may be a suppression of 10^{-2} - 10^{-6}.[5] Finally if the interior of the white dwarf is hot enough, neutrino emission may increase in importance. However, the possibility of monopoles keeping white dwarfs hot is interesting, as it may answer the old astrophysical question "where are the dim degenerates?"

There is another limit on monopole catalyzed nucleon decay from everyone's favorite astrophysical object -- the earth.[23] Monopoles in the earth release energy through catalysis at a rate

$$L_M = n_N \sigma_0 |v|$$

$$= 5 \times 10^4 \ \sigma_0 |v| \ \mathrm{erg \ s^{-1} \ monopole^{-1}} \quad , \tag{15}$$

for $\rho \sim 3 \ \mathrm{g \ cm^{-3}}$. The limit on the luminosity of the earth $L_\oplus \lesssim 10^{20} \ \mathrm{erg \ s^{-1}}$, implies $N_M \sigma_0 |v| \lesssim 2 \times 10^{15}$. The number of monopoles captured by the earth is given by Eqs. (4) and (5) using the earth radius and mass:

$$N_M = \pi F_M A_c \tau f_\oplus$$

$$= 2 \times 10^{20} \ F_{16} f_\oplus, \tag{16}$$

where $f_\oplus$ is the fraction of incident monopoles captured in the earth. In neutron stars, f is 1 but in the earth only monopoles with velocities less than about $3 \times 10^{-5} \ (\mathrm{m}/10^{16} \ \mathrm{GeV})^{-1}$ will be trapped[23], assuming $dE/dx = 30 \ \mathrm{GeV \ cm^2 \ g^{-1}} \ \beta\rho$, where β is the monopole velocity and ρ is the density in the earth.[24] If the local flux of monopoles has a velocity typical of the galactic virial velocity, $v_M = 10^{-3}$, then $f_\oplus \cong 10^{-6}(\mathrm{m}/10^{16} \ \mathrm{GeV})^{-4}$. If the local flux of monopoles has a velocity typical of the escape velocity from the sun, $v_M \simeq 10^{-4}$, then $f_\oplus \simeq 10^{-3} \ (\mathrm{m}/10^{16} \ \mathrm{GeV})^{-4}$. Therefore the limits from earth heat are[23]

$$F_{16} \ \sigma_0 |v| \lesssim 10^{-5}/f$$

$$\lesssim \begin{cases} 10(\mathrm{m}/10^{16} \ \mathrm{GeV})^4 & (v_M = 10^{-3}) \\ 10^{-2}(\mathrm{m}/10^{16} \ \mathrm{GeV})^4 & (v_M = 10^{-4}) \end{cases} \quad . \tag{17}$$

A similar analysis has been done for Jupiter,[23] with the result $F_{16}\sigma_0|v| \lesssim 5 \times 10^{-3}$ for $v_M = 10^{-3} - 10^{-4}$. Finally, it should be remembered that this limit may be susceptible to the low velocity suppression of the cross section.

The conclusions of the talk are given in Table 1. I think it would be worthwhile to repeat the uncertainties in the individual limits. It also should be remembered that quoted flux limits (except for the earth limit) are for typical galactic fluxes. If one believes the local flux is enhanced, it is necessary to make a correction to get the limit on the local flux. It is also dangerous to compare limits with different relative velocities.

Table 1. Astrophysical Limits on Monopole Catalysis

OBJECT	$\lvert v \rvert$	L^Y erg s^{-1}	LT/LY	$N_M \sigma_o \lvert v \rvert$	$F_{16} \sigma_o \lvert v \rvert$	Uncertainties
Old Neutron Stars Serendipitous[11]	0.3	$< 10^{31}$	10^2	$< 10^{14}$	$< 2.5 \times 10^{-7}$	Annihilation ; L^T/L^Y; n*
Old Neutron Stars Background[13]	0.3	$< 3 \times 10^{30}$	1	$< 3 \times 10^{11}$	$< 7.5 \times 10^{-10}$	ISM; n*; L^T/L^Y; Annihilation
Old Neutron Stars Background[11]	0.3	$< 2 \times 10^{32}$	5×10^3	$< 10^{17}$	$< 2.5 \times 10^{-4}$	ISM; n*; L^T/L^Y; Annihilation
Old Neutron Stars Background[15]	0.3	$< 10^{32}$	10^3	$< 10^{16}$	$< 2.5 \times 10^{-5}$	ISM; n*; L^T/L^Y; Annihilation
Young Neutron Stars (1929+10, etc)[18]	0.3	$< 3 \times 10^{30}$	1	$< 3 \times 10^{11}$	$< 2.5 \times 10^{-6}$	Age; distance; L^T/L^Y; Annihilation
Neutron Stars Main-Sequence Capture[20]	0.3	$< 3 \times 10^{30}$	1	$< 3 \times 10^{11}$	$< 3.5 \times 10^{-11}$	Above, plus main sequence evolution of neutron star
White Dwarfs[21]	10^{-3}	$< 10^{30}$	1	$< 4 \times 10^{19}$	$< 2 \times 10^{-4}$	Annihilation; L^T/L^Y; internal structure
Earth[23]	10^{-5}	$< 10^{20}$	---	$< 2 \times 10^{15}$	< 10	Monopole mass; low velocity; capture calculation; nuclear suppression
Jupiter[23]	10^{-3}–10^{-4}	$< 10^{20}$	---		$< 5 \times 10^{-3}$	nuclear suppression

All neutron star limits have the uncertainty of possible $M\text{-}\overline{M}$ annihilations reducing the number of monopoles. All the neutron star limits have used the π^a equation of state to correct for neutrino emission. If one uses the π^b equation of state, the L^T/L^γ correction would be about 10^3 for all the neutron star limits. The serendipitous limit is very sensitive to n_*, and if n_* is an order of magnitude smaller than expected, it is useless. The background limits are <u>linear</u> in n_*, but depend on absorption by the interstellar medium $\overline{(\text{ISM})}$. The limit in Ref. 13 did not correct for absorption, and the limit on L^γ is probably too severe. The limits from young neutron stars depend on estimates of their age and distance. The limit on the monopole flux if we include the main sequence capture is very good, but it assumes an evolutionary history for the neutron star progenitor, it assumes that monopoles do not annihilate and it assumes that monopoles were not expelled in the formation of the neutron star. The limit from white dwarfs requires more work before the uncertainties are settled. The limits from the Earth and Jupiter are sensitive to the monopole mass. If the mass is much larger than 10^{16} GeV, it is harder to capture the monopoles, and the flux limit is relaxed.

Terrestrial detection of monopole-catalyzed nucleon decay would be a remarkable discovery. It seems impossible in the near future if $F_{16}\sigma_0|v|$ is much less than one. In Table 1 are several astrophysical arguments that suggest that $F_{16}\sigma_0|v|$ is <u>much</u> less than one. Although there are uncertainties that might reduce any of the limits in Table 1 by a factor of 10 or so, it would seem unlikely that the independent astrophysical uncertainties conspire to remove <u>all</u> the limits to allow $F_{16}\sigma_0|v| \gtrsim 1$.

ACKNOWLEDGEMENTS

I would like to thank Stirling Colgate, Jeff Harvey, David Seckel, and Michael Turner for many conversations on the subjects in this talk. This work was supported by NASA and the DoE.

REFERENCES

1. C. Callan, in this volume, and references therein.
2. V.A. Rubakov, Nucl. Phys. <u>B203</u>, 311 (1982).
3. Recently Goldhaber (A.S. Goldhaber, in proceedings of the 4th Workshop on Grand Unification) has suggested that monopole catalyzed decay is more complicated than Eq. (1) indicates. He has discussed the possibility that the reaction proceeds through a metastable state, and although the cross section might be large, the catalysis rate might be small.
4. For a review of neutron star properties, see G. Baym and C.

Pethick, Ann. Rev. Ast. and Astro. 17, 415 (1979).

5. J. Arafune and M. Fukugita, Phys. Rev. Lett. 50, 1901 (1983).

6. J.A. Harvey, in this volume.

7. For monopoles of mass $\gtrsim 10^{16}$ GeV, the ratio of the gravita-tional force to the magnetic force at the surface of the neutron star is much greater than one. This prevents magnetic deflection of the incident monopole.

8. Assuming the only energy loss mechanism is through catalysis, Bais, et al. [F.A. Bais, J. Ellis, D.V. Nanopoulos, and K. Olive, Nucl. Phys. B219, 189 (1983)] have shown that it is possible for the monopole to pass through the neutron star. However I don't think it is reasonable to assume the only energy loss mechanism is via catalysis.

9. J. Ostriker, M. Rees, J. Silk, Ast. Lett. 6, 179 (1970); D. Lamb, F. Lamb, D. Pines, Nature 246, 52 (1973); J. Hills, Ap. J. 219, 550 (1978).

10. J.T. Stocke, et al., Ap. J. 273, 458 (1983).

11. E.W. Kolb, S.A. Colgate, J.A. Harvey, Phys. Rev. Lett. 49, 1373 (1982).

12. F.A. Cordova, E.W. Kolb, D.L. Tubbs, HEAO-2 Guest Investigator Proposal.

13. S. Dimopoulos, J. Preskill, F. Wilczek, Phys. Lett. 119B, 320 (1982).

14. D. McCammon, D.N. Burrows, W.T. Sanders, W.L. Kraushaar, Ap. J. 269, 107 (1983).

15. E.W. Kolb and M.S. Turner, Fermilab report (1983).

16. References to the original papers for the different equations of state may be found in K.A Van Riper and D.Q. Lamb, Ap. J. 244, L13 (1981) for π^a, PS, and BPS; M.B. Richardson, et al., Ap. J. 255, 624 (1982) for π^b; and S. Tsuruta in Canadian Journal of Physics 44, 1863 (1966) for I, II, IIB, III.

17. K.A. Olive and D.N. Schramm, Phys. Lett. 130B, 267 (1983).

18. K. Freese, M.S. Turner, and D.N. Schramm, Phys. Rev. Lett. 51, 1625 (1983).

19. A recent determination of the distance to PSR 1929+10 (≈ 250 pc) was made by D.C. Backer and R.A. Sramek, Ap. J. 260, 512 (1982). The distance determination is uncertain, and a good guess might be 100±100 pc.

20. This has been suggested by Bais et al. (Ref. 8) and Freese, Turner, and Schramm (Ref. 18).

21. K. Freese and R. Kron, University of Chicago preprint (1983).

22. For a review of White Dwarfs, see J. Liebert, Ann. Rev. Ast. and Astrophys. 18, 363 (1980).

23. M.S. Turner, Nature 302, 804 (1983).

24. S. Ahlen and K. Kinoshita, Phys. Rev. D26, 2347 (1982).

EXPERIMENTAL LIMITS ON MONOPOLE CATALYSIS OF NUCLEON DECAY

Steven M. Errede

Randall Laboratory of Physics
The University of Michigan
Ann Arbor, Michigan 48109

INTRODUCTION

During the past year since the Wingspread Monopole Workshop[1], perhaps the most significant and novel experimental advance in the now more than half-century "quest" for the detection of magnetic monopoles has come from terrestrial experiments searching for evidence of monopole catalysis of nucleon decay (MCND). A large class of grand and/or super-unified theories which predict nucleon decay and the existence of super-heavy magnetic monopoles of mass $M_m \sim 10^{16}$ GeV/c^2 as specific features of such theories[2], also predict that monopoles in the presence of hadronic matter may catalyze baryon number violating processes with cross sections typical of that for the strong interactions[3]!

Experiments designed to search for evidence of "spontaneous" nucleon decay are sensitive to monopole-catalyzed nucleon decay. Several of these underground detectors have now been in operation for well over a year; it is quite natural, and in fact important that these same experiments simultaneously search for evidence of monopole catalysis of nucleon decay.

Soon after the theoretical discovery[3] that monopole catalysis of nucleon decay may happen it was realized[4] that neutron stars could provide stringent astrophysical limits on the flux of super-heavy magnetic monopoles, because once captured by a neutron star, a monopole could literally "eat" grams of nuclear matter per second in the ultra-high density environment of the interior of a neutron star ($\rho_{ns} \sim 3 \times 10^{14}$ gms/cm^3), thereby releasing ~ 1 GeV per catalyzed nucleon, which in turn would heat up the star, causing it to radiate large amounts of thermal energy in the form

251

of X-rays and neutrinos. From limits on the X-ray luminosities of such objects, $L_{ns} < 10^{33}$ ergs/s, monopole flux limits of order $F_m < 10^{-22}$ cm^{-2} s^{-1} sr^{-1} were obtained[4] for catalysis cross sections $\sigma_c \sim 10^{-27}$ cm^2. If MCND truely exists, then monopole fluxes this small are hopelessly out of reach for any present or future monopole search experiment, whether it be via induction, dE/dX techniques, and especially MCND experiments.

However, we would like to stress that some of the assumptions made in obtaining monopole flux limits from neutron stars and white dwarfs may not necessarily be valid in an absolute sense, since the interior dynamics and structure of such objects are not truely known in a direct experimental manner. These assumptions also have inherent uncertainties to which the flux limits from neutron stars are sensitive, but not by many orders of magnitude[5]. There exist also many theoretical uncertainties associated with the details of the monopole catalysis reaction. A number of possible scenarios exist[6] in which these rather discouraging limits may be significantly ameliorated, making it sensible for all forms of terrestrial monopole searches to be carried out.

It is important to search specifically for terrestrial evidence of MCND because it is one of only a very few experimental "windows" in which grand or super-unified theories may be subjected to direct experimental test under laboratory (low energy) conditions. Because the experimental background for multiple catalysis interactions is very low, the observation of a single unambiguous example of multiple catalyzed nucleon decay would provide strong evidence for the existence of magnetic monopoles as well as nucleon decay, which in turn would imply the existence of some form of grand or super-unification of the forces of nature, as well as far-reaching implications for the early universe. Knowledge of the modes of monopole-catalyzed decay would yield information about generalized Cabibbo-Kobayashi-Maskawa (CKM) mixing in grand unified theories.

EXPERIMENTAL ASPECTS AND CONSIDERATIONS FOR MCND

Experimentally, the spatial-temporal requirements for the detection of magnetic monopoles are substantially different from those associated with the "prompt" physics studied in most high energy and cosmic ray experiments, because the expected velocities for super-heavy magnetic monopoles lie within the range $10^{-4} < \beta_m < 10^{-2}$. Exemplary transit times across a detector with path lengths of L = 3m, 30m for different monopole velocities are given in Table I.

Thus, the detection of magnetic monopoles via dE/dX or MCND requires specialized electronics which are "live" for rather long

Table 1. Transit times for a monopole passing through a detector.

Velocity	L = 3	L = 30m
$\beta = 1.0$	$\Delta t = 10$ ns	$\Delta t = 100$ ns
$\beta = 10^{-1}$	$\Delta t = 100$ ns	$\Delta t = 1.0$ μs
$\beta = 10^{-2}$	$\Delta t = 1.0$ μs	$\Delta t = 10$ μs
$\beta = 10^{-3}$	$\Delta t = 10$ μs	$\Delta t = 100$ μs
$\beta = 10^{-4}$	$\Delta t = 100$ μs	$\Delta t = 1.0$ ms
$\beta = 10^{-5}$	$\Delta t = 1.0$ ms	$\Delta t = 10$ ms

periods of time, up to 10 ms/event compared to conventional experiments which have live-times of order 100 -200 ns per event. Typical nucleon decay experiments have live times of order 5 - 10 μs for the purpose of detecting electrons from $\mu \rightarrow e\nu\bar{\nu}$ decay, with detector dead times of ~ few ms after this period for readout of the detector.

There are a broad range of possibilities for MCND with which the experimentalist must be concerned. For example, nucleon decay experiments may have poor detection efficiency for MCND if the catalysis cross section, σ_c and monopole velocity, β_m conspire in such a way that multiple catalysis interactions occur over a long period of time. The first interaction may trigger the detector and be observed as an "entering" event, while all subsequent catalysis interactions in the detector may not be observed due to detector dead time, as shown in Fig. 1. Such events could easily be missed, due to the ever-present background of cosmic ray muons and their not-infrequent interactions in the neighboring rock surrounding underground nucleon decay detectors.

An easy solution for covering this portion of β_m - σ_c "phase-space" is simply to install additional electronics to extend the detector live time, or at least to cover the dead time of the detector while the "prompt" electronics is read out, thereby making the experiment essentially "dead-timeless".

Another possible experimental scenario for MCND is the case where the catalysis cross section may be small. In this situation, it is likely that only single MCND interactions will be observed in a given detector. We shall see that single MCND interactions may bear a strong resemblance to atmospheric neutrino interactions having energies $E_\nu \sim 1$ GeV and also possibly spontaneous nucleon decay.

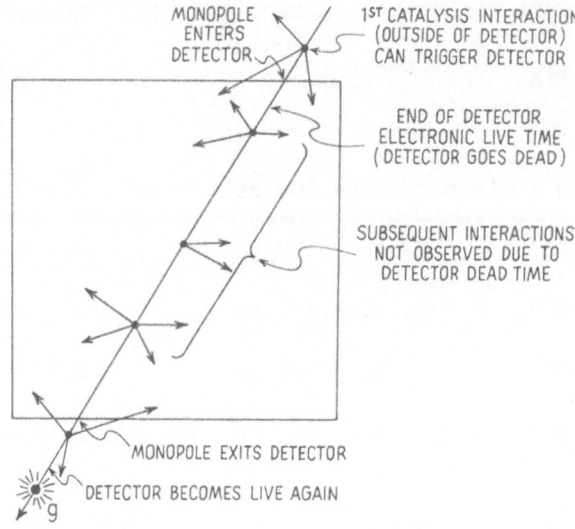

Fig. 1 Experimental signature of monopole-catalyzed nucleon
        decay(s) in a detector.

Catalysis Cross Sections and Interaction Lengths

Experimentalists need also concern themselves with the range
of expected catalysis cross sections, σ_c. A first-order estimate
of the catalysis cross section is given by[7]

$$\sigma_c \; \simeq \; \frac{1}{\beta}\Big[\frac{\sigma_o}{E_o^2}\Big](hc/2\pi)^2 \quad \text{for } \beta \ll 1 \tag{1}$$

where $1/\beta$ is the usual flux factor with β representing the relative velocity between the monopole and nucleon. $E_o \simeq M_p c^2 \sim 1$ GeV
is a scale parameter, σ_o is the dimensionless reduced cross
section, whose value is estimated[7] to be $10^{-6} < \sigma_o < 1$, with
$\sigma_o \sim 10^{-4}$ preferred. It can be seen that catalysis cross sections
comparable to that for the strong interactions are predicted.

Perhaps of more concern to the experimentalist is the expected range of interaction lengths, or mean free path between catalysis interactions, λ_c. If the interaction length is short compared
to the detector size, i.e. $\lambda_c \ll d$ corresponding to large catalysis cross sections, then multiple interactions in the detector
will dominate. However, if the interaction length is large compared to the detector size, $\lambda_c \gg d$, then likely only single
interactions will be observed. For a detector composed entirely of
one type of material, e.g. iron, the catalysis cross section may

be defined in terms of the experimentally measureable interaction length, λ_c, as

$$\sigma_c = \frac{1}{n\lambda_c} \tag{2}$$

where n is the nucleon number density of the detector. In terms of Eq. (1), the interaction length λ_c (cm) is given by

$$\lambda_c = \frac{4300\beta}{\rho\sigma_0} \tag{3}$$

where ρ (in g/cm^3) is the mass density of detector material.

However, we note that most detectors used to search for evidence of MCND are typically composed of more than one type of element to make up their bulk mass. For example, water Cherenkov detectors obviously are composed of H_2O. The Soudan-I detector is composed of taconite-loaded concrete interspaced with proportional tubes. Still other detectors such as the NUSEX and KGF detectors are made up of thin sheets of iron interspaced with various types of gas tracking detectors. The use of Eqs. (2)-(3) for detectors with composite materials is justifiable only if the catalysis cross section for nucleons on different nuclei are the same. In this situation, n is understood to be the average nucleon number density, ρ the average mass density, and σ_c the average catalysis cross section per nucleon. We shall see shortly that several phenomena complicate this issue considerably.

Velocity-Dependent Suppression/Enhancement Factors for MCND

There may exist important quantum mechanical effects which dramatically suppress or enhance the catalysis cross section for different nuclei. To understand this phenomenon more clearly, consider first the classical case of a nucleus of charge Ze incident on a monopole at rest, fixed at the origin. Classically, the motion of the nucleus is that of a particle spiralling inward on a right-circular cone of radius $r(z = \infty) = b$ where b is the impact parameter, as shown in Fig. 2. One can solve the equations of motion[8] to find $\dot{r} = \pm v\sqrt{1 - (b/r)^2}$. The distance of closest approach occurs when r = b, the initial impact parameter. Since all values of b are allowed, monopoles can easily penetrate to within distances of order b ~ 1 fermi, the size of the nuclear radius. Thus, it can be seen that within in the context of classical mechanics, no suppression (or enhancement) of the catalysis process will occur.

Quantum mechanically it has been shown[8,9] that there exists an angular momentum barrier due to the EM field angular momentum associated with the monopole-electric charge system, which is non-vanishing even in the lowest total angular momèntum J = 0 state, the total angular momentum operator (squared) defined as

$$J^2 = \left[\vec{r} \times (\vec{p} - Ze\vec{A})\right]^2 - 2q\vec{s}\cdot\hat{r}(1 - \kappa) \tag{4}$$

where $q \equiv Zeg/4\pi$ and κ = nuclear anomalous magnetic moment.

The solutions to the Schrödinger equation for this problem are of the usual form

$$\psi(r) = \sum_{\nu,m} C_{\nu m}\, \Phi_{\nu m}(\theta,\phi)\, R_{E\nu}(r) \tag{5}$$

Note that here $\psi(r)$ is <u>not</u> an incident plane wave. The $\Phi_{\nu m}(\theta,\phi)$ are angular eigenfunctions with eigenvalues of the total angular momentum-squared operator J^2 and Z-component of the total angular momentum operator, J_Z given by

$$J^2\Phi_{\nu m}(\theta,\phi) = \nu(\nu + 1)(h/2\pi)^2\, \Phi_{\nu m}(\theta,\phi) \tag{6}$$

$$J_z\Phi_{\nu m}(\theta,\phi) = mh/2\pi\, \Phi_{\nu m}(\theta,\phi) \tag{7}$$

The functions $R_{E\nu}(r)$ are solutions of the radial Bessel equation, specifically, they are spherical Bessel functions of the first and second kind, i.e. $j_\nu(kr)$, $n_\nu(kr)$, with $k = \sqrt{2ME}/h/2\pi \simeq \beta Mc^2/hc/2\pi$. Note that since $j_\nu(r \rightarrow 0) \rightarrow 0$, $n_\nu(r \rightarrow 0) \rightarrow \infty$ for k fixed, $k > 0$, the solutions $n_\nu(kr)$ are not allowed, and are therefore discarded. For low velocities, $kr \ll 1$ (valid for $r \sim 1$ fm, $\beta \sim 10^{-3}$) then

$$j_\nu(kr) \sim \frac{(kr)^\nu}{\Gamma(2\nu + 1)} \tag{8}$$

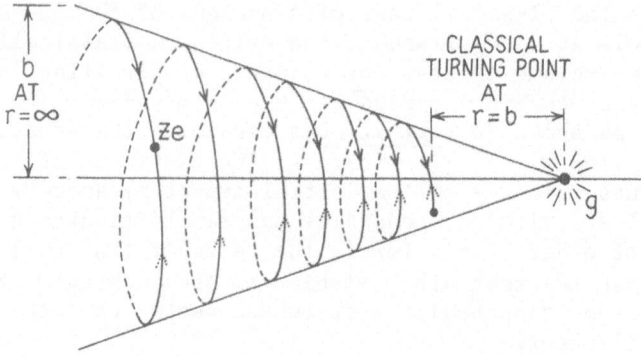

Fig. 2 Motion of particle of charge Ze incident on a monopole.

For the lowest <u>orbital</u> angular momentum state, when $L = |q| - s$ the total angular momentum operator (squared) becomes

$$J^2 = |q|\{1 - 2(1 + \kappa)s\} \tag{9}$$

$$= |q| \quad \text{for } s = 0 \quad \text{nuclei}$$

$$= -|q|\kappa \quad \text{for } s = 1/2 \text{ nuclei}$$

$$= \nu(\nu + 1)h/2\pi^2 \quad . \tag{10}$$

Note also that the order ν is not necessarily an integer, as in the case of more familiar problems. Furthermore, ν need not even be real!

$$\nu = \sqrt{Z/2 + 1/4} - 1/2 \quad \text{for } s = 0 \quad \text{nuclei} \tag{11}$$

$$\nu = -1/2 \pm i \sqrt{3/4} \quad \text{for } s = 1/2 \text{ nuclei with } \kappa = +1 \tag{12}$$

Now Arafune and Fukugita[9] have obtained a first-order estimate of the suppression/enhancement factor $F(\beta)$ for catalysis in nuclei, defined as:

$$F(\beta) \equiv |\text{wave function overlap}|^2 = |\psi(r)|^2 \quad . \tag{13}$$

The length scale is taken to be the nuclear radius. The effective catalysis cross section is obtained simply by multiplying the original "bare" catalysis cross section by $F(\beta)$:

$$\sigma_c = F(\beta)\sigma_c^{\text{bare}} \quad . \tag{14}$$

The suppression/enhancement effects for monopole catalysis were considered for the case of the lowest partial wave with minimal orbital angular momentum, because this is where the nucleus can penetrate closest to the monopole, and thus where the probability for the catalysis reaction to take place is the highest.

Approximating $F(\beta) = |\psi(r)|^2$ by using the argument kr of the small argument expansion of the Bessel function $j_\nu(kr)$, $(kr \ll 1)$ and ignoring all other numerical factors, they define $F(\beta)$ as

$$F(\beta) \equiv (kr)^{2\text{Re}\nu} = (\beta/\beta_0)^{2\text{Re}\nu} \tag{15}$$

where $\beta_0 \equiv hc/2\pi (r_0 M_n c^2 A^{4/3})$, $r_0 = 1.2$ fm, $M_n c^2 = $ nucleon mass, and $A = $ atomic number. The physical meaning of the (cutoff) parameter β_0 is the velocity threshold below which the catalysis reaction is suppressed/enhanced, and above which no such effects are observed.

As examples, consider the suppression/enhancement factors $F(\beta)$ for oxygen and hydrogen nuclei[11]:

Oxygen $F(\beta) \sim 10^{-2}$ at $\beta = 10^{-3}$ $Re(\nu_{Oxy}) = 1.56$, $\beta_0 = 0.1750$

$ \sim 10^{-5}$ at $\beta = 10^{-4}$

Hydrogen: $F(\beta) \sim 200$ at $\beta = 10^{-3}$ $Re(\nu_H)$ $= -0.50$, $\beta_0 = 0.0043$

$ \sim 10^4$ at $\beta = 10^{-4}$

In terms of the catalysis cross section, $\sigma_c = F(\beta) \cdot \sigma_c^{bare}$,

$$\sigma_c \simeq (\beta/\beta_0)^{2Re\nu} \frac{1}{\beta}\left[\frac{\sigma_0}{E_0^2}\right](hc/2\pi)^2 \tag{16}$$

The catalysis cross sections for oxygen and hydrogen are given by

Oxygen: $\sigma_c^{Oxy} \simeq 0.4(\beta/\beta_0)^{3.12} \frac{1}{\beta}\sigma_0$ (mb) $\simeq 13(\beta^{2.1})\sigma_0$ (mb)

Hydrogen: σ_c^H $\simeq 0.4(\beta/\beta_0)^{-1.00} \frac{1}{\beta}\sigma_0$ (mb) $\simeq 92(1/\beta^2)\sigma_0$ (mb)

The ratio of hydrogen to oxygen catalysis cross sections is $\sigma_c^H/\sigma_c^{Oxy} \sim 7.0/\beta^{4.1}$ which becomes very large at low velocities.

Thus, suppression of the catalysis reaction for higher Z elements with spin 0 nuclei, and enhancement of the catalysis reaction for elements with spin 1/2 nuclei and positive anomalous magnetic moment $\kappa > 0$ at low monopole velocities are expected. Note these numbers are sensitive to the value used for $r = r_0 A^{1/3}$.

One may well ask, whether or not there may exist similar enhancement or suppression effects for MCND associated with larger quantum mechanical systems, such as atoms and molecules? This is as yet a largely unexplored area of physics, although some theoretical work has already been done. S. Parke has shown[10], using a variational electron wave function in the Born-Oppenheimer approximation, that there exists no barrier to penetration of an isolated hydrogen atom by a slow magnetic monopole.

The dependence of the velocity threshold parameter β_0 as a function of atomic number A for various nuclei is shown in Fig. 3. Figure 4 shows the dependence of the order ν as a function of nuclear charge Z. Figure 5 shows contours of constant velocity for the suppression factor $F(\beta)$ as a function of nuclear charge Z.

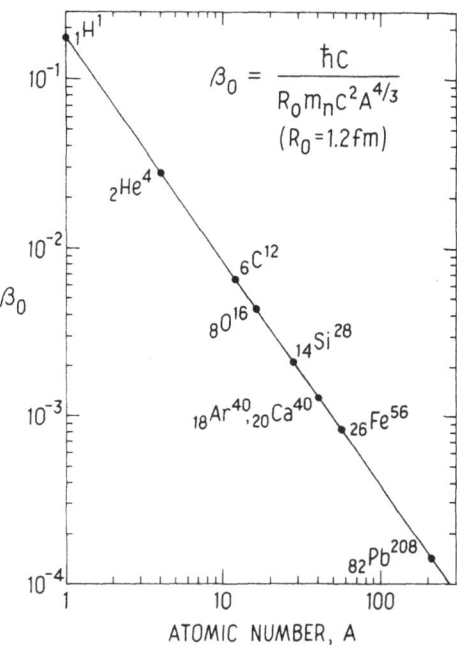

Fig. 3 Velocity threshold parameter β_0 vs. atomic number A for different nuclei.

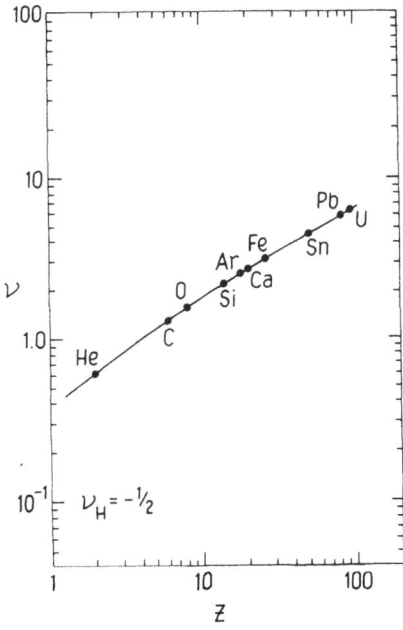

Fig. 4 Dependence of the order ν of the spherical Bessel function $j_\nu(kr)$ vs. the nuclear charge Z for different nuclei.

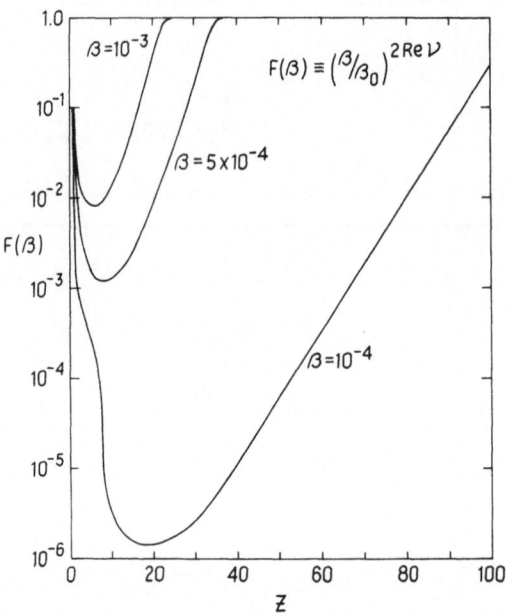

Fig. 5   Suppression factor $F(\beta)$ as a function of nuclear charge Z
(Contours of constant monopole velocity, β).

Monopole Catalysis Event Rate for Terrestrial Experiments

The number of expected monopole-catalyzed events observed in
a given detector is given by

$$N_0 = F_m \times A \times \Omega \times \Delta t \times \varepsilon(g, \beta_m, \lambda_c) \times \varepsilon_d \qquad (17)$$

where

F_m = monopole flux
A   = effective cross-sectional area of the detector for an
      isotropic flux of monopoles
Ω = solid angle = 4π
Δt = detector live time
$\varepsilon(g, \beta_m, \lambda_c)$ = efficiency for observing = 1, $\geqslant$ 2, $\geqslant$ 3,...,
      interactions in the detector. Function of: geometrical
      cuts (e.g. fid. vol. requirement); monopole velocity, β_m
      - time window associated with detector electronics
      $\Delta t_e \simeq$ 5 - 10μs, with deadtime for event data readout.;
      catalysis interaction length λ_c - (Poisson) probability
      for single (=1) or multiple ($\geqslant$ 2) interactions in
      detector within electronics time window.
ε_d = inclusive detection efficiency for triggering on MCND
      event, and passing event through analysis programs.

EXPERIMENTAL SIGNATURES OF MONOPOLE CATALYSIS OF NUCLEON DECAY

Monopole Catalyzed Decay Modes and Final States.

The most general theoretical expectations for baryon-violating catalysis reactions predict[3,7,11]:

$$M + N \rightarrow M' + e^+ + \text{pion(s)}, \text{ where } N = p \text{ or } n \qquad (18)$$

The production of pions in the final state is due to the excretion of isospin charge from the monopole core. The process of monopole catalysis of nucleon decay is shown in Fig. 6.

In a broader context, for the case of monopole catalysis in nuclei of atomic number A, the monopole catalysis reaction is

$$M + A(Z,N) \rightarrow M' + A'(Z',N') + e^+ + \text{pion(s)} \qquad (19)$$

Theoretical predictions differ somewhat on specific catalyzed decay modes and branching ratios for final states, although there is general agreement that the above reactions are essentially independent of the particular GUT, in that even for SUSY theories, the MCND final state will likely contain $N \rightarrow e^+ + \text{pion(s)}$[12], in contradistinction to spontaneous nucleon decay in such theories, where modes such as $N \rightarrow \bar{\nu} K$, or, for Higgs-dominated SU(5) theories, modes such as $N \rightarrow \mu^+ K$ are expected to comprise a major fraction of the total spontaneous nucleon decay rate[13]. However, for MCND reactions, final states with muons and/or kaons or vector mesons ρ, ω, K^* are expected to be suppressed by phase space factors and/or generalized CKM mixing angles in MCND reactions[7].

Surrounding the core of the magnetic monopole is a "plasma" of two, four, six, ... fermion "condensates" (zero-energy bound states). These condensates of fermions, in the context of SU(5) give rise to the following effective low-energy interactions[7]:

2 Fermion:

$$\bar{d}_b\, d_b, \quad \bar{u}_g\, u_g, \quad \bar{u}_r\, u_r, \quad e^+\, e^-, \quad \cdots$$

4 Fermion:

$$u_g\, d_b\, u_r\, e^+, \quad u_r\, e^-\, u_g\, d_b, \quad u_g\, s_b\, u_r\, \mu^+, \quad u_r\, \mu^-\, u_g\, s_b, \quad \cdots$$

6 Fermion:

$$u_g\, d_b\, u_r\, e^-\, \mu^+\, \mu^-, \quad u_r\, e^-\, u_g\, s_b\, e^+\, \mu^-, \quad \cdots$$

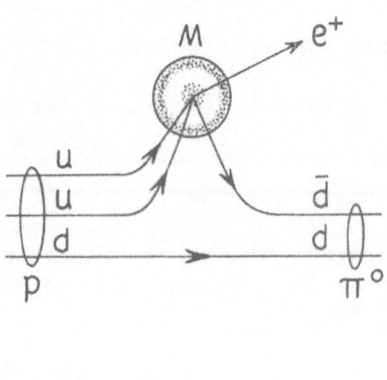

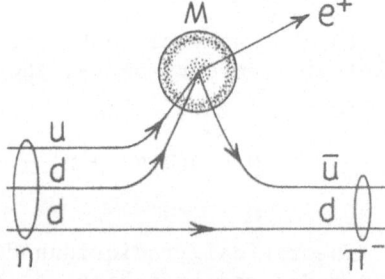

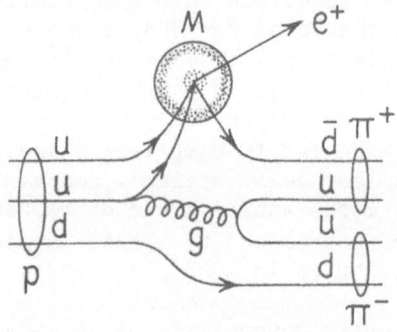

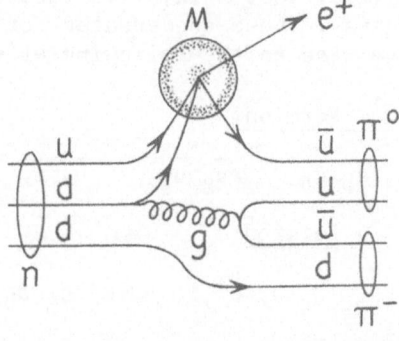

Fig. 6.   A schematic   representation of the monopole   catalysis of
          nucleon decay reaction.

Note that the 2-fermion interactions have $\Delta B = 0$, i.e. they conserve baryon and lepton number, while the 4 and 6-fermion interactions have $\Delta B = -\Delta L = \pm 1$ (but conserve B-L). The $\Delta B \neq 0$ condensates are expected to exist in the region surrounding the monopole core out to distances of $r_c \sim h/2\pi/M_{fc} \sim 1$ fermi, the spatial extent or confinement radius of the $SU(3)_c$ chromo-magnetic field of the monopole. It is for this reason that the catalysis cross section is expected to be of the order of that for conventional strong interactions, one of the 4 or 6-fermion interactions taking place when the monopole overlaps with a baryon during a collision.

From a historical point of view, it is interesting to note that the concept of a "plasma" of fermion condensates surrounding the "bare" monopole originated with the work of Kazama and Yang in 1977 when they considered the interaction of a spin-1/2 particle of charge Ze with a Dirac monopole[14]. They discovered that there likely existed a plasma of e^+e^- pairs in such zero-energy bound states surrounding the Dirac monopole. Thus, in the context of grand unified theories, it does not seem so un-natural that there should also likely exist a plasma of fermion condensates (comprised of both quarks and leptons) which violate baryon and lepton number in their interactions with normal matter.

The hierarchy of catalyzed decay modes in SU(5) is obtained from analysis of the 4, 6, ... fermion interactions[15], such that e^+ + pion(s) is expected to be the dominant catalyzed decay mode, being "Cabibbo"-favored; $\mu^+ K$ is expected to be suppressed by phase space; catalyzed modes such as $e^+ \mu^+ \mu^-$ + pion(s) or $\mu^+ e^+ e^-$ + K are expected to be suppressed by condensate factors; modes such as $\bar{\nu} \pi$, $\bar{\nu} K$ are forbidden (the neutrino has no magnetic charge); modes such as $e^+ K$ and $\mu^+ \pi$ are "Cabibbo"-unfavored/suppressed.

One particularly distinctive alternate possibility, as suggested by C. Schmid[16], is catalysis of $p \to e^+$ (or μ^+) to single prongs, where the positron (or muon) is ejected with essentially the full rest mass energy of the parent proton. This "decay" mode is a rather clean experimental signature for MCND, one which is forbidden to occur in spontaneous nucleon decay by energy and momentum conservation. Note that an interesting feature of this reaction is that it is not "isospin symmetric", unless catalysis reactions involving neutrons, $n \to \nu$ also occur. Unfortunately, the catalysis reaction involving $n \to \nu$ is likely to be unobservable in terrestial detectors, with the exception of the search for MCND neutrinos emitted from dense astrophysical sources such as neutron stars, or perhaps from local sources, such as the sun.

There are also many other possibilities. A.S. Goldhaber has suggested[17] that catalysis reactions may be of a much more subtle

and complicated nature, involving a multi-step process whereby the
proton (or neutron) is first captured by a monopole, after which
many low energy e^+e^- pairs may be ejected from the "plasma"
surrounding the monopole, with a final "burst" of e^+ + pion(s) as
the proton (or neutron) is catalyzed-- a process which occurs at a
weak decay rate and with commensurate lifetime $\tau_c \sim 100$ ns - 1μs.
If the catalysis reaction is indeed governed by an inherently weak
decay rate (rather than a strong rate with $\tau_c \sim 10^{-23} - 10^{-21}$ s)
as Goldhaber suggests, then there are obvious implications for
monopole flux limits obtained from neutron stars, while such decay
rates would have little impact on terrestrial searches, so long as
the detector's time window is much larger than the inherent
catalysis time, i.e. $\Delta t \gg \tau_c$.

Recently, monopole catalysis of nucleon decay in the context
of the Skyrme model has been investigated[18]. In this model, the
nucleon is treated as a soliton; the process of monopole catalysis
of skyrmion decay (MCSD) "unwinds" the topological nature of the
nucleon, also with a large catalysis cross section. However, the
catalyzed decay modes in this model differ from the standard MCND
modes, in that final states of a single lepton, (e^+) accompanied
by a large multiplicity of low-energy pions are expected.

A more exotic possibility for monopole catalysis may occur in
manifest left-right symmetric (LRS) theories of grand unification.
J. Van der Velde has suggested[19] the possibility that the reaction

$$M + n \rightarrow M' + \bar{n} \text{ (in nuclei)} \tag{20}$$

might occur in such theories, again with rather large catalysis
cross sections expected. The experimental signature for monopole
catalysis of neutron oscillation (MCNO) would be a sequence of
anti-neutron annihilations along the trajectory of a monopole,
with up to ~ 2 GeV of visible energy release per catalyzation.

Another possible scenario, suggested by H.J. Lipkin[20], is
monopole catalysis of nuclear β-decay, in which the reaction

$$M + A(Z,N) \rightarrow M' + A'(Z',N') + e^\pm + \overset{(-)}{\nu}_e \tag{21}$$

may occur, due to level mixing of the nuclear wave functions.
The signature of a sequence of low energy (MeV) β-rays along the
trajectory of a slow monopole, possibly accompanied by one or more
X or γ-rays from de-excitation of the final state nucleus is more
challenging to detect experimentally, but certainly feasible.
To date, no experiments have been performed specifically to search
for this type of catalysis reaction[21].

Yet another possibility to be considered is that if magnetic
monopoles are ultra-heavy, i.e. $M_m > M(\text{Planck})$, as predicted in

SUGRA[22] (supergravity) and also Kaluza-Klein theories[23], will quantum gravity play a role in catalysis? Will there be bizzare gravitationally-induced catalysis decay modes? Note that on this mass scale (or higher), magnetic monopoles may (or may not) be "micro" black holes.

The point to be stressed is the following. There exist many theoretical and experimental uncertainties surrounding the issue of magnetic monopoles. Since no conclusive evidence for the existence of magnetic monopoles has ever been shown, the experimentalist must face the dilemma of whether or not this lack of evidence is due to a vanishingly small flux of monopoles, or due to a fundamental lack of understanding as to the true nature and properties of magnetic monopoles and their interactions with matter.

Because of these uncertainties, experimentalists must be open-minded and sensitive to all experimentally realizable scenarios when designing or performing experiments to search for monopoles, especially for MCND experiments. The catalysis cross section is not well-defined and a broad range of possible catalyzed decay modes exist. Faced with all these uncertainties, and given that the background for MCND is so low, experimentalists can and should look for evidence of any and all conceivable decay modes with final state products such as single e^+, μ^+; as well as two or multiple-body final states such as e^+, μ^+, ν + meson(s), so as not to "close the door" on such possibilities.

Note also that each of the different experimentally possible catalyzed decay modes may have different trigger/detection efficiencies for a given detector, an experimental aspect which must be carefully considered and well understood, in terms of sensitivity to the detection of specific catalyzed decay modes, and/or setting limits on the monopole flux, etc.

THE KINEMATICS OF MONOPOLE CATALYSIS OF NUCLEON DECAY

One important aspect of MCND is the kinematics associated with the catalysis reaction. Understanding of this aspect of monopole catalysis is important for single interactions, because of the inherent background from atmospheric neutrino interactions in a given detector. The kinematics of MCND is strikingly distinct from that of ordinary elementary particle interactions or decay kinematics because in this baryon number violating reaction, the monopole is an active participant in the process and cannot be neglected.

There are three basic kinematic classes of monopole catalysis reactions:

1.) Two body final state/single-prong reactions:

$$M + p \rightarrow M' + \underline{e}^+, \mu^+ \qquad \text{(free proton)}$$
$$M + n \rightarrow M' + \bar{\nu}_e, \bar{\nu}_\mu, \bar{\nu}_\tau \qquad \text{(free neutron)}$$

2.) Three body final state/two-prong reactions:

$$M + p \rightarrow M' + e^+, \mu^+ + \pi^\circ, K^\circ \qquad \text{(free proton)}$$
$$M + n \rightarrow M' + e^+, \mu^+ + \pi^-, K^- \qquad \text{(free neutron)}$$

3.) Multibody final state/multi-prong reactions:

$$M + p \rightarrow M' + e^+, \mu^+ + \text{(mesons)}^\circ \qquad \text{(free proton)}$$
$$M + n \rightarrow M' + e^+, \mu^+ + \text{(mesons)}^- \qquad \text{(free neutron)}$$
$$M + A \rightarrow M' + A' + e^+, \mu^+ + \text{mesons} \qquad \text{(nuclei)}$$

We include the case of free neutrons merely as an illustrative example of a possible MCND reaction. Where allowed by charge conservation, final states with e^-, μ^- and/or ν_e, ν_μ, ν_τ are also experimental possibilities. "Mesons" generically means π, K, ρ, K^*, ω, etc. allowed by charge and energy conservation. Let us consider first the simplest case, that of the two-body catalyzed final state, or single-prong catalysis reaction.

The kinematical aspects of MCND are most clearly and easily understood if one first makes a Lorentz transformation into the monopole-free proton center-of-mass system, as shown in the inset of Fig. 3. If we consider the effect of the mass of the outgoing single-prong on the catalysis kinematics, using the e, μ lepton π, K meson and p, n nucleon masses as examples, setting aside for the moment the requirements of angular momentum conservation in the catalysis reaction, we can obtain a first-order estimate of the kinematical aspects of MCND.

The four-momentum transfer squared (Q^2) vs. CM scattering angle for two body/single prong catalysis reactions are shown in Fig. 7 for various final states. Points of interest and importance to note about this figure are the following:

1.) $Q^2 \equiv 2M_p E_x - M_p^2 - M_x^2$ decreases with increasing final state single prong rest mass M_x, along with final state single-prong lab momentum. For example, for single e^+ in the final state, the MCND reaction is highly exothermic in nature, in the sense that for the reaction $M + p \rightarrow M' + e^+$, $P_p^{lab} = 0$ GeV/c, whereas for the e^+, $P_{e^+}^{lab} \sim M_p c$ GeV/c. As the rest mass M_x of the outgoing particle increases, this "exothermicity" monotonically decreases to the elastic scattering (billiard-ball) limit, as in the limiting case of $M + p \rightarrow M' + p'$ in which the final state lab momentum of the outgoing proton is very low, $P_p' \simeq \beta_m M_p c$ GeV/c.

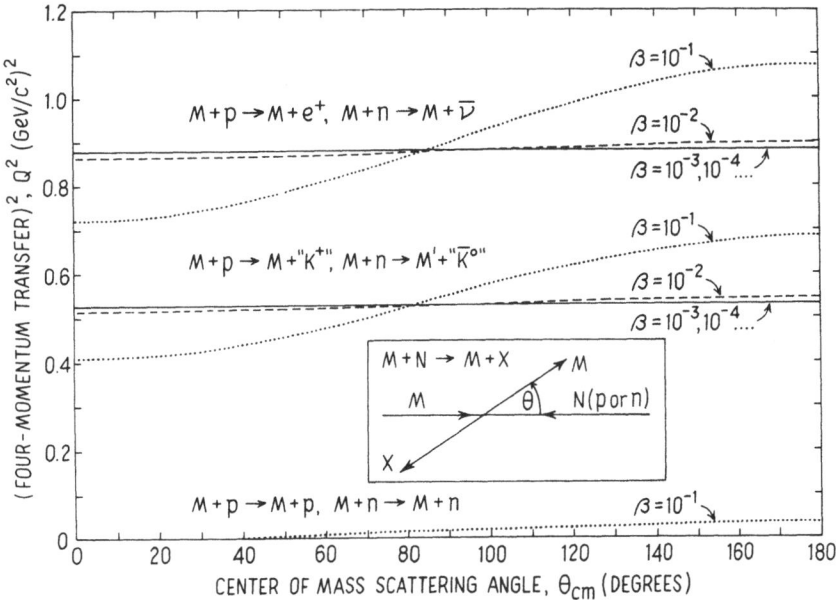

Fig. 7. Q^2 vs. CM scattering angle Θ_{cm} for two-body MCND reactions

2.) Note that the outgoing final state particle can be emitted backwards in the lab for any and all $M_x < M_p$ (or M_n).

3.) The monopole cannot easily transfer energy at low velocities, being kinematically limited to less than $\Delta E_T < 1/2 \ \beta_m^2 M_p \sim 0.5$ keV at $\beta_m \sim 10^{-3}$. However, the monopole can transfer 3-momentum, of up to $\Delta P_T \sim 1$ GeV/c for $p \to e^+$. The transfer of 3-momentum in the CM frame can be seen as merely a consequence of the monopole recoiling against the outgoing single-prong. Thus, four momentum transfers (squared) of $Q^2 \sim 1$ (GeV/c)^2 are possible, even at low monopole velocities.

4.) In an experimental situation for monopole catalysis where the monopole is not directly observed but only the catalyzed decay products, then a seemingly apparent "violation" of energy and momentum conservation may be observed.

5.) Note that Q^2 vs. Θ_{cm} is almost flat at low β. Q^2_{max} occurs for CM scattering angle $\Theta_{cm} = 180°$ ("bounce back"), Q^2_{min} occurs at $\Theta_{cm} = 0°$ ("diffractive" scattering).

6.) The lab angle, θ_{lab} of the outgoing single-prong particle reflects the CM scattering angle, Θ_{cm} very accurately, i.e. $\theta_{lab} \simeq \Theta_{cm}$ for all Θ_{cm}, except for relativistic monopoles. The maximum transverse 3-momentum, $P_\perp^{max} \simeq M_p c$ (for catalysis of $p \to e^+$) occurs at $\Theta_{cm} \simeq 90°$ in the monopole-nucleon center of mass system.

7.) The scattering angle of the monopole in the lab is folded forward from the CM scattering angle, essentially undeflected from its original lab direction, $\Theta_m \sim P_\perp/P_\parallel < (1\ \text{GeV/c})/(10^{13}\ \text{GeV/c})$, which is $\Theta_m \sim 1/10$ picoradian for $M_m \simeq 10^{16}\ \text{GeV/c}^2$, $\beta_m \simeq 10^{-3}$.

8.) It is possible to obtain some understanding of the catalysis kinematics associated with multi-body decay modes, e.g. the reaction $M + p \rightarrow M' + e^+ \pi^\circ$ from the single-prong reactions. For example, if we let the effective invariant mass of the $e^+\pi^\circ$ system $M_{inv}(e^+\pi^\circ)$ be set equal to the kaon mass M_K, then the net momentum of the $e^+\pi^\circ$ system will have $P(e^+\pi^\circ) = P_K \simeq 700\ \text{MeV/c} \neq 0$. If we then "decay" the "K" $\rightarrow e^+\pi^\circ$, it can be seen that the opening angle of the $e^+\pi^\circ$ system is obviously not $\Theta_{op} = 180^\circ$, i.e. not back-to-back, and manifestly different than that for spontaneous two-body nucleon decay. Only in the elastic scattering limit ($Q^2 \simeq 0$), when $M_{inv}(e^+\pi^\circ) \simeq M_p$ will $\Theta_{op}(e^+\pi^\circ) \simeq 180^\circ$.

9.) For multi-body catalyzed decay modes, the three momentum transfer ΔP_T might be expected to be of the order for that for conventional strong interactions at low energies, or, from another point of view, of the order of the "primordial", or "intrinsic" momentum associated with the valence quark(s) inside the nucleon bag which participate in the catalysis reaction, i.e. $\Delta P_T \sim k_T \sim 300\ \text{MeV/c}$. The three momentum transfer and invariant mass of the outgoing lepton + meson(s) system are obviously not independent, one determining the other at low monopole velocities, where the energy transfer is small.

10.) Similarly, for MCND in nuclei the residual final state nucleus must receive some fraction of the available 3-momentum, both the residual nucleus and monopole recoiling against the outgoing catalyzed decay product(s).

11.) Because of the kinematical aspects of MCND, for the thin-detector regime of the catalysis cross section ($\lambda_c \gg d$) where single catalysis interactions dominate, it is unlikely that MCND interactions will be distinguishable from atmospheric neutrino interactions. Since the "structure", or form of the catalysis "current" J_c (if one many call it that) is at present a theoretical unknown, also we have no justification for excluding back-to-back events as possible MCND candidates either! Hence even "spontaneous" nucleon decay forms a background to MCND (and vice-versa)! Similarly, one must reason so for the case of n n̄ oscillations. Therefore, in determining 90% C.L. monopole flux limits for single catalysis interactions, the most conservative approach is to include all observed events with $E_{vis} < M_p$ or M_n (or in the case of n n̄, $E_{vis} < 2 M_n$).

There are additional "higher-order" effects and phenomena associated with MCND which must also be considered:

12.) As in the case of spontaneous nucleon decay in nuclei, experimentalists cannot neglect nuclear re-interaction effects of the outgoing final state mesons with the nucleons in the residual nucleus[24]. For example, for pions produced in spontaneous nucleon decay within an oxygen nucleus, there is ~ 40% chance of the outgoing pions undergoing some form of hadronic interaction via charge exchange, inelastic scattering, absorbtion, etc. before escaping the nucleus. This aspect is also expected to play a role in MCND in nuclei.

13.) Are "final state" interactions with the monopole also important? The outgoing final state meson(s) must pass through the $SU(3)_C$ portion of the monopole. It is reasonable to expect that even in the case of free proton catalysis, some form of final state hadronic interactions with the magnetic $SU(3)_C$ part of the monopole and with the condensate/plasma of "ordinary" fermion anti-fermion pairs may have important and observable consequences.

14.) For monopole catalysis of nucleons in nuclei, the effects of Fermi motion of individual nucleons, both on the catalysis cross section, σ_c and on the final state kinematics cannot be neglected. (N.B. P_f(Oxygen) ~ 235 MeV/c). The $1/\beta$ factor in the catalysis cross section is the relative velocity between nucleon and monopole, which can be sizeable ($\beta \sim 0.2$) even at very low monopole velocities.

The most important point in this discussion of the kinematics of MCND is that the monopole catalysis reaction must conserve both energy and momentum.

Due to the nature of the kinematics associated with MCND, it can be seen that the measurement of the monopole flux from single catalysis interactions, or the monopole flux limits for single catalysis interactions that can be set by a given experiment will be dominated by the ever-present background from atmospheric neutrino interactions. No improvement will be made for any length of detector live time Δt, modulo statistical fluctuations, since $F_m \leqslant N_{obs}/[\ A \cdot \Omega \cdot \Delta t \cdot \varepsilon(g,\beta_m,\lambda_c) \cdot \varepsilon_d]$; the numerator, N_{obs} is also a linear function of the live time Δt.

This situation can be improved somewhat for single MCND interaction flux limits by subtracting the expected atmospheric neutrino background, since it is known to exist and neutrino interactions in these detectors will occur. However, the neutrino flux is only just now being measured in these same detectors (modulo background from spontaneous nucleon decay and MCND!). Furthermore, the calculated ν-flux at low energies is uncertain to factors of ~ 2. When this subtraction is performed, little improvement on single interaction MCND flux limits occurs due to the systematic error on the expected number of events for the

90% C.L. on the difference of the number observed minus the number
expected events.

Only if it can be shown that there is a statistically
significant excess of single interaction events above the inherent
ν-background, for example, in a particularly distinctive channel
(such as p → e⁺, μ⁺) or events with particularly distinctive
kinematic features, will there ever be any hope of discriminating
between catalysis reactions and ν-interactions, spontaneous nucleon decay, etc.

It would be extremely useful for the experimentalists to
obtain more detailed information with regard to theoretical
expectations for the Q^2 distribution for multi-prong interactions
as shown in Fig. 8. It would also be helpful to obtain information on the energy sharing of outgoing leptons and pions for multi
body reactions. We can define an energy sharing variable, Z_{ℓ^+} as:

$$Z_{\ell^+} \equiv E_{\ell^+}/M_p \quad \text{with} \quad 0 < Z_{\ell^+} < 1 \tag{22}$$

Hopefully, theorists could also determine a corresponding MCND
"fragmentation function" $D_c(Z_{\ell^+})$, which represents the probability
that the outgoing lepton receives a fraction Z_{ℓ^+} of the total
available energy in the final state, as shown in Fig. 9.

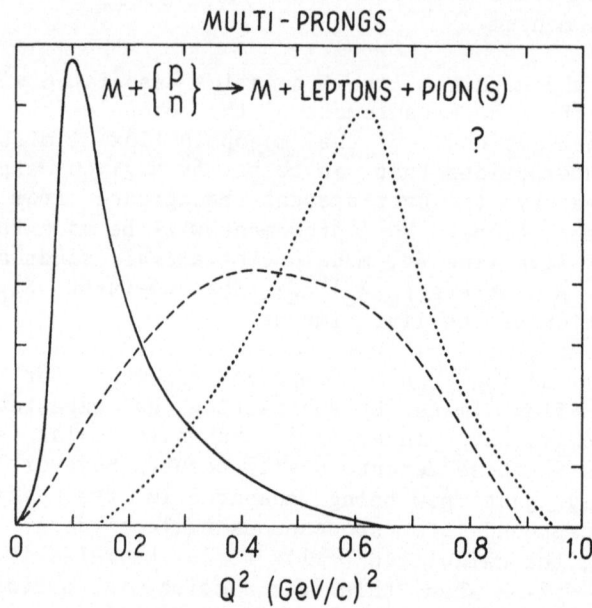

Fig. 8 Possible Q^2 distributions for multi-prong MCND reactions.

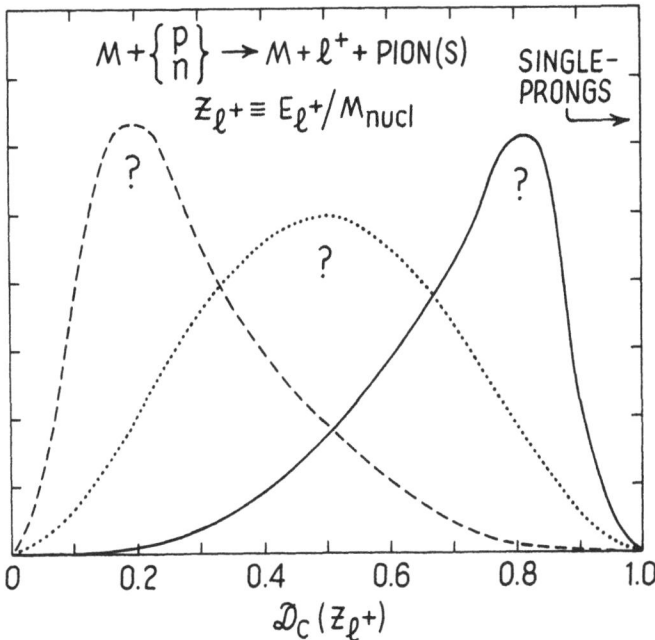

Fig. 9 MCND fragmentation function $D_c(Z_{\ell^+})$ for the outgoing lepton ($\ell^+ = e^+$, μ^+) in multi-prong MCND reactions.

Such distributions would be extremely useful for Monte Carlo simulation of MCND in terms of determining efficiencies, flux limits, etc. and also in terms of potential capabilities for discerning monopole catalyzed events from background atmospheric ν-interactions, spontaneous nucleon decay, $n\bar{n}$ oscillations, etc.

EXPERIMENTAL RESULTS - TERRESTRIAL SEARCHES FOR EVIDENCE OF MCND.

There are five experiments reporting on searches for MCND at this conference. They are the AHT, NUSEX, SOUDAN-I, KAMIOKANDE and IMB experiments. As implied by the title of this talk, none of these experiments have observed any events which even remotely resemble (multiple) catalysis interactions in their detectors. Hence, each experiment has been able to set limits on the flux of monopoles,

$$F_m < N_0/[A \cdot \Omega \cdot \Delta t \cdot \varepsilon(g, \beta_m, \lambda_c) \cdot \varepsilon_d] \qquad (23)$$

which for MCND is a function of both monopole velocity, β_m and catalysis interaction length, λ_c. Thus, in general, the limits on monopole flux via MCND form a 3-dimensional surface in $F_m-\beta_m-\lambda_c$ space, the flux limits presented are merely 2-dimensional slices through this surface. We now discuss each of these experiments and their results.

The AHT (Aachen-Hawaii-Tokyo) Experiment

This experiment is unique in the sense that it is a sea-level MCND search experiment, in contrast to the other four which are all deep underground experiments.

The AHT experiment[25] consists of a cylindrical tank of 17T of purified water, 3.0m ϕ × 2.4m high in, which Cherenkov light from relativistic charged particles traversing the tank is detected by 6 - 8" hemispherical phototubes located on the bottom of the tank. The trigger for this detector was such that whenever a 4-fold PMT coincidence within 30 ns occurred, a series of 8 successively delayed time windows were opened, each of ~ 280 ns duration for a total time window Δt = 2.2 µs. Three (or more) 4-fold 30 ns PMT coincidences within any of the eight time windows after the initial 4-fold coincidence consititued a trigger, the event being recorded in 8 parallel TDC's. Eight digital counters were also gated on after the initial 4-fold coincidence, each for a succesively longer time for the purposes of unfolding the time structure information in each of the 8 time windows of the event.

No triggers were observed in a live time of 290 hours (12 days) of running. The experimenters claim good sensitivity for detecting MCND interactions occurring from any point within the detector. The AHT detector's sensitivity for MCND as a function of β_m and λ_c, and corresponding limits on the monopole flux, F_m are shown in Figs. 10 and 11.

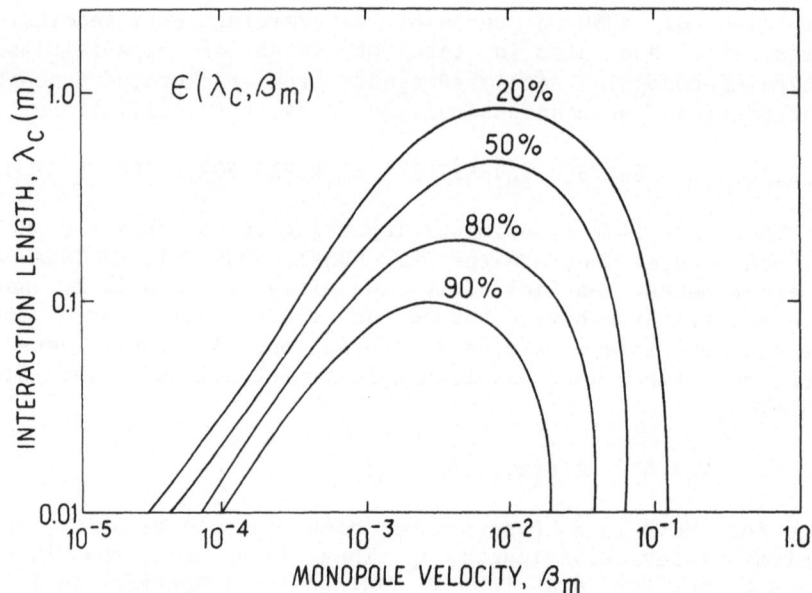

Fig. 10 AHT detection efficiency for MCND vs. catalysis interaction length, λ_c.

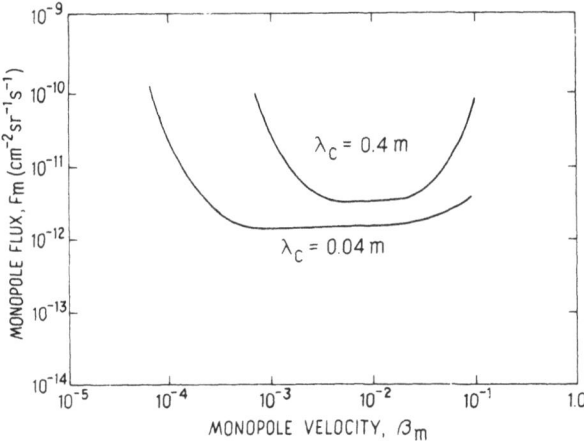

Fig. 11   AHT 90% C.L. monopole flux limit contours for MCND.

The AHT detector is sensitive to MCND only in the regime of relatively short interaction lengths, $\lambda_c \leqslant 0.5$ m. Their detection efficiency decreases for $\beta_m > 10^{-2}$, when the transit time of a fast monopole through their detector becomes comparable to the 30 ns 4-fold PMT coincidence window. For $\beta_m < 10^{-4}$, the detection efficiency for 4 or more catalysis interactions in the detector decreases due to the maximum time window of $\Delta t = 2.2$ μs. The detection efficiency beyond $\lambda_c > 0.5$ m also decreases rapidly because of the decreasing probability for 4-fold interactions to occur within the detector. The limit on the monopole flux for interaction lengths, $\lambda_c \leqslant 0.4$ m is $F_m < 3.0 \times 10^{-12}$ cm^{-2} sr^{-1} s^{-1} (90% C.L.) over the velocity range $5 \times 10^{-3} < \beta_m < 5 \times 10^{-2}$ for 12 days of detector live time.

The NUSEX Nucleon Decay Experiment

The NUSEX detector[26] consists of a 3.5 m cube of 150T of 1 cm thick iron plates interspaced with 134 planes of plastic streamer tubes with bi-dimensional readout. The experimenters search for magnetic monopoles via two techniques: the dE/dX loss signature from "conventional" grand unified magnetic monopoles with no baryon number-violating catalysis interactions, and MCND with or without the dE/dX loss signature from the primary monopole track. Since M. Price[27] will shortly discuss the Mt. Blanc results of their monopole search via dE/dX, I will confine myself primarily to the discussion of their MCND results.

The NUSEX experiment has searched for evidence of both single and multiple MCND interactions in their detector. Their event topology requirement for multiple MCND interactions is $\geqslant 2$ interactions with tracks emanating from catalyzation vertices along the trajectory of the monopole within the detector time window of

5.0 μs; or sequential events with single (or multiple) catalysis interactions per event, separated by the detector electronics dead time of 20 ms. The observation of the primary track of the monopole is not necesarily required in either case, for multiple catalysis interactions, due to the extremely low background from cosmic ray muons, and/or atmospheric neutrino interactions in their detector.

Note that because of the use of limited streamer tubes, the NUSEX detector has good sensitivity for the detection of very small ionization losses, from either direct (ionization) or indirect (e.g. level-mixing) electromagnetic processes at the single electron level, i.e. $dE/dX > {}^{1}/100$ minimum ionizing. This corresponds to a velocity threshold of $\beta_m > 10^{-4}$. Note also that at velocities above $\beta_m > 10^{-3}$, monopoles are expected to become more and more highly ionizing; because of the use of limited streamer tubes and digital (hit/no hit) readout, no direct pulse height information is obtained from the detector.

No candidates for multiple (> 2) interaction MCND events were observed within a period of 317 days of detector live time. The NUSEX 90% C.L. limit on the monopole flux, F_m, as a function of interaction length, λ_c is shown in Fig. 12. Their MCND flux limit is essentially independent of monopole velocity above $\beta_m > 10^{-4}$. Below this velocity, the capability of the NUSEX detector to observe the primary monopole track via dE/dX loss decreases very rapidly, and also because of the rapid decrease in the fficiency for observing multiple catalysis interactions in their detector due to the finite (5 μs) electronics time window. Below monopole velocities of $\beta_m < 10^{-6}$ the detection efficiency for multiple MCND interactions rapidly increases again due to the detector being re-enabled after the 20 ms deadtime.

However, in this ultra-low velocity regime, MCND interactions will appear in separate, sequential events with no associated primary track, with single (or multiple) interactions in the detector per event. Note that it is unlikely that monopoles will have velocities this low, unless they are extremely massive and also gravitationally bound to the earth.

For catalysis interaction lengths of $\lambda_c > 1.0$ m, the NUSEX detection efficiency for multiple MCND interactions decreases due to the decreased probability for 2 or more interactions to occur within the detector. Below interaction lengths of $\lambda_c < 1.0$ m their detection efficiency is essentially unity, and they obtain a limit on the monopole flux, F_m for > 2 interactions within their detector of $F_m < 2.3 \times 10^{-14}$ cm^{-2} sr^{-1} s^{-1} (90% CL) for velocities $10^{-6} < \beta_m > 10^{-4}$ and $\lambda_c < 1.0$ m.

The NUSEX experiment has also set limits on the flux of monopoles for single MCND interactions in their detector. Single interactions are an important aspect of MCND, because as stated earlier, if the catalysis cross section is small, the effective interaction length, λ_c is larger than the size of the detector, and it is likely that only single interactions will be observed.

There are two single interaction flux limits to be presented. The first is for monopole velocities above $\beta_m > 10^{-4}$, where the monopole track is expected to be observed in conjunction with the single catalysis interaction. This single interaction flux limit is shown in Fig. 12. The 90% C.L. monopole flux limit for single ($\equiv 1$) MCND interactions has a local mimimum at $\lambda_c \sim 4.0$ m, where $F_m < 1.1 \times 10^{-13}$ cm^{-2} sr^{-1} s^{-1}, independent of monopole velocity.

The single interaction flux limit increases at short interaction lengths, due to the increased probability of multiple (> 2) interactions occurring within the detector live time of 5.0 µs. At short interaction lengths, single interactions in the NUSEX detector will occur only in the short monopole path length regime, i.e. monopole trajectories confined solely to the edges of the detector, a region which must be avoided because it is fraught with neutron and K^0_L-induced background (see below). The single MCND interaction flux limit increases at large interaction lengths simply due to the decreased probability for single catalysis interactions to occur within the detector volume.

The second single interaction flux limit is for the situation where the monopole velocity is below $\beta_m < 10^{-4}$, where the primary

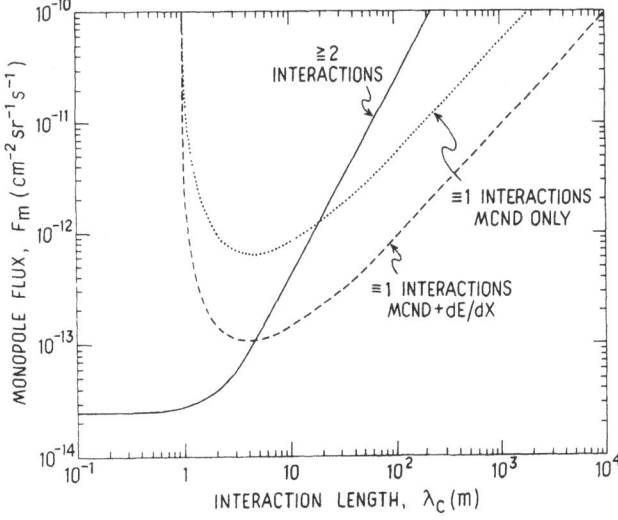

Fig. 12   NUSEX 90% C.L. monopole flux limits for MCND.

track from the monopole is not observed. In this regime, only the single catalysis interactions are detected. Just as in the case for spontaneous nucleon decay, it is necessary to make fiducial volume cuts to remove potential problems associated with neutron and K^0_L background(s) which originate from (unobserved) cosmic ray muon–nucleus interactions in the neighboring rock surrounding the detector. The muon-induced neutron and K^0_L background predominantly occurs within the first few interaction lengths of penetration into the detector volume. The fiducial mass of the NUSEX detector is $\sim$ 100 T.

The Mt. Blanc detector has observed 10 fully-contained events within the same period of 317 days live time. Of these 10 events, 8 are kinematically consistent with single MCND interactions. Thus for this situation, all 8 events should be included in obtaining single interaction flux limits for MCND. The 90% single-sided C.L. on 8 observed events is 13. The single ($\equiv$ 1) interaction MCND flux limit again has a local minimum at $\lambda_c \sim 4.0$ m, with $F_m < 6.1 \times 10^{-13}$ cm^{-2} sr^{-1} s^{-1} (90% C.L.) as shown in Fig. 12. Note also that, strictly speaking, this limit is not independent of monopole velocity, because it is coupled to the $\geqslant 2$ interaction MCND monopole flux limit, due to the search for sequential catalysis events beyond the 20 ms dead time. In any case, it is likely to be only weakly dependent on the monopole velocity for $10^{-6} \leqslant \beta_m \leqslant 10^{-4}$.

The SOUDAN-I Nucleon Decay Experiment

The SOUDAN-I experiment[28], located in northern Minnesota, consists of 31.4T of taconite-loaded concrete interspaced with proportional tubes. The spatial dimensions of the detector are 2.9 m $\times$ 2.9 m $\times$ 2.0 m. The SOUDAN-I dectector has an electronics time window of $\Delta t = 6.5$ μs, with a dead time of 800 ms for digitization and readout of the detector data. Since J. Bartelt[29] will shortly be discussing their results, I will be brief. The SOUDAN experimenters search for monopoles in three ways. Since their detector obtains pulse height information from each of the proportional tubes, at high monopole velocities, $\beta_m \geqslant 10^{-1}$, they search for the signature of highly-ionizing ($\geqslant 16\times$ m.i.) particles traversing their detector without the use of relative time information from the proportional tubes. In this same velocity regime, they also search for evidence of MCND with additional tracks emanating from points along the monopole track, either from within their detector or the rock surrounding their detector.

In the medium velocity regime, $10^{-3} \geqslant \beta_m \geqslant 10^{-1}$, the same method for searching for monopoles is used as above, but now with the addition of timing information from the proportional tubes, performing a velocity fit to tracks.

In the low monopole velocity region, $\beta_m \leqslant 10^{-3}$, only multiple MCND events with at least one interaction occurring within the detector volume and at least one other event, either inside the detector or in the neighboring rock are searched for. A catalyzed decay occurring in the rock may precede or follow a catalyzation within the detector volume. Since the dE/dX threshold for their proportional tubes is $\sim 1/3$ minimum ionizing, it is unlikely that the monopole will be directly observed in this velocity region.

The SOUDAN-I experiment has observed no monopole candiate events, either from dE/dX, timing or MCND for a detector live time of 285 days. From their search via dE/dX and timing, they obtain a 90% C.L. limit on monopole flux of $F_m < 1.3 \times 10^{-13}$ cm^{-2} sr^{-1} s^{-1} for $\beta_m \geqslant 10^{-3}$. For monopole flux limits from MCND, rather than parameterize their results in terms of contours of constant interaction length, λ_c, they chose to use the parameterization

$$\sigma_c = \frac{\sigma_0(hc/2\pi)^2}{\beta(M_pc^2)^2} \tag{24}$$

where β is the relative velocity between nucleon and monopole, σ_0 is the dimensionless cross section. Their flux limits are essentially unchanged for $10^{-2} \leqslant \sigma_0 \leqslant 10^1$ for $\beta_m \geqslant 10^{-3}$.

Fermi motion of nucleons in their (complex) target material is included. No velocity suppression/enhancement factors $F(\beta)$ for the catalysis cross section on different target nuclei were assumed. They find that because of the $1/\beta$ factor in the catalysis cross section parameterization, at low monopole velocities essentially all of MCND events would arise from catalysis of protons in atomic hydrogen in their detector (and/or the surrounding rock), which comprises approximately $\sim 1\%$ of the total detector (or rock) mass.

The 90% C.L. limit on the monopole flux for 285 days of SOUDAN-I detector live time, for multiple ($\geqslant 2$) catalysis interactions is $F_m < 1.5 \times 10^{-13}$ cm^{-2} sr^{-1} s^{-1} for $\beta_m \geqslant 10^{-3}$, with dimensionless cross sections in the range $10^{-2} \leqslant \sigma_0 \leqslant 10^1$ as shown in Fig. 13. Below monopole velocities of $\beta_m < 10^{-3}$, their detection efficiency decreases due to the electronics time window of $\Delta t = 6.5$ μs, and the loss of the ability to use the dE/dX signature of the primary monopole track. The flux limits below this velocity are dependent on the value of the dimensionless cross section parameter, σ_0.

The KAMIOKA Nucleon Decay Experiment

The KAMIOKANDE detector[30] has been operational since July of 1983 and already much physics has been accomplished. Their

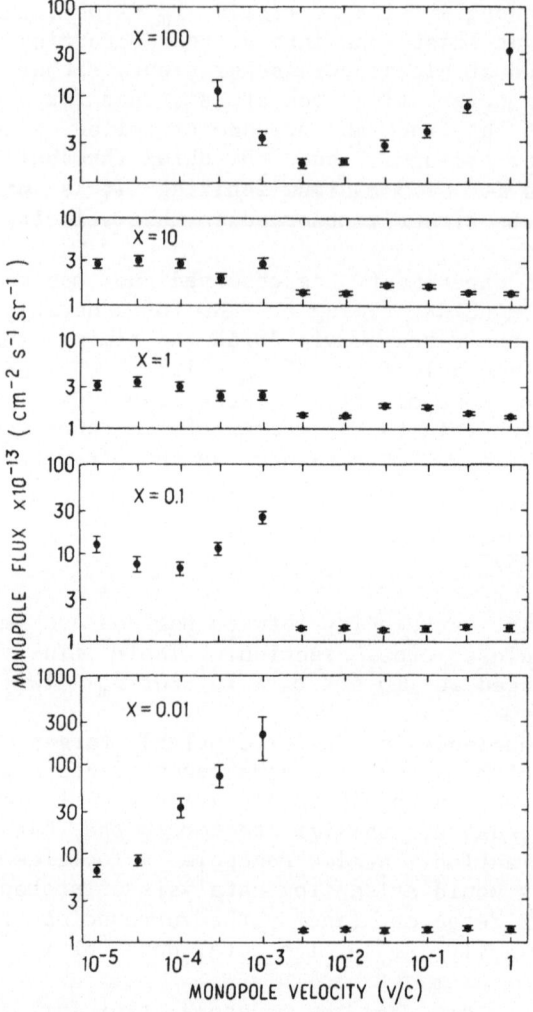

Fig. 13   SOUDAN-I 90% C.L. monopole flux limits for MCND.

ring-imaging water Cherenkov detector consists of a cylindrical
steel tank 15.5 m φ × 16.0 m high containing 3000T of purified
water and 1000 20" photomultiplier tubes lining the walls of the
detector, with ~ 20% photocathode coverage of the total surface
area of the detector, a factor of 10× more coverage than the IMB
detector. The signal from each of the PMT's is split into two, one
portion of the signal is sent to a channel of LeCroy 2285A ADC
system, the other to a 12-channel linear fan-in. The total summed
signal of the detector, after 3 stages of linear fan-in, is used
to obtain a total energy trigger. The detector is triggered
whenever the summed detector analog signal exceeds a threshold
voltage corresponding to a deposition of > 140 photoelectrons
(p.e.'s) or an energy threshold of approximately ~ 40 MeV. The

detector has no high resolution timing, in the sense that the time for lit phototubes is known to within the ADC gate, Δt = 550 ns. The summed detector signals are also sent to two Tektronix transient digitizers, one of which covers a time scale of 5 μs; the other a time scale of 100 μs, for the purposes of detecting $\mu \rightarrow e\nu\bar{\nu}$ decay and MCND.

The experimenters have searched[31] for evidence of multiple catalysis interactions occurring in their detector over a time period of Δt = 88 μs with an analyzed live time of 56.6 days. In order to sift through the data sample to find such events, they impose the following requirements on each event:

1.) < 5000 p.e.'s be observed in the initial event triggering the detector. This cut should be compared with expectations for various "spontaneous" nucleon decay modes in their detector, where < 4000 p.e.'s are expected for even the high Cherenkov light-yield modes such as $p \rightarrow e^{+}\pi^{\circ}$, from which typically ∿ 3000 p.e.'s are expected. The < 5000 p.e. cut reduces the initial raw data sample by a factor of ∿ 78%, from 1.3×10^{6} events to 2.9×10^{5} events.

2.) The pulse-height on the long-time scale transient digitizer is required to be ≳ 310 p.e.'s, allowing them to search for delayed events in which all catalyzed decay modes other than $p \rightarrow \bar{\nu}K^{+}$, $(K^{+} \rightarrow \mu^{+}\nu_{\mu})$ and $n \rightarrow \bar{\nu}K^{\circ}$, $(K^{\circ} \rightarrow \pi^{+}\pi^{-})$ may be observed with high efficiency. The time window used is from 1 μs to 88 μs after the main trigger. The minimun time of 1 μs after the trigger is due to the inability to resolve two waveforms of comparable magnitude as separate entities on the transient digitizer if they are closer in time than this amount. The cut requirement on the transient digitizer removes ∿ 99.8% of the events passing the first cut requirement, with 693 events remaining.

3.) Each transient digitizer waveform of the remaining 693 events is scanned by physicists to remove events which are consistent with PMT dark noise, i.e. events with low pulse-height waveforms, the time integral of which exceeds the 310 p.e. requirement. The physicist scan results in a further reduction of 96.8%, leaving 22 events. Of these 22 events, 8 events have > 1000 p.e.'s in the second event observed on the transient digitizer. For a detector live time of 56.6 days, 9.3 ± 3.0 2-fold coincidence events are expected from random accidental coincidences of cosmic ray muons, one of which triggers the detector and is observed via the primary data acquisition electronics, the second event recorded on the transient digitizer.

4.) The final cut requirement demands that for each event, that either there be (i) more than 2 delayed events OR that (ii) at least one delayed event be observed and the first event originate within the KAMIOKANDE detector fiducial volume (1000T mass). None of the 22 events remain after these cuts are made.

The expected background rate from random accidentals due to cosmic ray muons yielding ≥ 2 delayed events is $\sim 1 \times 10^{-2}$ events/yr. The estimated background rate for a neutrino interaction inside the KAMIOKANDE detector fiducial volume and ≥ 1 delayed events is somewhat higher, $\sim 3 \times 10^{-2}$ events/yr. Thus, a total background rate of ~ 0.04 events/yr satisfying the above event selection criteria are expected from known cosmic ray sources.

Since no candidate events for MCND were found, 90% C.L. limits were obtained on the monopole flux vs. monopole velocity in terms of contours of constant catalysis cross section/interaction length for 56.6 days of detector live time, as shown in Fig. 14.

We express the KAMIOKANDE flux limit contours in terms of both the catalysis cross section, σ_c and interaction length, λ_c.

For catalysis cross sections $\sigma_c \geq 10$ mb (i.e. $\lambda_c \leq 1.7$ m) the 90% C.L. limit on the monopole flux, for multiple catalysis interactions is

$$F_m < 3.8 \times 10^{-14} \text{ cm}^{-2} \text{ sr}^{-1} \text{ s}^{-1} \text{ for } 10^{-4} < \beta_m < 10^{-3}$$

For catalysis cross sections $\sigma_c \geq 100$ mb (i.e. $\lambda_c \leq 17$ cm) the 90% C.L. limit on the monopole flux, for multiple catalysis interactions is

$$F_m < 1.8 \times 10^{-14} \text{ cm}^{-2} \text{ sr}^{-1} \text{ s}^{-1} \text{ for } 10^{-5} < \beta_m < 10^{-3}$$

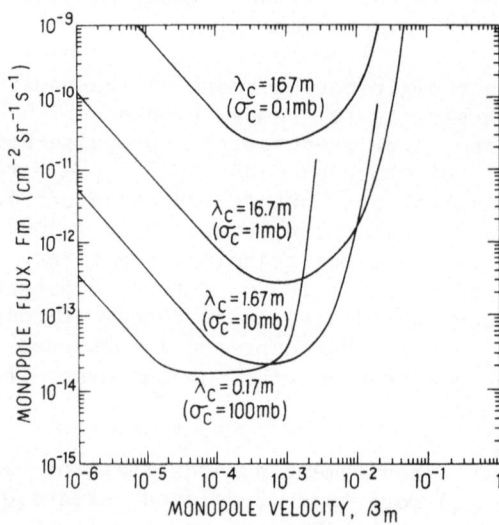

Fig. 14 KAMIOKANDE 90% C.L. monopole flux limits for MCND.

The shape of the flux limit contours for the KAMIOKANDE multiple interaction MCND results are due to factors associated with the detection efficiency for MCND, which, in addition to the inherent properties of magnetic monopoles themselves, are governed by the electronic and spatial-temporal properties of the detector, the time window associated with the transient digitizer and the above event selection criteria.

For example, all of the KAMIOKANDE flux limit contours become less stringent for monopole velocities above $\beta_m > 10^{-3}$ due to the minimum time requirement of $\Delta t \geqslant 1$ µs between the first and subsequent catalysis interactions. In this same velocity region, the 90% C.L. flux limit contours for relatively short interaction lengths, $\lambda_c < 1$ m are less stringent at fixed β_m than for larger interaction lengths of $\lambda_c \sim 1\text{-}10$ m because at such velocities and short interaction lengths, the first catalysis interaction, which triggers the detector, increasingly occurs outside the detector fiducial volume, and is likely to be interpreted as an entering event. Subsequent multiple catalysis interactions will likely not be observed for monopole velocities above $\beta_m > 10^{-3}$, because the mean transit time of the monopole through the detector will be increasingly less than the 1 µs mininumum time requirement for the transient digitizer as the monopole velocity, $\beta_m \rightarrow 1$.

The flux limits become less stringent at low velocities, $\beta_m < 10^{-4}$ because the mean time between catalysis interactions becomes increasingly larger than the maximum time window of the transient digitizer, $\Delta t = 88$ µs.

The flux limits become less stringent at large interaction lengths (for fixed β_m), simply due to the decreasing probability for two or more interactions to occur within the detector.

The IRVINE-MICHIGAN-BROOKHAVEN (IMB) Nucleon Decay Experiment

The IMB detector[32,33,34] is an 8000T water Cherenkov ring-imaging detector with 2048 5" hemispherical PMT's facing inward on a surface grid of ~ 1 m lattice spacing. The volume within the PMT planes has dimensions of 22.8 m × 16.8 m × 17.5 m. The effective cross-sectional area of the detector is 550 m^2, with a mean path length through the detector of 12.8 m for an isotropic flux of monopoles. The readout electronics independently records for each tube the time of arrival (T1) and the amount (Q) of Cherenkov light received. The T1 time scale spans a period of 512 ns with 1 ns resolution. The time resolution of the PMT's is $\sigma_t \sim 4.6$ ns for PMT operation at the 1 photo-electron light level, the regime expected for nucleon decay in our detector.

The IMB detector is triggered whenever $\geqslant 3$ PMT's in a group of 8 × 8 tubes fire within 50 ns and any two or more groups of tubes fire within 150 ns; a second unbiased trigger requires $\geq$ 12 PMT's to fire within 50 ns. These trigger requirements correspond to energy thresholds of 25 and 50 MeV, respectively. These should be compared with expectations for spontaneous nucleon decay, e.g. for $p \rightarrow e^+\pi^\circ$ we expect 170 ± 25 PMT's to fire with an energy deposition of ~ 940 MeV. For nearly all other nucleon decay modes, > 40 PMT's are expected, with the mean number of lit tubes for the majority of decay modes > 100 PMT's.

Immediately following a trigger on the T_1 time scale, the electronics activates for each tube a second time scale (T_2) for an additional 7.5 μs with 15 ns resolution, enabling the detection of $\mu \rightarrow e\nu\bar{\nu}$ decays for muons which stop in the detector. For monopoles, the T_1-T_2 timing enables the detection of a sequence of multiple catalyzed nucleon decays occurring over a time period of 8 μs. At the end of this period, the detector goes dead for a period of 3 ms while the event data is read out from the detector electronics into a PDP 11/34 computer.

We search for evidence of multiple MCND interactions in the IMB detector[33] by requiring at least one event with $\geq$ 30 PMT's in the T_1 time scale and at least one additional event with $\geq$ 50 PMT's firing within any 300 ns time window in the T_2 time scale. We measure a rate of 4.6 ± 0.3 coincidences/day, consistent with the expected rate of 4.7 events/day due to random two-fold coincidences from 2.7 cosmic rays/s through the detector. No ($\geq$ 3)-fold coincidences have been observed.

We then perform a double analysis of the remaining events. The analysis bifurcates into two independent methods, one of which is totally free of human interaction, except for the final stage; the other method relies on physicists interactively scanning each coincidence event.

For the computer analysis chain for searching for multiple MCND events, since we have observed only two-fold coincidences, one event in each of the two time scales, we require the number of tubes lit from Cherenkov light in each of the T_1 and T_2 time scales to be $\leq$ 300 PMT's. More than 85% of all coincidence events and a negligible number of two-fold catalysis interactions are eliminated by this requirement. In addition to, and independently of the above PMT requirement, we require at least one of the events to originate within the detector fiducial volume, 2 m inside the PMT planes. The event occurring on the T_1 time scale in each of the coincidence events is reconstructed using the same track-fitting algorithms as used for the IMB spontaneous nucleon decay event search. Event vertex locations and track directions are reconstructed to an accuracy of $\sigma_x \sim 0.7$ m, $\sigma_\theta \sim 8°$, respectively. None of the T_1 events were reconstructed as originating inside the fiducial volume. From computer simulations, the detection efficiency for single track events in the T1 time scale occurring in the fiducial volume, e.g. for monopole catalyzed $p \rightarrow e^+$ or $p \rightarrow \nu\pi^+$ is $\varepsilon_d(T1) > 70\%$ and for $\geq$ 2-track events , e.g. for monopole catalyzed $p \rightarrow e^+\pi^\circ$, the detection efficiency is $\varepsilon_d(T1) > 90\%$. Because the electronic time resolution of the second time scale (T2) is $\sigma_t = 15$ ns, the event occurring on the T2 time scale in each of the coincidence events is reconstructed using a purely geometrical algorithm rather than temporal one. This algorithm rejects entering tracks by a factor of > 250, while saving events which originate within the fiducial volume with $> 90\%$ efficiciency. Three events which passed the T2 algorithm were carefully scanned and interactively reconstructed by physicists. All three of the coincidence events were found to be due either to random coincidences of downward going cosmic ray events which clipped the corner(s) of the IMB detector, and/or entering muons which stopped in the detector. None of the events is even remotely consistent with expectations for a broad class of event topologies for monopole-catalyzed nucleon decay inside the fiducial volume of the detector.

Because of the broad range of experimental possibilities and many uncertainties associated with monopole catalysis, in terms of catalysis cross section, catalyzed decay modes, etc., physicists independently scanned every coincidence event to determine whether or not candidates for MCND existed in the data sample. Each event was interactively scanned, events which were obvious cosmic ray events entering the detector were easily rejected, whereas events which bore even the remote possibility as being candidates for MCND were interactively reconstructed using the IMB color graphics program. For these events, the results from interactive fitting were compared with computer reconstruction and found to be in good agreement. None of the coincidence events analyzed in this manner were found to be candidates for MCND, with the requirement that at least one event occur within the detector fiducial volume. Furthermore, none of the observed 2-fold events is consistent with entering MCND interactions occurring outside the fiducial volume of the detector, expected in the large σ_c region.

We have analyzed 200 days of data and found no events passing the event selection criteria consistent with multiple interactions from MCND. The overall detection efficiency for multiple MCND, from the use of both of the analysis chains is conservatively estimated to be $\varepsilon_d = 90 \pm 10\%$ for the inclusive decay mode $N \rightarrow e^+ + \text{meson(s)}$, where $N = p$ or n. The flux limits for multiple interactions will be based on this detection efficiency, and are also valid for catalysis modes involving the emission of single charged e^+, μ^+, and π^+, as well as nn annihilations.

Figure 15 shows the 90% C.L. upper limits on the monopole flux for multiple interactions in our detector as a function of

monopole velocity, expressed in terms contours of constant catalysis interaction length. The flux limits are less stringent in the
high velocity region because of the increased probability of all
catalysis interactions occurring only on the $T1$ time scale, and in
the low-velocity region due to the increased probability of the
second interaction occurring after the end of the $T2$ time scale.

In the region of low monopole velocities, $\beta_m \leqslant 10^{-3}$ and short
catalysis interaction lengths, $\lambda_c \leqslant 1.0$ m, the requirement of at
least two interactions, at least one of which must occur in the
fiducial volume of the IMB detector actually hurts the MCND
detection efficiency as the monopole velocity decreases, because
the IMB detector electronics goes dead for 3.0 ms before the
monopole ever gets into the fiducial volume of the detector.
However, "veto-ing" multiple catalysis interactions can be easily
overcome by simply requiring the logical OR of the above event
selection criteria with the requirement of three or more catalysis
interactions on a trajectory through the IMB detector fiducial
volume. The probability of observing $\geqslant 3$ catalysis interactions in
this manner for such interaction lengths is essentially unity, and
is very free from cosmic ray background. We expect less than
2×10^{-2} events/yr. of detector live time due to random 3-fold
coincences of cosmic rays within the $T1$-$T2$ time scale. No $\geqslant 3$-fold
coincidences have been observed for 200 days of live time.

The 90% C.L. limit on the monopole flux, F_m for multiple
catalysis interactions, for $\lambda_c \leqslant 10$ m ($\sigma_c \geqslant 1.7$ mb) is

$$F_m < 8.5 \times 10^{-15} \text{ cm}^{-2} \text{ sr}^{-1} \text{ s}^{-1} \text{ for } 5 \times 10^{-3} \leqslant \beta_m \leqslant 5 \times 10^{-2} \ .$$

The 90% C.L. limit on the monopole flux, F_m for multiple
catalysis interactions, for $\lambda_c \leqslant 1.0$ m ($\sigma_c \geqslant 17$ mb) is

$$F_m < 3.4 \times 10^{-15} \text{ cm}^{-2} \text{ sr}^{-1} \text{ s}^{-1} \text{ for } 10^{-3} \leqslant \beta_m \leqslant 10^{-1} \ .$$

Monopole flux limits presented in terms of contours of
constant interaction length, λ_c, are merely slices through a
three-dimensional surface, as shown in Fig. 16. For any of the
various forms of catalysis cross section parametrization, such as
σ_c or σ_0, the monopole flux limits, expressed in terms of these
cross section parameters will have contours which lie on this
three-dimensional surface in F_m-β_m-λ_c space.

When the catalysis interaction length λ_c becomes larger than
the mean path length $\langle L \rangle$, only single catalysis interactions are
likely to be observed in the detector. During our independent
search for spontaneous nucleon decay, 169 single-interaction
events within the fiducial volume of the detector were found in
200 days of detector live time. The event rate and characteristics

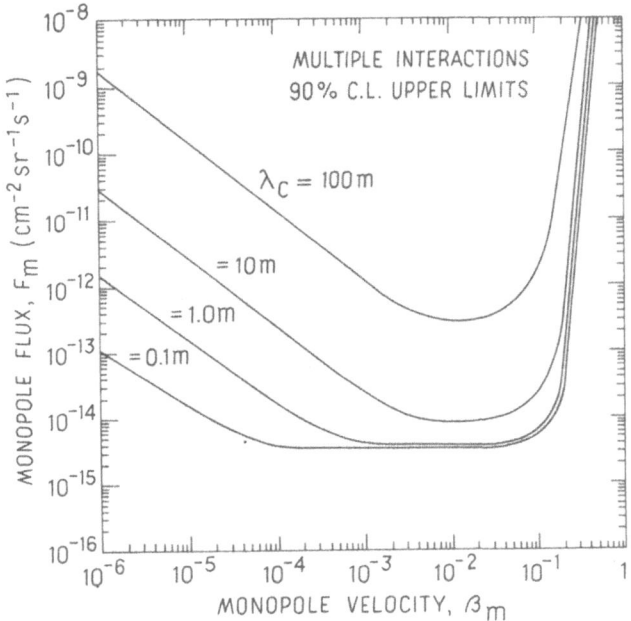

Fig. 15   IMB 90% C.L. monopole flux limits for multiple MCND
          interactions -contours of constant interaction length.

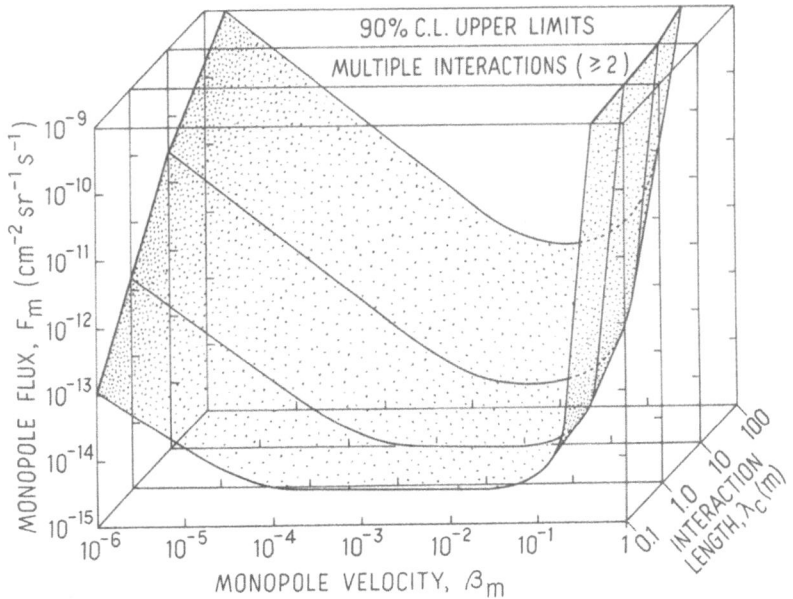

Fig. 16   IMB F_m-β_m-λ_c "surface" for multiple MCND interactions.

of these events appear to be consistent with neutrino interactions in the detector. Of the 169 events, 163 have energies below twice the nucleon rest mass, M_n which we use to obtain conservative limits for single MCND interactions without further event selection, to incorporate the broadest range of possibilities for MCND. The inclusive detection efficiency ε_d, for $N \to e^+ +$ meson(s) is conservatively taken as $\varepsilon_d = 70 \pm 10\%$. The 90% C.L. upper limit on the monopole flux for single interactions in the fiducial volume of the detector are velocity independent, and is shown in Fig. 17. The 90% C.L. upper limit on the monopole flux, F_m for single interactions in the fiducial volume has a mimimum at $\lambda_c \sim 16.7$ m ($\sigma c \sim 1.0$ mb) of $F_m < 1.7 \times 10^{-12}$ cm^{-2} sr^{-1} s^{-1}.

We have recently improved our sensitivity to MCND from the installation of additional electronics in the IMB experiment which allows us to search for multiple interactions occuring over a period of 8 ms. This system is live during the main detector electronics dead time of 3 ms. In the near future, we hope to anaylze the data from this electronics system to search for additional evidence of MCND in our detector. We will also analyze data obtained via an on-line algorithm which presently saves all events occurring within 10 ms of each other. We can also improve our single interaction MCND flux limits somewhat by considering specific exclusive channels for MCND, e.g. $p \to e^+$.

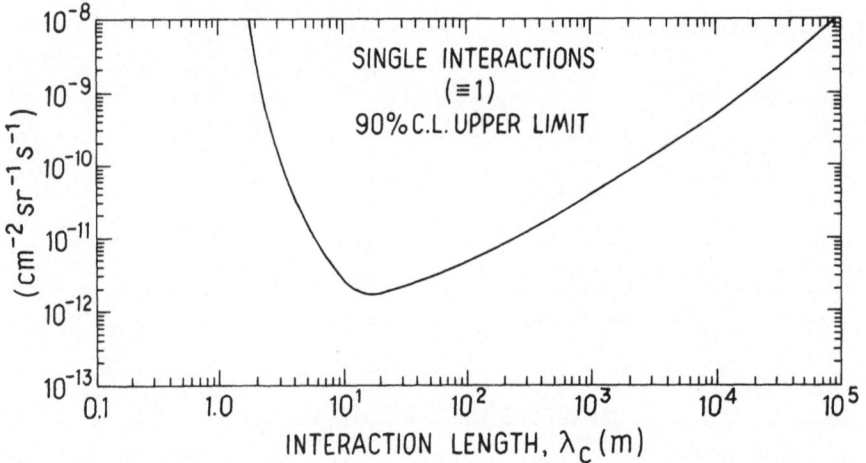

Fig. 17 IMB 90% C.L. flux limit for single MCND interactions.

SUMMARY OF DETECTOR PARAMETERS FOR MCND SEARCH EXPERIMENTS

The detector parameters for each of the five experiments searching for evidence of MCND are summarized in Table 2.

Table 2. Detector Parameters for Experiments Searching for Evidence of MCND

	AHT	Mt. Blanc	SOUDAN-I	KAMIOKANDE	IMB
Depth (mwe)	0	5200	1800	2700	1500
Detector Mat'l	H_2O	Iron	Taconite Cement	H_2O	H_2O
Avg. Density (gm/cm^3)	1.0	3.5	1.9	1.0	1.0
Detector Mass (Ton) Tot/F.Vol	17 (17)	150 (100)	31 (16)	3000 (1000)	8000 (3300)
Detector Dim. (meters)	$3\phi \times 2.4h$	$(3.5)^3$	$3 \times 3 \times 2$	$15.5\phi \times 16h$	$23 \times 17 \times 18$
Eff. Area $\langle A \rangle$ * of detector(m^2)	7.0	18	10	220	550
Avg. Path Length $\langle L \rangle$ (m) *	1.9	2.3	1.7	10.3	12.8
Electronic Live Time (μsec)	2.2	5.0	6.5	88	8.0
Electronic Dead time (msec)	2.0	20	800	3.0	3.0

---

* The effective area $\langle A \rangle$ and average path length $\langle L \rangle$ for an isotropic flux of monopoles are given for each detector. These quantities are purely geometric in origin, and may not necessarily reflect the actual collecting power of each detector. Spatial/-temporal requirements for the detection of MCND specific to each detector are not included.

SUMMARY OF EXPERIMENTAL LIMITS ON MCND

We summarize the 90% C.L. monopole flux limits from the five terrestrial experiments which have searched for evidence of MCND:

1.) The AHT (Aachen-Hawaii-Tokyo) MCND Experiment (Water)
$\geqslant 4$ interactions within $\Delta t < 2.2$ µs, 12 d live time:
$F_m < 3.0 \times 10^{-12}$ cm^{-2}sr^{-1}s^{-1} for $\lambda_c \leqslant 0.4$ m ($\sigma_c \geqslant 40$ mb)
    and the velocity range $5 \times 10^{-3} \leqslant \beta_m \leqslant 5 \times 10^{-2}$.

2.) The NUSEX (Mt. Blanc) Nucleon Decay Experiment (Iron)
$\equiv 1$ interactions (MCND + dE/dX)   317 d live time:
$F_m < 1.1 \times 10^{-13}$ cm^{-2}sr^{-1}s^{-1} for $\lambda_c \sim 4$ m, $\beta_m > 10^{-4}$.

$\equiv 1$ interactions (MCND only)   317 d live time:
$F_m < 6.1 \times 10^{-13}$ cm^{-2}sr^{-1}s^{-1} for $\lambda_c \sim 4$ m, $\beta_m > 10^{-4}$.

$\geqslant 2$ interactions within $\Delta t = 5.0$ µs, 317 d live time:
$F_m < 2.3 \times 10^{-14}$ cm^{-2}sr^{-1}s^{-1} for $\lambda_c \leqslant 1$ m and $\beta_m \geqslant 10^{-4}$.

Electromagnetic (dE/dX > 1/100m.i.+ timing), 120 d live time:
$F_m < 1.3 \times 10^{-13}$ cm^{-2}sr^{-1}s^{-1} for $5 \times 10^{-4} \leqslant \beta_m \leqslant 5 \times 10^{-2}$.

3.) The SOUDAN-I Nucleon Decay Experiment (Taconite Concrete)
$\geqslant 2$ interactions within $\Delta t = 6.5$ µs, 285 d live time:
$F_m < 1.5 \times 10^{-13}$ cm^{-2}sr^{-1}s^{-1} for $10^{-2} \leqslant \sigma_0 \leqslant 10^2$ and $\beta_m \geqslant 10^{-3}$.

Electromagnetic (dE/dX and/or timing), 285 d live time:
$F_m < 1.3 \times 10^{-13}$ cm^{-2}sr^{-1}s^{-1} for $\beta_m \geqslant 10^{-3}$.

4.) The KAMIOKANDE Nucleon Decay Experiment (Water)
$\geqslant 2$ interactions within $\Delta t < 88$ µs, 56.6 d live time:
$F_m < 3.8 \times 10^{-14}$ cm^{-2}sr^{-1}s^{-1} for $\lambda_c \leqslant 1.7$ m ($\sigma_c \geqslant 10$ mb)
    and the velocity range $10^{-4} \leqslant \beta_m \leqslant 10^{-3}$.

$\geqslant 2$ interactions within $\Delta t < 88$ µs, 56.6 d live time:
$F_m < 1.8 \times 10^{-14}$ cm^{-2}sr^{-1}s^{-1} for $\lambda_c \leqslant 17$ cm ($\sigma_c \geqslant 100$ mb)
    and the velocity range $10^{-5} \leqslant \beta_m \leqslant 10^{-3}$.

5.) The IMB Nucleon Decay Experiment (Water)
$\equiv 1$ interactions within $\Delta t < 8.0$ µs, 200 d live time:
$F_m < 1.7 \times 10^{-12}$ cm^{-2}sr^{-1}s^{-1} for $\lambda_c \sim 16.7$ m ($\sigma_c \sim 1.0$ mb)

$\geqslant 2$ interactions within $\Delta t < 8.0$ µs, 200 d live time:
$F_m < 8.5 \times 10^{-15}$ cm^{-2}sr^{-1}s^{-1} for $\lambda_c \leqslant 10$ m ($\sigma_c \geqslant 1.7$ mb)
    for the velocity range $5 \times 10^{-3} \leqslant \beta_m \leqslant 5 \times 10^{-2}$.

$\geqslant 2$ interactions within $\Delta t < 8.0$ µs, 200 d live time:
$F_m < 3.4 \times 10^{-15}$ cm^{-2}sr^{-1}s^{-1} for $\lambda_c \leqslant 1.0$ m ($\sigma_c \geqslant 17$ mb)
    for the velocity range $10^{-3} \leqslant \beta_m \leqslant 10^{-1}$.

CONCLUSIONS

Five experiments have searched for and found no terrestrial evidence of monopole catalysis of nucleon decay within a combined live time of over 870 days. The cross section and velocity dependent limits on the monopole flux from these experiments encompass significantly wider velocity range and are considerably lower than, or comparable to, experiments which are sensitive to the electromagnetic interactions of magnetic monopoles with matter. However, we stress that such limits can place no useful constraint on the monopole flux if the catalysis cross section is vanishingly small. The present MCND experiments, in some sense are just beginning. If improvements in each of the detectors are made, with regard to enlarging Δt_e, the electronics time window per event and coverage of the electronics dead time during readout of the main detector electronics, then significantly improved detection efficicency for MCND will result. Because the presently operating nucleon decay experiments are expected to be taking data for an extended period of time, we can expect that these experiments will begin to probe the monopole flux region of the mass-independent part of the Parker Bound within 2-3 years.

MCND EXPERIMENTS IN THE FUTURE

Due to the current great interest in spontaneous nucleon decay, magnetic monopoles and also particle astrophysics, it is reasonable to expect that much larger experiments than at present will be proposed (and hopefully funded) in the forseeable future. Such detectors, because of their large area and mass will be also very useful for searching for evidence of MCND. Thus, from the beginning design stages of such detectors, sensitivity to physics occurring over long time scales should be incorporated into the design. It would be wise to consider the use detector materials which contain substantial amounts of free protons, such as water, to maximize the sensitivity to MCND, because of the expected enhancement at low velocities of the catalysis cross section.

Several new nucleon decay tracking-calorimeter experiments are expected in the near future, FREJUS and SOUDAN-II. Deep water large area experiments for muon and neutrino astrophysics are planned, e.g. DUMAND and a similar experiment at Lake Baikal. Future monopole experiments with large collection areas of order 1000 m^2 or more are presently being discussed by many people, for potential future experiments here in the U.S., in Europe and also in Japan. Such experiments will have sensitivity for probing the monopole flux region of $F_m \sim 10^{-16} \rightarrow 10^{-17}$ cm^{-2}sr^{-1}s^{-1}, well below the Parker bound.

ACKNOWLEDGEMENTS

I wish to thank many people for many rewarding, interesting
and illuminating discussions regarding the plethora of issues and
aspects surrounding monopole catalysis of nucleon decay. They are
S. Ahlen, C. Akerlof, R. Akhoury, B.C. Barish, J. Bartelt,
P. Bosetti, C.G. Callan, Jr., N.S. Craigie, S. Dawson, M. Einhorn,
J. Ellis, G.W. Ford, P. Galeotti, C. Goebel, A.S. Goldhaber,
M. Goldhaber, C. Hill, T.W. Jones, G.L. Kane, E.W. Kolb,
J.G. Learned, H.J. Lipkin, M.J. Longo, J.M. LoSecco, D. O'Connor,
S.J. Parke, E.A. Peterson, M. Price, M. Rollier, S. Rudaz,
E.L. Shumard, J.L. Stone, L.R. Sulak, G. Tarle, Y. Tomozawa,
W.P. Trower, J.C. Van der Velde, M. Veltman, C.N. Yang, and Y.P.
Yao.

Special thanks are due to M. Koshiba and Y. Tatsuka for their
help and cooperation in obtaining the Kamiokande MCND results for
this conference.

I wish also to thank the members of the IMB collaboration[34]
for their support, and gratefully acknowledge their individual and
collective efforts in our experiment.

Last, but by no means least, I also wish to thank Jim Stone
for his efforts in holding a wonderful and well-organized conference. Thanks go to the advisory committee as well as the local
organizing committee. Special thanks are also due to the conference secretaries, Sue Streicher and JoAnne Sulak, whose hard
work and diligence also helped to make this conference a success.

REFERENCES

1. Magnetic Monopoles (Proceedings of the Wingspread Magnetic
   Monopole Workshop), ed. R.A. Carrigan and W.P. Trower,
   (Plenum, New York, 1983).
2. H. Georgi and S.L. Glashow, Phys. Rev. Lett. 32, 438 (1974);
   H. Georgi, H.R. Quinn and S. Weinberg, Phys. Rev. Lett. 33
   451 (1974); C. Dokos and T. Tomaras, Phys. Rev. D 21, 2940
   (1981); P. Langacker, Physics Reports 72, 185 (1981); E.J.
   Weinberg, these proceedings. For an alternate view of grand
   unification, see J.C. Pati and A. Salam, Phys. Rev. Lett. 31,
   661 (1973); Phys. Rev. D 8, 1240 (1973); and Phys. Rev. D 10,
   275 (1974).
3. V.A. Rubakov, Nucl. Phys. B203, 311 (1982), and Pis'ma Zh.
   Eksp. Teor. Fiz. 33, 658 (1981) [JETP Lett.33, 644 (1981)]
   F. Wilczek, Phys. Rev. Lett. 48, 1146 (1982); C.G. Callan
   Jr., Phys. Rev. D 26, 2058 (1982) and 25, 2141 (1982), and in
   Magnetic Monopoles, (see ref. 1) p. 97.

4. E.W. Kolb, S.A. Colgate and J.A. Harvey, Phys. Rev. Lett. 49, 1373 (1982); S. Dimopoulos, J. Preskill, and F. Wilczek, Phys. Lett. 119B, 320 (1983); J.A. Harvey, these proceedings. See also: K. Freese and M.S. Turner, Phys. Lett. 123B, 293 (1983); K. Freese and R. Kron, Univ. Chicago preprint, (1983); K. Freese, these proceedings.

5. E.W. Kolb, these proceedings. See also J. Harvey, these proceedings; M.S. Turner, these proceedings.

6. F.A. Bais, J. Ellis, D.V. Nanopoulos, and K.A. Olive, Nucl.-Phys. B219, 189 (1983); V.A. Kuzmin and V.A. Rubakov, ITCP Report No. IC/83/17, 1983 (to be published); A.S. Goldhaber, in Magnetic Monopoles (see ref. 1) p. 1, and in The Fourth Workshop on Grand Unification, edited by H.A. Weldon, P. Langacker, and P.J. Steinhardt, (Birkhauser Boston, Mass. 1983) p. 115; see also refs 4-5.

7. J. Ellis, D.V. Nanopolous and K.A. Olive, Phys. Lett. 116B, 127 (1982); J. Ellis, in Magnetic Monopoles, (see ref. 1), p. 17.

8. A.S. Goldhaber, Phys. Rev. 140, B 1407 (1965), and private communication; G.W. Ford, private communication(s).

9. J. Arafune and M. Fukugita, Phys. Rev. Lett. 50, 1901 (1983).

10. S.J. Parke, private communication; see also S. Drell, N. Kroll, M. Mueller, S.J. Parke and M. Ruderman, Phys. Rev. Lett. 50, 644 (1983); G. Tiktopoulos, Phys. Lett. 125B, 156 (1983); N. Kroll, V. Ganapathi, S.J. Parke, and S. Drell, this conference.

11. F.A. Bais, J. Ellis, D.V. Nanopoulos, and K.A. Olive, Nucl. Phys. B219, 189 (1983).

12. S. Dawson and A.N. Schellekens, Phys. Rev. D 27, 2119 (1983); D. London, J. Rosner and E. Weinberg, Enrico Fermi preprint EFI 83-39-Chicago, (1983); S. Dawson, these proceedings; A.N. Schellekens, these proceedings; D. London, these proceedings.

13. S. Dimopoulos and H. Georgi, Nucl. Phys. B193, 150 (1981); N. Sakai, Z. Phys. C 11, 153 (1982); S. Dimopoulos, S. Raby and F. Wilczek, Phys. Lett. 112B, 133 (1982); J. Ellis, D.V. Nanopoulos and S. Rudaz, Nucl. Phys. B202, 43 (1982).

14. Y. Kazama and C.N. Yang, Phys. Rev. D 15, 2300 (1977); see also Y. Kazama, C.N. Yang and A.S. Goldhaber, Phys. Rev. D 15, 2287 (1977); Y. Kazama, Phys. Rev. D 16, 3078 (1977).

15. See references 13, 9, 3 and also reference(s) 14.

16. C. Schmid, Phys. Rev. D 28, 1802 (1983).

17. A.S. Goldhaber, in Magnetic Monopoles (see ref. 1) p. 1, and in The Fourth Workshop on Grand Unification, edited by H.A. Weldon, P. Langacker, and P.J. Steinhardt, (Birkhauser, Boston, Mass. 1983) p. 115; A.S. Goldhaber, private communication.

18. C.G. Callan, Jr., these proceedings; N.S. Craigie, these proceedings; C.G. Callan, Jr., private communication; N.S. Craigie, private communication; see also M. Rho, A.S.

Goldhaber and G.E. Brown, Phys. Rev. Lett. 51, 747 (1983);
A.D. Jackson and M. Rho, Phys. Rev. Lett. 51, 751 (1983).

19. J.C. Van der Velde, private communication; see also S. Errede
et al., Phys. Rev. Lett. 51, 245 (1983).

20. H.J. Lipkin, these proceedings, and Phys. Lett. 113B, 347
(1983); see also L.W. Alvarez, LBL internal report (unpublished); J.S. Trefil et al, Nature 302, 111 (1983).

21. After this talk was given, it came to the author's attention
that a search for monopole catalysis of nuclear β-decay had
already been carried out, with potentially interesting results. See S.N. Anderson, J.J. Lord, S.C. Strausz and
R.J. Wilkes, Phys. Rev. D 28, 2308 (1983).

22. F.A. Bais, these proceedings; P. van Baal, F.A. Bais and P.
van Nieuwenhuizen, to be published in Nucl. Phys. B; P. van
Baal and F.A. Bais, to be published in Phys. Lett. B.

23. M.J. Perry, these proceedings; R. Sorkin, these proceedings;
Q. Shafi, these proceedings.

24. T.W. Jones, et al., in the Proceedings of the 1982 Summer
Workshop on Proton Decay Experiments, ed. D.S. Ayres,
ANL-HEP-PR-82-24, and references therein.

25. P.C. Bosetti, et al., Phys.Lett. 133B, (1983); P.C. Bosetti,
J.G. Learned and D. O'Connor private communication(s).

26. G. Battistoni, et al., Phys. Lett. 118B, 461 (1982); G.
Battistoni, et al., to be published in Phys. Lett. B, (1983);
P. Galeotti, M. Price and M. Rollier, private communica-
tion(s).

27. M. Price, these proceedings.

28. J. Bartelt, et al., Phys. Rev. Lett. 50, 651 and 655 (1983);
J. Bartelt and E.A. Peterson, private communication(s).

29. J. Bartelt, these proceedings.

30. K. Takahashi, in the Third Workshop on Grand Unification,
P.H. Frampton, S.L. Glashow and H. van Dam, ed. (Birkhauser
Boston, Mass. 1982) p. 56; K. Arisaka, et al., contributed
paper # 103 at the 1983 International Symposium on Lepton and
Photon Interactions at High Energies, ed. D.G. Cassel and
D.L. Kreinick, published by F.R. Newman Laboratory of Nuclear
Studies, Cornell University, Ithaca, New York (1983).

31. M. Koshiba and Y. Tatsuka, private communication(s).

32. R.M. Bionta, et al., Phys. Rev. Lett. 51, 27 (1983); J.C. van
der Velde, et al., in the Proceedings of the International
Conference on Neutrino Physics and Astrophysics: Neutrino
'81, Maui, Hawaii, 1981, ed. R.J. Cence, E. Ma, and A.
Roberts (Univ. of Hawaii Press, Honolulu 1982), Vol. 1, p.
205; R. Bionta et al., in the Proceedings of the XVIIth
Recontre de Moriond, 1982, ed. J. Tran Thanh Van (Editions
Frontieres, Gif-sur-Yvette, France, 1982).

33. S. Errede, et al., Phys. Rev. Lett. 51, 245 (1983); S.
Errede, et al., in the Proceedings of the XVIIIth Recontre de
Moriond, 1982, ed. J. Tran Thanh Van (Editions Frontieres,
Gif-sur-Yvette, France, 1983).

34. The members of the Irvine-Michigan-Brookhaven collaboration
     consist of: R.M. Bionta, G. Blewitt, C.B. Bratton,
     D. Casper, R. Claus, B.G. Cortez, S. Errede, G.W. Foster,
     W. Gajewski, K. Ganezer, M. Goldhaber, T.J. Haines,
     T.W. Jones, D. Kielczewska, W.R. Kropp, J.G. Learned,
     E. Lehmann, J.M. LoSecco, H.S. Park, F. Reines, J. Schultz,
     S. Seidel, E. Shumard, D. Sinclair, H.W. Sobel, J.L. Stone,
     L.R. Sulak, R. Svoboda, J.C. Van der Velde, and C. Wuest.

EXCITATION OF SIMPLE ATOMS BY SLOW MAGNETIC MONOPOLES

Norman M. Kroll and Venkatesh Ganapathi

Department of Physics
University of California - San Diego
LaJolla, CA 92093

Stephen J. Parke[a] and Sidney D. Drell

Stanford Linear Accelerator Center
Stanford University
Stanford, CA 94305

ABSTRACT

We present a theory of excitation of simple atoms by slow moving massive monopoles. Previously presented results for a monopole of Dirac strength on hydrogen and helium are reviewed. The hydrogen theory is extended to include arbitrary integral multiples of the Dirac pole strength. The excitation of helium by double strength poles and by dyons is also discussed. It is concluded that a helium proportional counter is a reliable and effective detector for monopoles of arbitrary strength, and for negatively charged dyons.

INTRODUCTION

As discussed by Drell et al.[1] (hereafter referred to as DKMPR), massive Dirac monopoles have a large effect on atomic energy levels, and can cause degeneracy or near degeneracy between the ground state and excited states of the atom. This phenomenon leads to greatly enhanced excitation cross sections for slow moving monopoles. For simple atoms the effect can be reliably calculated, thus providing the possibility of reliable low $\beta = v/c$ detection. What is believed to be a quite accurate calculation has been carried out for a minimum strength monopole in hydrogen,

295

and a calculation of uncertain accuracy has been carried out for such a monopole in helium.

In the following we shall review the method and results of DKMPR, discuss some further investigation of the hydrogen case which includes an extension to monopoles of arbitrary multiples of the Dirac charge, and describe work in progress which should lead to comparably accurate predictions for helium. We shall also discuss the interaction of double strength poles and dyons with helium. Finally, we shall conclude with a few comments about the potential utility of other noble gas atoms as monopole detectors.

THE BASIC MECHANISM AND COMPUTATIONAL STRATEGY

In the following we treat the monopole as infinitely massive, and, until we discuss recoil effects, the nucleus as infinitely massive. We use non-relativistic theory and assume the normal Dirac magnetic moment for the electrons. The Hamiltonian for an atom may then be written

$$H = \sum_{i=1}^{z} \frac{(p_i - A_i)^2}{2m} - \frac{e}{2m} \vec{\sigma}_i \cdot \vec{B}_i + U(r_1 \ldots r_z). \tag{1}$$

In Eq. (1) we have

$$\vec{B}_i = \frac{g\hat{r}_i}{r_i^2} \tag{2}$$

$$\vec{A}_i = g \,\hat{\phi}_i \, \frac{(\pm 1 - \cos \theta_i)}{r_i \sin \theta_i}. \tag{3}$$

The two signs in Eq. (3) refer to alternate gauges[2] and (r_i, θ_i, ϕ_i) are spherical coordinates for the ith electron relative to the pole. Finally, U represents the electrostatic interaction between the electrons and between the electrons and nucleus.

The angular momentum operator for this system may be written[3]

$$J = \sum_i \left(\vec{r}_i \times (\vec{p}_i - e\vec{A}_i) + \frac{1}{2} \vec{\sigma}_i - q\hat{r}_i \right) \tag{4}$$

where $q = eg$ must be an integer or half integer. The term $\vec{r}_i \times (\vec{p}_i - e\vec{A}_i) = m\vec{r}_i \times \dot{\vec{r}}_i$ is the mechanical orbital angular momentum while $-q\hat{r}_i$ represents the angular momentum associated with the electron's electrostatic field crossed with magnetic field of the pole.[4]

Because of the coordinates we have used in the above equations, and also because the pole is assumed to be much more massive than the atom, it is convenient to work in a reference frame in which the pole is at rest. If the atom impinges on the monopole with zero impact parameter so that the monopole passes through it (i.e., the nuclear coordinate z_N goes from $-\infty$ to $+\infty$) J_z is conserved. Since, however, the z component of field angular momentum changes from $+qZ$ to $-qZ$ in this process, the mechanical angular momentum and spin, which we identify with the atomic angular momentum when the atom is outside the range of the pole's magnetic field, must change so as to compensate. Thus J_z (atom) increases by $2qZ$ as a result of the collision. Let us suppose that the incident atom has atomic spin zero, that is to say the atomic ground state has zero angular momentum (as would be the case for a noble gas atom). Then no matter how slowly the collision takes place, after it is over the atom must be in a state of angular momentum greater or equal (typically equal) to $|2qZ|$, and hence in an excited state. Even if J (atom) is not zero for the ground state, some of the magnetic substates must become excited. For example, if the incident atomic spin is $1/2$ then half the atoms are excited for $|2qZ| = 1$, and all for any larger value.

To see what happens when the impact parameter is nonzero, we first note that when the pole and nucleus coincide, all three components of J commute with the Hamiltonian and hence the states may be characterized as eigenstates of $\vec{J} \cdot \vec{J}$ with eigenvalue $\sqrt{J(J+1)}$ and a $2J+1$ degeneracy. Thus the atomic spin zero incident atom finds itself in a state of degeneracy of at least (typically, precisely) $|2qZ|+1$ as it passes over the nucleus. In the case of nonzero impact parameter it is convenient to use a time dependent coordinate system in which the nucleus is on the negative z axis. In that case the instantaneous states are still characterized by the eigenvalues of J_z. Furthermore, for adiabatic motions transition between states of different J_z will be very improbable unless degeneracy or near degeneracy occurs. (Note that constant J_z here means that the atom remains in its ground state. This differs from our previous discussion because here J_z refers to a z axis which has a different direction before and after the collision.) The previous discussion, however, tells us that for small impact parameters the spin zero incoming state will become nearly degenerate with the $|2qZ|$ partners which it would have at the center. Quasi-adiabatic transfers to these states become probable and lead to excitation. Additional degeneracies will occur if the $z_N = 0$ state to which the incident ground state connects is not the ground state of the $z_N = 0$ Hamiltonian. These additional degeneracies, which we refer to as off center level crossings, occur for a range of impact parameters, and can provide an additional source of excitation. They may enhance the excitation cross sections for double strength poles on He, and they are expected to be important for the heavier noble gas atoms. For non-zero incident spin the situation is similar but different magnetic substates must be treated separately as they connect with different $z_N = 0$ states, and the degree of degeneracy of the $z_N = 0$ state varies with magnetic substate.

With the above picture in mind, a general strategy for calculating excitation cross sections may be described.

(1) The $z_N = 0$ low lying energy level system must be established. This is a relatively simple task for one and two electron atoms. A substantial effort would be involved to do it reliably for such things as Ne and A. Connections between these states and the states at $z_N = \pm\infty$ can be established by assuming that states having the same eigenvalues of J_z do not cross. These connections are sufficient to establish whether or not there are off center crossings and to determine which states will be excited.

(2) Excitation transitions via the central multiplet can be calculated by means of a simple extension of the Landau-Zener theory.[5] To obtain quantitative results for a given distance of closest approach, and velocity at that distance, one requires only a knowledge of the central multiplet splitting which occurs at distances of that order. We assume here that these transitions are probable only if the splitting is small, and hence linear in the separation. It can therefore be obtained by applying first order perturbation theory to the central multiplet, provided a sufficiently accurate form for the wave function of the central multiplet can be obtained.

(3) Excitation via off center crossings are significant only if $\Delta J_z = \pm 1$ for the crossings, a circumstance which has, among the cases we have studied, occurred only for double strength poles on He. Such crossings are also expected to occur in more complex systems such as Ne and A. They are most likely to lead to transitions when the distance of closest approach is equal to the crossing distance. Hence the determination of this distance and the wave function is needed, as well as an appropriate treatment of the differential equation which couples the states.

(4) Since we are treating the orbit of the incoming atom classically, we can use the dependence of the energy level on z_N to determine an effective potential between the atom and the pole. For excitation via the central multiplet, one is primarily interested in relating the distance of closest approach and the atomic velocity at that distance to the velocity and impact parameter at $z_N = -\infty$. This can be determined to a good approxima-

tion from the information already determined in Eqs. (1) and (2), and together with Eqs. (1) and (2) provides a practical procedure for determining the excitation cross sections. More detailed information about the energy as a function of z_N is likely to be required to discuss excitation via off center crossings.

The general procedure described above will be illustrated and further explained in the applications to be described in subsequent sections. Before doing so, however, we explain the $\Delta J_z = \pm 1$ selection rule for off center crossings. We first imagine the eigenvalues and wave functions of the Hamiltonian with the nucleus on the negative z axis to be known. The states are also eigenstates J_z, and as mentioned before, states of the same J_z will not cross so that a unique energy $E_n(z_N)$ can be defined for each state n. We next consider the case of the nucleus moving in some orbit in the x = 0 plane, and define a primed coordinate system whose x′ = 0 plane coincides with the x = 0 plane (i.e., x′ = x) and whose z axis points from the pole to the nucleus. The Hamiltonian in these primed coordinates has the same form as Eq.(1) with nuclear coordinate $\vec{r}'_N = (0,0,-z_N)$. In these variables the Hamiltonian has an explicit time dependence due to the variation of z_N with time and an implicit time dependence due to the time dependence of y′ and z′ with reference to a fixed y,z. A similar remark applies to the wave function. Now write

$$\psi = \sum C_m(t)\psi_n(\vec{r}',z_N)e^{-i\int E_n dt},$$

and insert in the time dependent Schroedinger equation to obtain[6]

$$\frac{dC_n}{dt} = - \sum_m \left(\psi_n, \frac{d\psi_m}{dt}\right) e^{-i\int(E_m-E_n)dt} C_m. \tag{5}$$

The quasi adiabatic approximation[7] consists of restricting the values of n,m and refer to the pair of levels which are crossing. With the natural choice of phase $(\psi_n, d\psi_n/dt)$ always vanishes so that the sum in Eq. (5) reduces to a single term, and we have a pair of coupled equations for C_n and C_m. Next we observe that

$$\frac{d\psi_m}{dt} = \left(-i\frac{\vec{v}\cdot\hat{y}'}{z_N} L'_x - \hat{z}' \cdot \vec{v} \frac{d}{dz_N}\right) \psi_m$$

where

$$L'_x = \sum_j^z \vec{r}'_j \times \vec{p}'_j \cdot \hat{x}'.$$

Now $(\psi_n, d\psi_m/dz_N)$ vanishes unless $\Delta J' = 0$, which never holds for crossing states, while $(\psi_n, L_x \psi_m)$ vanishes unless $\Delta J_z = \pm 1$. The selection rule for off center crossings is thus explained.

THE HYDROGEN ATOM

The simplest illustration of the preceding discussion is provided by the hydrogen atom. For notational simplicity, we assume the pole strength parameter q to be positive, and, of course, Z = 1.

The energy levels at $z_N = 0$, where the pole and charge coincide, is given by,[8] $E = Ry/n^{*2}$ where $n^* = 1+n_r+\sqrt{(J+1/2)2-q2}$, $J = q - 1/2$, $q = 1/2$, ... and a linearly independent set with $n^* = n_r+\sqrt{(J+1/2)2-q2}$, $J = q + 1/2$, $q + 3/2$, ... In both cases, $n_r = 0,1,...$ is the number of radical nodes. Since for large J

$$\sqrt{(J+1/2)^2 - q^2} \approx J + \frac{1}{2} - \frac{q^2}{2J+1}$$

and thus n^* depends primarily upon n_r+J, the pattern of J degeneracy and multiplet structure is similar to that of the pole free case.

The ground state, with $E = -Ry$, $J = q - 1/2$; and the first excited state, with $E = -Ry/(2q+1)$, $J = q + 1/2$ are the two $z_N = 0$ states which connect to the ground state at $z_N = -\infty$, and correspond to the central multiplets referred to in the previous section. The state with J_z (atom) = $-1/2$ at $z_N = -\infty$ has $J_z = q - 1/2$. If the collision is at zero impact parameter, it connects to the lowest state with J_z (atom) = $2q - 1/2$ at $z_N = +\infty$. For $q = 1/2$ this is simply the $+1/2$ component of the ground state and no excitation is involved. For larger q excitation must occur. For collisions in which the impact parameter is nonzero but sufficiently small, the 2q components of the ground state never actually cross but come sufficiently close together to allow quasi adiabatic transfer to occur among them, leading to a distribution of excited final states with $2q - 1/2 \geqslant J_z$ (atom) $\geqslant 3/2$. The state with J_z (atom) = $+1/2$ at $z_N = -\infty$ has $J_z = q + 1/2$ and hence must connect to the $J = q + 1/2$ state. Thus as z_N varies from $-\infty$ to zero, its energy increases from $-Ry$ to $-Ry/(2q+1)$. At the same time there are $2q-1$ states which descend from $z_N = -\infty$ excited states to the $z_N = 0$ ground state with energy $-Ry$ and hence experience off center crossings with the $J_z = q+1/2$ state. These off center crossings are with states having $q - 3/2 \geqslant J_z \geqslant -(q - 1/2)$. Thus a transfer to any of these states involves $\Delta J_z \geqslant 2$, and hence do not occur in the quasi adiabatic regime. Hence we confine our attention to excitation via the central multiplet.

In order to discuss excitation via the central multiplet, we first write the Hamiltonian as

$$H = H_0 + H_e$$

where

$$H_0 = \frac{(p-eA)^2}{2m} - \frac{q}{2m} \frac{\vec{\sigma} \cdot \hat{r}}{r^2} - \frac{e^2}{r} \qquad (6)$$

$$H_e = \frac{-e^2}{r^2} \hat{r} \cdot \vec{r}_N. \qquad (7)$$

We are assuming the pole to be fixed at the origin and the distance r_N of the proton from the origin to be sufficiently small to allow us to represent the change in the electrostatic potential by the electric dipole approximation.[9] Furthermore, we write

$$\vec{r}_N = v_0 t \, \hat{z} + b_0 \hat{x} \qquad (8)$$

where b_0 is the distance of closest approach and v_0 is the velocity of the nucleus when it is at that distance. Thus we are neglecting orbit curvature and velocity variation near the point of closest approach. Degenerate first order perturbation theory applied to the subspace formed by the central multiplet can be written

$$i \frac{d\psi}{dt} = \gamma(v_0 t J_z + b_0 J_x)\psi \qquad (9)$$

where

$$\gamma = -\frac{e^2}{J(J+1)} \left\langle \frac{\hat{r} \cdot \vec{J}}{r^2} \right\rangle = -\frac{e^2}{J(J+1)} \left\langle \left(\tfrac{1}{2} \hat{r} \cdot \vec{\sigma} - q \right)/r^2 \right\rangle \qquad . \qquad (10)$$

In Eq. (9) ψ is a $2J+1$ component column matrix whose components are the amplitudes of the J_z eigenstates, and J_z and J_x are angular momentum matrices in the standard J representation.

Setting

$$\psi = \begin{bmatrix} C_J \\ \cdot \\ \cdot \\ C_M \\ \cdot \\ \cdot \\ C_{-J} \end{bmatrix} \tag{11}$$

we obtain

$$i\dot{C}_M = \gamma v_0 t M C_M + \frac{\gamma b_0}{2} \left(C_{M+1}\sqrt{(J-M)(J+M+1)} + C_{M-1}\sqrt{(J+M)(J-M+1)} \right)$$

$$-J \leqslant M \leqslant J \tag{12}$$

The Ansatz

$$C_M = \left[\frac{(2J)!}{(J-M)!(J+M)!} \right]^{1/2} u^{J+M} v^{J-M} \tag{13}$$

solves these equations provided

$$\dot{u} = -\frac{1}{2} i\gamma v_0 tu - \frac{1}{2} i\gamma b_0 v \tag{14}$$

$$\dot{v} = \frac{1}{2} i\gamma v_0 tv - \frac{1}{2} i\gamma b_0 u. \tag{15}$$

Equations (14) and (15) are equivalent to Zener's Eq. (4).[5] We seek a solution for which $C_J = 1$, $C_M = 0$ ($M \neq J$) at $t = -\infty$. This is obtained by solving Eqs. (14 and 15) with $u(-\infty) = 1$, $v(-\infty) = 0$. These are just the Zener boundary conditions, so that we can adopt his solutions. Since we only need the $t = +\infty$ value, we write

$$|u^2(+\infty)| = \exp(-\pi\gamma b_0^2/(2v_0)) \tag{16}$$

$$|v^2(+\infty)| = 1 - |u^2(+\infty)| \tag{17}$$

which yields at $t = +\infty$

$$|C_M|^2 = \frac{(2J)!}{(J-M)!(J+M)!} \exp \frac{-\pi\gamma b_0^2(J+M)}{2v_0} \left(1 - \exp \frac{-\pi\gamma b_0^2}{2v_0}\right)^{J-M} \tag{18}$$

It will be shown below that the impact parameter $b = \lambda b_0$ and the incoming velocity $v = v_0/\lambda$, where λ is a function of v which we shall determine. Assuming this to be the case here, we obtain the partial excitation cross sections

$$\sigma_M = \pi \int db^2 |C_M|^2 = \lambda^3 \frac{2v}{\gamma} \int dx e^{-(J+M)x}(1-e^{-x})^{J-M} \frac{(2J)!}{(J-M)!(J+M)!}$$

$$= \frac{2v\lambda^3}{\gamma(J+M)} \, , \quad J \geqslant M \geqslant -J+1 \tag{19}$$

It should be clear that the above theory of excitation via a central multiplet applies to any atom. The quantitative problem in the case of complex atoms arises in the evaluation of γ and λ. For the case of hydrogen, we find

$$\gamma = 4\alpha/(2q+1) \tag{20}$$

for the $J = q - 1/2$ state, and

$$\gamma = 4\alpha q/\left[(2q+3)(2q+1)^2\left(2q+1 - \frac{1}{2}\sqrt{2q+1}\right)\right] \tag{21}$$

for the $J = q + 1/2$ state. Here α is the fine structure constant and in our units γ is an inverse length squared, so that the units are inverse Bohr radii squared.

To determine λ, we assume the motion of the proton may be described classically and that it is determined by studying motion in the potentials $V_- = E(q - 1/2, \vec{r}_N)$ and $V_+ = E(q + 1/2, \vec{r}_N)$ where $E(q + 1/2, \vec{r}_N)$ is the minimum eigenvalue of the Hamiltonian with $\vec{r}_N$ held fixed and J_z with eigenvalue $q\pm 1/2$. Here J_z refers to the component of angular momentum along an axis directed from the pole to the proton. Since the potential depends only on $|\vec{r}_N|$, angular momentum is conserved in the motion, yielding $vb = v_0 b_0$. The ratio of v to v_0 is given by energy conservation. In order to obtain a b independent form for λ, we have assumed $V_\pm(b_0) = V_\pm(0)$. With these assumptions, $\lambda = 1$ for the lower state and

$$\lambda = \left(1 - \frac{2q\alpha^2}{(2q+1)v^2}\right)^{1/2} \tag{22}$$

for the upper state. Evidently λ depends only upon the excitation energy of the central multiplet state so that its evaluation for complex atoms depends only upon a determination of the excitation energies.

The total excitation cross section is thus given

$$\sigma = \frac{1}{2}(\sigma_{-1/2} + \sigma_{1/2})$$

$$\sigma_{-1/2} = (q+1/2) \frac{\beta}{\alpha} \left(1 + \frac{1}{2} + \dots + \frac{1}{2q-1}\right) a_0^2, \quad q > 1/2 \qquad (23)$$

$$\sigma_{1/2} = \frac{\beta}{2\alpha q} \left(1 - \frac{2q\alpha^2}{(2q+1)\beta^2}\right)^{3/2} (2q+3)(2q+1)^2$$

$$\qquad (24)$$

$$\times \left(2q + 1 - \frac{1}{2}\sqrt{2q+1}\right)\left(1 + \frac{1}{2} + \dots + \frac{1}{2q+1}\right)a_0^2$$

These last expressions have been written in standard units so that v has been replaced by $\beta = v/c$ and the Bohr radius factors have been put in explicitly. The factors $(1 + 1/2 + \dots)$ come from the sum over partial cross-sections, and in the form written it is assumed that all are above threshold. For example, for $q = 1$, $\sigma_{-1/2}$ has only one term and represents excitation to the $n = 2$ level with $(v/c) = (\alpha/2)\sqrt{3m_e/m_p}$ at threshold. $\sigma_{1/2}$ has three terms, representing excitation to an $n = 2$ level and two excitations to $n = 3$ levels with a slightly higher threshold. For fixed n, λ determines the threshold when the energy of the relevant central multiplet exceeds that of the final state, that is when $2q+1 > n^2$. The excitation cross sections increase quite rapidly with q, and our approximations lose their validity as cross sections become of order a_0^2. As was emphasized in DKMPR, the cross section is quite large even for $q = 1/2$.

THE HELIUM ATOM

The simplest system of practical interest to which our theory applies is the helium atom. We obtain results in this section for the $q = 1/2$ and $q = 1$ case.

The q = 1/2 Case

The lowest two hydrogenic states for $z_N = 0$ have $j = 0$ and 1. Thus in the shell model approximation the ground state will be an antisymmetrized product of $j = 0$ and 1 orbitals yielding a ground state of $J = 1$. This state, which provides the central multiplet through which transitions occur, has a binding energy which we have variationally estimated to be 8.49 eV, so that the He atom sees a repulsive barrier of height 16.09 eV. This leads to a threshold factor $\lambda = \left(1 - \beta_c^2/\beta^2\right)^{1/2}$ with $\beta_c = 9.29 \times 10^{-5}$. The excitations are to the lowest $^3S_1$ and $^3P_2$ states. These occur in the ratio 2:1 with thresholds at $\beta \times 10^4 = 1.03$ and 1.06, respectively. Applying Eq. (19), the excitation cross section is given by $\sigma = \frac{2v\lambda^3}{\gamma}(1+1/2)$. The critical problem remaining is the determination of γ.

A preliminary determination has been made using a product wave function whose spin and angle dependence is the same as that of the hydrogenic orbitals and whose radial parts take the form

$$f_0 = e^{-\alpha_0 r/a_0}$$

$$f_1 = r^{\sqrt{2}-1} e^{-\alpha_1 r/a_0\sqrt{2}} \left(1 - t\frac{r}{a_0} e^{-wr/a_0}\right)$$

In these expressions, α_0, α_1, t, and w are parameters determined by minimizing the energy subject to the constraint[9] that the electric and magnetic dipole forms of the perturbation energy give the same result. While they are guaranteed to give the same result if the perturbation is computed using exact wave functions of the Hamiltonian, they are in general not the same for approximate wave functions. The value obtained for γ is 4.66 γ_H where $\gamma_H = \alpha/4(4-\sqrt{2})a_0^2$. The result reported in DKMPR, 2.35 γ_H, was obtained from the magnetic dipole form, using a cruder wave function. The cruder wave function with the electric dipole form yields 3.53 γ_H. While we believe the new value to be the more reliable, the results above have convinced us that a more elaborate and systematic approach is desirable. The computational problem is quite similar to that involved in computing the fine and hyperfine (He_3) structure of the helium P states, which has been done to great accuracy, and we are in the process of adopting the methods used there[10] to this problem. It should be possible to assess the accuracy of a given level of approximation for the wave function by comparison with the helium hyperfine structure results and by comparing its results for the electric and magnetic dipole forms of the interaction. For the present, however, we consider our best value to be

$$\sigma = 2.5 \times 10^{-18} (\beta/10^{-4})(1 - \beta_c^2/\beta^2)^{3/2} \text{ cm}^2$$

with $\beta_c = 9.29 \times 10^{-5}$.

The q = 1 Case

The lowest hydrogenic states for $z_N = 0$ have $j = 1/2$ and $3/2$. Thus the shell model ground state is obtained by putting both electrons in the ground state, yielding a $J = 0$ ground state. This state connects with the lowest $^3P_2$ state at $z_N = \pm\infty$. Because the ground state at $z_N = -\infty$ has $J_z = 2$, the central multiplet which it connects is the $J = 2$ level formed by the $(1/2)(3/2)$ configuration. This configuration also has a $J = 1$ level and it is important to know which of the two lies lower. A simple variational calculation seems to provide convincing evidence that the $J = 1$ level is lower by about 0.9 eV. There are therefore four off center crossings by the $J_z = 2$ state as z_N varies from $-\infty$ to zero. The first, which occurs at $z_N \approx a_0$, is with the $J_z = 0$ state which connects the $^3P_2$ state at $z_N = -\infty$ to the $J = 0$ ground state at $z_N = 0$. There is no quasi adiabatic transfer at this crossing because $\Delta J_z = 2$. The other three crossings are with the $J_z = 1, 0, -1$ components of the $J = 1$ central multiplet. The $J_z = 1$ crossing, which satisfies the $\Delta J_z = \pm 1$ rule, occurs at $z_N \approx 0.1 \ a_0$ and is a potential additional source of excitation. More careful examination of the spectrum at small z_N ($z_N \approx 0.5 \ a_0$ and less) reveals a more complex situation than we have discussed previously. The coupling of the electric dipole perturbation to the inner electron is so strong that it breaks the quartet-doublet coupling for $z_N \gtrsim 0.003 \ a_0$. Since such distances make a negligible contribution to the cross section, one can confine one's attention to the "Paschen-Bach" region in which the inner electron has $j_{1z} = 1/2$ and the outer electron has $j_{2z} = 3/2$. The levels which contribute to the excitation process are just the $j_z = -3/2, -1/2, 1/2$ levels of the outer electron. The electrostatic interaction between the inner and outer electrons does, with the z_N range we are considering, cause a departure from the uniform spacing of the level, and in particular reduces the spacing between the $j_{2z} = 3/2$ and $j_{2z} = 1/2$ components. As mentioned above, these two levels actually cross at $z_N \approx 0.1 \ a_0$. We have not worked out the details of the excitation process for this more complex situation but it is clear that application of the central multiplet theory to this excited electron will yield a lower limit for the cross sections. Comparing γ values for $q = 1/2$ and $q = 1$ and using Eq. (21), we see that the $q = 1$ monopole is at least a factor 2.3 more effective than the $q = 1/2$ monopole. This factor will be enhanced not only by the electrostatic narrowing of the level separations but also by the fact that the screening will be stronger and the fact that there is an additional level. The excitations are to $^3S_1$, $^3P_2$, and $^3D_3$ with $\beta \times 10^4$ thresholds at 1.03, 1.11, and 1.11, respectively. The relative intensities will surely be altered from the 3:3/2:1 of the central multiplet, with $^3S_1$ strongly enhanced and $^3D_3$ strongly suppressed.[11] While one might imagine that the threshold factor λ^3 might be smaller at threshold due to the fact that the central barrier is somewhat higher, it is likely that the tighter binding of the inner $j_{1z} = 1/2$ electron at small negative z_N overcomes this effect. We conclude, therefore, that He is a very effective detector for $q = 1$ monopole and is very likely to be so for higher q's as well.

As discussed in Ref. 1 and by several speakers at this conference, the preferred method of detecting the excitation is by collisional ionization of the $^3S_1$ states with a doping gas such as CO_2 or CH_4. All of the higher triplet states quickly decay either directly or by cascade to the metastable $^3S_1$ state.

DYON INTERACTIONS WITH HELIUM

In this section we consider the interaction of a dyon with helium. The dyon is assumed to have q = 1/2 and plus or minus one unit of electric charge. Such a charge on a monopole could arise during the production mechanism in the early universe or in the case of positive charge by the subsequent capture of a proton. The size of the monopole-proton bound state is approximately 10 fm and is therefore very small compared to atomic dimensions and hence will be considered point like.

The positively and negatively charged dyons need to be considered separately, and we first consider the positively charged case. Following the discussion of previous sections, the eigenstates of the dyon-helium atom system are first identified at large separation. Apart from the usual states of the helium atom, an electron can form bound states with the positively charged dyon with energies $-\frac{1}{2} m\alpha^2/(n+\mu)^2$ where n = 0, 1, ... and $\mu = \sqrt{j(j+1)}$; j = 0, 1, 2, ... being the angular momentum. The ground state of the dyon is 13.6 eV below the continuum and thus the dyon will pick up an electron while travelling through matter. Metal surrounding or making up a detector would be an ideal source of such electrons. Therefore we need to consider a dyonic atom and a helium atom colliding, i.e., a three electron system.

Next, we consider the dyon and helium nucleus on top of one another. In doing this, we must ignore for the time being the Coulomb repulsion between the dyon and the nucleus. This interaction, which turns out to be quite important, will be taken into account later.

The relevant z_N = 0 central multiplet of the three electron system is in a $(0)(1)^2$ configuration formed from the two lowest hydrogenic states and because of the Pauli principle, must have J = 1. Because one of the electrons is already bound to the dyon (with J = 0) when the collision begins, the J_z value which connects to the z_N = $-\infty$ configuration is J_z = 1. This shows that the ground state z_N = 0 central multiplet described above is indeed the relevant central multiplet and there are no off center crossings. The states which connect to the J_z = 1 and J_z = 0 states of the central multiplet when $z_N \rightarrow +\infty$ are, as discussed in a previous section, the minimum energy J_z = 1 and J_z = 0 states with z_N = $+\infty$. One sees by inspection that these are states with He in its ground state and with one electron bound to the dyon and in excited J = 2 or J = 1 states with excitation energies of 11.3 and 6.8 volts, respectively.

Following the method of the Hydrogen atom calculation, we may calculate the transition probabilities as a function of distance of closest approach b_0 Eq. (18). Again the interaction parameter

γ (call it γ_D here) must be calculated. Following the previous calculation for helium, this is performed variationally using the same wave functions as in the previous section but with the extra electron in a $j = 1$, $m_j = 0$ monopole harmonic. The parameters of the variational wave functions are determined so as to minimize the energy subject to the constraint that the electric and magnetic dipoles as discussed earlier give the same result. The interaction parameter γ_D is found to be approximately 20 times the hydrogen value due mainly to the extra charge (partially shielded) at the origin, which comes in as the third power. Thus we find

$$\gamma_D \approx 1.4 \times 10^{-2}/a_0^2.$$

In order to compute the partial cross sections we use Eqs. (16), (17), and (18). For this case, we obtain

$$|c_0|^2 = 2(e^{-\Gamma} - e^{-2\Gamma})$$

$$|c_1|^2 = e^{-2\Gamma}$$

where

$$\Gamma = \pi\gamma_D b_0^2/2\beta_0 = \pi\gamma_D b_0^3/2b\beta.$$

We have used angular momentum conservation of the nuclear orbit in the last equality. In order to compute a cross section, we need to express b_0 in terms of b and β. This can easily be done taking account of the Coulomb repulsion between the dyon and the nucleus, the attractive force of the electrons, and angular momentum conservation. Thus we have

$$\frac{1}{2} Mv^2 = \frac{1}{2} M \frac{b^2 v^2}{b_0^2} + \frac{2\alpha^2}{b_0} - \Delta \tag{25}$$

where Δ is the net increase in binding energy of the electrons when the dyon and nucleus are separated by distances small compared to a_0. Here b_0 and b are in units of a_0 and $\Delta \approx 180$ eV. The cross sections may be evaluated by computing

$$\sigma_M = 2\pi \int |c_M^2(b_0)| b\,db$$

numerically, where b_0 is computed from Eq. (25). It is convenient to express the result as

$$\sigma(6.8) = (\beta/10^{-4})\Lambda_1 \, 2.0 \times 10^{-19} \, cm^2$$

$$\sigma(11.5) = (\beta/10^{-4})\Lambda_2 \, 1.0 \times 10^{-19} \, cm^2$$

Here $\sigma(6.8)$ represents the excitation cross section for the $J = 1$ excited state of the dyon, which has an excitation energy of 6.8 eV. It corresponds to $J = 1$, $M = 0$ in Eq. (18). Similarly, $\sigma(11.5)$ corresponds to $J = 1$, $M = 1$ in Eq. (18) and represents excitation of the dyon to its $J = 2$ state with excitation energy 11.5 eV. The parameters Λ_1 and Λ_2 represent threshold factors, analogous to the λ^3 which appears in Eq. (19), and are determined by the numerical integration described above. Representative values are listed below:

$\beta/10^{-4}$	Λ_1	$\beta/10^{-4}$	Λ_2
2	1.4×10^{-3}	3	7.1×10^{-4}
3	8.5×10^{-2}	4	4.4×10^{-2}
4	.48	5	.236
5	.86	6	.48
6	1.02	8	.77
10	1.04	10	.88
	1.00	20	.99
			1.00

The difference in the behavior of Λ_1 and Λ_2 arises from the fact that $|C_1^2|$ peaks at $b_0 = 0$ where the Coulomb suppression is complete, while $|C_0^2|$ vanishes at b_0 and peaks at a finite value. The fact that Λ_0 slightly exceeds unity is a consequence of the fact that the force becomes attractive at large distances.

The excitation may be detected via the 6.8 eV and 4.5 eV photons which would be emitted by these states. The small cross section, the strong Coulomb suppression for $\beta < 4 \times 10^{-4}$, and the requirement that one detect photons rather than ionization makes He rather unattractive as a detector of positively charged dyons.

For a negatively charged dyon the eigenstates for a widely separated dyon and atom are just the eigenstates of the atom. When the dyon and helium atom are on top of one another we are

trying to bind two electrons to a charge one center with a
monopole. This situation has similarities to the cases of a
hydrogen minus ion which has one bound state with energy -0.75 eV
and a helium minus ion which does not exist as a bound state. Let
us compare the dyon of charge one and the hydrogen minus ion. The
first electron goes into a state with energy -13.6 eV in both
cases, whereas the second electron for the dyon case goes into a
triplet state with energy -6.8 eV (ignoring shielding) and into
the n = 1 state with energy -13.6 eV (ignoring shielding) for
hydrogen ion. Also the shielding for the dyon case will be larger
than for the hydrogen ion so that the monopole will be bound less
than 0.75 eV and there is probably no bound state. The details
here are not important as a dyon at a velocity of 10^{-4} c can cause
transitions of a few tenths of an electron volt.

Thus the cross section for ionization of helium is just π
times the square of the distance at which binding becomes a few
tenths of an electron volt. This distance is estimated to be of
order 1/3 a_0, giving a cross section of 3×10^{-17} cm^2. Of course,
the Coulomb attraction between the dyon and the nucleus will
enhance this effect by drawing the nucleus towards the dyon. The
added energy in the electrons due to the reduction in charge near
the center (< 55 eV) is less than the Coulomb attraction between
dyon and nucleus at a distance of a_0 and is therefore negligible.
The energy loss, neglecting the attraction between dyon and
nucleus will be of order 100 MeV cm^2 g^{-1}. This is about 50 times
minimum ionizing and thus an uncertainty in our estimate of the
cross section by a factor of 4 or more is not important. Also
there is no need for a quenching gas like CO_2 or CH_4 for this case
as the dyon ionizes the helium atom directly. The threshold for
this process is 1.2×10^{-4} c and there is no reduction in cross
section near threshold. Below threshold, the ionization cross
section probably continues to be high but accompanied by capture
of the α particle and electron by the dyon. This process
ultimately leads to a tightly bound α particle dyon system which
behaves like a positive dyon.

In summary, helium would be a very efficient detector for
negatively charged dyons down to a β of 1.2×10^{-4}, but appears to
be much less promising as a detector for positively charged dyons.

HEAVIER NOBLE GAS ATOMS

The primary limitation of He as a slow monopole detector is
its threshold at $\beta = 1.03 \times 10$-4. One would prefer to be sensitive at least to 0.37×10^{-4}, the velocity of escape from the
earth. Because our arguments relied heavily upon the spherical
symmetry which is obtained when the monopole and nucleus coincide,
it is natural to consider heavier noble gas atoms. Such atoms are

extensively used in proportional counters and are not subject to the annoying leakiness of He proportional counters. Assuming the threshold to be determined by the excitation energy, which we take to be of the order of the ionization energy one finds (β_t = threshold β):

Ne	A	Kr	Xe

$$10^4 \beta_t = \quad 0.48 \quad 0.29 \quad 0.19 \quad 0.14$$

It is not, however, clear that we have used a proper measure of the threshold. We recall that the monopole presents a 16 eV central barrier to the impinging He atom, a fact which gives rise to the λ^3 threshold factor which appears in Eq. (19). This barrier provides a threshold factor, rather than the principal determinant of the threshold, because it is lower than the required excitation energy. We have already noted that the existence of a Coulomb barrier in the He-positive dyon problem more than doubles the effective threshold.

In order to assess the barrier height question in a preliminary way, we first consider the diamagnetic repulsion which the atom experiences when it is far from the monopole. Using experimental values for the diamagnetic susceptibility, we find for $a_0/z_N \ll 1$,

$$E(\text{diamagnetic}) = H(a_0/z_N)^4 \tag{26}$$

with

He	Ne	A	Kr	Xe

$$H(eV) = \quad 1.3 \quad 4.8 \quad 130 \quad 265 \quad 423$$

While Eq. (26) obviously gives a gross overestimate as $z_N \to 0$, it should be noted (see Fig. 1 of DKMPR) that for He at $z_N = a_0$ it is roughly a factor ten too small. While it would be unwarranted to assume ten is a universal factor, it does seem very likely that the actual barrier at $z_N = a_0$ is several times the value given by Eq. (26). We further comment that an argon atom with $\beta = 1.03 \times 10^{-4}$ the He threshold, cannot penetrate beyond a 200 eV barrier height, and on the basis of the above discussion, seems unlikely to achieve a z_N as small as a_0. It may therefore be out of the range at which quasi adiabatic excitation takes place. If this is the case, the quasi adiabatic threshold for argon would be higher than that of He.

Once β is sufficiently large to allow barrier penetration, it is certain that degeneracies and near degeneracies will be encountered, and hence quasi adiabatic excitation will take place. The question of the magnitude of γ factors, which depend primarily on the extent to which the last shell electrons are screened for the state which connects to the ground state, is then of crucial importance in determining quantitative cross sections. A contribution from off center crossings may also prove to be important, and distortions of the multiplet structure of the sort found in the q = 1 case for He may further complicate matters.

In summary then, the probability that the threshold for quasi adiabatic excitation can be pushed below $\beta = 10^{-4}$ by using heavy noble gas atoms seems to be low. More work is required both to determine what the threshold is and how large the cross sections are. It seems likely, however, that they become satisfactorily large somewhere between $\beta = 10^{-4}$ and $\beta = 10^{-3}$.

CONCLUDING REMARKS

Despite some quantitative uncertainties about cross section magnitudes, it seems certain that He proportional counters will satisfactorily detect massive Dirac monopoles of arbitrary charge in the $\beta = 10^{-4}$-10^{-3} velocity range. The threshold increases slowly with magnetic charge but will remain below the ionization limit at $\beta = 1.2 \times 10^{-4}$. Such counters will also detect negatively charged dyons at this threshold.

APPENDIX

Because our use of the magnetic dipole perturbation has received some criticism,[12] we add some technical comments about the hydrogen energy level problem here.

Considering the energy as a function of z_N, the term linear in z_N near $z_N = 0$ can be obtained from either the point electric or point magnetic dipole. Here we restrict the perturbation Hamiltonian to terms linear in z_N, which means the δA^2 term is omitted in the magnetic dipole case. As explained in footnote 9, the two forms of the Hamiltonian give the same result.

One might also consider carrying out first order perturbation theory with real dipoles, that is two point charges (electric or magnetic) of opposite sign separated by a distance z_N. In this case, our argument for the equivalence of the two cases no longer applies. First order perturbation theory for the real electric dipole can be readily carried out, and one finds a correction to our point dipole result of order z_N/a_0. Higher order perturbation theory would, of course, be expected to yield terms of similar order. First order perturbation theory for the real magnetic dipole case is, however, singular. To be specific, we consider the case in which the dipole is aligned along the z axis, with the original magnetic pole displaced from the origin in the positive direction. Then for the $J_z = 1$ state, first order perturbation theory can be carried out, and a correction to the point dipole result of order $(z_{Na_p})^{2\sqrt{2}-2}$ is found. However, for the $J_z = 0$ and $J_z = -1$ cases, the perturbation integrals diverge. If one first performs the radial integrals, which do converge, but carries out the small z_N expansion before completing the angular integration, one finds that the linear terms agree with the point dipole result, but the angular integral for the order $(z_N/a_0)^{2\sqrt{2}-2}$ correction diverges. If the poles are displaced to the negative side, it is the $J_z = -1$ state rather than the $J_z = 1$ state which is well behaved. (Note that the difficulty does not depend upon the choice of string direction, which only affects the ϕ dependence of the states.) The formal origin of the difficulty lies in the structure of the differential equation and has been discussed in the context of variational calculations by Tiktopoulos.[6] To explain the origin physically we first recall that the ordinary spherical harmonics vanish along the z axis for $m \neq 0$. This vanishing can be understood as resulting from the effect of the repulsive centrifugal potential associated with the $m^2/r^2\sin^2\theta$ term in the Hamiltonian. In the case of monopole harmonics, the magnitude of the centrifugal potential changes as one switches from one sign of the z axis to the other, and for the $q = 1/2$ case, may vanish on one side or the other for $|J_z| = 1$ or 0. Accordingly, the wave functions do not vanish for such states along the entire z axis. Shifting the position of the pole without changing the states, which is what one does when one carries out first order perturbation theory, can shift the repulsive and singular centrifugal potential into a region where the wave function does not vanish, leading to the divergent expectation values described above. Thus an attempt to apply first order perturbation theory to a real magnetic dipole using numerical integration techniques[12] is bound to lead to unreliable and confusing results.

As a final remark, it may be of interest to mention that we have carried out numerical variational calculations for the $J_z = \pm 1$ excited states and the $J_z = 0$ ground state all the way from $z_N = -\infty$ to $z_N = +\infty$ using wave functions which guarantee consistency with perturbation theory at the $|z_N| = 0$ and infinity limits. (Our wave function for the ground state is the same as that used by Tiktopoulos[6] and our results agree with his.) These have been done accurately enough to allow us to numerically confirm the value of the slope at the origin to 0.2 percent. The calculations are useful because they indicate that the departure from linearity is small, linearity being valid within ~ 5 percent out to a distance $0.5\ a_0$. This result supports the approximations we have used in calculating excitation probabilities. Analogous calculations have also been carried out for $q = 1$.

ACKNOWLEDGEMENTS

Work supported in part by the Department of Energy, contracts
DE-AC03-76SF0015 (SLAC) and DE-AT03-S1ER40029 (UCSD). Fermilab
is operated by the Universities Research Association, Inc. under
contract with United States Department of Energy.

REFERENCES

a. Now at Fermi National Accelerator Laboratory, Batavia, IL.
1. S.   Drell, N. Kroll, M. Mueller, S. Parke, and M. Ruderman,
    Phys. Rev. Lett. $\underline{50}$, 644 (1983).
2. T.T. Wu and C.N. Yang, Nuclear Phys. B $\underline{107}$, 365 (1976).
3. M. Fierz, Helv. Phys. Acta $\underline{17}$, 27 (1944).
4. For divergence free magnetic fields and vector potentials,
    this field angular momentum is given by $e\vec{A}$, so that the
    canonical angular momentum $\vec{r}_i \times \vec{p}_i$ represents the sum.
5. C. Zener, Proc. Roy. Soc. A $\underline{137}$, 696 (1932).
6. G. Tiktopoulos, Phys. Lett. B $\underline{125}$, 156 (1983). Compare with
    Eq. (15) of the above paper. We are including our time
    dependent Hamiltonian and hence do not have the $d\Omega/dz_0$ term
    which appears there.   While we think that such a term should
    not be included in our treatment, to be on the safe side we
    have verified that it also satisfies the selection rule.
7. We are using the phrase "quasi adiabatic" to refer to a
    situation in which relative velocities are slow enough to
    allow us to assume that excitations are dominated by transi-
    tions to levels whose energy separation from the incident
    state is small compared to the minimum excitation energy
    which would be required for the isolated atom.   The transi-
    tions are actually non-adiabatic.
8. W.V.R. Malkus, Phys. Rev. $\underline{83}$, 899 (1951).
9. In DKMPR we discussed this problem in a reference frame in
    which the nucleus was fixed at the origin.   This made it
    natural to use as the perturbation Hamiltonian the change in
    H due to the addition of a point magnetic dipole instead of a
    point electric dipole.   That is to say, in place of H_e we
    used

$$H_m = -\frac{-q}{2mr^3}\, \vec{r}_N \cdot [2\vec{r} \times (\vec{p}-e\vec{A}) - \vec{\sigma} + 3\hat{r}\vec{\sigma}\cdot\hat{r}]$$

That the two must give the same result follows from

$$H_e - H_m = \vec{r}_N \cdot \vec{\nabla}H = i\vec{r}_N \cdot [\vec{P},H]$$

and $(\psi_n, [\vec{P},H]\psi_n) = 0$ for any eigenstate of H.
10. C.   Schwartz, Phys. Rev. $\underline{134}$, A1181 (1964); J.D. Prestage,
     E.A. Hinds and F.M.J. Pichanick, Phys. Rev. Lett. $\underline{50}$, 828
     (1983).

11. There are four additional excitation states, including the $^3F_4$ state to which all excitation occurs at zero impact parameter, which are not included in the lower Paschen-Bach multiplet discussed in the text. Their total contribution to the cross section is much smaller.

12. R. Hagstrom, Theory of Excitation and Ionization of Matter by Monopoles, paper presented at this conference, Oct. 6, 1983.

THE COUPLING BETWEEN MAGNETIC CHARGES AND MAGNETIC MOMENTS

Gianni Fiorentini

Dipartimento di Fisica
Istituto Nazionale di Fisica Nucleare
Sezione di Pisa, I-56100, Pisa

ABSTRACT

I discuss several consequences of the coupling of magnetic charges with the magnetic moments of electrons and nucleons. Particularly I discuss the formation of bound states of monopoles and nuclei, and its implications on the Rubakov-Callan effect. I also report on recent calculations of the energy loss of very slow magnetic monopoles and on the possibility that magnetic monopoles induce nuclear reactions.

INTRODUCTION

What is new with magnetic monopoles is that their existence is nowadays predicted by several field theories, all more or less directly aiming to a unified description of the basic forces of nature. As we heard at this conference, the predictions about the monopole mass, m_M, span several orders of magnitude: from $m_M \simeq 10^{10}$ GeV up to $m_M \simeq 10^{20}$ GeV. The mass being so large, the only hope for finding a monopole is to look at the relics of the Big Bang. Such heavy particles are expected to be very slow, the arrival velocity onto Earth being in the range $v/c \sim 10^{-5} - 10^{-3}$. This low velocity implies non trivial interactions with matter, which are to be understood in view of planning experiments on the search of magnetic monopoles. Concerning this problem, I will discuss a few points, on which there have been recent developments.

First I will discuss the possibility that slow monopoles can form bound systems with atomic nuclei[1-5]. It has been found that the electromagnetic interaction gives rise to bound systems with binding energies in the range 1-100 KeV and with typical linear sizes of the order of 20 Fermi. The scales of these systems being approximately the same as of mesonic atoms, the name "monopolic atoms" seems appropriate. The existence of these states has interesting implications on the monopole catalysis of proton decay, the so called Rubakov-Callan effect.

Next, I will discuss other kinds of bound states, where monopoles and atomic nuclei are bound together by electrons, in a way similar to the chemical binding of molecules.[6,7] In these systems monopoles and nuclei are typically 1 Ångstrom apart and the dissociation energy – i.e. the energy needed to break the system into a monopole plus an (ordinary) atomic system – is in the eV range. These systems, which I refer to as "monopolic molecules", are interesting in connection with the Rubakov-Callan effect.

The third point I will consider is the energy loss of very slowly moving magnetic monopoles, $v/c < 10^{-3}$, particularly the transfer of kinetic energy through elastic collisions with atoms. We will see that this mechanism originates energy losses of order 100 MeV/cm in a solid,[8] thus implying that monopoles reaching Earth with $v/c < 10^{-4}$ are stopped. On these grounds one can derive bounds on the flux of very slow monopoles in the solar system.

Finally I would like to mention the possibility that a monopole, passing close to a Uranium nucleus, induces nuclear fission, an effect which could provide an interesting signature for monopoles.

THE COMMON SOURCE

The four processes I just mentioned seem to be unrelated. Actually, they all have a common origin, which I would like to explain with an example. Consider a monopole, its magnetic charge q_M being given by the Dirac condition,

$$qM = hc/4\pi e \tag{1}$$

At a point $\underline{R}$ one has a magnetic field:

$$\underline{B} = qM\underline{R}/R^3 = hc\underline{R}/4\pi eR^3 \tag{2}$$

Consider now an electron in the point $\underline{R}$, the relative monopole-electron velocity being $\underline{v}$. The electron feels the Lorentz force:

$$\underline{F}_L = \frac{-e}{c} \underline{v} \times \underline{B} \to F_L \sim vh/4\pi R^2 \quad . \tag{3}$$

Since the electron has a magnetic moment $\mu_e = -eh/4\pi m_e c$ and the field is not homogeneous there is an additional force:

$$\underline{F}_M = -\underline{\nabla} (\underline{\mu}_e \times \underline{B}) \to F_M \sim h^2/8\pi^2 m_e R^3 \quad . \tag{4}$$

Now let us see which is bigger between F_L and F_M. From Eqs. (3) and (4) one gets:

$$F_L/F_M \sim (v/\alpha c)(R/a_o) \quad , \tag{5}$$

where $a_o \simeq 0.5 \times 10^{-8}$ cm is the electron Bohr radius. Therefore, at small velocities ($v < \alpha c$) and/or small distances ($R < a_o$) the force arising from the coupling between the magnetic charge and the magnetic moment is dominant.

Recall that one encounters an opposite situation in ordinary atomic physics, where spin coupling provides fine and hyperfine structures, negligible in a first approximation, and let me add that the above example can be generalized to any other particle of mass m provided that in Eq. (4) m_e is replaced with m and the appropriate gyromagnetic factor is introduced.

It is also interesting to have an estimate of the interaction energy arising from the coupling of the magnetic moment, in order to understand which effects can be induced by this coupling. For a magnetic moment

$$\mu = k \, eh/4\pi mc \tag{6}$$

the interaction energy with the monopole is:

$$U_{dip} = -(\underline{\mu} \times \underline{B}) \sim kh^2/16\pi^2 mR^2 \quad . \tag{7}$$

Note that the electron charge disappears, since "e" in Eq. (6) cancels with the factor "1/e" in Eq. (2), as typical of the magnetic monopole interaction.

Consider now (See Fig. 1a) a monopole at a distance $R \sim a_o$ from an electron ($m = m_e$, $k = 1$). The interaction energy is

$$U_{dip} \simeq h^2/16\pi m_e a_o^2 \simeq 7 \text{ eV} \quad , \tag{8}$$

a value which is quite efficient for a strong deformation of an atomic system.

If we consider (See Fig. 1b) a monopole at a distance

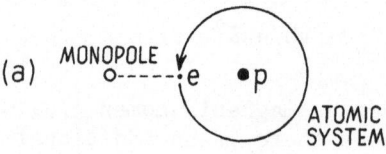

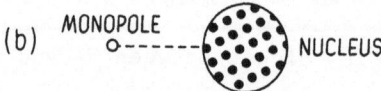

Fig. 1   Interaction of a monopole M with an atomic system, a), and
         with a nucleus, b).

$R = R_o = 1$ fermi from a nucleon ($m = m_p$, $k \sim 2$),   the interaction
energy is

$$U_{dip} \simeq h^2/8\pi^2 m_p R_o^2 \simeq 20 \text{ MeV}, \tag{9}$$

a value comparable to the binding energy of nucleons in nuclei, so
that one can expect significant modifications of the nucleus in
the presence of the monopole.

MONOPOLIC ATOMS

Consider a nucleus $_Z N^A$ with spin $\underline{s}$ and magnetic moment:
$$\underline{\mu}_N = (eh/4\pi m_N c) \ k \ \hat{s}. \tag{10}$$

k is the gyromagnetic factor, which, for many nuclei, is

$$k \approx A . \tag{11}$$

The monopole-nucleus dipole Hamiltonian is:
$$U_{dip} = - (\underline{\mu}_N \times \underline{B}) = - (\hat{s} \times \hat{R}) \ k \ h^2/16\pi^2 m_N R^2 . \tag{12}$$

For a suitable spin orientation U_{dip} is attractive.   Can this
result in monopole-nucleus bound states?   For this to occur, U_{dip}
has to overcome the centrifugal barrier,

$$U_{cen} \approx (J + 1/2)^2 \ h^2/8\pi^2 m_N R^2 \tag{13}$$

The comparison of Eqs. (12) and (13) shows that for k large
enough and for small angular momenta the effective potential,
$U_{dip} + U_{cen}$, is attractive at any R, thus suggesting that bound
states can exist. Note that only the gyromagnetic factor k enters
in the discussion, and k is large for many nuclei.

It is worth observing that potentials like U_{dip}, which behave as $-R^{-2}$ near the origin, are pathological. Classically this means that one has to fall onto the center of the potential. In quantum mechanics the spectrum of the Hamiltonian is not bounded from below. Actually, these pathologies are unphysical, in that the dipole approximation does not make any sense for distances smaller than the nucleus radius,

$$R_N \simeq 1.2 \times A^{1/3} \text{ Fermi.} \tag{14}$$

The monopole nucleus interaction U_{MN} is really unknown at distances $R < R_N$. However, it is possible to use a safe prescription in order to get lower bounds on the binding energies. Let me take, as in Ref. 2:

$$U_{MN} = \begin{cases} U_{dip} \text{ for } R > R_N \\ + \infty \text{ for } R < R_N. \end{cases} \tag{15}$$

This prescription is safe in that the true potential, whatever it is, is more attractive than I am assuming and consequently the true binding energy is larger than I will find.

Thus, by using Eq. (15) one generally underestimates the binding energy. The important point is that, if a nucleus is found to bind to a monopole in this approximation, a fortiori, it will do so in the real world.

In Ref. 2 Bracci and myself have studied the Schrödinger equation for spin 1/2 nuclei resulting from the Hamiltonian:

$$H = (\underline{p} - Ze/c \ \underline{A}_M)^2/2m_N + U_{MN} \quad , \tag{16}$$

where $\underline{A}_M$ is the monopole vector potential and U_{MN} is taken as in Eq. (15). In summary, our results are the following:

i) Either zero or infinite bound states exist, depending on the value of the gyromagnetic factor k.

ii) The bound states are grouped in families of given total angular momentum. Inside each family, the ratio of the binding energies of two consecutive levels,

$$C = E_n/E_{n-1} \quad \text{,is approximately constant.} \tag{17}$$

iii) There are many stable spin 1/2 nuclei that can bind to monopoles. They are listed in Table 1, where one sees that the binding energies of the ground states are in the range

TABLE 1. Binding energies of stable spin 1/2 nuclei $_Z N^A$ to magnetic monopoles with charge $q_M = hc/4\pi e$ and infinite mass. $\tilde{\mu}$ is the nucleus magnetic moment in units of the nuclear magneton, E_b is the binding energy of the most tightly bound state and C is the ratio between the energies of two consecutive levels.

Nucleus	$\tilde{\mu}$	E_b(keV)	C
$_1H^1$	2.79	15.1	4×10^{-4}
$_1H^3$	2.98	112	2.2×10^{-2}
$_2He^3$	-2.13	13.4	3.85×10^{-3}
$_6C^{13}$	.7	1.8	4.2×10^{-3}
$_9F^{19}$	2.63	383	.25
$_{15}P^{31}$	1.13	49.2	.134
$_{48}Cd^{113}$	-.62	6.3	.14
$_{50}Sn^{115}$	-.92	29.6	.3
$_{50}Sn^{117}$	-1.0	38.6	.33
$_{50}Sn^{119}$	-1.05	43.9	.35
$_{52}Te^{123}$	-.74	14.6	.23
$_{54}Xe^{129}$	-.78	28.3	.26
$_{70}Yb^{171}$	.49	1.5	9.1×10^{-2}
$_{78}Pt^{195}$	.61	7.6	.24
$_{80}Hg^{199}$	.50	2.2	.13
$_{81}Tl^{203}$	1.61	94.5	.57
$_{81}Tl^{205}$	1.63	99.4	.57
$_{82}Pb^{207}$	.59	6.9	.24

$$E_b \approx 1 - 100 \text{ KeV} \quad , \tag{18}$$

the typical size of these states being

$$L \approx 20 \text{ Fermi.} \tag{19}$$

Note, in particular, that protons can bind to monopoles with a binding energy of about 15 KeV in this approximation. Other authors, by using different approximations for the interaction at distances $\leqslant 1$ Fermi, have found stronger binding. I recall that this is consistent with our results, since these latter are to be interpreted as lower bounds.

The possibility that nuclei bind to monopoles has several phenomenological consequences. I would like to discuss in some detail the implications on the Rubakov-Callan effect, starting with a pedagogical discussion of the nuclear capture of slow negative muons:

$$\mu^- + N \to \nu_\mu + N' \quad . \tag{20}$$

This process has a typical weak interaction cross section,

$$\sigma_W \approx (c/v_\mu) \, 10^{-40} \text{ cm}^2 \quad , \tag{21}$$

where v_μ is the muon velocity. The capture rate,

$$\lambda_{cap} = \sigma_W \times v \times (\text{target density}) \quad ,$$

is at most of order 10^{-6} s^{-1}, absolutely negligible in comparison with the muon decay rate, $5 \times 10^{+5} \text{ s}^{-1}$. However, if the target is a solid, heavy material, practically any muon undergoes reaction in Eq. (20). The point is that there is an intermediate stage,

$$\mu^- + N \to (\mu^- N) \to \nu_\mu + N' \quad , \tag{22}$$

where a muonic atom $(\mu^- N)$ is formed. This system is formed with a cross section of typical atomic dimensions, $\sigma_{for}^{(\mu)} \approx 10^{-16} \text{ cm}^2$, when the muon is sufficiently slow, say $v_\mu \leqslant 10^{-3}$ c. This corresponds to a formation rate:

$$\lambda_{for} \approx 10^{14} \text{ s}^{-1} \quad , \tag{23}$$

i.e. is a very fast process. The muonic atom is very compact, the typical size L_μ being a few Fermi for a heavy nucleus. Thus the effective density, i.e. the overlap probability between the muon and the nucleus per unit volume, is:

$$\rho_{eff} \approx L^{-3} \approx 10^{38} \text{ cm}^{-3} \quad . \tag{24}$$

Inside the muonic atom, muon capture occurs at a rate:

$$\lambda^{(\mu^- N)}_{cap} = \sigma_W \ v_\mu \ \rho_{eff} \approx 10^8 \ s^{-1} \quad , \tag{25}$$

which is larger than the muon decay rate, thus accounting for the observation of muon capture. One sees that the formation of the bound system considerably enhances the process of muon capture. Similarly it can occur that the monopolic atom formation enhances the Rubakov-Callan effect. In addition to the catalysis in flight, the equivalent of Eq. (20),

$$M + N_{B=A} \rightarrow M + e^+ + (\dots)_{B=A-1} \quad , \tag{26}$$

one can have a two step process:

$$M + N_{B=A} \rightarrow (M \ N_{B=A}) \rightarrow M + e^+ + (\dots)_{B=A-1} \quad . \tag{27}$$

If the cross section for forming the monopolic atom, $\sigma^{(M)}_{for}$, is known, from results of proton decay experiments one can get information on the flux ϕ of monopoles impinging onto Earth and/or on the cross section σ_{cd} for catalyzed decay. In Fig. 2, I show as an example the information one gets about monopoles with $v \simeq 10^{-3}$ c.[2] The variable w along the horizontal axis is the ratio of the true catalysis cross section to the estimate of Rubakov and Callan $w = \sigma_{cd}/\sigma_{RC}$, (28). Curve a) denotes the boundary of the allowed region by considering proton decay catalysis in flight only, i.e. Eq. (26). Curves b) and c) denote the boundary when monopolic atom formation is taken into account. They are obtained, respectively, by assuming

$$\sigma^{(M)}_{for} = 10^{-26} \ cm^2 \quad ; \quad \sigma^{(M)}_{for} = 10^{-23} \ cm^2 \quad .$$

These are two conservative estimates of the cross section for the formation of monopolic atoms, which actually can be quite larger.

It is remarkable that by taking into account monopolic atom formation one gets rather strict bounds on the monopole flux, even if the cross section for catalyzed decay is orders of magnitude smaller than predicted by Rubakov and Callan. The bounds found are comparable to the Parker bound (see Ref. 9 and the talk by Professor Parker at this conference) and compete with the bound derived from analysis of neutron stars if $w \lesssim 10^{-6}$ (see Ref. 10 and the talk by Professor Kolb at this conference).

Let me observe that the values of $\sigma^{(M)}_{for}$ which have been used were derived by using some rough approximation. We plan to perform in the future a more refined analysis of the formation process of monopolic atoms. Some work in this direction has been recently done by Olaussen et al.[11]

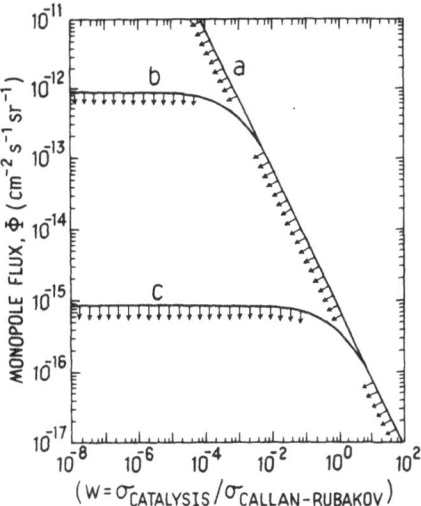

Fig. 2 Bounds on the monopole flux φ and on the probability of
        catalyzed proton decay, $w=\sigma_{c,d}/\sigma_{RC}$.   Curve a) corresponds
        to catalysis in flight only.   In curves b) and c) the
        formation of monopolic atoms has been taken into account,
        assuming, respectively, 10^{-26} cm^2 and 10^{-23} cm^2 as the
        formation cross section.

        In conclusion, the formation of bound states can strongly
affect the Rubakov-Callan effect.   Monopolic atoms, however, are
not the end of the story in this respect. We will see in the next
section that other kinds of bound states can be interesting too.

MONOPOLIC MOLECULES

        Consider as a prototype the three body system consisting of a
monopole, an electron and a proton.   The electron is attracted by
the proton through the Coulomb interaction and can be attracted by
the monopole for a suitable spin interaction.   Can the three
bodies stay bound together? The affirmative answer can be guessed
just by looking at the large distance behavior of the (ep)- M
interaction.   What matters is the coupling of the electron
magnetic moment $\mu_e = eh/4\pi m_e c$ with the monopole magnetic field.
From Eq. (7) we get:

$$U_{(ep)-M} \approx - h^2/16\pi^2 m_e R^2 \quad , \tag{29}$$

which has to be compared with the centrifugal barrier,

$$U_{cen} \approx (J + 1/2)^2 \ h^2/8\pi^2 m_p R^2 \quad . \tag{30}$$

Clearly Eq. (29) is stronger than Eq. (30) as long as $J < (m_p/m_e)^{1/2}$, which suggests that there are many families of bound states.[2] The size and the binding energy of the ground (M-e-p) state can be derived by resorting to a Born-Oppenheimer calculation of the proton-monopole potential V_{pM} in the presence of the electron, what a chemist would call the molecular term. The result of the calculation[7,12] is presented in Fig. 3. Curve a) corresponds for R → ∞ to a hydrogen atom in the 1s state, the electron magnetic moment being parallel to the monopole magnetic field. In this effective potential the proton can oscillate around the equilibrium position, $R \simeq a_0$, the zero oscillation mode corresponding to a binding energy of about 1.6 eV.[7] Thus one has a configuration resembling an ordinary molecule.

It is worth observing that the formation of monopolic molecules can inhibit the catalysis of proton decay. Indeed, in these systems the probability of having the proton at the same site as the monopole is exponentially small, since R = 0 lies in the classically forbidden region. For this reason it is necessary to study in detail the competition between the formation of monopolic atoms, which enhances the monopole-proton overlap, and the formation of monopolic molecules, where on the contrary the overlap is suppressed. I plan to investigate this point in the future.

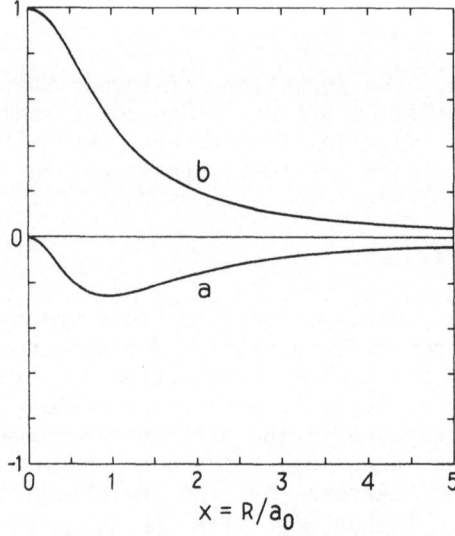

Fig. 3 Molecular terms for the system (M-e-p), in units of $m_e \alpha^2 c^2/4$. In the limit R → ∞ curve a (b) corresponds to a hydrogen atom in the 1s state, the electron magnetic moment being parallel (antiparallel) to the monopole magnetic field.

ENERGY LOSS OF VERY SLOWLY MOVING MAGNETIC MONOPOLES

It is again the coupling of the electron magnetic moment,

$$U_{dip} = -\underline{\mu}_e \times \underline{B} = (\hat{s} \times \hat{R})\, h^2/16\pi^2 m_e R^2 \quad , \tag{31}$$

which accounts for the energy loss of monopoles colliding with atoms. As long as v/c exceeds 10^{-4}, energy is lost by the monopole in excitation of atomic levels through a mechanism of adiabatic transitions between Zeeman levels discussed in Refs. 2 and 13 and reported at this conference by Professor Kroll. I would like to discuss the energy loss arising from elastic collisions with atoms, which is the only mechanism available for $v/c < 10^{-4}$, i.e. below the threshold for atomic excitations.

A rough estimate of the stopping power can be derived quite simply. The energy lost by the monopole per unit path is proportional to the number of atoms per unit volume, ρ, to the energy in the center of mass of the collision, E_{cm}, and to some cross section σ, which is actually the transport cross section:

$$dE_M/dx = \rho\, E_{cm} \sigma \quad . \tag{32}$$

σ can be estimated by using dimensional considerations. For the motion in the potential Eq. (31) at an energy E_{cm} the only quantity with dimensions of $(length)^2$ is $h^2/4\pi^2 m_e E_{cm}$, so we take:

$$\sigma \approx h^2/4\pi^2 m_e E_{cm} \quad . \tag{33}$$

By substituting Eq. (33) into Eq. (32) one gets:

$$dE_M/dx \approx h^2/4\pi^2 m_e \quad , \tag{34}$$

which corresponds to an energy loss of about 100 MeV/cm for typical solid density. The result of a more precise calculation[8] for the energy loss in liquid hydrogen due to elastic collisions is reported in Fig. 4, curve (a). In the same figure are shown for comparison, the energy losses arising from adiabatic excitations, curve (b), from Ref. 13, and from impact ionization, curve (c), from Ref. 14.

One sees also that at very low velocity the energy loss is quite large and could be detectable. It is worth noticing that the energy lost in elastic collisions is released to the medium in the form of elastic vibrations and/or infrared radiation.

There are several consequences of this large energy loss. In particular, monopoles with $v/c < 10^{-4}$ are trapped when impinging onto Earth or onto the Moon. Unless some annihilation mechanism occurs, these monopoles contribute to the magnetic field of the

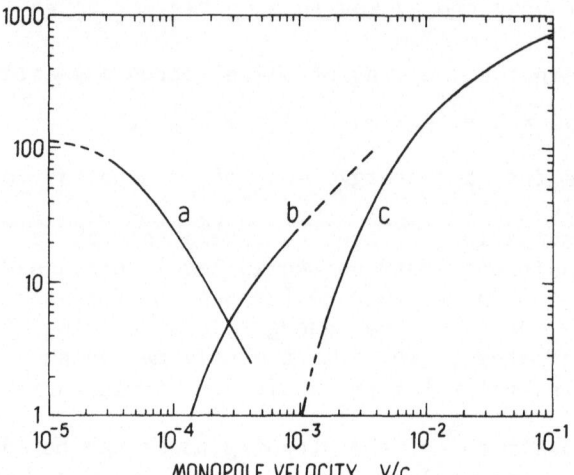

Fig. 4 The various contributions to the energy loss of monopoles
in liquid hydrogen. Curve (a) corresponds to elastic
scattering.[8]   Curve (b) is the effect of atomic excitation.[13]   Curve (c) is the energy loss through ionizing
collisions.[14]

celestial bodies where they are stopped.   The absence of magnetic
field on the Moon surface ($B_{Moon} < 10^{-3}$ Gauss) implies a bound on
the monopole flux:

$$\phi \ (v/c < 10^{-4}) < 10^{-13} \ cm^{-2} \ s^{-1} \ sr^{-1} \quad . \tag{35}$$

Otherwise one can assume, following Carrigan[15], that monopoles and antimonopoles annihilate in the interior of Earth when
the Earth's magnetic field reverses, which occurs typically once
every half million years. The energy released in the annihilation
contributes to the flux of heat from Earth's interior.   From data
on heat flow one deduces again:

$$\phi \ (v/c < 10^{-4}) \lesssim 10^{-13} \ cm^{-2} \ s^{-1} \ sr^{-1} \quad . \tag{36}$$

In conclusion, by using quite different assumptions on the
fate of monopoles trapped inside Earth and the Moon one gets the
same bound on the flux of monopoles.   This bound is complementary
to the information so far obtained from ionization experiments,
which are only sensitive to monopoles with v/c $> 10^{-3}$ and which
provide for these higher velocities a bound of comparable
strength, see Ref. 16 and the final talk by Professor Giacomelli
at this conference.

NUCLEAR REACTIONS INDUCED BY MAGNETIC MONOPOLES

From the discussion of "The Common Source", and particularly from Eq. (9), it follows that the monopole-nucleon interaction is so strong that a monopole passing close to a nucleus could induce some nuclear reactions. This is a point which is worth investigating, since it could result in effects which provide clean signatures of monopoles (see also the talk by Professor Lipkin at this conference). Particularly it is worth investigating exothermic reactions, other reactions being presumably inhibited since the collision energy, $E_{cm} \simeq m_N v^2/2$, is presumably too small for slow monopoles.

The possibility that monopoles induce fission of fissile nuclei, such as ^{238}U for example, is appealing. The idea (see Fig. 5), is that when a monopole passes close to a nucleus the nucleons which are closer orient their magnetic moments in the direction of the monopole, whereas the nucleons on the opposite side are almost unperturbed, since they feel a much weaker magnetic field. Thus the nucleus gets locally polarized and the nucleons close to the monopole are attracted by it. The nucleus

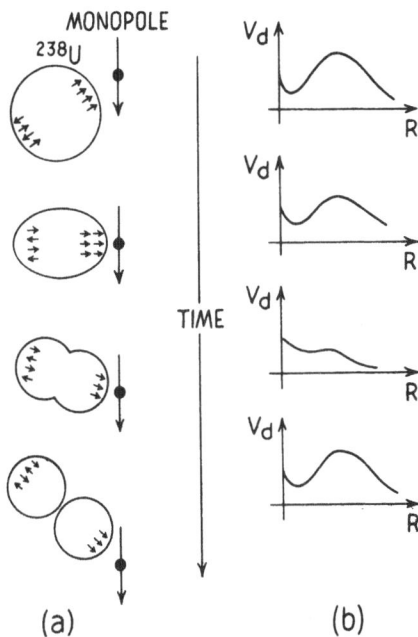

Fig. 5 Sketch of a monopole nucleus interaction that can result in nuclear fission. (a)Shape of the nucleus and distribution of nucleons' magnetic moments. (b)The total energy of the nucleus V_d as a function of a nuclear deformation parameter R, in the presence of the magnetic monopole.

therefore becomes elongated, by a sort of tidal effect. This deformation can in principle result in nuclear fission. In other terms, the curve representing the energy of the nucleus as a function of a deformation parameter can be modified by the presence of the monopole in such a way that for a short time the barrier preventing fission disappears (Fig. 5). Some work on monopole induced fission has been done in Pisa by L. Bracci, M. Rosa-Clot and myself but thus far we do not have a quantitative description of the process.

If it really occurs, monopole induced fission could be useful for the detection of monopoles. One could use several separated layers of ^{238}U and look for fissions occurring in coincidence along a straight line, which corresponds to the monopole trajectory.

CONCLUSIONS

There is a lot of atomic and nuclear physics which has to be studied in order to understand the fate of slow monopoles in matter. This physics is amusing. At the same time it is necessary in order to understand how known effects -the Rubakov-Callan effect for example - work in practice. Also this physics is interesting since it is possible to discover new effects which can provide clean signatures of magnetic monopoles. The problems of monopolic atoms and molecules, their energy spectra and formation processes, are a world to be explored. The same has to be said about nuclear reactions induced by magnetic monopoles.

There are so many and so amusing things monopoles can do, provided only that they exist.

ACKNOWLEDGEMENTS

I am grateful to Professor Giacomelli for several discussions about magnetic monopoles and it is a pleasure to thank my friends Bracci and Tripiccione, with whom most of the results presented have been accomplished. Also I would like to express my appreciation to the organizers of this nice conference.

REFERENCES

1. L. Bracci and G. Fiorentini, Phys. Lett. 129B, 29 (1983).
2. L. Bracci and G. Fiorentini, Phys. Lett. 124B, 493 (1983) and Pisa preprint IFUP TH 83/2, to appear in Nucl. Phys. B.
3. K. Olaussen, H.A. Olsen, P. Osland and I. Overbo, DESY preprint 83-041.

4.   J. Makimo, M. Maruyama and O. Miyamura, Osaka preprint
     OUAM-82-11-2.
5.   J.S. Trefil et al. Nature 302, 11 (1983).   See also the talk
     by Professor Goebel at this conference.
6.   L. Bracci and G. Fiorentini, Pisa preprint IFUP TH-83/2, to
     appear in Nucl. Phys. B.
7.   Th. W. Ruijgrok, T.A. Tjon and T.T. Wu, DESY preprint 83-026.
8.   L. Bracci, G. Fiorentini and R. Tripiccione, Pisa preprint
     IFUP TH-83/26.
9.   E.N. Parker, Astrophys. Journal 160, 383 (1970).   See also
     M.S. Turner, E.N. Parker and T.J. Bogdan, EFI preprint N.
     92-18.
10. E.W. Kolb, S.A. Colgate and J.A. Harvey, Phys. Rev. Lett. 49,
     1373 (1982).
11. K. Olaussen, H.A. Olsen, P. Osland and I. Overbo, "Proton
     Capture by Magnetic Monopoles", Trondheim Univ. preprint
     (1983).
12. G. Tiktoupolos, "Atomic Excitation and Ionization by Slow
     Magnetic Monopoles", Tech. Univ. Athens preprint (1983).
13. S.D. Drell, N.M. Kroll, M.T. Mueller, S.J. Parke and M.A.
     Rudermann, Phys. Rev. Lett. 50, 644 (1982).
14. S. Geer and W.G. Scott, CERN pp note (1981).
15. R.A. Carrigan Jr., Nature 288, 348 (1980).
16. G. Giacomelli, Invited paper at the "Magnetic Monopole Work-
     shop", Wingspread, Racine, Wisconsin, October 1982.

BINDING OF NUCLEI TO MONOPOLES

Charles J. Goebel

Department of Physics
University of Wisconsin-Madison
Madison, WI

The Hamiltonian describing the motion of a nucleus in the magnetic field of a monopole is

$$H = \frac{1}{2M} \left[p_r^2 + \frac{\Lambda^2 - \nu^2 - \eta \vec{S} \cdot \hat{r}}{r^2} \right] \tag{1}$$

where

$$\vec{\Lambda} = \vec{r} \times \left(\vec{p} - Q\vec{A}(\hat{r}) \right) - \nu \hat{r} \tag{2}$$

is orbital plus field angular momentum, with eigenvalues $\lambda = \nu$, $\nu+1,\ldots$; $\vec{A}$ is the vector potential of the monopole's field. M, $Q = Z_e$, and $\vec{S}$ are the mass, charge, and spin of the nucleus. The parameter ν is the product of the monopole strength and the nuclear charge. It has quantized values, and if the monopole has the minimal strength $G = 1/2e$, then

$$\nu = GQ = Z/2, \tag{3}$$

[for convenience, we take G and therefore ν positive]. Finally,

$$\eta \vec{S} = 2M \, G \, \vec{M}', \quad \text{or} \quad \eta = g\nu \tag{4}$$

where $\vec{M}'$ is the magnetic dipole moment of the nucleus, and g is its gyromagnetic ratio.

The second term in (1) includes centrifugal and Lorentz forces, and the force of the inhomogeneous magnetic field on the dipole moment.   Since these   all have the same r   dependence, the Schrodinger equation nicely separates, and each eigenstate   of the operator $\Lambda^2 - \nu^2 - \eta \vec{S} \cdot \hat{r}$ (these are of course eigenstates of $\vec{J} = \vec{\Lambda} + \vec{S}$) has an effective radial potential of the form $-C/2Mr^2$.   It can be shown that if $g > 0$ and $s \leqslant \nu$ the most attractive state has

$$j = j_{min} = \nu - s, \quad C = \eta s - \nu. \tag{5}$$

Hence, if $\eta s > \nu$ there is   a long   range $(1/r^2)$ attraction in the radial   wave equation in   at least   one angular state.   The table gives two examples:

Table I

	ν	s	η	j_{min}	C	B_o	g.s. BE
proton	.5	.5	2.79	0	.90	.0008	.01 Mev
$A\ell_{27}$	6.5	2.5	19.5	4	42.2	12.7	.56 Mev

For the proton, $j = j_{min} = 0$ is the   <u>only</u> angular state   in which $C > 0$.   For $A\ell_{27}$ there are 6 more; in order of decreasing strength of attraction the angular states with $C > 0$ are

Table II

j	4	5	6	7	5	6	8
C	42.2	35.1	26.5	16.2	14.9	6.4	4.4

Of   course the Hamiltonian Eq. (1) is valid   only   when the nucleus can be considered as a point   charge and   with a fixed magnetic moment, i.e., at sufficiently large r.   However, because of the   long-range $(1/r^2)$ nature of the   interaction, meaningful lower limits can be put on binding energies and radiative capture cross sections if one assumes that   Eq. (1) is nearly correct beyond   some radius r_o.   If one takes the "worst case" that the effective interaction is $+\infty$ ("hard core") for $r < r_o$, then the ground state energy is of the form

$$E_{g.s.} = - \frac{B_o(C)}{2Mr_o^2} \tag{6}$$

where B_o is the square of the largest root of the Hankel function $K_{i\sqrt{C-.25}}(x)$; the smaller roots give the excited states.   Here is a

brief   table   of $B_j(C)$ for   the   g.s.,   first,   and   second excited
states

Table III

C	B_0	B_1	B_2
.25	0	0	0
.5	$.5 \times 10^{-5}$	1.8×10^{-11}	$.64 \times 10^{-16}$
1.0	1.4×10^{-3}	1×10^{-6}	$.7 \times 10^{-9}$
3.0	.083	1.8×10^{-3}	$.42 \times 10^{-5}$
10.0	1.2	.15	.021
30.0	7.7	2.2	.68
100.0	41.8	19.8	10.0

In Table III above, the values of B_0 are given for the proton
and $A\ell_{27}$ and, in the last column, the estimate for the g.s.
binding energy if one takes $r_0 = 1.4\ A^{1/3}F_m$ in Eq. (6).

Half an Mev is a respectable binding energy, and an $A\ell_{27}$
nucleus bound by that much to a monopole passing through matter at
a moderate velocity is not going to be knocked off. For instance,
if the monopole has a "cosmic" velocity, $v/c = 10^{-3}$, in a
collision with an iron nucleus the c. of m. energy is only
$0.5(56)(.93\ Mev)(10^{-3})^2 = 25$ KeV. Further, when the nuclei are
close together the Coulomb repulsion between the nuclei reduces
the available energy, so that the cross section for knock off is
not large even if energetically allowed. A reasonable way to
estimate the knock off cross section would be to integrate the
rate of tunneling of the bound nucleus out of the $1/r^2$ well under
the polarizing effect of the struck nucleus during the Coulomb
collision. I estimate very crudely that the c. of m. energy of
the iron nucleus would have to be of the order of 100 Mev to make
the cross section appreciable.

The most obvious way for a nucleus to get bound to a monopole
is radiative capture. The binding energy is radiated away as a
photon emitted by the electric dipole of the transition of the
nucleus from a continuum state to the bound state. The dipole
matrix element is not very sensitive to the short range part of
the interaction; the big contribution to the radial integral comes
from the region of the bound state's turning point. For the

radiative capture of Al_{27} to the $j = 4$ ground state from a $j = 4$ state of small positive energy I find

$$\sigma_{Cap} \approx \frac{8\pi}{5} (13e)^2 \frac{B.E.}{27m_p} \times \pi/k^2$$

$$= 1.4 \times 10^{-4} \left(\frac{B.E.}{.56 \text{ Mev}}\right) \times \pi/k^2 \qquad (7)$$

$$= .3 \text{ mb, if B.E.} = .56 \text{ Mev and } v/c = 10^{-3}$$

where B.E. is the ground state binding energy, and k is the c. of m. (= monopole rest frame) nucleus momentum. This result puts a lower limit on the total capture cross section.

The radiative capture of a proton by a slow monopole is suppressed by the no 0-0 transition rule. All proton-monopole bound states have $j = 0$, and $j = 1$ to $j = 0$ transition matrix elements have an additional factor of $k \cdot {}^{52}$ from the centrifugal barrier.

A final remark on the factor $1/k^2$ of Eq. (7): This is due to the long-range $(1/r^2)$ attractive potential. The same factor occurs in exothermic absorptive cross sections when there is an attractive Coulomb potential in the initial channel. The factor comes about as follows: The transition matrix element is of the form $M_{fi} = \int d^3r \, \psi_f \, \vec{r} \, \psi_i$. Suppose that for r greater than some value R, the WKB approximation to ψ_i is valid. Then for $r > R$ we have

$$\psi_i = \sin\left(\int^r p(r')dr'\right)/r\sqrt{k \, p(r)} \qquad (8)$$

where $p(r)$ is the local wave number. This ψ_i is properly normalized as

$$\psi_i \xrightarrow{r \to \infty} \sin(kr+\delta)/kr,$$

since $p(r) \xrightarrow{r \to \infty} k$. At a fixed r, everything in the expression for ψ_i, Eq. (8), has a finite zero energy limit, except for the factor $1\sqrt{k}$. The same is true for ψ_i at $r \leqslant R$ which is matched to the WKB ψ_i at $r = R$. Hence $M_{fi} \sim 1/\sqrt{k}$ as $k \to 0$, and so $\sigma_{Cap} \sim (1/\sqrt{k})^2(1/k) = 1/k^2$ (the factor $1/k$ is the flux factor). The normal result in the limit $k \to 0$, namely $M_{fi} \sim$ const., so $\sigma_{Cap} \sim 1/k$, comes about because for initial-channel attractive potentials of shorter range than $1/r^2$ the WKB approximation fails. For instance, the reflection coefficient from an exponential attractive tail $V \sim -e^{-\mu r}$ is $R = e^{-4k/\mu}$; the place where the reflection occurs (i.e., where the WKB approximation fails) is roughly where the magnitudes of potential and kinetic energies are equal, $r = 2\ell n(1/k)/\mu + \cdot const, \xrightarrow{k \to 0} \infty$. So as $k \to 0$, ψ_i at short range is suppressed by the factor $\sqrt{T} = \sqrt{1-R} \approx \sqrt{4k/\mu}$, resulting in the normal result $M'_{fi} \xrightarrow{k \to 0}$ constant.

ACKNOWLEDGEMENTS

This research was supported in part by the University of Wisconsin-Madison Graduate School Research Committee with funds granted by the Wisconsin Alumni Research Foundation, and in part by the Department of Energy under contract DE-AC02-76ER00881.

VACUUM ANTI-SHIELDING OF MONOPOLES

Charles J. Goebel

Department of Physics
University of Wisconsin-Madison
Madison, Wisconsin 53706

At last year's Monopole Conference, R. Iengo gave an informal talk on his formalism for treating the interaction of the electromagnetic field simultaneously with electric and magnetic charges, and his deduction from it that the vacuum polarization effect of virtual monopole-pair creation on charge renormalization (Z_3) would be opposite to the effect of virtual charge-pair creation. That is, virtual charges shield charge, but virtual monopoles antishield charge.

This is a nice result because it sheds light on an old problem first found by Schwinger.[1] As described by Coleman[2] in his Erice '81 talk, the assumption that the vacuum is "self dual" under the interchange of electricity and magnetism leads to the conclusion that the renormalization (i.e., vacuum shielding) of charges and poles should be the same. So if Dirac's quantization condition eg = integer/2 is to hold for both the bare and physical couplings, Z_3 must be a rational number. A stronger conclusion follows from the more general consideration of running coupling constants: If the product of e(R) and g(R) is to stay quantized, they cannot run at all and Z_3 must equal 1. The joker in this is that a "self dual" vacuum must have the same content of electrical and of magnetic fields, that is, for every kind of charge-carrying particle there must be a particle with the same properties except that its "charge" is pole strength. As Iengo argues, the contribution of the virtual charges and poles to vacuum shielding would then just cancel and so the shielding of charges and poles would be indeed nil.

But the vacuum of the real world is not self dual; there are many kinds of charge carrying particles much lighter than monopoles, and consequently the vacuum is much more electrically than magnetically polarizable. (In fact, henceforth I neglect virtual monopole pairs; the vacuum is taken to be a vacuum of charged fields only. This is justified at scales $>1/$ Mpole. This also avoids having to worry how to treat virtual composite particles.) Coleman[2] goes on to give an argument (of remarkable opaqueness compared with the Coleman standard) that by contrast to Dirac point monopoles, topological monopoles enjoy a running magnetic strength which runs just opposite to charge, so that the product of pole strength and charge, $e(R)g(R)$, is the same at all scales and Dirac quantization always holds. But vacuum polarization effects at distances greater than, say, 10^{-2} Fm certainly can't tell a GUT topological monopole from a Dirac point monopole, so there cannot be such a contrast. In fact, the dual of Iengo's statement that virtual monopoles antishield charge is that virtual charges anti-shield pole strength. Consequently $e(R)$ and $g(R)$ run oppositely. A corollary to the running of $g(R)$ to smaller values at smaller distance is that the force between two poles should weaken at short distances, compared to inverse square, in contrast to the vacuum polarization effect on the force between two charges.

Can Iengo be understood in a simple minded way? My first simple minded thought was that virtual charges could not affect a monopole's strength at all, because the combination of spherical symmetry and $\nabla \cdot B = 0$ everywhere (except at the monopole) ensures that the monopole's $\vec{B}$ field cannot be altered. But I gradually realized that the $\vec{H}$ field is something else. I was raised to believe that the only real magnetic field was $\vec{B}$, while $\vec{H}$ was merely a construct to help one design magnets. But to a monopole $\vec{H}$ (likewise $\vec{D}$) is real! The force on a monopole of strength g is $\vec{F} = g(\vec{H} - \vec{v} \times \vec{D})$. This is a fact which everyone who properly learned classical electromagnetism knows (even though the "mono-poles" are merely the ends of bar magnets), and a deduction from the stress tensor can be found in Landau and Lifshitz. Not only is $\vec{H}$ just as real as $\vec{B}$ but the relation between them in the vacuum is not just the jejune $\vec{B} = \mu_0 \vec{H}$ of MKS units. If the permittivity ε is wave number and frequency dependent as a consequence of vacuum polarization the same is true of the permeability μ. In fact $\varepsilon\mu = c^{-2}$, which is a constant; the velocity of light c is not changed by vacuum polarization.

A final comment on the recognition by a monopole of $\vec{H}$ and $\vec{D}$ as the force fields is that this only seems peculiar because we live in an electrical world. If there were an appreciable density of monopoles, either real or virtual, the force on a charge e would not be given by $\vec{E}$ and $\vec{B}$, but would be

$$e[\vec{E} + \vec{M}_M + \vec{v} \times (\vec{B} + \vec{P}_M)] \tag{1a}$$

where $\vec{P}_M$ is magnetic polarization (i.e., the magnetic dipole moment density due to the polarization of monopoles--not to be confused with $\vec{M}$, the magnetic dipole moment density due to induced currents of charge) and $\vec{M}_M$ is the "electricalization" (the electric dipole moment density due to induced currents of pole strength.) This is parallel to the expression for the force on a monopole g, namely

$$g[\vec{B} - \vec{M} - \vec{v} \times (\vec{E} + \vec{P})] = g[\vec{H} - \vec{v} \times \vec{D}] \qquad (1b)$$

as stated above. [Note that if both electric and magnetic polarizability exist, the fields $\vec{E}$, $\vec{D}$, $\vec{B}$ and $\vec{H}$ would not suffice to write Maxwell's equations in terms of free charge and pole strength.]

There are two problems one faces in calculating the vacuum polarization (more precisely magnetization, $\vec{M}$) induced in the vacuum by a monopole. One is that the usual Lagrangian formalism and consequent Feynman rules don't apply (this is what Iengo coped with). The other is that the interaction strength between the monopole and a charge, eg, is not small (unless it vanishes exactly) because it is quantized to half integer values; thus perturbation theory is not applicable. These problems are solved by using the old fashioned method of calculating a vacuum expectation value as a sum over single particle modes. The modes of a charged particle in a monopole field are known exactly, and so no expansion is made in the strength of the monopole-charge interaction. The Wichmann-Kroll calculation[3] is an example of this technique; they used exact states of an electron in the Coulomb field of nucleus of charge Ze in order to get the vacuum polarization exactly to all orders in Ze. In these calculations the interaction between the particles (order α) is neglected; in modern terminology they are one (charged particle) loop calculations.

But the first problem resurfaces when we come to write the expression for the quantity of interest, namely magnetism, $\vec{M}$. In the Wichmann-Kroll calculation the quantity of interest is charge density, for which there is a well known expression. But for magnetization there is no direct expression known, at least to me. One has $\vec{J}^{tot} = \vec{J}^{free} + \vec{\nabla} \times \vec{M}$, but this is not enough to determine $\vec{M}$; in particular for the monopole the spherical symmetry forces $\vec{M}(\vec{r})$ to have the form $f(r)\hat{r}$, and hence $\vec{\nabla} \times \vec{M} = 0$. But there is a quantity which has a known expression and which involves the magnetization, namely the energy density

$$E_V = \frac{\vec{B} \cdot \vec{H}}{2} = \frac{\vec{B}^2}{2\mu} \,. \qquad (2)$$

Before finding E_V in the monopole field, I shall first find E_V in a uniform $\vec{B}$ field; this is not only easier but also relevant, because $\mu(\vec{r})$ in the monopole field will turn out to be nearly a local function of $\vec{B}$. I shall consider only <u>spinless</u> charged particles; this simplicity will be important in the monopole case. The expression for E_V for a uniform $\vec{B}$ field in the vacuum of a spinless charged field is

$$E_V(B) = \frac{1}{2} B^2 + \frac{eB}{4\pi^2} \sum_{n=0}^{\infty} \int_{-\infty}^{\infty} dk\sqrt{m^2 + k^2 + eB(2n + 1)} \qquad (3)$$

to one loop order. The first term is the Faraday "tree term"; the second term is the one-charged-particle-loop term, otherwise known as the zero point energy density of the charged field, i.e., $V^{-1}\sum \frac{\omega}{2}$ where the sum is over modes and V is the volume of space.

The sum $\sum_{n=0}$ over the Landau levels can be dealt with by the Sommerfeld-Watson trick:

$$\sum_{n=0}^{\infty} f(n) = \int_C dy \frac{f(y)}{1-e^{2\pi iy}} =$$

$$\int_{-1/2}^{\infty} dy\, f(y) + 2 \int_0^{\infty} \frac{dx}{1+e^{2\pi x}} \operatorname{Im} f\left(-\frac{1}{2} + ix\right) \qquad (4)$$

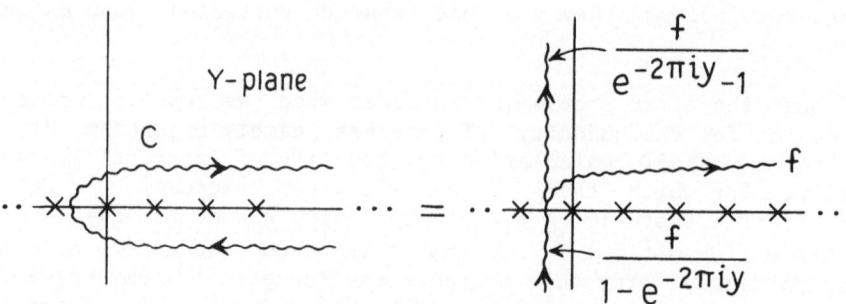

The first equality is obvious by evaluating the integral by residues. For the second, we have rotated the contours; note that the factor $1/(1-e^{2\pi iy})$ decreases exponentially into the ℓ.h.p. (Imy → -∞), and approaches 1 exponentially into the u.h.p. (Imy → +∞). [We have assumed that f(y) has no singularities in the way of the rotations.] So the sum is expressed as the integral approximant and a correction term which is a rapidly convergent integral. (If f were expanded in power series in x and the correction integral done term by term one would get a form of the Euler–Maclaurin formula.)

The integral approximant to the zero point energy term is independent of B (in fact it is just the free particle zero point energy density); this term is dropped. As for the correction term, do the $\int_{-\infty}^{\infty}$ dk first; putting in a cutoff Λ, the integral is

$$\int_{-\Lambda}^{\Lambda} dk \sqrt{M^2 + k^2} = 2\Lambda^2 + M^2[\ln(\Lambda/M) + \text{const}] + O(M^4/\Lambda^2).$$

where $M^2 = m^2 + 2ieBx$. The first term contributes nothing to Imf in Eq. (3). The remaining divergence is only logarithmic, and contributes only a B^2 term to E_V. Separating the part of E_V going like B^2 from the part of higher order in B (recall that parts of zero order in B have been dropped) we have

$$E_V(B) = \frac{1}{2} B^2 (1 + \frac{\alpha}{6\pi} L) + G(eB)$$

where

$$G(eB) = -\frac{m^4}{12\pi^2} \int_0^\infty \frac{dt}{e^{(\pi m^2/eB)t}+1} \left[\frac{t}{2} \ln(1 + t^2) + \tan^{-1}t - t\right] \quad (5)$$

where $L = \ln(\Lambda/m) + \text{const}$. This result was first given to order B^4 by Euler and Heisenberg,[4] and to all orders by Schwinger.[5] Comparing Eq. (5) to $E_V(B) = B^2/2\mu(B)$ gives the vacuum permeability $\mu(B)$. In the weak field limit $B \to 0$ we have

$$\mu(0) = \left[1 + \frac{\alpha}{6\pi} L\right]^{-1} \quad (6)$$

This will be observed to be just (to order α) the reciprocal of $\varepsilon(0)$, A.K.A. Z_3^{-1} where

$$Z_3 = 1 - \frac{\alpha}{6\pi} L \quad (\text{scalar QED}) \quad (7)$$

So $\mu(0)\varepsilon(0) = 1$, in agreement with the invariance of the velocity of light. Since $L = \infty$, at vanishing wave number the charged-field vacuum is a perfect dielectric, $\varepsilon = \infty$ (charge is totally screened) and also a perfect diamagnet, $\mu = 0$ (monopole strength is totally

antiscreened).   Consequently, for finite physical charges, bare
electric charge must be infinite, bare pole strength vanishing. As
we shall see later, the product of the permeability $\mu(R)$ at a
distance R from a pole, and the permittivity $\varepsilon(R)$ at a distance R
from a charge, is at least roughly constant, $\mu(R)\varepsilon(R) \approx 1$; hence
the running charges defined from $\mu(R)$ and $e(R)$ have a constant
product, $g(R)e(R) \approx$ const.   [The statement of the penultimate
sentence that at short distance the weak becomes strong and the
strong weak is not to be taken literally, at least not in this
world.   Before $e^2 = \alpha$ has run up to 1/2, to equal $g^2 = 1/4\alpha$ which
has run down to 1/2, one has reached the GUT scale.]

As a final step, before passing on to the monopole, let me
express $E_V(B)$ in terms of renormalized quantities. Define

$$B^2/\mu(0) = B_R^2$$

$$eB = e_R B_R,$$

(8)

hence $e^2 = e_R^2/\mu(0) \equiv e_R^2/Z_3$ as already mentioned.   In terms of
these renormalized quantities, the energy density $E_V(B)$, Eq. (5),
becomes

$$E_V(B) = \frac{1}{2} B^2 + G(eB)$$

(9)

where we have dropped the subscript R.   In the strong field limit
this becomes

$$E_V(B) = \frac{B^2}{2} \left(1 - \frac{\alpha}{6\pi} \left[\ln(\sqrt{eB/m}) - .7994\right]\right), \quad eB \gg m^2.$$

(9a)

In the monopole field, $\vec{B} = g\vec{r}/r^3$, the expression for E_V is

$$E_V(r) = \frac{1}{2} B^2 + \frac{1}{8\pi r} \sum_{j=\nu} (2j + 1) \int_o \frac{kdk}{\omega} \left[(\omega^2+m^2)J_p^2(kr)\right.$$

(10)

$$\left. + \frac{k^2}{4p} \left\{(2p - 1)J_{p-1}^2(kr) + (2p + 1)J_{p+1}^2(kr)\right\}\right],$$

where $\omega = \sqrt{m^2 + k^2}$, $\nu = |eg|$, and $p = \sqrt{(j + 1/2)^2 - \nu^2}$. This is more elaborate than the expression for E_V in the uniform $\vec{B}$ field; it is position dependent and the $\int dk$ cannot be done in closed form. However, by proper use of the asymptotic expansion for J_p and various known formulae, $E_V(r)$ can be found[6] in the limits $mr \gg 1$ and $mr \ll 1$. At large r expressed in terms of $B = g/r^2$, we have

$$E_V(r) \rightarrow \frac{1}{2} B^2 \left(1 + \frac{\alpha}{6\pi} L\right) \qquad mr \gg 1. \tag{11}$$

Not surprisingly, this is identical to the uniform-field result, Eq. (5), in the limit $B \rightarrow 0$.

At small r, the renormalized E_V can be written

$$E_V(r) = \frac{B^2}{2} \left(1 - \frac{\alpha}{6\pi} \left[\ln(\sqrt{eB}/m) - .7994 + \frac{F(\nu)}{\nu}\right]\right), \quad mr \ll 1 \tag{12}$$

where $F(\nu)$, known only numerically, is not far from a constant, varying monotonically from 1.303 at $\nu = 1/2$ to 1.152 at $\nu = \infty$. One sees that $E_V(r)$ differs from the strong uniform field result, Eq. (8), only by the term $F(\nu)/\nu$. In the limit $\nu \gg 1$, $E_V(r)$ is thus given exactly locally by the uniform-B $E_V(B)$. This can be understood by noting that the scale on which the monopole's B field varies is r and the length scale of the B-field wave functions (radius of lowest Landau level) is $R = 1/\sqrt{eB} = r/\sqrt{\nu}$; hence if $\nu \gg 1$, $R \ll r$. In the opposite limit, $\nu \rightarrow 0$, the reason for the behavior of the term $F(\nu)/\nu$, namely $\approx 1.5/\nu$, is not evident; of course only half integral values of ν are meaningful. In any case, the leading correction to $\mu(r)$ at small r, the $\alpha \ln r$ term, is correctly given by the uniform-B result.

This result can be used to give an estimate of the force between two poles at a separation R. The force is given by an integral of the stress tensor, $\sim B^2/\mu$, on a surface which is conveniently taken as the plane whose points are equidistant from the two poles. Because μ is slowly varying with position, we can estimate as follows: The dominant contribution to the integral comes from points on the plane whose distance from each monopole is of order R; thus the result has a factor $1/\mu(R)$ where $\mu(R)$ is the permeability at a distance of order R from each monopole. This permeability is in the field of the two poles; it is not easy to find because the wave equation for the modes of a charged particle in the $\vec{B}$ field of two poles does not separate. But we found that $\mu(\vec{r})$ in the field of one pole was nearly a local function of $\vec{B}$, and it seems reasonable that this be true more generally. Hence the factor in the force is $\approx 1/\mu(R)$ where $\mu(r)$ is given by Eq. (12). It can be concluded that the "vacuum polarization" effect of the virtual creation of pairs of a spinless particle of charge e and mass m on the force between two poles G_1 and G_2 at a separation R is to multiply the $G_1 G_2/R^2$ zero order force by a factor which is less than 1, and for separations R $\ll$ m^{-1} is roughly

$$1 - \frac{\alpha}{6\pi} \ln(1/mR). \tag{13}$$

ACKNOWLEDGEMENTS

This research was supported in part by the University of Wisconsin-Madison Graduate School Research Committee with funds granted by the Wisconsin Alumni Research Foundation, and in part by the Department of Energy under contract DE-AC02-76ER00881.

REFERENCES

1. J. Schwinger, Phys. Rev. 151, 1048 and 1055 (1966).
2. S. Coleman, in International School of Subnuclear Physics (19th) A. Zichichi, ed., Plenum (1983).
3. E.H. Wichmann and N.M. Kroll, Phys. Rev. 101, 843 (1956).
4. W. Heisenberg and H. Euler, Z. Physik 98, 714 (1936); W. Weisskopf, Egl. Danske Videnskab. Selstabs. Mat.-fys. Medd. 14, no. 6 (1936).
5. J. Schwinger, Phys. Rev. 82, 664 (1951).
6. More details will appear in a forthcoming paper, coauthored with T. Thomaz, now of the University of Pernambuco, Brazil.

MONOPONUCLEOSIS - THE WONDERFUL THINGS THAT MONOPOLES CAN DO TO NUCLEI IF THEY ARE THERE

Harry J. Lipkin

Department of Nuclear Physics
Weizmann Institute of Science
Rehovot, Israel

and

Physics and High Energy Physics Divisions
Argonne National Laboratory
Argonne, Illinois 60439

and

Fermi National Accelerator Laboratory
Batavia, Illinois 60510

In the proceedings of the 1965 Coral Gables Conference, Behram Kursonoglu included appropriate folk tales about the Turkish folk hero, Nasreddin Hoja, before every talk. One of these tales seems particularly appropriate for this conference.

One morning a wood cutter saw Hoja by the edge of a lake, throwing quantities of yeast into the water. "What the devil are you doing, Hoja?" he asked. Hoja looked up sheepishly and replied, "I am trying to make all the lake into yogurt." The woodcutter laughed and said, "Fool, such a plan will never succeed." Hoja remained silent for a while, and stroked his beard. Then he replied, "But just imagine if it should work!"

A modern version of this story would have Hoja sitting by the edge of a monopole detector saying, "But just imagine if the monopoles are there!" In this talk we consider the wonderful things that they can do to nuclei by examining nuclear physics in strong magnetic fields. We have seen that monopoles can bind nuclei.[1-3] We shall investigate the following other possible processes:

1. Mixing of singlet and triplet states of deuteron-like positronium.
2. Production of a new kind of nuclear matter with nucleon moments oriented in the field.
3. Catalysis of nuclear fission.
4. Catalysis of nuclear fusion (with implications for solar neutrinos).
5. Enhancement of forbidden decays like triplet positronium, e.g. fission products.

MIXING OF DEUTERON TRIPLET AND SINGLET STATES

The magnetic energy of a nuclear spin in the magnetic field produced by a monopole at a distance of several fermis is of the order of nuclear binding energies and level spacings.[4] This field can produce appreciable mixing of nuclear wave functions and may give observable effects. For a rough estimate consider a deuteron in a very strong magnetic field. In the same way that a field splits and mixes the triplet and singlet spin states of positronium, the spin triplet deuteron ground state is split into three energy levels and its central member is mixed with the (unbound) singlet spin state to produce the eigenstates $|n{\uparrow}p{\downarrow}\rangle$ and $|n{\downarrow}p{\uparrow}\rangle$. The splitting between the two states by the field at a distance of $\underline{r}$ from a monopole of charge g is

$$\Delta E = g(1.91 + 2.79)(eh/2\pi Mc)(1/r^2) \qquad (1)$$

where M is the proton mass. If we set $2\pi eg/hc = 1$, the splitting is equal to the binding energy of the deuteron, 2 MeV when

$$hc/2\pi r = 20 \text{ MeV.} \qquad (2a)$$

$$r \simeq 10 \text{ fermis} \qquad (2b)$$

This suggests that a monopole might break up a deuteron at a distance of 10 fermis.

However, a slowly-moving heavy monopole does not have sufficient kinetic energy in the monopole-deuteron center mass system to break up a deuteron, and the two nucleons either remain bound to the monopole or escape as a normal deuteron. This differs from positronium, where both the triplet and singlet states are metastable and decay by annihilation. The triplet decay is inhibited by selection rules and is third order in α (3γ) while the singlet is second order (2γ). An external magnetic field catalyzes the decay of the triplet state by mixing in a singlet component for which the second order decay is allowed.

The nuclear analog of positronium is a mestastable nuclear state whose decay can be enhanced or catalyzed by the presence of the monopole magnetic field; e.g. beta unstable odd-odd nuclei like Al^{26} whose ground states have the proton spin j_p and the neutron spin j_n coupled to the maximum possible spin $J = j_p + j_n$ and whose beta decay to a $J = 0$ even-even nucleus is highly forbidden because of the large spin change. A strong magnetic field decouples j_p and j_n and mixes in all lower spin couplings down to $J = |j_p - j_n|$. As in triplet positronium, decays from the admixed states have a much lower order of forbiddenness and the decay rate is enhanced.

MONOPOLE NUCLEAR MATTER

If a monopole is placed at the center of a nucleus, the nucleons will gain energy if their magnetic moments are oriented parallel to the field. Changing the spin orientation will lose nuclear binding energy. However, if the magnetic energy gained is greater than the nuclear binding energy lost, the magnetically polarized nucleus will be the ground state of the system and there will be a different kind of nuclear matter.

For a crude estimate of this effect, let us consider a monopole at the center of a sphere of nuclear matter of radius R with nuclear density $\rho(R)$. The magnetic energy gained by orienting the magnetic moments of all nucleons parallel to the field is

$$E_{mag} = \int_{o}^{R} \frac{\bar{\mu}egh}{2\pi Mcr^2} \quad \rho(r)dr \times 4\pi r^2 \tag{3}$$

where $\bar{\mu}$ is the mean value of the magnitude of the nucleon magnetic moment. For systems with equal numbers of neutrons and protons, we take $\bar{\mu} = 2.35$ nuclear magnitons. For a sphere of uniform density containing A nucleons

$$\rho(r) = \frac{A}{\frac{4}{3}\pi R^3} \tag{4a}$$

It is convenient to parameterize the radius R as

$$R = \frac{e^2}{mc^2} \frac{A^{1/3}}{\xi} = 2.8 \frac{A^{1/3}}{\xi} \text{ fermis} \tag{4b}$$

where ξ is a parameter of order unity and is about 2 for conventional densities. Then, for a monopole strength g given by the minimum Dirac value

$$\frac{2\pi eg}{hc} = \frac{1}{2} \tag{4c}$$

Equation (3) becomes

$$\Delta E_{mag} = \frac{3}{2} \; \bar{\mu} \; \frac{h^2 c^2}{4\pi^2 e^4} \; \frac{m^2 c^2}{M} \; \xi^2 \; A^{1/3} \approx 18 \; \xi^2 A^{1/3} MeV \tag{5a}$$

$$\frac{\Delta E_{mag}}{A} \approx 18 \; \zeta^2 \; A^{-2/3} \; MeV \quad . \tag{5b}$$

We see that for $\zeta = 2$, Eqs. (5) give magnetic energies which are comparable to nuclear binding energies. Whether or not monopole nuclear matter is stable compared to normal nuclear matter cannot be determined by such a crude calculation. It is necessary to calculate also the effects of reorienting the nucleon spins on the nuclear interaction and to include the changes in kinetic energy if the density is varied to give a minimum energy.

Note also that even if a magnetic field is not strong enough to change the polarization states of an entire nucleus, the states in the higher shells are more likely to be affected than the inner shells. A monopole field can split the degeneracy of the states in the spherical shell model and cause the levels whose magnetic moments are oriented parallel to the field to move downward in energy while those with magnetic moments oriented anti-parallel to the field move upward. In this way some of the levels in the highest filled shell move upward while some of the levels in the lowest unfilled shell move downward. At some value of the magnetic field strength, these levels will cross and the ground state configuration for the nucleus will change. One might consider the analogue of the Nilsson model used to consider the effects of deformation on nuclear levels. Instead of plotting the level energies as a function of deformation, they can be plotted as a function of the external magnetic field. Thus, an equilibrium may be reached in which the center of the nucleus is normal nuclear matter, whereas the outer shells have become monopole nuclear matter.

CATALYSIS OF SPONTANEOUS FISSION

The rearrangement of nuclear levels produced by a magnetic monopole may make the nucleus much more susceptible to spontaneous fission. Consider a nucleus which could gain energy by splitting into two nuclei because the energy gained from the Coulomb repulsion is greater than the energy lost from the nuclear attraction. However, because the nuclear force is a short range force, a potential barrier is created as the nucleus is deformed. A small deformation reducing the Coulomb energy only slightly increases the nuclear energy very sharply. However, if an appreciable part of the binding energy no longer comes from the nuclear force, but comes instead from the magnetic interaction with the monopole, this situation can change. The magnetic interaction has a longer range than the nuclear interaction and the deformation of the nucleus will cost less when part of the energy is magnetic than when all of the energy is nuclear. Thus, the fission barrier may be reduced appreciably and nuclei which do not fission spontaneously in the absence of monopoles may fission rapidly when monopoles are present.

CATALYSIS OF NUCLEAR FUSION BY MAGNETIC MONOPOLES

If two nuclei with magnetic moments are near a monopole, the attraction of both nuclei by the monopole can compensate for the Coulomb repulsion between the nuclei and greatly reduce the potential barrier which inhibits nuclear fusion. Such an effect can catalyze the He^3-He^3 reaction in the sun while not affecting the He^3-He^4 reaction because He^4 has no magnetic moment. Since the He^3-He^4 reaction leads to the Li-Be-B chain which produces the high energy solar neutrinos investigated in Davis' experiment, monopole catalysis could explain Davis' failure to observe solar neutrinos.[5,6]

Quantitative estimates to this effect are difficult because barrier penetration factors are exponential and very sensitive to small effects, while the three-body problem of two nuclei and a monopole is not easily solved to the precision required. For the thermonuclear reactions in the sun, the reaction rate depends on an integral over nuclear kinetic energies of the product of a Boltzmann factor which decreases exponentially with <u>increasing</u> energy and a barrier penetration factor which decreases exponentially with <u>decreasing</u> energy. The maximum of the integrand, called the Gamow peak, occurs at energies ten times larger than thermal energy where the Boltzmann factor is e^{-10} and barrier penetration factors of e^{-60} are common.[7] The problems arising are illustrated by the following simple example.

We assume that a bound state of nucleus of charge z and a monopole of magnetic charge g exists and consider a collision between this bound state and another nucleus of charge Z, mass number A and magnetic moment μ in nuclear magnetons. At distances large compared to the size of the bound state the interaction between the two bodies at a distance r is the sum of the Coulomb interaction and the magnetic interaction

$$V = \frac{Zze^2}{r} - \frac{\mu eh}{2\pi M_p c} = \frac{Zze^2}{r} - \frac{\mu h^2}{8M\pi^2 r^2} \tag{6}$$

where we have assumed that the magnetic moment is oriented parallel to the magnetic field to give an attractive interaction and used the Dirac value, Eq. (4c).

At large distances the potential is the normal Coulomb repulsion. However at small distances the attractive magnetic interaction takes over and the potential goes through zero and becomes attractive at the distance

$$R_0 = \frac{\mu}{2} \frac{h^2}{4\pi^2 MZze^2} \qquad . \qquad (7a)$$

The maximum value of the interaction, Eq. (6), occurs at the distance

$$R_{max} = 2R_0 = \frac{\mu h^2}{4\pi^2 MZz^2} = \frac{\mu h^2}{4\pi^2 MZz^4} \frac{mc^2}{} \times \left(\frac{e^2}{mc^2}\right) = \frac{(137)^2}{1840} \times \frac{\mu}{Z} \left(\frac{e^2}{mc^2}\right). \qquad (7b)$$

Then

$$V(R_{max}) = \frac{Zze^2}{2R_{max}} = \frac{Z^2z^2}{2\mu} \left(\frac{2\pi e^2}{hc}\right)^2 M c^2 \qquad . \qquad (7c)$$

The cross section for a reaction between the two systems includes a barrier penetration factor $e^{-\gamma}$ where γ is given by the usual Gamow expression

$$\gamma = \frac{4\pi}{h} \sqrt{2AM} \int_R^b (V-T)^{1/2} dr \qquad . \qquad (8)$$

The limits of the integral R and b are the two points where the integrand vanishes, the classical turning points, and T is the kinetic energy.

For the case where the monopole is absent, the dominant contribution to the result in Eq. (8) comes from the upper limit and the lower limit can be taken as zero for a first approximation. Let us write

$$\gamma = \gamma_0 - \gamma_R \qquad (9)$$

where γ_0 denotes the value of the expression, Eq. (3), with R = 0 and γ_R denotes the correction due to the finite value of R. For the case where there is no magnetic monopole present, the values of γ_0 and γ_R are given to a good approximation as

$$\gamma_0 = \frac{4\pi^2 Zze^2}{hv} \qquad (10a)$$

$$\gamma_R = \frac{8\pi}{h} (2Zze^2 AM R)^{1/2} \qquad (10b)$$

where v is the relative velocity.

To include the effect of the magnetic monopole, we must choose the value of R to be the point where the integrand vanishes in the presence of the monopole potential; i.e. a value greater than R_0 given by Eq. (7a). There will also be a considerable correction in the integrand in the region between R_0 and distances several times this radius. We can give a rough estimate of these two effects by setting $R = R_{max} = 2R_0$ in Eq. (10b). In this case we obtain

$$\gamma_{2R_0} = 4\sqrt{2\mu A} \tag{11}$$

For the case of a He^3 nucleus relevant to fusion in the sun, $Z = 2$, $A = 3$ and $\mu = 2.13$. With these values we obtain γ_R approximately equal to 14. This means that the cross section is enhanced by a factor e^{14}.

A more refined calculation which evaluates the integral in Eq. (8) explicitly gives an approximate result with the factor 4 in Eq. (11) replaced by 2π. This changes the enhancement factor from e^{14} to e^{22}. With factors like these arising easily one can understand that large effects are possible and that it is necessary to be very careful before drawing quantitative conclusions.

Another estimate of this effect is obtained by noting that the barrier exponent γ must vanish when the kinetic energy T is equal to the barrier height,

$$T = \frac{1}{2} AM_p v^2 = V(R_{max}) = \frac{Z^2 z^2}{2\mu} \left(\frac{2\pi e^2}{hc}\right)^2 M c^2 \approx 190 \text{ KeV} \tag{12a}$$

Then

$$v = \frac{2\pi Z z e^2}{h\sqrt{\mu A}} \tag{12b}$$

But the barrier exponent γ_0 in the absence of the monopole at this energy is given by Eq. (10a) as

$$\gamma_0 = 2\pi\sqrt{\mu A} \tag{13}$$

For the He^3 nucleus, $\gamma_0 = 2\pi\sqrt{6.39} \sim 16$. Thus a barrier penetration factor of e^{-16} is present at this energy in the absence of the monopole and completely disappears when the monopole attraction is included.

With such large factors present which are sensitive to details of the calculations, and many unknown factors, it is very difficult to obtain quantitative results. One example of such an unknown factor is the effect of a condensate of electron-positron pairs that must be created in the monopole field at distances of the order of the electron Compton wavelength. The magnetic energy of an electron-positron pair in the magnetic field of a monopole at a distance r is

$$\frac{2eh}{2\pi mc} \times \frac{g}{r^2} = mc^2 \left(\frac{h}{2\pi mcr}\right)^2. \tag{14}$$

Thus at distances of the order h/2πmc the vacuum seems to become unstable against creating pairs and orienting the moments parallel to the field. This naive picture is not correct. More sophisticated treatments of the charge density around a monopole are given elsewhere.[8]

The electron-positron pairs cannot screen a magnetic charge, because of Gauss' law applied to magnetic charges. However, the cloud of pairs could very well screen the Coulomb repulsion between the nuclei.

At the distance $r = h/2\pi\sqrt{2}mc$ which makes the magnetic energy of a pair, Eq. (9), equal to its rest energy, the Coulomb barrier has the value

$$\frac{Zze^2}{r} = \frac{Zze^2 2\pi\sqrt{2}\ mc}{h} = \frac{Zz\sqrt{2}\ mc^2}{137} \tag{15}$$

This is about 20 keV for He^3.

If the pair condensate screens off the Coulomb barrier at this point, then there is no barrier penetration factor for energies above 20 keV, while at lower energies the penetration is enhanced by the factor $e^{\gamma R}$ with

$$\gamma_R = 4\left(\frac{2\pi\sqrt{2}\ Zze^2}{hmc}\ AM\right)^{1/2} = 4(19ZzA)^{1/2}. \tag{16}$$

For He^3, $\gamma_R = 60$, which is enormous.

The essential physics underlying these numbers is that the conventional calculation of barrier penetration includes enormous contributions to the Gamow integral, Eq. (8), from distances smaller than (h/2π√2 mc. Any effect which reduces this contribution in the exponent produces a large enhancement of the cross sections.

What is this fermion charge around a monopole? Is it observable? Can you polarize the distribution by putting it between condenser plates? Is it a dielectric? Is it a conductor? Clearly, a better understanding of the underlying physics is needed in order to obtain reliable estimates of fussion catalysis by monopoles.

ENHANCEMENT OF FORBIDDEN β DECAYS

Spontaneous electromagnetic mixing has been considered[10,11] as a radiative correction to ordinary beta decay and found to be much too small to produce an observable effect. The induced mixing due to a monopole is similar to this radiative mixing, but the transition matrix element is of order unity instead of order α. The transition probability is thus increased by the large factor of α^2.

The enhancement factor in the transition matrix element M_B for the magnetically induced beta decay over the ordinary decay matrix element M_O for various combinations of electromagnetic transitions and one ordinary beta decay is given by standard perturbation theory as

$$\frac{\langle f|M_B|i\rangle}{\langle f|M_O|i\rangle} = \frac{\langle f|V|A\rangle}{\langle f|V|i\rangle} \times \frac{\langle A|H_B|i\rangle}{(E_A-E_i)} \tag{17a}$$

$$\frac{\langle f|M_B|i\rangle}{\langle f|M_O|i\rangle} = \frac{\langle B|V|i\rangle}{\langle f|V|i\rangle} \times \frac{\langle f|H_B|B\rangle}{(E_B-E_f)} \tag{17b}$$

$$\frac{\langle f|M_B|i\rangle}{\langle f|M_O|i\rangle} = \frac{\langle B|V|A\rangle}{\langle f|V|i\rangle} \times \frac{\langle f|H_B|B\rangle}{(E_B-E_f)} \times \frac{\langle A|H_B|i\rangle}{(E_A-E_i)} \tag{17c}$$

where V denotes the transition operator for ordinary beta decay, H_B denotes the electromagnetic transition operator, $|i\rangle$, $|f\rangle$, $|A\rangle$ and $|B\rangle$ denote the initial and final nuclear ground states and the intermediate excited states and E_i, E_f, E_A, and E_B denote their energies.

Particularly interesting cases might be odd-odd nuclei with the odd proton and odd neutron in the same L-shell and coupled to a spin of 2 or greater. The beta decay to a 0^+ ground state then involves recoupling the angular momenta of the two nucleons and is forbidden because orbital factors are needed for a change in total angular momentum larger than one. However, the monopole could break the couplings of the proton and neutron spins. In perturbation theory this appears as a cascade of M1 transitions via the other states of the same configuration down to the 1^+ state, from which the beta decay is an allowed GT transition. Thus very long-lived highly forbidden transitions might go much more rapidly via the magnetic transition through several intermediate states. The relevant matrix elements for such transitions can be crudely estimated using shell model wave functions and experimental values of magnetic moments and Gamow-Teller matrix elements within the same configurations.

Consider, for example Al^{26}, which has $J^P = 5^+$ and decays to the excited 2^+ state of Mg^{26} with a lifetime of 7.2×10^5 years and a log ft of 14.2. Al^{26} also has excited states with $J^P = 4^+$, 3^+, 2^+ and 1^+ at excitation energies of 2 MeV, 0.4 MeV, 1.8 MeV and 1.1 MeV respectively. The beta decays from the 1^+ state of Al^{26} to the 0^+ ground state of Mg^{26} and from the 3^+ and 4^+ states of Al^{26} to the 2^+ and 3^+ excited states of Mg^{26} are both allowed GT transitions. Thus a monopole-induced fifth-order transition via four intermediate states or a second or third-order transition via one or two intermediate states might have a much shorter lifetime than the observed decay.

The transition matrix elements for the fifth and third order transitions are

$$\langle 0^+|M_B|5^+\rangle = \langle 0^+|v|1^+\rangle \frac{\langle 1^+|H_B|2^+\rangle}{E(1^+)-E(5^+)} \times \frac{\langle 2^+|H_B|3^+\rangle}{E(2^+)-E(5^+)}$$

$$\frac{\langle 3^+|H_B|4^+\rangle}{E(3^+)-E(5^+)} \times \frac{\langle 4^+|H_B|5^+\rangle}{E(4^+)-E(5^+)} \tag{18a}$$

$$\langle 2^+|M_B|5^+\rangle = \langle 2^+|v|3^+\rangle \times \frac{\langle 3^+|H_B|4^+\rangle}{E(3^+)-E(5^+)} \times \frac{\langle 4^+|H_B|5^+\rangle}{E(4^+)-E(5^+)} \, . \tag{18b}$$

The matrix elements $\langle 0^+|v|1^+\rangle$ should be approximately equal to that of the mirror transition from the 0^+ ground state of Si^{26} to the 1^+ state of Al^{26}, which has an experimentally measured log ft of 3.5. The matrix element $\langle 2^+|v|3^+\rangle$ cannot be taken directly from another transition like $\langle 0^+|v|1^+\rangle$. Reasonable estimates are obtained by using log ft values of the neighboring decay so the 3^+ ground state of Na^{26} to the same 2^+ state of Mg^{26} with a log ft of 4.7 and of the 3^+ ground state of Al^{28} to the 2^+ state of Si^{28} with a log ft of 4.9.

Rough quantitative estimates of the expression (18) are obtainable from the shell model description of the states in Al^{26} as a neutron and a proton in the $d_{5/2}$ shell coupled to spins 1, 2, 3, 4 and 5. We can use the experimental magnetic moments of the $5/2^+$ ground states of the nuclei Mg^{25} and Al^{25}; namely -0.9 n.m. and $+3.6$ n.m. respectively, as values for the effective magnetic moments of the $5/2^+$ neutron and the $5/2^+$ proton configurations in

Al^{26}. We therefore need assume only that the states of spins 1-5 in Al^{26} are described by different couplings of the neutron configuration of Mg^{25} and the proton configuration of Al^{25}, without assuming a particular model like a single-particle description for either.

The electromagnetic transition operator H_B and its relevant matrix elements can then be written

$$H_B = (g_p j_{pz} + g_n j_{nz})B_z = (g_p + g_n)J_z B_z/2 + (g_p - g_n) \qquad (19a)$$
$$(j_{pz} - j_{nz})B_z/2$$

$$\langle J|H_B|J+1\rangle = \langle J|j_{pz} - j_{nz}|J+1\rangle \{0.9(eh/4\pi Mc)B_z\} \qquad (19b)$$

where g_p and g_n denote the gyromagnetic ratios of the proton and neutron configurations, j_{pa} and j_{nz} the z-components of the angular momenta of these configurations, B_z the magnetic field strength, chosen to be in the z-direction and J_z the z-component of the total angular momentum. The values of g_p and g_n were taken from the experimental moments,

$$(g_p - g_n) = (2/5)(3.6 + 0.9)(eh/4\pi Mc) = 0.9(eh/2\pi Mc) \quad . \quad (19c)$$

The angular momentum matrix elements are easily evaluated by standard methods. Assuming equal populations for the 11 J_z states and using the values log ft = 3.5 and 4.9 respectively for the two beta transitions we obtain

$$\log ft(5^+ \to 0^+) = 7.4 + 16 \log r \qquad (20a)$$

$$\log ft(5^+ \to 2^+) = 6.8 + 8 \log r \qquad (20b)$$

where r is in fermis.

The $(5^+ \to 0^+)$ transition is seen to have log ft values of 7.4 and 12 for values of r of 1 and 2 fermis respectively. The $(5^+ \to 2^+)$ transition (9b) has log ft values of 6.8, 9.2 and 10.6 for values of r of 1, 2 and 3 fermis respectively. These should be compared with the log ft of 14.2 for the competing observed decay.

These very crude estimates only indicate orders of magnitude. Better calculations can be made with time-dependent magnetic fields to account for the passage of a monopole by a nucleus, and with more complicated nuclear wave functions, but these are probably not worth the effort until more information is available about monopoles.

ACKNOWLEDGEMENTS

It is a pleasure to thank D. Schramm for pointing out the significance of the He^3-He^3 reaction and J.D. Bjorken, E. Kolb and I. Talmi for stimulating discussions. This work was supported in part by the Israel Commission for Basic Research, and in part by the U.S. Department of Energy under contract W-31-109-ENG-38.

REFERENCES

1. Dennis Sivers, Phys. Rev. D2, 2048 (1970).
2. C. Goebel, these proceedings.
3. G. Fiorentini, these proceedings.
4. Harry J. Lipkin, Fermilab Pub. 83/64-THY, Physics Letters in press.
5. L.W. Alvarez, Lawrence Berkeley Laboratory, Physics Note (unpublished).
6. J.S. Trefil, et al., Nature 302, 111 (1983).
7. Donald D. Clayton, "Principles of Stellar Evolution and Nucleosynthesis," McGraw-Hill, New York (1968) p. 302.
8. C.G. Callan, these proceedings.
9. B. Grossman, these proceedings.
10. Harry J. Lipkin, Phys. Rev. 76, 567 (1949).
11. Eugen Merzbacher, Phys. Rev. 81, 942 (1951).

COHERENT, CO-OPERATIVE ASPECTS OF MONOPOLE AND MATTER INTERACTION

Timir Datta

Physics and Astronomy Department
University of South Carolina
Columbia, South Carolina 29208

Since no highly ionizing monopoles are observed,[1] the local velocity of the monopole must be $v < 10^{-4}c$. Also because the sensitivity of the inductive superconducting[2] detection technique used by Cabrera[3] is independent of the monopole speed, the Cabrera result will be consistent with the negative results elsewhere, if it is agreed that the interesting monopole speeds lie in the rather restrictive low range $10^{-5}c < v < 10^{-4}c$. Therefore, it is of considerable interest to identify new mechanisms by which a slowly moving monopole may interact with a material medium.[4] This may be useful in the design of future monopole experiments.

In this paper, we report on a cooperative effect between a magnetic medium and a monopole slowly ($10^{-5}c < v \lesssim 10^{-3}c$) moving through it. As such this is a rather involved microscopic problem and recently Drell[5] and collaborators and Ahlen and Kinoshita[6] have reported on monopole matter interaction mechanisms. However, our aim is to discuss the coherent-collective aspects of the phenomenon, and we will be interested in the long wave length or hydrodynamic (HD) models of the collective oscillations of this magnetic system. The physical principle behind this mechanism is that any object passing through a medium at a speed faster than the speed at which the medium carries wave excitation will generate a cone of shock waves.

Physically a HD picture is applicable if each spin in the magnetic medium directly interacts with at least its immediate neighbors, since in this way any local disturbance will be collectively shared in turn by the entire chains of spins giving rise to a cooperative or HD motion. Curiously HD is also appropriate in the opposite limit. That is, even if there is no direct scattering between individual spins, HD modes still appear if the individual spins are coupled to an average molecular or collective field in the medium. In general, these are modes where the spins interact with each other via a field produced jointly by the coherent cooperative motion of the spin excitation (Magnon) and the moving monopole, and not due to individual interactions. In terms of the order parameter[7] of this problem $S \equiv S_o e^{i\phi}$, a quantitative criterion for the validity of this hydrodynamic description is provided by the correlation length which is a measure of the range of fluctuations of S. These local fluctuations will be meaningful only if they are spatially uniform over a distance scale larger than ξ. Thus if this spatial variation is characterized by a wave vector k, then a HD description is appropriate if $|k|\xi \ll 1$, or $\xi \ll 1/|k|$. Furthermore, we are aiming at a finite temperature case, where the magnetic order is incomplete but local equilibrium is reached within a relaxation time which is microscopically short. So these renormalized modes satisfy $\omega\tau \ll 1$.

Let a monopole be moving with a velocity v along the x direction, through a ferromagnetic material. The spin equation or the equation of motion for S in the presence of the monopole[8] is

$$\frac{\partial S}{\partial t} - iD\nabla^2 S = -g\delta(x-vt) \tag{1}$$

Although the transport coefficient D in Eq. (1) is calculable microscopically, at the macroscopic level presently contemplated it will be convenient to treat D as a phenomenological macroscopic parameter characteristic of the particular magnetic material. By the same token the source or the inhomogeneous term g is determined by the microscopic details of the scattering processes, the coupling strength and the nature of the medium material. So at least in principle, g may also be calculated ab initio. However, once again by restricting our interest to the collective aspects of this process, we will regard g to be given and to simply characterize the strength of the spin wave or magnon excitation due to the monopole.

Notice this is a very general level of describing the phenomena, such that the physical nature of the solution will not be affected by any microscopic information on D or g.

We solve Eq. (1) by obtaining the Green function G for the homogeneous equation. Let us first perform a Fourier transformation in ω, i.e.,

$$S_\omega(r) = \frac{1}{2\pi} \int_{-\infty}^{\infty} S(r,t)\, e^{i\omega t} dt \tag{2}$$

and similarly g'_ω is obtained by transforming $g\delta(x-vt)$.   With S_ω, g_ω and Eq. (1), we obtain[9]

$$\nabla^2 S_\omega + k^2 S_\omega = -g_\omega \qquad\qquad (3)$$

The wave vector k is determined by the dispersion relation $\omega = Dk^2$. These $\omega \to 0$, as $k \to 0$ Goldstone magnons are a consequence of the breaking of the rotational symmetry of the microscopic Hamiltonian of the spin system.[10,11]   The equation for G is

$$(\nabla^2 + k^2)G(r,r') = -\delta(r-r') \qquad\qquad (4)$$

so

$$G = \frac{e^{\pm ik\cdot(r-r')}}{4\pi|r-r'|}. \qquad\qquad (5)$$

The retarded physically meaningful solution is given by

$$S_\omega^{(r)} = \int \frac{g_\omega(r')e^{-ik\cdot|r-r'|}}{|r-r'|} \, d^3r' \qquad\qquad (6)$$

For magnon phase velocity $V_{ph} = (\omega/k) < V$ we obtain the well known Cherenkov[12,13] situation, with the cone angle $\theta_c = \cos^{-1}(V_{ph}/V)$.

The singularity is not physical, because (i) in any real magnetic material the maximum saturation magnetization is finite and (ii) the presence of dispersion, i.e., the magnon phase velocity is frequency dependent as given by the dispersion relation.   However, dispersion is present even in the case of ordinary optical Cherenkov radiation.   Hence, this complication may be incorporated in a manner analogous to the optical case.[14] Such a procedure does not change the qualitative description, but quantitative results require calculating each ω component in Eq. (2) separately and determining the contribution to the radiation in the range $\omega \pm \Delta\omega$. In general, in the inverse integral of Eq. (2) the magnon radiation will be operative only within frequency range $\omega_0 < \omega < \omega_L$ where the cut off frequency ω_L is determined by the shock condition $V(\omega_L)_{ph} = V$. ω_0 may be obtained from the lower bound k_0 of k. k_0 is determined by the anisotropy and a typical value is $\sim 10^{-2}a^{-1}$, where a is the lattice size. The width of the shock front may be estimated by equating it to the distance from the monopole when the monopole field equals the field present in the ferromagnetic medium, i.e., $q/L^2 \sim 10^6 G$, with $q = 3.3 \times 10^{-8}$ G cm^2 and $L \sim 10$Å.   For a typical value of $D \sim 10^3$ MeV Å^2 the $V_{ph} \sim 10^{-5}c$.   So for a monopole arriving out of a 10 μm thick magnetic sheet, the radius of the intersected Cherenkov cone is 3.6 μm.

The criterion for sustaining such a spinwave shock front[15] is $l_a/l_c > 1$. Where l_a is the magnon annihilation mean free path and the creation distance l_c is given by, $l_c = 1/n\sigma$. Here n is the volume density of the spins in the medium and σ is the magnon creating cross section. For neutrons σ ~ one barn, so, for 1 spin per unit cell or 1 Å^3, i.e., $n = 10^{24}$ $spin/cm^3$, we obtain $l_c \sim 10$ cm. However, at room temperature in a polycrystalline medium $l_a < 10$ cm. Hence, for neutron $l_a/l_c \ll 1$. Similar arguments hold for most other types of particles. This is because most particles are (i) non-penetrating, highly interacting such as electrons; (ii) penetrating but moving very rapidly compared to the magnon velocity, viz., cosmic ray particles or (iii) like neutrons, too weakly interacting.

A magnetic monopole on the other hand has all the right characteristics. It has extremely large inertia, even at ultra subrelativistic speeds. Also the monopole charge is very large, so that the spin wave generation cross section[16] σ (monopole) $> 10^{12}\sigma$ (neutron). Hence for monopoles: $l_c < 1 \text{Å}$ and $l_a/l_c \gg 1$. Thus, the observation of a slow moving highly penetrating agent producing a magnon shock cone may be uniquely associated with a transient monopole. These cones may be employed to reconstruct the trajectory, velocity, energy and energy loss of the monopoles.

Figure 1 is a schematic of a prototypical magnetic, solid state, monopole detector based on the coherent spin wave excitation. Notice, this is a magnetograph or non-volatile magnetic "photograph" of the magnetic disarray left in the wake of the monopole. The material for the magnetic medium is determined by the velocity of the targeted monopole. Iron oxide is an attractive possibility, because of the large room temperature value of the spin wave stiffness parameter D, discussed earlier, and also because the electrical conductivity is low. The tracks can be read electromagnetically or optically by scanning the magneto optical film, coated on the detector face. For a film with the Verdet constant of 0.5 G^{-1} cm^{-1} the Faraday rotation in a μm thick film is ~ 5×10^{-3} deg, which is well within the present day magnetic bubble technology.[17] For supportive confirmatory evidence, a kaptan type track detector may be attached on the under side or the ram side of each disk.

In short, we have shown that along with the incoherent individual scattering processes in a material medium a slow moving monopole can also excite collective modes.[18] These modes may be employed to detect such monopoles.

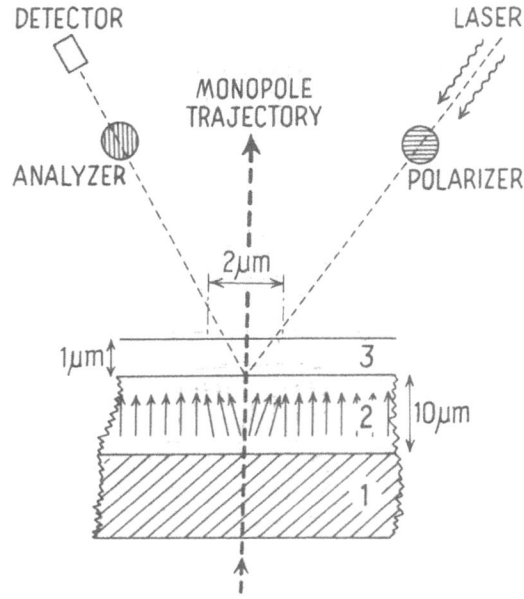

Fig. 1   The schematic of a "magnetic shock wave" monopole detector.   If a monopole passes through the magnetic medium
(2) the moments over the cone region get collectively
disturbed from their initial ordered position.   This is
very much like storing information in a bubble memory
device.   The only difference is that here the monopole
furnishes the magnetizing field, and the collective process enlarges the action radius to the cone size.   For an
oblique trajectory the patch on the detector face is a
conic section.   The direction of the local magnetization
is determined by the polarity of the monopole charge
(shown north seeking).   This patch may be located by an
electromagnetic pickup head or optically as shown.   This
may be done by scanning a layer of magneto-optical film
(3) on the detection medium.   The size of this patch
(~2μm) is determined by the thickness of the sensitive
layer (~10μm), the spin wave stiffness D and the monopole
velocity v.   The substrate (3), may be constructed from
any suitable non-magnetic material as in commercial memory
disks.

ACKNOWLEDGEMENTS

    The author would like to thank Professors C.P. Poole, and
H.A. Farach for helpful discussions, Professor C.N. Yang for
pointing out the need to understand slow monopole-matter inter-

actions, Professors B.C. Barish and G. Giacomelli for some of their recent preprints. This work was partially supported by a grant from NSF (no. ISP-80-11451) and a NASA/JPL grant (no. RE-152/344.

REFERENCES

1. For $v \geqslant 10^{-3}e$ the monopole flux F_m is $< 4.1 \times 10^{-13}$ cm^{-2} sr^{-1} s^{-1} and $v \geqslant 10^{-4}e$ $F_m < 5 \times 10^{-12}$ cm^{-2} sr^{-1} s^{-1}, see (a) J. Bartelt, H. Courant, K. Heller, T. Joyce, M. Marshak, E. Peterson, K. Ruddick, and M. Shupe, D.S. Ayrus, J.W. Dawson, T.H. Fields, E.N. May and L.E. Price, Phys. Rev. Letts. 50, 655 (1983); (b) D.E. Groom, E.C. Loh, H.N. Nelson and D.M. Ritson, Phys. Rev. Letts. 50, 573 (1983). Also, see reports in these conference proceedings.

2. L.J. Tassie, Nuovo Cimento 38, 1935 (1965).

3. B. Cabrera, Phys. Rev. Letts 48, 1378 (1982).

4. T. Datta, Nuovo Cimento, to appear (1983).

5. S.D. Drell, N.M. Kroll, M.T. Muller, S.J. Parke, M.A. Ruderman, Phys. Rev. Letts. 50, 644 (1983).

6. S. Ahlen and K. Kinoshita, Phys. Rev. D, 26, 2347 (1982).

7. (a) B.I. Halperin and P.C. Hohhenberg, Phys. Rev. 188, 898 (1969) and 177, 952 (1969); for more details see (b) Dieter Forster, Hydrodynamic Fluctuations, Broken Symmetry and Correlation Functions. W.A. Benjamin Inc., Reading (MA) (1975).

8. Here we have expressed the spin wave equation with the imaginary (i) part explicit, this makes the analogy with the Cherenkov problem more obvious as in Eq. (3). For the ferromagnetic case being considered, Eq. (1) is identical with the time dependent Schrödinger equation with a source term. In the case of antiferromagnetic spin waves, Eq. (1) will have to be trivially replaced by an inhomogeneous wave equation and will lead to a generation of Cherenkov antiferromagnetic magnons. S displays this wave like behavior because it is a conserved quantity. A local deviation of S cannot relax locally (hence rapidly), but equilibrates slowly and collectively over the whole system.
    Also $S \sim M_\perp$ where $M_\perp$ is the magnetization perpendicular to the internal magnetization M. A discontinuity in $M_\perp$ will accompany the magnon shock wave given by Eq. (1).

9. Where $g_\omega \equiv (i/D)g'_\omega$.

10. Robert M. White, in Quantum Theory of Magnetism, 2nd edition, Springer-Verlag, NY (1983).

11. See Ref. 7(b).

12. J.D. Jackson, Classical Electrodynamics, John Wiley and Sons, NY, 2nd ed (1975).

13. A.I. Akhiezer, V.G. Baryaktar and S.V. Peletminskii, Phys. Letts. 4, 129 (1963).

14. H. Motz and L.I. Schiff, Am. Journ. Phys. $\underline{21}$, 258 (1953).

15. T. Datta, H. Farach and C.P. Poole to appear in Bul. Am. Phys. Soc. (1983).

16. T. Datta, USC preprint (1983).

17. J.F. Dillon Jr., in Physics of Magnetic Garnets, A. Paoletti, ed., North-Holland, Amsterdam, 379 (1978).

18. Such collective excitations will create acoustic Mach shock waves as well. In Aluminum V_{sound} 5km/s, so with $V = V_e$, $\theta_c = 34°$.

THE PHYSICS OF MONOPOLE DETECTION

Barry C. Barish

Physics Department
California Institute of Technology
Pasadena, California 91125

ABSTRACT

In this review, I discuss the techniques and challenges involved in the detection of Grand Unified Monopoles. The only positive evidence remains the original candidate of Cabrera[1] from an induction experiment in a superconducting loop. Further work with both induction and non-induction techniques is being vigorously pursued. The main ideas in these experiments are reviewed with emphasis on prospects for future large scale detectors.

INTRODUCTION

Detection of magnetic monopoles has presented a major challenge for experimental physicists for many years. The early interest was motivated by the apparent symmetry between Electric and Magnetic fields in Maxwell's equations. However, due to the lack of abundance of free magnetic charges compared to electric charges, they were not included in the final formulation of those equations.

Renewed interest in monopoles developed after Dirac showed in 1931 that the existence of free magnetic charges (Dirac monopoles) could provide reason for the quantization of electric charge $(nh/2\pi c)$.[2,3] This provided motivation for generations of experiments looking for production of monopoles on each new accelerator and in cosmic rays.

In trying to find such monopoles several experimental

features, making the signatures unique, were used. First, Dirac
monopoles are quantized to 137/2 electric charge and therefore are
expected to give a great deal of ionization going through matter.
This assumes that the mass of these monopoles is similar to the
masses of other elementary particles (e.g., $m = m_{proton}$ or
$m = 137 m_{proton}$, etc.), meaning these monopoles could be produced
prolifically in high energy accelerations. Another consequence of
the high ionization is that Dirac monopoles could come to rest in
matter. A variety of searches in different matter (e.g.,
sediment, moon rocks, etc.) where they might be abundant have been
performed.

All of these possibilities have been sensitively investigated
and the final results are that there exists no evidence for Dirac
monopoles. There was one event of a possible Dirac monopole
reported from the cosmic rays[4] but that proved incorrect.

In recent years, the possibilities of Grand Unified Theories
have become more attractive. After the success of Electro-Weak
Unification (e.g., $SU_2 \times U_1$), further Unification of the Strong
Interactions seems likely. One consequence of Grand Unifications
is the possibility of proton decay and many experimental efforts
are underway investigating that possibility. Another consequence
of Grand Unification might be motivation for the existence of
magnetic monopoles. G. 't Hooft and Polyakov[5,6] showed that such
monopoles exist as solutions in many non-abelian gauge theories.
The possibilities of the existence of these monopoles has
stimulated the recent theoretical and experimental interest in the
subject. The one feature of these grand unified monopoles that
makes the problem very different experimentally is that they are
extremely heavy ($M \sim 10^{16}$ GeV). This means that essentially all
the previous searches for Dirac monopoles that relied on large
ionization losses are insensitive to such monopoles. These
monopoles will not stop and be collected in matter, are too heavy
to be produced from accelerators, and the fluxes in cosmic rays
are expected to be extremely small and require different experimental techniques.

GUT monopoles are so massive that they will necessarily be
non-relativistic ($\beta \simeq 10^{-3}$) if they are galactic in origin. At
such low β a monopole has great penetrating power and easily
passes through the Earth. These facts result in new challenges
for experiment, and I review here the major attempts at detection.

FLUXES IN COSMIC RAYS

Before discussing the actual experiments, I review the
expected fluxes of GUT monopoles in cosmic rays. As stated above,

the existence of Grand Unified Monopoles is a general consequence of Grand Unification. However, even if they "exist", is there a detectable flux in cosmic rays? At present, this question cannot be answered. The usual picture is that monopoles would have been produced in the "Early Universe" at the phase transition where the unifying gauge symmetry (e.g. SU_5) breaks down ($\sim 10^{15}$ GeV). Combining this with the standard Big Bang model leads to a problem. Too many monopoles are produced ("a glut") and the number estimated exceed the limit from the mass density of the Universe (Hubble). Several theorists have attacked this problem and have suppressed monopole production by various versions of a "New Inflationary Universe". The general idea is that reheating creates the baryons we see today, that is the phase transition is postponed to a later time. Unfortunately, this results in negligible monopole production ("a famine").

The final conclusion on monopole production from cosmology is very unclear. For experiment, we must conclude at this point that we have little guidance and do not know from cosmology what the production of monopoles in the early Universe might have been and how many are around today. The problem is illustrated schematically in Fig. 1.

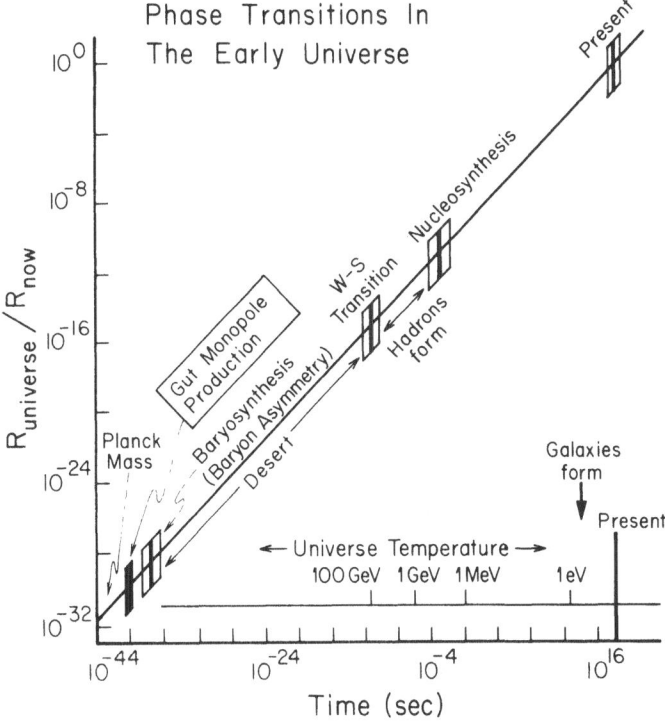

Fig. 1 Monopoles in cosmic rays; production in the Early Universe.

The best that can be done with our present knowledge is to use some constraints on the possible fluxes of monopoles in cosmic rays from astrophysics for guidance. These constraints come from two considerations: (1) the limit on the total mass of the Universe; and (2) the existence of galactic magnetic fields. Below I briefly review these constraints.

Mass of the Universe

The most straightforward astrophysical limit comes from assuming that magnetic monopoles account for the unseen mass of the Universe.[7]

The mass density contained in galaxies ρ_G accounts for about $.02 \rho_c$, the critical density to close the Universe. This implies that the number of monopoles,

$$n_M \lesssim \frac{10 \times \left[(\text{Galactic Mass}/\text{Monopole Mass}) \right]}{(\text{Distance between Galaxies})^3}$$

$$n_M \lesssim 4 \times 10^{-20} \ \text{cm}^{-3} \quad .$$

For comparison, the number of nucleons is

$$n_N \sim 4 \times 10^{-6} \ \text{cm}^{-3}$$

and

$$\frac{n_M}{n_N} \lesssim 10^{-14} \ \frac{\text{monopoles}}{\text{nucleon}} \quad .$$

From this, one obtains a flux limit

$$F \lesssim 5.4 \times 10^{-18} \ \text{cm}^{-2} \ \text{sr}^{-1} \ \text{s}^{-1} \ m_{19}^{-1} \left[v/10^{-3} \ c \right] \quad .$$

So, for example, this gives $F \lesssim 5 \times 10^{-15} \ \text{cm}^{-2} \ \text{sr}^{-1} \ \text{s}^{-1}$ for $\beta \sim 10^{-3}$ and $m_M \sim 10^{16}$ GeV. A more optimistic flux limit is obtained if we use the fact that monopoles could cluster like mass in our galaxy (10^{12} solar masses within 20 kpc). This results in a revised flux limit (clustering):

$$F_c \lesssim 3 \ 10^{-13} \ \text{cm}^{-2} \ \text{sr}^{-1} \ \text{s}^{-1} \ m_{19}^{-1} \left[v/10^{-3} \ c \right] \quad .$$

Survival of Galactic Magnetic Fields

The usual understanding of the existence of galactic magnetic fields is that they are due to persistent currents, ($\nabla \times B_{gal} \neq 0$). If now magnetic monopoles exist and move along

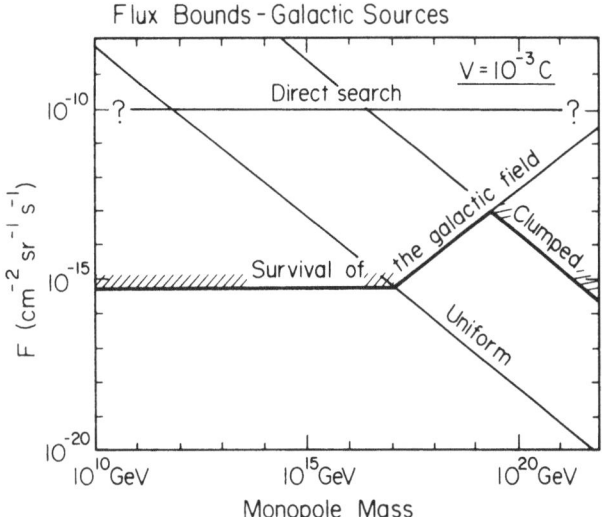

Fig. 2 Bounds on the flux of monopoles from astrophysics considerations. These bounds are orders of magnitude below the Cabrera level.

these field lines, they will gain energy at the expense of the field. In order for the fields to survive, the field energy must not be dissipated more rapidly than the currents can be regenerated by dynamo action ($t_{regeneration} \cong 10^8$ yrs). This requirement places a limit on the fluxes of monopoles, usually called the Parker Bound,[8] viz.,

$$F \lesssim 10^{-16} \text{ cm}^{-2} \text{ sr}^{-2} \text{ s}^{-1} \qquad \text{(Parker Bound)} \quad .$$

This bound has been reexamined by Turner, Parker and Bogdan using the monopole mass, velocity distributions, etc. They obtain a less restrictive bound for $M_M > 10^{16}$ GeV. Figure 2 shows these bounds combined with the flux bounds from the mass density of the Universe. The level is $F \lesssim 10^{-15} \text{ cm}^{-2} \text{ sr}^{-1} \text{ s}^{-1}$ for $M_M < 10^{17}$ GeV and the maximum possible level is $F_{MAX} \lesssim 10^{-12} \text{ cm}^{-2} \text{ sr}^{-1} \text{ s}^{-1}$ for $M_M \sim 10^{19}$ GeV and $\beta \sim 3 \times 10^{-3}$ and assuming clustering.

The conclusions and guidance from cosmology and astrophysics are that the production of monopoles in the early Universe is uncertain, and that the flux of monopoles from astrophysical arguments appears bounded at $F \lesssim 10^{-15} \text{ cm}^{-2} \text{ sr}^{-1} \text{ s}^{-1}$ (Parker Bound). This is for Galactic monopoles with $\beta \sim 10^{-3}$. This means that very large scale experiments, much larger than the present day, will be necessary for detection. These detectors will need to be $\sim 10^4$ m^2 (larger than a football field) to record several events/year at the Parker Bound.

Below I describe the principal techniques being used at smaller scales and the attempts to develop larger detectors.

INDUCTION EXPERIMENTS

The "best" technique that has been used to search for GUT monopoles is by the electromagnetic induction from passage of a magnetic monopole through a closed superconducting loop. This technique has a unique signature in a predicted amount of induced current which is independent of the monopole mass, velocity, electric charge, etc.

The technique was first used to search for the presence of magnetic monopoles in various materials.[10,11] These searches were negative using bulk matter varying from moon rocks to sea water sediment.

Cabrera[1] extended these ideas to the sensitivity needed to detect the passage of a single free GUT monopole through a superconducting loop. He has applied the combination of a SQUID (Superconducting Quantum Interference Device), and a very well magnetically shielded loop.

A schematic view of this detection method is shown in Fig. 3. From Maxwell's equation,

$$\vec{\nabla} \times \vec{E} + \frac{1}{c}\frac{\partial \vec{B}}{\partial t} = - \left[\frac{4\pi}{c}\right] \vec{I}_m \ ,$$

integrating over the plane of the ring, one obtains

$$\int_\Gamma \vec{E} \times \vec{d1} + \frac{1}{c}\frac{\partial \phi}{\partial t} = \frac{-4\pi g}{c} \delta(t)$$

E vanishes along the path Γ, therefore,

$$\phi(t) = -4\pi g \ \Theta(t)$$

$$= \phi_g + \phi_s,$$

where ϕ_g is from the monopole, ϕ_s is the induced supercurrent, and $\phi_s = -I(t)L$ (the self inductance of the ring). Note that ϕ changes by $2\phi_0$, where $\phi_0 = hc/2e$ is the flux quantum in a superconductor and $4\pi g = hc/e$.

To get an idea about the sensitivity of this technique, $\Delta I = 2\phi_0/L \sim 10^{-9}$ amps; $\Delta E = 8 \times 10^{-30}/L$ joules. So consider the example of a loop of 5 mil wire with $D \sim 1$ meter.

$L \sim 2\mu H$

$\Delta E = 4 \ 10^{-24}$ joules (2 fluxons) .

For comparison, a commercial SQUID has a sensitivity of $\Delta E \sim 10^{-27}$ joules or 10^{-3} fluxons! Therefore, we can see that the combination of a SQUID and a superconducting loop are ideal. The signal/noise is excellent and the observed "jump" in current must agree with the 2 fluxons expected of a magnetic monopole.

The real limitation of this technique comes from the problem of shielding the loop magnetically. The change in flux due to a monopole ($2\phi_0$) is 4×10^{-7} Gauss-cm^2, while the Earth's magnetic field is $\sim .01 - .1$ Gauss, so the Earth's field is 10^6 fluxons/cm^2, which represents a formidable background to shield.

In the Cabrera experiment, a superconducting lead shield of $\sim 10^{-8}$ Gauss is used. This gives the necessary sensitivity, however this technique is extremely difficult to extrapolate to larger areas. For example, a 1 m^2 detector would require the field to be shielded to $\sim 10^{-12}$ Gauss.

The Cabrera experiment has been improved by installing a 3 coil detector. This gives a redundancy as the monopole crosses more than one loop. Also, accelerometers, etc. have been added to reduce the possibility of false signals.

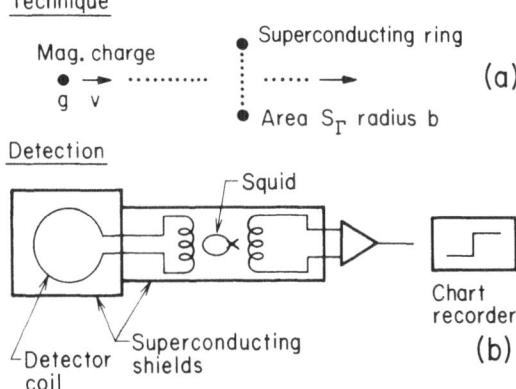

Fig. 3 The induction technique used in the Cabrera experiment is shown. The flux change from passage of a monopole is $2\phi_0$ (the flux quantum in a superconductor).

There are several efforts to duplicate Cabrera's result and to extend the techniques to larger areas. There are several groups (e.g., IBM) who are working to stabilize the ambient field (.01-.5 Gauss), rather than work at $\sim 10^{-8}$ Gauss, like Cabrera. These techniques, (e.g., high order gradiometer configurations) may well lend themselves to extensive multiplexing. This might mean that a future induction detector as large as, say, 100 m^2 might be realizable. However, the possibility of reaching the area ($\sim 10^4$ m^2) necessary to challenge the Parker bound seems remote.

It is worth noting that attempts are underway to use induction techniques in non-superconducting loops (e.g. Price at CERN and D. Morris at LBL). If the formidable signal/noise problems can be solved, this may simplify the experimental problems of building a large scale detector by removing the cryogenic requirements.

IONIZATION DETECTORS

The other major technique being used in searches for GUT monopoles is use of detectors relying on detecting ionization loss from the passage of monopoles. A monopole passing through matter will lose energy by ionization loss. The amount of energy loss for charged particles (or monopoles) is well understood for relativistic particles, however, the ionization loss for very slow particles is much less understood. In fact, discrepancies in the various calculations for slow monopoles ($\beta \sim 10^{-3}$) have plagued the interpretation of scintillator and gas ionization experiments and have made it difficult to evaluate the use of scintillators for very large arrays. Recent work has gone a long way to clarify this situation.

Determination of ionization loss at low β requires calculations of the atomic collisions and excitation which can be quite complicated. At relativistic energies, the impulse approximation can be used and the problem is simplified. However, at low β, a detailed model of the atom must be used in order to understand the dynamics and this requires various approximations. Ahlen and Kinoshita[12] have done a detailed calculation of this energy loss in analogy to calculations of Lindhard[13] for protons. The results of this calculation as a function of β are shown in Fig. 4.

The possibility of an enhancement in the energy loss at low β has been discussed by Drell et al.[14] They have shown that energy losses due to Zeeman splittings, diamagnetic shifts and crossings of the energy levels caused by the interaction of the atomic electrons with the monopole magnetic field, is substantially enhanced in simple atoms. This energy loss for atomic hydrogen is

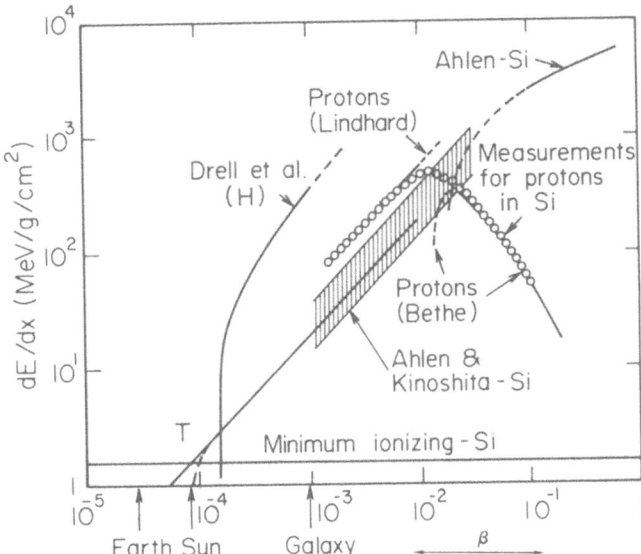

Fig. 4   Ionization loss for slow magnetic monopoles. The calculations for Ahlen and Kinoshita at low β have been used to
analyze scintillator experiments.   Also shown is the
energy loss in hydrogen due to Zeeman splitting, which
might be exploited in future detectors.

also shown in Fig. 4. It is not entirely clear what these effects
are for a complicated organic material like plastic scintillator,
but they certainly indicate the possibility that a scintillating
material for low β monopole experiments could be chosen that will
display these enhancements.

A large number of experiments[15-21] have already been performed searching for heavy, slow monopoles using ionization loss
techniques.   Typically, the detectors consist of multiple layers
of ionization detectors (e.g., scintillators) set to respond to
low light levels.   This threshold level, combined with the energy
loss calculations discussed above, determine the lowest detectable
β for each experiment.   The actual "signal" for a monopole is
determined by recording the time of passage of each layer and
thereby searching for very slow moving penetrating particles.
This is illustrated in Fig. 5.

The present searches are negative and the various bounds are
shown in Fig. 6. Each measurement differs in terms of solid angle
acceptance, ionization materials, minimum ionization threshold,
signals electronically recorded and the minimum β accepted.   Note
that none of the results yet approach the astrophysics bounds and
a much larger array is needed to reach such sensitivities.

Special requirements are necessary in order for scintillator experiments to be sensitive to very low β. In particular, this requires electronics that will respond to the low level signals, and stretched out train of pulses resulting from the passage of a very slow particle. (At β ~ 10^4, it takes a monopole hundreds of nanoseconds just to pass through a single layer.) Two experiments[17,18] have taken the required special measures and are sensitive to lower β. These experiments can be more directly compared with the results from induction experiments. These

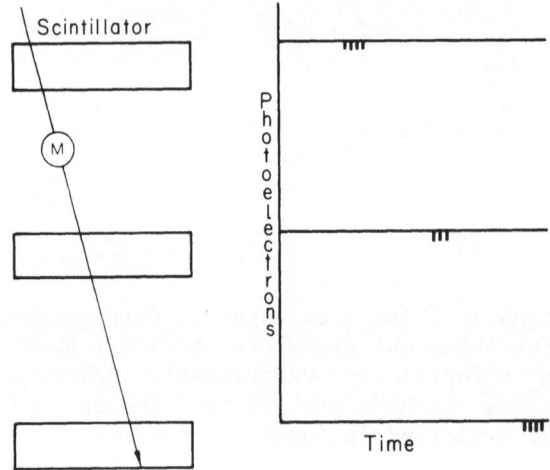

Fig. 5 Monopole signal from slow moving penetrating particles.

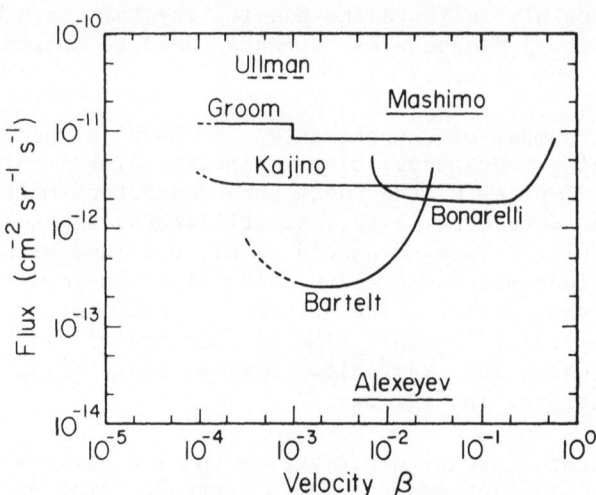

Fig. 6 Summary of limits versus β for various scintillation experiments. The limits are well below the Cabrera level but do not approach the astrophysics bounds.

results are particularly important since they are significantly below the Cabrera experiment and are sensitive to local sources of monopoles ($\beta \sim 10^{-4}$), as well as galactic monopoles ($\beta \sim 10^{-3}$).

Extending these techniques to a very large detector, capable of sensitivities below the Parker Bound, seems quite possible. Before building such a detector array, many questions will need to be answered, however. For example, should the array be built at sea level, which is inexpensive and has easy access, or is it necessary to work in a mine to reduce cosmic ray backgrounds? What is the optimal and most practical scintillating material? What electronic recording and calibrations are necessary? How many layers are needed, and are tracking as well as scintillation layers desirable? Despite these many questions, some variation of this technique seems likely and possible for a very large array capable of challenging the Parker Bound.

OTHER NON-INDUCTION TECHNIQUES

Several alternate methods for monopole detection are worth consideration:

Eddy Current Losses in a Conductor

A second mechanism for energy loss exists for monopoles passing through a conductor. The energy loss by this mechanism has been calculated by Martem'yanov and Khakimov[22] and more recently by Ahlen and Kinoshita.[23] Ahlen obtains

$$\frac{dE}{dx} = \frac{4\pi^2 N_s e^2 g^2}{m_s c^2} \times \frac{\beta c}{v_F} \approx 1 \text{ GeV/cm}$$

for aluminum and $\beta \sim 3\times10^{-3}$. This is considerably larger than ionization loss, but also varies as β making detection more difficult for slow moving monopoles. It seems tempting to develop techniques to detect these Eddy current energy losses.

One possibility is to pick up the thermal-acoustic shock wave; this has been looked at extensively. Akerlof[24] has done a theoretical study of the intrinsic limit due to acoustic noise and has concluded that the signal/noise will not allow detection. At Caltech, we have built a prototype acoustic detector using aluminium as a conductor and specially built 10 MHz transducers. The measured S/N $\approx$ 1/50 has been observed, indicating a large factor is needed before detection by this technique would be feasible. Various possibilities, like acoustic focusing, use of magnetostrictive or ferromagnetic material, and improved receivers and electronics, are being investigated.

Searches in Old Iron Ore

The idea is that relic poles are bound to ferromagnetic domain boundaries by image forces. A group from Kobe[25] and another from Wisconsin[27] are searching in many kg of iron ore ~10^7 yrs old. This will, in principal, be equivalent to searching with sensitivity of ~10^{14} cm^{-2} sr^{-1} s^{-1} for very slow monopoles $\beta < 10^{-4}$. This is a region investigated no other way.

Detection of Monopoles by Etching in Mica

In this case, low velocity monopoles cause damage in the mica. The tracks are then revealed by etching. Price[28] has done such an experiment for mica with $\tau \sim 4.6 \times 10^8$ yr. He sees no evidence for monopoles with $4 \times 10^{-4} < \beta < 2 \times 10^{-3}$. For this range of velocities, this yields a flux limit of $F \leq 10^{17}$ cm^{-2} sr^{-1} s^{-1}. This technique can be improved significantly in the future.

PROTON CATALYSIS

There has been a great deal of interest recently in the possibility that monopoles can catalyze proton decay. The work of Dokos and Tomaras,[29] Rubakov,[30] and Callan[31] has lead to the possibility of detecting the passage of monopoles through a proton decay detector by the detection of proton decay by the process

$$M + N \rightarrow M + e^+ + \text{mesons}$$

where N = proton or neutron. If these catalyzed decays occur with cross sections typical of the strong interaction ($\sigma \sim 10^{-26}$ cm^2), several decays will occur along the trail of the monopole path through the detector. These will be separated by time characteristic of the β of the monopole.

The IMB proton decay detector has already been used for such a search.[32] The effective cross sectional area of the detector is 550 m^2 for an isotropic flux of monopoles, with area times solid angle of 6900 m^2-sr. The mean path length through the detector is 12.8 m for an isotropic flux.

The most sensitive way to search involves looking for multiple interactions, since a single interaction gives events that are kinematically difficult to distinguish from the ~1 event/day of cosmic ray neutrino interactions depositing ~940 MeV. The back-to-back kinematical constraint applied to the "natural" $p \rightarrow \pi^0 e^+$ decay mode is not applicable here.

For multiple interactions, the first 100 days of data were analyzed looking for two proton decay candidates within 8 μs. No events of this type have been observed and therefore a 90% C.L. upper limit on the product of the monopole flux times the cross section has been determined. For σ_c > 10 mb, this corresponds to F_m(min) = 7.2 × 10^{-15} cm^{-2} sr^{-1} s^{-1} within a velocity window 10^{-4} < βm < 10^{-1}. The 90% C.L. upper limit on the product F_m × σ_c has a min value $(F_m × \sigma_c)_{min}$ = 3.6 10^{-41} sr^{-1} s^{-1} at σ_c = 1 mb and monopole velocity. These results are summarized in Fig. 7. The significance of this result clearly depends heavily on the catalysis cross section. This is very uncertain, and active theoretical work is in progress to better determine this cross section.

CONCLUSIONS

The expected small flux from astrophysical considerations, coupled with the slow velocity (β ~ 10^{-3}), makes the detection of GUT monopoles a demanding challenge. A variety of techniques are being used and larger detectors are under development. It should be possible to build an inductive detector approaching 100 m^2 in the future, and a non-inductive detector large enough to challenge the Parker Bound is definitely feasible.

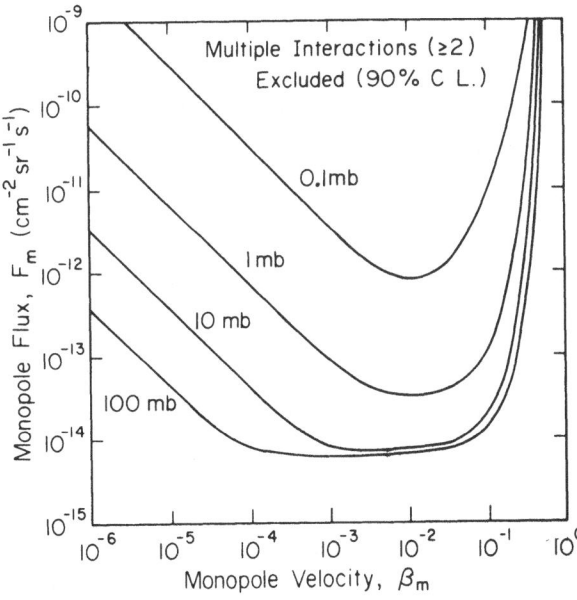

Fig. 7 Limits on monopole flux from proton catalysis search for multiple decays along the monopole track. This result is for 250 days of running on the IMB detector.

REFERENCES

1. B. Cabrera, Phys. Rev. Lett. 48, 1378 (1982).
2. P.A.M. Dirac, "Quantized Singularities in the Electromagnetic Field", Proc. Royal Soc. London A233, 60 (1931).
3. P.A.M. Dirac, "Theory of Magnetic Poles", Phys. Rev. 74, 817 (1948).
4. P.B. Price, E.K. Shirk, W.Z. Osborne, and L.S. Pinsky, Phys. Rev. Lett. 35, 487 (1975).
5. G. 't Hooft, "Magnetic Monopoles in Gauge Theories", Nucl. Phys. B79, 276 (1974).
6. A. Polyakov, "Particle Spectrum in Quantum Field Theory", Pis'ma Zh. Eksp. Teor. Fiz. [JETP Lett.] 20 [20], 430 [194] (1974).
7. M.J. Longo, "Massive Magnetic Monopoles: Indirect and Direct Limits on Their Number Density and Flux", Phys. Rev. D25, 2399 (1982).
8. E.N. Parker, Astrophys. J. 163, 225 (1971).
9. M.S. Turner, E.N. Parker, and T.J. Bogdan, Phys. Rev. D26, 1296 (1982).
10. L.W. Alvarez, Lawrence Radition Laboratory Physics Note 470 (1963).
11. L.W. Alvarez, P.H. Eberhard, R.R. Ross, and R.D. Watt, Science 167, 701 (1970); Phys. Rev. D4, 3260 (1971); Phys. Rev. D8, 698 (1973); and P.H. Eberhard, R.R. Ross, J.D. Taylor, L.W. Alvarez, and H. Oberlock, Phys. Rev. D11, 3099 (1975).
12. S.P. Ahlen and K. Kinoshita, Phys. Rev. D26, 2347 (1982).
13. J. Lindhard, Mat. Phys. Medd. Dan. Vid. Selsk. 28, 8 (1954).
14. S.D. Drell, N.M. Kroll, M.T. Mueller, S.J. Parke, and M.A. Ruderman, Phys. Rev. Lett. 50, 644 (1983).
15. J.D. Ullman, Phys. Rev. Lett. 47, 289 (1981).
16. T. Mashima, K. Kawagoe, and M. Koshiba, J. Phys. Soc. Japan 51, 3067 (1982).
17. D.E. Groom, E.C. Loh, H.N. Nelson, and D.M. Ritson, Phys. Rev. Lett. 50, 573 (1983).
18. F. Kajino, S. Matsuno, Y.K. Yuan, T. Aoki, T. Kitamura, K. Mitsui, Y. Ohashi, and A. Okada, Inst. Cosmic Ray Research, Univ. of Tokyo Preprint (1983).
19. R. Bonarelli, P. Capiluppi, I. D'Antone, G. Giacomelli, G. Mandrioli, C. Merli, and A.M. Rossi, Phys. Lett. 112B, 100 (1982).
20. J. Bartelt, H. Courant, K. Heller, T. Joyce, M. Marshak, E. Peterson, D.S. Ayres, J.W. Dawson, T.H. Fields, E.N. May, and L.E. Price, Phys. Rev. Lett. 50, 655 (1983).
21. E.N. Alexever, M.M. Boliev, A.E. Chadakov, B.A. Makoev, S.P. Mikheyev, and Y.V. Sten'kin, Lett. Nuova Cimenta 35, 413 (1982).
22. V.P. Martem'yanov and S. Kh. Khakimov, "Slowing-Down of a Dirac Monopole in Metals and Ferromagnetic Substances", Zh.

Eksp. Teor. Fiz. [Sov. Phys. JETP] $\underline{62}$ [$\underline{35}$], 35 [20] (1972).

23. See Reference 12.
24. C.W. Akerlof, Phys. Rev. $\underline{D26}$, 1116 (1982).
25. B.C. Barish, R.G. Cooper, C.E. Lane, and G. Liu, Caltech Preprint CALT-68-988 (1982) and proceedings, Monopole '82.
26. T. Ebisu and T. Watanabe, in this volume.
27. D. Cline et al., in this volume.
28. B. Price et al., in this volume.
29. C.R. Dokos, T.N. Tomaras, Phys. Rev. $\underline{D21}$, 2940 (1980).
30. V.A. Rubakov, Nucl. Phys. $\underline{B203}$, 311 (1982); JETP $\underline{33}$, 644 (1981).
31. C. Callan, Jr., Phys. Rev. $\underline{D26}$, 2058 (1982); and Phys. Rev. $\underline{D25}$, 2141 (1982).
32. S. Errede et al., Phys. Rev. Lett. $\underline{51}$, 245 (1983).

A SEARCH FOR GUT MONOPOLES WITH MICA TRACK ETCH DETECTORS

Steven P. Ahlen[a], P.B. Price[b], S. Guo[b], and
R.L. Fleischer[c]

[a]Physics Department
Indiana University
Bloomington, Indiana 47405

[b]Physics Department
University of California Berkeley
Berkeley, California 94720

[c]General Electric Co.
Schenectady, New York

INTRODUCTION

Given the astrophysical bounds on monopole flux implied by survival of the Galactic magnetic field[1], it is quite difficult to carry out meaningful searches for GUT monopoles. To improve significantly upon the Parker limit with direct searches using conventional techniques (superconducting loops, scintillation counters, He-CH_4 proportional counters) would require detector areas of the order of 10^4 m^2, and collection times in excess of several years. An attractive alternative scheme which will be described in the present paper, would be to use an unconventional particle detector of area 10^{-3} m^2 which has been in operation for over 10^8 years.

TRACK ETCH DETECTORS

It is well known[2,3] that virtually any transparent solid will record the passage of charged particles by means of the formation of submicroscopic damage trails tens of Angstroms in diameter. It was the discovery by the authors of Ref. 2 that such damage trails could be amplified by chemical etching that enabled the track etch technique to be of enormous value in such diverse fields as anthropology, geology, microbiology, astrophysics and nuclear physics. It has been found that the effective thresholds for track registration, the nature of the formation process of the latent track, and the optical etching procedures are quite dependent on the type of detector being used. For example, organic polymers are the most sensitive materials with the most sensitive of these (CR-39, the most commonly used material for plastic lenses) being capable of recording the track of a fast particle with stopping power greater than ~ 70 MeV cm^2 g^{-1}. At the other extreme are the inorganic glasses and crystals which may require nearly 100 times greater energy deposition to form a track which can be amplified to macroscopic dimensions by chemical etching. It is not surprising that these two classes of detectors have such disparate sensitivities in view of the difference in the damage production mechanisms. The latent damage trail of an organic solid is believed to be dominated by disrupted molecules while that of an inorganic crystal is predominantly due to a trail of crystal defects formed by the displacement of atoms from their equilibrium positions. It is the smallness of the mass of the electron compared to the nucleus which results in plastics being more sensitive to high velocity particles than are crystals.

A great deal of work has been done with the track recording properties[2,4] of the mineral muscovite mica. This is one of the most common of the various types of mica[5] and can be found in a wide variety of geological environments. Its chemical composition is given by the formula $K_2Al_4[Si_6Al_2O_{20}](OH,F)_4$. Muscovite is among the most useful naturally occurring minerals for track work due to its outstanding optical characteristics. Since geological formation of micas typically occurred 10^8-10^9 years in the past, it is a natural choice for a monopole detector. In the next section we summarize the conditions under which this application is possible.

MICA AS A DETECTOR OF GUT MONOPOLES

In order to evaluate the possibility of using mica as a detector of monopoles it must be determined if the damage trail is large enough to be observable. Information on the amount of damage required is available from studies of tracks induced in mica by low energy heavy ions. It is important to distinguish between two classes of energies of heavy ions. For ion velocities greater than 10^{-2} c, most of the energy lost by an ion goes into electronic excitation and ionization. For velocities smaller than 10^{-2} c, the energy lost goes predominantly into recoils of the atoms of the absorbing material. This energy loss is commonly referred to as the "nuclear" or "atomic" component as opposed to the "electronic" component which dominates at large velocity.

Track formation in mica ultimately depends on a sufficient linear
density of crystal defects.   At low velocity this is easily
related to nuclear stopping power.   At high velocity this situation is more complex due to the ion explosion spike mechanism[2]
which couples electronic energy loss to the formation of crystal
defects due to charge imbalance.   However since GUT monopoles are
expected to have velocities $\sim 10^{-3}$ c, it is not necessary to
consider such effects in evaluating mica response.

The studies of Ref. 4 have shown that etchable tracks are
produced in mica which is irradiated with very low velocity
(5×10^{-4} c to 2.5×10^3 c) ions ranging from Ne to Th.   Furthermore the track etch rate V_e, which is the velocity at which the
etchant (hydrofluoric acid is the best choice for mica) eats along
the particle track, is found to be proportional to the nuclear
stopping power, S_n.   This latter quantity is determined according
to the expressions given in Ref. 6 which have a sound theoretical
basis and which have been well fit to experimental data.   The data
in Ref. 4 and the formulae of Ref. 6 imply that

$$V_e = 0.012(um/hr)S_n(GeVcm^2/g)$$

for mica etched in 40% HF at 25 C.   In considering particle
detectability it is necessary to be aware of two other types of
etch rate: V_o is the rate at which the mica surface is removed
even in the absence of radiation damage while V_p is the rate at
which the edge of the mica is removed (stated more precisely, V_o
is the etch rate perpendicular to the 001 cleavage plane while V_p
is the etch rate parallel to the cleavage plane).   For the
conditions stated above, $V_o = 0.027$ um/hr and $V_p = 1.36$ um/hr.
The difference between V_o and V_p is due to the highly anisotropic
structure of mica.   In order for a penetrating particle having
zenith angle θ to leave a detectable track it is necessary that
$V_e \cos \theta > V_o$; otherwise the surface etches more rapidly than track
etching proceeds.   It follows from this that the nuclear stopping
power must exceed 2.25 GeV cm^2 g^{-1} to produce an etchable track.
If one naively assumes atomic nuclei to be point charges with spin
0, and if one neglects any effects of electron clouds, it can be
shown[7] that the nuclear stopping power of a Dirac monopole
(g = 137 e/2) is far too small to form a track.   However, a number
of authors[8-12] have concluded that monopoles will form bound
states with nuclei through magnetic dipole-magnetic monopole
interactions.   The interactions of such a composite system with
matter would be dominated by the electric charge of the bound
nucleus (this follows from the observation that the scattering
cross section of an electron is 4 times smaller for a Dirac
monopole scattering center than for a proton scattering center).
One could calculate S_n for such a system by replacing the nuclear
mass of the projectile with ∞ (assuming a supermassive GUT
monopole companion) and using the expression for S_n in Ref. 6.

When this is done it is found that for a monopole bound to ^{27}Al, a particularly abundant nucleus in the earth's crust having large magnetic moment, $S_n > 2.5$ GeV cm^2 g^{-1} for velocities from 2×10^{-4} c to 2×10^{-3} c. Thus it seems likely that if monopoles capture nuclei in the earth's crust, they will record tracks in mica located beneath the capture point. It is important to note that if monopoles capture nuclei in the crust they will probably be stopped within the earth, whereas they are unlikely to come to rest in the earth if they carry zero or one unit of electric charge.

Estimates of radiative capture cross sections[9],[13] and of nuclear abundances in the earth's crust yield typical capture mean free paths of ~ 10 km. Furthermore, estimates of mean burial depths over the lifetime of samples of mica currently near the earth's surface are ~ 5 km. Thus, a substantial fraction of monopoles would be detected. An experiment using the above approach to look for monopoles is reported in Ref. 13. The particular sample used had a measured age of 460 million years and an area of 13 cm^2 was scanned for monopole tracks. None were observed and the limit shown in Fig. 1 was obtained. Also shown are Cabrera's latest limit,[14] experimental limits from other groups,[15] and indirect limits based on Galactic magnetic field arguments.[16] The mica limit is well below the most pessimistic indirect limit, and it covers the most significant velocity interval (3×10^{-4} c to 2×10^{-3} c). If the scenario required for the mica technique can be unequivocally demonstrated it would seem that there is little more to be done in searching for monopoles. It is to this important consideration that I turn my attention in the next section.

CAN THE MICA LIMIT BE EVADED

In this section I will consider mechanisms which could reduce the significance of the mica result. It will be seen that there continues to be strong motivation for undertaking large area monopole searches but that the mica result is useful in reducing the necessary parameter phase space covered by such searches.

Track Fading

Although it is well known that elevated temperatures can obsterate latent tracks,[2] this cannot be used to reduce the significance of the mica experiment. This is because the age obtained for the mica with Rb-Sr dating of $(4.90 \pm 0.20) \times 10^8$ y agrees well with the age determined by fission track dating $\left((4.41 \pm 0.43) \times 10^8 \text{ y}\right)$. Furthermore, study of recoil tracks in mica due to alpha decay of U and Th (typical recoil energy ~ 0.4 keV/amn) has shown that the ratio of density of recoil tracks to density of fission tracks is the same within 30% in samples of widely different fission track ages, implying comparable resistance of the two types of tracks to thermal fading at ambient temperatures typical for mica. Since the track produced by a monopole-nucleus system would have a damage distribution similar to that of an alpha decay recoil track, the fission track age is a reliable measure of the monopole collection time.

Collisional Disruption of the Monopole-Nucleus System

If the nucleus bound to the monopole were jarred loose through a collision with a nucleus in the Earth's crust the monopole would not record a track in underlying mica. However, such disruption is not possible for monopole velocities smaller than $\sim 3 \times 10^{-3}$ c since available CM energies in such collisions (even in collisions with ^{238}U) are less than ~1 MeV, the typical monopole-nucleus binding energy.[9]

Catalysis of Baryon Decay

For Grand Unified Theories which predict proton decay, Rubakov and Callan[17] argue that GUT monopoles strongly catalyze baryon decay, making it likely that monopole-nucleus bound states would be short lived. However, there is no proof yet that baryon number violating reactions occur.[18] In addition, it has been argued that SU(5) GUT monopoles would not catalyze baryon decay,[19] and that in some GUTs, baryon number violating proton decay does not occur. In any case, limits on monopole flux derived under the assumption of catalyzed proton decay[20] or neutron decay[21] form a useful complement to the mica experiment.

Reduction of Nucleus Capture Cross Section

Arafune and Fukugita[22] have argued that various effects can reduce the probability of close collision of a monopole with a nucleus and that these effects would thereby reduce the probability of induced baryon decay. Clearly such would also be the case for capture of a nucleus by a monopole. The mechanism discussed most thoroughly in Ref. 22 involved a long range force between the monopole and nucleus due to extra angular momentum carried by a monopole-electric charge system. Depending on the sign of the anomalous magnetic moment of the nucleus, this effect could result in enhanced or reduced capture cross sections. For the case of ^{27}Al, which has a positive anomalous moment, the cross section would actually be enhanced so that the limit of Fig. 1 would be conservative in that this effect was not considered.

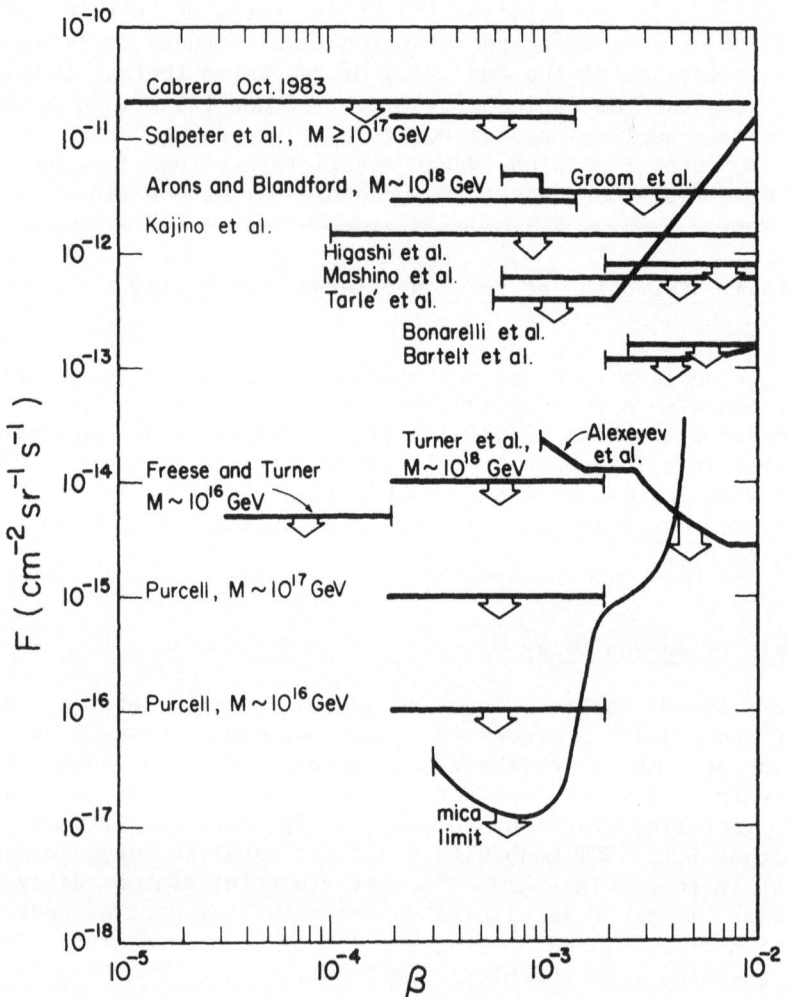

Fig. 1 Limits on monopole flux. See text for discussion.

Diamagnetic Repulsion

It was suggested in Ref. 22 that diamagnetic effects involving atomic electrons would prevent monopole-nucleus contact at
very low velocities. In particular, it was pointed out that the
energy difference between the ground state of a free monopole and
free helium atom and that of a helium atom including a monopole
bound to the nucleus is 16 eV. Thus, for monopoles with
velocities less than 10^{-4} c it would be kinematically impossible
to closely approach a helium nucleus. The situation is somewhat
worse for Al for which the energy difference is estimated[23] to be

1400 eV. The threshold velocity for nuclear capture would be ~ 3×10^{-4} c in this case. One would expect a rapidly rising cross section near threshold (note that at 5×10^{-4} c, the CM energy is more than twice the threshold energy). This effect was included in Fig. 1 by imposing a low velocity cutoff at 3×10^{-4} c. No attempt was made to correct for above threshold cross sections.

Monopole Carrying Positive Electric Charge

It seems that the only way to avoid the mica flux limit is to suppose that almost all monopoles enter the earth with a net positive electric charge. This would almost certainly prevent a monopole from capturing an ^{27}Al nucleus due to Coulomb repulsion. Although this is unlikely if monopoles are created as dyons (one might suspect equal numbers of positive and negative dyons), Bracci and Fiorentini have suggested[10] that monopoles may capture hydrogen nuclei during travel through the interstellar medium if an Auger capture mechanism has a sufficiently large cross section. Furthermore, depending on the binding energy and capture cross section, it may be possible to form proton-monopole bound states in the early universe through a pure radiative process, much as several of the light isotopes are believed to have been formed.

CONCLUSION

As seen above, the only significant challenge to the mica flux limit comes from the possibility of a large probability for forming proton-monopole states. It is important that this issue be clarified theoretically. It is also important that large area experimental searches be undertaken which do not rely on the incoming monopole to be in any particular electric charge state.

REFERENCES

1. E.N. Parker, these proceedings.
2. R.L. Fleischer, P.B. Price and R.M. Walker, Nuclear Tracks in Solids: Principles and Applications, University of California Press, Berkeley (1975).
3. S.P. Ahlen, P.B. Price and G. Tarle, Physics Today, 34, No. 9 (1981).
4. J. Borg, J.C. Dran, Y. Langevin, M. Manrette and J.C. Petit, Rad. Effects 65, 133 (1982).
5. W.A. Deer, R.A. Howie and J. Zussman, Rock Forming Minerals, Vol. 3, Sheet Silicates, Longmans, Green and Co. Ltd., London (1962).
6. W.D. Wilson, L.G. Haggmark and J.P. Biersack, Phys. Rev. B15, 2458 (1977).

7. S.P. Ahlen and K. Kinoshita, Phys. Rev. $\underline{D26}$, 2347 (1982).

8. D. Sivers, Phys. Rev. $\underline{D2}$, 2048 (1970).

9. C. Goebel, Proc. Monopole Seminars, ed. D. Cline, Univ. of Wisconsin, p. 51 (1981).

10. L. Bracci and G. Fiorentini, Phys. Lett $\underline{123B}$, 493 (1983).

11. J. Makino, M. Maruyama and O. Miyamura, Prog. Theo. Phys. (Japan) 69, 1042 (1983).

12. K. Olaussen, H.A. Olsen, P. Osland and N. Ouerbo, DESY preprint 83-041.

13. P.B. Price, S. Guo, S.P. Ahlen and R.L. Fleischer, Phys. Rev. Lett., in press (1984).

14. B. Cabrera, M. Taber, R. Gardner and J. Bourg, Phys. Rev. Lett. $\underline{51}$, 1933 (1983).

15. E.N. Alexeyev et al., Proc. 18th Inter. Cosmic Ray Conf., Bangalore, India 5, 52 (1983); D.E. Groom, E.C. Loh, H.N. Nelson and D.M. Ritson, Phys. Rev. Lett. 50, 573 (1983); J. Bartelt et al., Phys. Rev. Lett. 50, 655 (1983); F. Kajino et al., Proc. 18th Inter. Cosmic Ray Conf., Bangalore, India 5, 56 (1983); S. Higashi, R. Bonarelli et al., Phys. Lett. $\underline{126B}$. 137 (1983); G. Tarle, S.P. Ahlen and T.M. Liss, Phys. Rev. Lett. $\underline{52}$, 90 (1984); T. Mashimo et al., Phys. Lett. $\underline{128B}$, 327 (1983).

16. M.S. Turner, E.N. Parker and T.J. Bogdan, Phys. Rev. $\underline{D26}$, 1296 (1983); E.M. Purcell in Magnetic Monopoles, ed., R.A. Carrigan and W.P. Trower (Plenum, New York, 1983), p. 141; J. Arons and R.D. Blandford, Phys. Rev. Lett. $\underline{50}$, 544 (1983); E.E. Salpeter, S.L. Shapiro and I. Wasserman, Phys. Rev. Lett. $\underline{49}$, 1114 (1982); S. Dimopoulos, S.L. Glashow, E.M. Purcell and F. Wilczek, Nature 298, 824 (1982); K. Freese and M.S. Turner, Phys. Lett. $\underline{123B}$, 293 (1983).

17. V. Rubakov, JETP Lett. 33, 644 (1981); C. Callan, Phys. Rev. $\underline{D26}$, 2058 (1982).

18. R.M. Bionta et al., Phys. Rev. Lett. $\underline{51}$ 27 (1983).

19. T.F. Walsh, P. Weisz and T.T. Wu, DESY preprint 83-022.

20. S. Errede et al., Phys. Rev. Lett. $\underline{51}$, 245 (1983).

21. S. Dimopoulos, J.P. Preskill and F. Wilczek, Phys. Lett. $\underline{119B}$, 320 (1982); E.W. Kolb, S.A. Colgate and J.A. Harvey, Phys. Rev. Lett. $\underline{49}$, 1373 (1982).

22. J. Arafune and M. Fukugita, Phys. Rev. Lett. $\underline{50}$, 1901 (1983).

23. S. Parke and N. Kroll, private communication (1983).

ON THE FEASIBILITY OF INFRARED PHOSPHORS IN

SUPER-SLOW PARTICLE SEARCHES

Ray Hagstrom and Anthony D. Rugari[a]

High Energy Physics Division
Argonne National Laboratory
Argonne Illinois 60439

INTRODUCTION

Monopoles are an inevitable byproduct of the process of symmetry breaking from a grand (semi-simple) symmetry to the sorts of symmetry required to explain results from mainstream elementary particle physics. These grand unified theories (GUT s) represent one of the most promising extrapolations of current theoretical thinking toward the shortest distances. It is more than usually important that we adopt very conservative standards in the interpretation of data on the monopole question because of the link between GUT's and monopoles. GUT monopoles are predicted to be heavy and slow; that is, slow enough that they may not have been detected, and heavy enough that they would neither stop in the earth (but would pass through) nor that they would be accelerated to relativistic speeds by existing magnetic fields in the galaxy.

The point has been made long before now that there is a possibility that GUT monopoles could actually have escaped detection if they were circulating throughout space as a flux at very slow speeds.[1] On the other hand, imposingly strong experimental limits already exist on the flux of fast monopoles[2] and on the abundance of monopoles at rest within matter.[3] It will, of course, always remain relevant to increase the sensitivity of these searches. We, on the other hand, propose that it is also crucial to close the gaps which are not examined by these existing techniques. Thus, we will temporarily adopt the working hypothesis that there are no monopoles stopped in the earth's crust and that whatever monopoles are passing through the Earth must be sufficiently slowly moving to have escaped detection via

normal techniques. This point of view places a lower bound of
3×10^{-5} c on the speeds of monopoles incident on the surface of
the earth.   There is a solid consensus of opinion that standard
detector techniques could have observed monopoles passing through
the Earth if their speeds had exceeded 1×10^{-3} c.   The remaining
niche which has not been thoroughly explored is precisely this
decade and a half of velocities (3×10^{-5} c to 1×10^{-3} c).   Once
this gap has been closed by some search technique which examines
fluxes as low as the present-day limits on faster monopoles, the
question of observing monopoles will be provisionally settled
without major exclusions and, assuming that the quarry has not
then been found, what will remain will be merely a question of
degrees of improvement of the limits on all fronts.

POSSIBILITIES FOR PRACTICAL DETECTORS

     We take the point of view that settling the problem of the
experimental observability of monopoles requires the most rigorous
exclusion of untested and imperfect theoretical arguments.   We
consider it to be most appropriate to construct detectors which
are known to have narrow bandgaps even in the absence of the
strong fields of a monopole.   If the presence of the magnetic
fields further diminishes these bandgaps, all the better.   One
might consider the possibility that the presence of the magnetic
fields of the monopole actually increases the bandgap in the
detector medium.   Even if this were to be the case, the "activa-
tion energy" should be considered to be the minimum value of the
bandgap during the course of the passage of the monopole.   Thus,
by adhering to the use of narrow bandgap detector media, we can
totally finesse questions of the detailed atomic physics of the
detector medium.   Of course, it is clear that without employing
narrow bandgaps, there is no hope of detecting slow projectiles
which merely carry electrical charges.   The data sets on total-
ionization rates show this without question.

     As a first step toward establishing a narrow bandgap detector
for very slow projectiles, a modest array of silicon ionization
detectors has been fabricated.[4]   The total area covered is about
1200 cm^2, with an effective area for redundant time-of-flight
identification on the order of 100 cm^2.   Here the analogy is to
the operation of ionization chambers: Electrons are promoted
across the bandgap by the electromagnetic fields of the moving
projectile.   The tiny currents caused by these electrons are
detected without gain within the device itself.   The very nature
of detecting electrons promoted above a narrow bandgap requires
that very pure detector media be used.   The economic prospect of
extending this sort of technique to larger areas is poor.

     As an alternative, let us explore the possibility of detecting photons which are emitted in the process of the decay of excitations which are initiated by the projectile. Here the analogy is to the operation of scintillators: A passive medium has electrons promoted across a bandgap and when the electrons de-excite they emit photons which may be detected by remote photo-detectors. We are, thus, considering how to build a "scintillator" which has a narrower bandgap in the work reported here than standard instruments.

There are two important subtleties in the practical operation of scintillators which will confront us here. First is the two-step nature of the process: a projectile promotes some electron across a bandgap; the electron subsequently de-excites by spontaneously radiating a photon whose wavelength is prescribed by the Planck relation and the energy difference. If the decay were from the bottom of the conduction band back to the valence band, the resulting photon would have sufficient energy to promote another electron across the same gap. This means the medium would be strongly absorbing to the photon and that signals so produced would be stifled before they could emerge to be detected. Thus, successful scintillator media are seeded with trace contaminants which allow the promoted electrons to de-excite to energies which lie within the bandgap of the bulk of the material and subsequently to revert to their ground state within the valance band. The resulting photons have energies which lie within the bandgap and the bulk of the medium and thus the medium is transparent to them. The second practicality is that the resulting photon must be sufficiently energetic to be detected within financial limitations.

DESCRIPTION OF DETECTORS BASED ON NARROW BANDGAP PHOSPHORS

Our approach, and the heart of the original investigations to be reported in this paper, is to examine the practicality of deploying detectors based upon a special class of modified Lenard phosphors[5]. These remarkable phosphors have the capability to convert infra-red perturbations into visible radiation with reasonable quantum efficiency and at feasible costs. The key to their performance is the phosphors' ability to store energy in the form of very long-lived meta-stable states and to release the stored energy under infra-red stimulation. Figure 1 shows a simplified history of the operation of such a phosphor.

We have examined two similar versions of these phosphors, one activated with Eu and the other activated with Ce. In the true ground state of the phosphor, the Eu or Ce ions have eligible electrons attached. When the phosphor is stimulated via ultra-violet "pumping" radiation, these electrons are promoted to metastable states attached to Sm ions present elsewhere within the

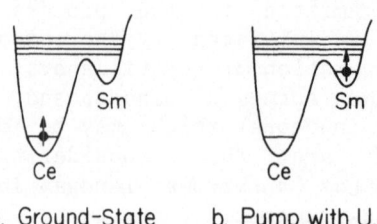

a. Ground-State   b. Pump with U.V.

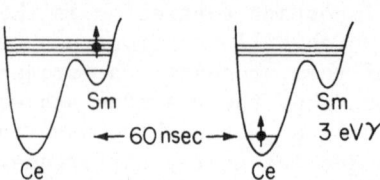

c. Stimulated with I.R. from Monopole
   Weizacker-Williams Field

Fig. 1   Schematic career of active electrons in Sr-S-K-Cl-Ce-Sn
         infrared-sensitive phosphor.   (a) Ground state configura-
         tion —electron attached to Ce activator.   (b) External
         stimulating radiation promotes the electron to attach to
         the Sm trap. (c) Stimulation by infrared radiation boosts
         the electron from the trap, whence it returns to the Ce
         activator with a decay time of 60 nsec. During this decay
         visible light is emitted.

lattice.   These metastable states have very long lifetimes for
spontaneous decays even at room temperatures (up to 10^7 seconds
according to Ref. 5).   The metastable states may be de-excited by
the influence of infra-red frequencies of stimulation, however,
leading to the emission of photons at a characteristic "line" of
the Eu or Ce ion.   Thus, infra-red perturbation frequencies yield
photons with wavelengths near 5800Å ("Orange", henceforth) for Eu
and near 4800Å ("Blue") for Ce.

     KODAK supplies the Eu-activated phosphors as a commercial
product.[6]   KODAK Research Laboratories has generously provided
samples of the Ce-activated versions.[7]   Of course, for the
Ce-based phosphors, emissions would be peaked in the blue-green, a
range well-suited to the 20% quantum efficiency attained by
photomultipliers made with standard 111 formulation phosphor. The
principal mode of spontaneous de-excitation for the active sites
is via emission of the same blue photons as are released by
infra-red perturbation frequencies.

     With a suitably chosen array of phosphorescent material it
would, in principle, be possible to detect the passage of slow

ionizing particles (see Fig. 2). By constructing a multi-layer array of sheets of the phosphor, one can observe light pulses using photomultipliers which surround the array. Then, by knowing the spacing of the layers and the time intervals between photo-multiplier pulses one can estimate the speeds of projectiles passing through. The unique signature which the array could produce would be time-of-flight measurements which yield anomalously slow velocity estimates for projectiles. The array would be optimized to identify projectiles with speeds below 0.01 c. The total range of normal ionizing particles starting with velocities of 0.01 c is generally less than 10^{-4} gm/cm^2. Hence, signals from highly penetrating particles such as super-heavy monopoles are easily distinguished from signals due to normal particles.

Let us consider a trial for such an array. A list of standard parameters to be used in sensitivity computations will be helpful at this point:

A = total area of infra-red phosphor per layer $\qquad$ 10^5 cm^2

H = total height of array $\qquad$ 300 cm

δ = thickness of one phosphor layer $\qquad$ 10^{-1} cm

L = # of layers in array

Y = time duration of data taking $\qquad$ 3×10^7 s

v = speed of monopole under consideration

v_{min} = minimum speeds of monopoles to be sought $\qquad$ 10^6cm/s

v_{max} = maximum speeds of monopoles to be sought $\qquad$ 3×10^7 cm/s

g = charge of monopole= $e/(2\alpha)$ = 68.5e = 3×10^{-8} esu

n_γ = mean # of blue photons emitted as monopole passes one layer

N_γ = # of blue photons which are stored in phosphor array

ρ_γ = # of blue photons stored in 1 cm^3 of phosphor $\qquad$ 10^{17}/cm^3

Ipmt = # of photodetectors for blue photons $\qquad$ 150

ω = angular frequency of equivalent γ in EM field of moving monopole

ω_0 = angular frequency of IR transition in phosphor $\qquad$ 1.6×10^{15}/s

b = impact parameter of monopole trajectory with respect to IR sensitive site

T = decay lifetime of phosphor spontaneously to emit blue photons $\qquad$ 3×10^6 s

τ = luminescence lifetime $\qquad$ 60×10^{-9} s

The mean number of spontaneous emissions within one layer of the detector array during one resolvable time period is $N\gamma\tau/TL$ and the mean number of stimulated emissions is $n\gamma$. In order to establish the signal to background significance we use normal statistics. We require that

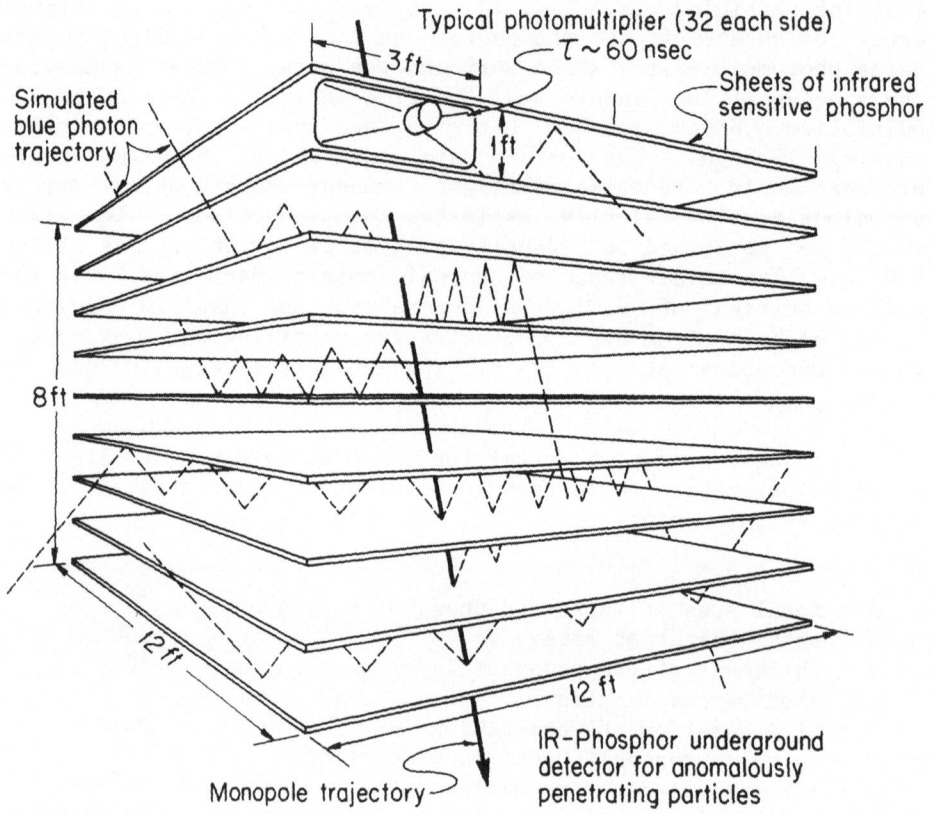

Fig. 2 Schematic of detector using infrared phosphors in 8
        sheets. This schematic is for purposes of discussion only.

$$n_\gamma > 5 \left(\frac{N_\gamma \tau}{TL}\right)^{1/2} \quad . \tag{1}$$

Then the probability of Poisson fluctuations simulating a monopole
within any given layer, within any given detector, within any
given time interval is very small ($\sim 10^{-6}$ with normal statistics).
Hence, the probability of noise simulating a monopole of specified
speed and a specified starting time is negligible $(10^{-6})^L = 10^{-54}$.
Then, the probability of noise simulating a monopole at any speed
at any time during the run is $10^{-6L}YH/(V_{min}\tau^2) \approx 10^{-35}$. These
estimates are clearly naive and are intended merely to allow
discussion.

EXPERIMENTAL DETERMINATION OF RELEVANT PROPERTIES OF NARROW
BANDGAP PHOSPHORS

Now that we have introduced the desired characteristics of
the IR phosphor, Fig. 3 shows a schematic of the test stand used
to evaluate the samples of the phosphors. We have a high
performance RCA C31034 photomultiplier. When its photocathode is
cooled to −25 C the rate of dark current pulses of the tube (above
0.3 photoelectrons) is less than 30 Hz. The tube is mounted in a
refrigeration chamber to facilitate the cooling of the photocathode. Atop the photo tube face sits a quartz-optics telescope
to focus incident light onto the 3 cm^2 Gallium-arsenide photocathode. Within the chamber viewed by the photomultiplier are a
pumping lamp for the phosphor, an infra-red (.93 μm) LED, the
phosphor sample itself, an alpha particle source, and several
calibration LED's. We have the capability of blocking the source
by means of interposing a brass absorber. An optical shutter can
block light incident on the photomultiplier. We also have the
capability of evacuating the test chamber to pressures on the
order of 50 μ Hg and of heating or cooling the phosphor sample by
means of another refrigerator external to the test chamber. The
temperature of the phosphor sample is monitored with a Constan-
tan/copper thermocouple mounted between the sample and the heat-
sink to which the sample is attached. The photomultiplier output
is analyzed with a 1000 channel pulse height analyzer and a six
digit scaler.

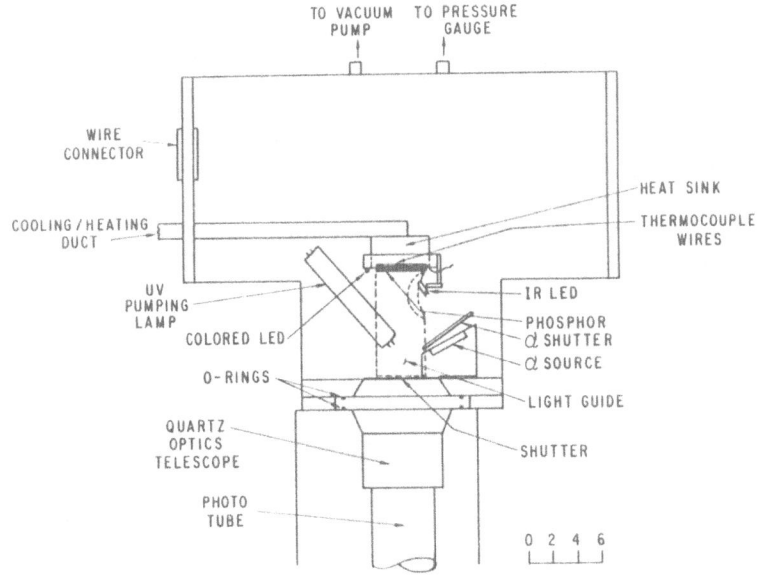

Fig. 3 Schematic of test stand used to evaluate phosphors.

We have calibrated the gain and offset of this pulse-height
analysis system by taking data with several different brightness
settings of the calibration LED's. A typical value of the gain is
0.225 photoelectrons/channel. The next task is to determine the
dark current of the phototube. Dark currents respond to environmental perturbations. We were able to achieve dark currents of
6 Hz. This is sufficiently quiet to allow us considerable
sensitivity to the tiny rates at which the phosphors glow.

Our UV optical pumping lamp was a potentially important
source of light even somewhile after the power to the lamp had
been removed. The glow rate decays via a power law so that the
counting rate is proportional to $t^{-1.25}$, where t is the time
elapsed from turning the lamp off. The glow is reduced to
1.5×10^{-5} times the original rate after two hours.

After these calibrations, we determine how the rate of
spontaneous emission of the phosphor sample varies with time.
This is determined by using an external light source to pump the
phosphor sample so that the early phases of the decay can be
observed without complications due to the UV pumping lamp. We
find that the decay of glow from the phosphors has two phases, a
bright, fast-decaying initial phase followed by a much longer-
lived phase. It was found that our phosphors decay exponentially
in the initial phase of their decay, with dependence as
$\exp(-t/30.4 \text{ min})$ for the Ce-activated (blue) phosphor and as
$\exp(-t/26 \text{ min})$ for the Eu-activated (orange).

We were able to show that increasing the temperature has
little influence on the initial phase of the decay. We explored
the temperature dependence of the constant spontaneous emission.
The sites within the phosphor which participate in this fast-
decaying initial phase of the decay are completely exhausted
overnight, but another class of sites still maintains the sensitivity of the phosphors to Infra-red stimulation. We allow the
phosphor sample to reach the "quiet phase" of its decay and then
we vary its temperature. After allowing the phosphor temperature
to stabilize, the spontaneous glow was measured as follows:
Counting rates were determined for two 60 s intervals with the
shutter open and then two 60 s intervals with the shutter closed
(to monitor the dark current). These intervals were averaged and
the dark current was subtracted off. The clock times when these
emission measurements were taken also were recorded to allow
adjustments to correct for long term drifts. Unlike the situation
for the initial phase of the phosphor glow, the quiet phase shows
strong temperature dependence of its glow rate. For the blue
phosphor the rate of glow depends as $\exp(\text{Temperature}/9.9 \text{ C})$ while
for the orange the rate of glow depends as $\exp(\text{Temp}/12.4 \text{ C})$.
Thus, it appears that at easily reachable temperature we can
effectively cut off the background glow.

OBSERVATION OF IONIZING PARTICLES USING NARROW BANDGAP PHOSPHORS

Ideally, we would use super-slow monopoles or super massive electrically charged particles as the source of ionization. Our most appropriate practical source of ionizing particles is an alpha-source. By comparison to the monopoles (even slow mono-poles) our alpha particles have no more than the ionization rates of monopoles even at the ionization maximum for the alpha particles while the total ranges of such slow alpha particles is very short (about 20 μm). Thus, unlike the situation with monopoles etc., no larger signal is generated by making the phosphor layer thicker than 20 μm. The encouraging side of this situation is that, because we can manage to see signals from the alpha particles, our confidence for detecting monopoles etc. is great.

The distance between source and phosphor in our test stand apparatus is ~ 7 cm; a consultation of standard range tables is sufficient to ensure that our alpha particles reach the phosphor at any pressures less than 0.5 atmosphere. We operate at pressures of 50 μm Hg so that the total energy lost to residual atmosphere is small (<500 eV). We estimate on theoretical grounds that the mean number of photoelectrons observed by the photo-multiplier should be 0.06 per alpha particle incident upon the sample. The time scales for data-taking runs were adjusted to match this expectation.

First we examined the orange emitting phosphor sample. In order to assure that the signals we observed were due to orange photons emitted by the phosphor (and not, for instance, ultra-violet photons emitted by scintillation of the alpha particles in the residual atmosphere of the chamber) we installed a filter in front of the photomultiplier.[8] This filter is a long wavelength-pass filter. Here, 10 runs of 100 s were made to monitor the constancy of the glow. Then, two runs of 1020 s were made to look for the alpha ionized photons. The two runs are completely consistent within statistical fluctuations. Figure 4 summarizes background-subtracted data from one of the 1020 s runs. Here we see the alpha ionized photons with a peak in energy at 0.95 photoelectrons with noticeable signals of 2, 3 and 4 photo-electrons.

The same procedure was then used to examine the blue phosphor. These data were taken with a blue/green glass filter in the system, this filter is a rather narrow-pass filter.[9] Once again, in Fig. 5, we see that with the blue phosphor installed there is a similar peak in the spectrum at about 0.9 photo-electrons. The same consistency is also apparent for the blue phosphor. The difference between spectra taken with the alpha shutter open and the alpha shutter closed is the same (within

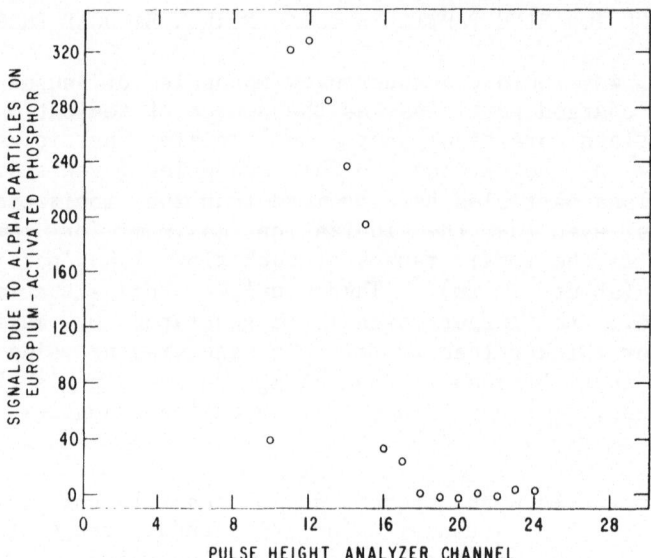

Fig. 4   Measurements of the pulse height histogram of alpha-induced optical emissions for the orange phosphor. The no-alpha background has been subtracted off. The peak in energy is at 0.945 photoelectrons.

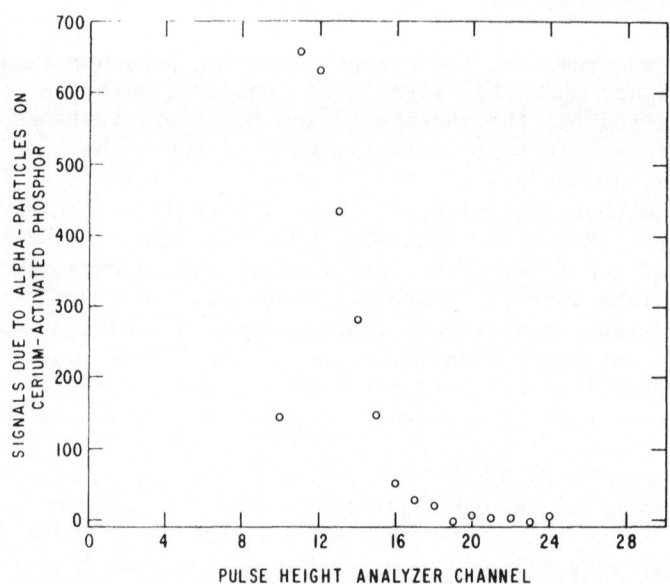

Fig. 5   Measurements of the pulse height histogram of alpha-induced optical emissions for the blue phosphor. The no-alpha background has been subtracted off. The peak in energy is at 0.9 photoelectrons.

statistics) for both 1020 s data taking runs.   This difference is also consistent with being 10 times the difference between the open and closed counts for the 100 s runs.   Convincingly to rule out scintillation of air we made several 100 s runs with no phosphor in the chamber (replacing the phosphor with ordinary paper, mimicking the backing sheet on our samples).   We found no significant difference between the alpha absorber open and closed count rates.

We are able to observe both the orange phosphor and the blue phosphor light emitted when the phosphors are exposed to slow ionizing particles.

Finally, we estimate the observed absolute light yields of our samples.   We know the absolute activity of our radioactive α-source.[10]   We can estimate the rate of α-particles incident on the phosphor samples. Including the quantum efficiency, transmittance of filters, etc., we obtain the number of photons/α-particle.   The largest uncertainty here is our estimate for the light collection efficiency of the system: We are, for instance, uncertain whether photons emerging from the phosphor in directions away from the photomultiplier are absorbed by the backing material or whether they are reflected back toward the photomultiplier. We suspect that the latter is the case.   In any event, we wish to quote our estimate of the photon yield / α-particle conservatively so that the latter case will be assumed.   We conclude that the Eu-activated phosphors produce ~ 20 γ/α while the Ce-activated phosphors produce ~ 30 γ/α.   If light is absorbed by the backing material, these estimates should be doubled.

DISCUSSION

Scaling up our observed photon yields from α-particles to monopoles and electrically charged super massive projectiles we conclude that the yield will be at least 10^4 γ/cm of track length. We are now able to decide the practicability of meeting our criterion for significance, Eq. (1).   We cast Eq. (1) in terms of a required spontaneous glow rate of the phosphor.

$$T > \frac{25A \; \rho_{\gamma\tau}}{\sigma\lambda^2} \tag{2}$$

Clearly we require a marked increase of the lifetime for spontaneous decay of the phosphor compared with room-temperature measurements of Ref. 5 in order to reduce the glow levels to allow our required significance. A factor of 500 improvement here would be appropriate.   Our data show that this sort of reduction merely requires reducing operating temperatures by about 60°C.

Let us summarize our results: We argue on the basis of well-known experimental data that super-slow projectiles will produce ionization signals in media with narrow bandgaps. We propose a specific choice of narrow bandgap phosphors which would be economically suitable for use as detectors. We have performed measurements of these phosphors which determine the light yield of the phosphors to slow projectiles. We have performed measurements on these phosphors which demonstrate that the temperature dependence of the glow of the phosphors is strong enough that a reduction of operating temperature by about 60 C could be sufficient to produce the desired signal to background ratio.

ACKNOWLEDGEMENTS

This work was performed at Argonne National Laboratory, a contract laboratory of the United States Department of Energy.

The work was supported in part by Department of Energy Outstanding Junior Investigator program (RH) and by the Summer 1983 Student Research Participation Program (AR).

REFERENCES

a) Present address: Department of Physics, Virginia Polytechnic Institute and State University, Blacksburg, Virginia 24061.
1. "Particle Physics far from the High Energy Frontier", S.L. Glashow, Bruges Multipart-Dym. 1980:701 and V.F. Mikhaylov, Akademiia Nauk Kazakhstan, Prepritn IFVZ 81-18.
2. "Experimental Status of Monopoles", G. Giacomelli in Magnetic Monopoles, edited by R.A. Carrigan Jr. and W.P. Trower (Plenum, New York, 1983).
3. R.R. Ross, P.H. Eberhard, L.W. Alvarez and R.D. Watt, Phys. Rev. D8, 689 (1973); Phys. Rev. D4, 3260 (1971) and references therein.
4. F. Bieser, M. Bronson, F. Crawford, D. Greiner, R. Hagstrom, P. Lindstrom, ANL-HEP 83 (unpublished).
5. Urbach, Pearlman, and Hemmendinger, Journal of the Optical Society of America 36, 372 (1946).
6. KODAK Special Products Division.
7. KODAK Research Lab.
8. KODAK Wratten No. 22 filter.
9. Schott # BG18 Blue-Green glass filter (red absorbing).
10. Pu239 (70,000 counts/min) courtesy of Ellis Steinberg, ANL-CHM.

A STUDY OF ACOUSTIC DETECTABILITY OF MAGNETIC MONOPOLES

Gang Liu

Department of Physics
California Institute of Technology
Pasadena, California 91125

ABSTRACT

This is an analysis of thermoacoustic detectability of magnetic monopoles. Thermodynamical arguments make the thermal noise calculation much simpler. The result shows that our prototype detector is far from being able to detect magnetic monopoles. The possibility of improvements made by focusing techniques is discussed.

INTRODUCTION

Astrophysical estimations and other information indicate that the flux of the GUT magnetic monopoles might be very small[1,4,5], so developing low cost detectors to cover a large area becomes important. It is for this purpose the acoustic method is suggested. According to the calculation of V.P. Martem'yanov[6] and S.P. Ahlen, even a slow monopole with $\beta \approx 3 \times 10^{-3}$ can have an energy loss as high as 1 GeV/cm in a good conductor[1]. Unfortunately, it seems the only thing one might be able to pick up from a good conducting metal is the acoustic radiation due to the thermal expansion of the monopole track. Whether or not this acoustic signal is detectable is an interesting question. This question has been studied by B. Barish[1] and C. Akerlof[2], and a prototype detector has been built in Caltech to study this question empirically. In this article, thermodynamics is used to simplify the noise calculation, a new S/N value for the prototype detector is obtained and the possible improvements made by focusing techniques are discussed.

S/N OF THE PROTOTYPE DETECTOR

The thermal noise at the output of the transducer is the sum of the contributions from the noises of the medium, the piezoelectric material, the resistance of the transducer etc. Calculating them individually is very complicated. However, a simple thermodynamical argument tells us that the total thermal noise output from the transducer is simply equal to the thermal noise of the real part of its impedance measured when the transducer is coupled to the medium and the whole system is kept at the same temperature.

When one concentrates on the output wires of the transducer, it is equivalent to a noise source V_1 with a complex impedance Z_1 (Fig. 1a). V_1 and Z_1 depend on the coupling with the medium, so they should be taken as is measured when the transducer is coupled to the medium the same way as when it is used to detect the signal.

If Z_1 is pure resistance and is independent of the frequency, we connect the transducer to a resistor R_2 which also contains a noise source[3], $V_2 = \sqrt{4kTR_2\Delta f}$ (Fig. 1b). If the system is kept at constant temperature, the net energy transfer between the transducer and R_2 should be zero, otherwise the second law of thermodynamics is violated. So

$$\left(\frac{V_1 R_2}{Z_1 + R_2}\right)^2 (R_2)^{-1} = \left(\frac{V_2 Z_1}{Z_1 + R_2}\right)^2 (Z_1)^{-1}$$

gives

$$V_1 = \sqrt{4kTZ_1\Delta f}$$

which is just the thermal noise of Z_1.

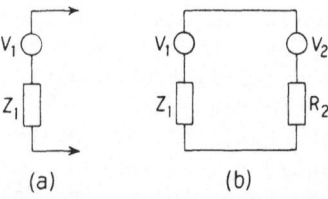

(a)   (b)

Fig. 1 (a) For noise analysis, the transducer V_1 is equivalent to a noise source with a complex impedance Z_1, (b) At thermal equilibrium, the noise power fed into R_2 by V_1 should be equal to the noise power fed into Z_1 by V_2.

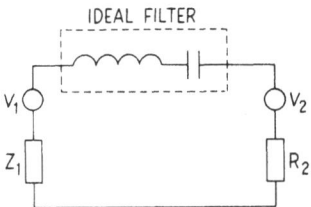

Fig. 2 Ideal noiseless filter is added to restrict the frequency range of the transfer between the transducer and R_2.

If Z_1 is complex and depends on the frequency, we use an "ideal filter" to restrict the frequency range of the energy transfer between the transducer and R_2 (Fig. 2). The ideal filter is made of ideal capacitance and inductance without any resistance. To make is noise free it's kept at very low temperature. Since there is no energy dissipation, to keep it cool doesn't require any energy. Again, the second law of thermodynamics requires

$$\int \left| \frac{V_1(f)R_2}{Z_1(f) + R_2 + Z_0(f)} \right|^2 \frac{df}{R_2} = \int \left| \frac{V_2 Z_1(f)}{Z_1(f) + R_2 + Z_0(f)} \right|^2 \frac{Re(Z_1(f))}{|Z_1(f)|^2} \, df \quad (1)$$

where $Z_0(f)$ is the impedance of the ideal filter, which is pure imaginary. Equation (1) holds for any arbitrarily made $Z_0(f)$, which means

$$|V_1(f)|^2 R_2 = |V_2|^2 Re(Z_1(f)) \quad .$$

Thus,

$$V_1^2(f) = 4k \, TRe(Z_1(f)) \quad . \tag{2}$$

So the thermal noise at the output of the transducer is just the thermal noise calculated for the real part of its impedance. This can be generalized to any sourceless system.

The signal is easily found to be

$$V_{sig}^2 = \frac{\eta A R_p}{Z_a} P_{sig}^2$$

where R_p is the input impedance of the preamplifier, P_{sig} is the signal pressure on the transducer, A is the area of the transducer, Z_a is its acoustic impedance and η is its efficiency. This gives

$$\frac{S}{N} = \frac{V_{sig}^2}{V_n^2} = \frac{\eta A P_{sig}^2 R_p}{4kTZ_a Re(Z_1)\Delta f} \qquad . \qquad\qquad (3)$$

For maximum energy transfer from the transducer to the pre-amplifier we should make $R_p \approx Re\ (Z_1)$. Taking $\eta = 1$ and using the same data as in Ref. 1, we have

$$\frac{S}{N} = 1.2 \times 10^{-4} \qquad .$$

This is the theoretical limit of our prototype detector. In practice the transducer's noise plus the noise of the preamplifier is measured to be 3 times larger in amplitude and η is about 70%. This means S/N is only 9.4×10^{-6}. We see that our prototype detector is far from being able to detect a monopole. This result agrees with the calculations of Akerlof[2].

FOCUSING TECHNIQUES

In Eq. (3), the only parameter we are free to chose to improve S/N is the area A. To increase S/N by 10^5 times we have to make $A = 10^5$ cm^2, or about three and half meters in diameter. The thickness of the piezoelectric ceramic has to be made about half a wavelength, or 0.3 mm. This is impossible to do in practice.

Another way to increase area is to use focusing techniques to converge a large area of acoustic wave to the small area of the transducer. The focusing lens system should keep the coherence of the acoustic wave and should convert the cylindrical wave from the monopole track into a plane wave when it reaches the transducer. To make such a focusing system, many serious technical problems may arise, but in principle it should be possible.

A question as to whether focusing techniques also focus the noise was raised when focusing was first suggested. With the thermodynamical arguments above, the answer is clearly "no", because when the transducer is in thermal equilibrium with the environment, the noise power that is focused into it is exactly equal to what it sends away, and even balanced in frequency. So the noise at the transducer should only depend on its own characteristics and the temperature. It doesn't matter whether or not a lens is put in the medium. Just like black body radiation, it doesn't matter how you heat the black body, once it is in thermal equilibrium with the environment, the energy density and frequency distribution of the radiation inside the cavity is only a function of its own characteristics and the temperature.

CONCLUSION

Thermodynamics makes the thermal noise calculation much simpler. When the whole system is in thermal equilibirum, only the noise of the transducer's impedance needs to be calculated. The calculation shows that our prototype detector is far from being able to detect the acoustic signal from a monopole. Focusing techniques may solve the problem, but many technical problems remain to be solved.

ACKNOWLEDGEMENT

I am especially grateful to Professor Barry Barish, without whose guidance and great help this work would be impossible. Also many thanks are due to Charles Lane, Ron Cooper and Mark Muldoon, with whom many helpful discussions were essential for my understanding of the subject.

REFERENCES

1. B.C. Barish, "Acoustic Detection of Monopoles", Magnetic Monopoles (edited by R.A. Carrigan, Jr. and W. Peter Trower), Plenum Press, 1983, p. 219. Also Caltech Preprint CALT 68-988 (1983).
2. C.W. Akerlof, "Limits on the Thermoacoustic Detectability of Electric and Magnetic Charges" Phys. Rev., D26, 1116 (1982).
3. D.K.C. MacDonald, Noise and Fluctuations, John Wiley & Sons, Inc.
4. M.S. Turner, E.N. Parker, and T.J. Bogdan, "Magnetic Monopoles and the Survival of Galactic Magnetic Fields." Phys. Rev., D26, 1296 (1982).
5. M.J. Longo, "Massive Magnetic Monopoles: Indirect and Direct Limits on Their Number Density and Flux", Phys. Rev., D25, 2399 (1982).
6. V.P. Martem'yanov and S.Kh. Khakimov, "Slowing-Down of a Dirac Monopole in Metals and Ferromagnetic Substances" Zh. Eksp. Teor. Fiz. 62, 35 (1972).

THE CALTECH PROTOTYPE SCINTILLATOR MONOPOLE DETECTOR

Charles E. Lane

Lauritson Laboratory
California Institute of Technology
Pasadena, California 91125

ABSTRACT

A scintillator detector for magnetic monopoles was built, emphasizing enough redundancy for unambiguous detection of monopoles if a signal is found. This detector, now in operation, is intended as a prototype for a much larger detector required to reach the Parker bound.

INTRODUCTION

Our goal is to build a detector capable of unambiguously detecting magnetic monopoles, and not just setting limits. A modestly instrumented detector might be able to set limits in the absence of candidate events while our more ambitious goal places more severe constraints upon the design of a detector. First, we need to keep in mind that the ultimate detector will probably have to cover a large area in order to get below the Parker Bound. This implies that the detector will have to be relatively inexpensive per unit area in order to be practical. The requirement of having a clear, unambiguous monopole signal requires one to collect as much information with as much redundancy about each candidate event as possible, and to design the detector in such a way that the signature of a monopole will stand out from any background signals. For example, the minimum ionization necessary for detection should be considerably lower than the theoretically predicted ionization for a monopole with a velocity in the range of interest. One wishes to have a detector that is rather quick to build, and hence a design that utilizes familiar techniques is to be preferred. It is the considerations of expense and

familiarity that leads one to the use of scintillator as a possible
detector of monopoles.

The Caltech prototype monopole detector was constructed with
these goals in mind, to test and study the scintillator techniques
for monopole detection in detail.   It is intended as a prototype
to determine the necessary elements of a much larger detector with
which one would be able to challenge the Parker Bound.

DESIGN CONSIDERATIONS

The primary consideration in the design of a scintillator
detector is the slowness of a GUT monopole. It is expected that a
'cosmic ray' monopole will have a β of order $10^{-3} \rightarrow 10^{-4}$.   The
penetration and low velocity of a monopole provides a unique
signature for detection in a scintillation array.   Below are
listed the time scales involved in monopole detection for a
monopole with β = 10^{-4}:

   i) muon   through 5 ft. detector   5   ns

   ii) monopole through 1 in. scintillator   80   ns

   iii) monopole through 5 ft. detector   50 μs   .

The timing of pulses from several parallel planes of scintillator provides the monopole signature required for detection.   As
a monopole passes through the scintillator planes, it produces
scintillation photons which are spread out over the transit time
of the monopole. (See Fig. 1).   These photons are detected, and
appear as a train of small (single photoelectron) pulses, unlike a
cosmic ray muon which will produce a sharp pulse with a larger
amplitude.   The delay between 'bursts' in several planes gives a
signature of a slowly moving particle, and may be detected by
timing. (See Fig. 1).

Since the monopole signal is based upon the timing of PMT
pulses from several planes of scintillator, it becomes important
to keep an event history of the PMT pulses for each event.   It is
desirable that this event history be long (longer than a monopole
transit time through the detector), and have good timing resolution.   Notice that the spread of the PMT pulses from one plane of
scintillator   gives an indication of the velocity of the particle
producing them, however, it is much better experimentally to see a
'delayed coincidence' between independent planes of scintillator
as confirmation of a monopole.

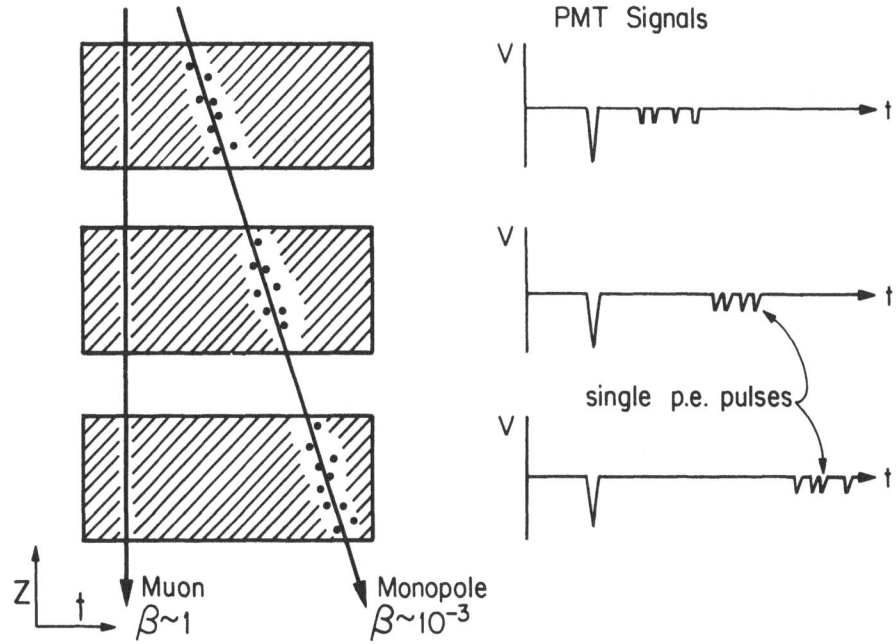

Fig. 1 Relative pulse heights and time distribution of muon and
monopole signals through 3 layers of scintillation.

DETECTOR DESIGN

Scintillator

The Caltech prototype monopole detector consists of 6 planes
of 1 inch NE 114 scintillator, measuring 5 by 10 feet (Fig. 2).
There are two pieces of scintillator in each plane, each of which
has two 56 AVP photomultipliers on it, with BBQ shifter bars used
to collect the light from the scintillator.   A cosmic ray muon
(minimum ionizing) passing through the detector produces an
average total of 12 p.e. in each plane of the detector.

Electronics

The signals out of the PMT's are amplified, discriminated at
a level of about 0.6 p.e., and sent to the trigger logic and the
data acquisition electronics.   The trigger logic and the data
acquisition functions are performed independently, with only an
'event' signal and 'trigger type' being sent from the trigger
logic to indicate an interesting event.   The triggers are
deliberately less selective than one would have normally, since it
is assumed that off-line analysis of the events will be used to
throw out events that do not satisfy the 'software' trigger
criteria.

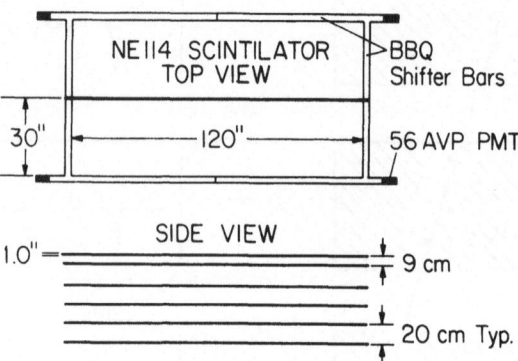

Fig. 2   The Caltech scintillator array.

Trigger Logic

The trigger logic contains four separate triggers which are combined to generate the event trigger. The triggers use signals from each plane that are the 'or' of the PMT pulses from that plane, since the triggers are based upon inter-plane timing and not the details of pulse timing within an individual plane.

The first trigger detects cosmic rays passing through the detector by means of a fast coincidence of three out of four of the center planes. This trigger is also used to veto the monopole trigger, since most of the 'false alarms' produced by the monopole trigger are a result of cosmic rays. The amount of dead time introduced by this cosmic ray veto is a few percent; it is of the same order as the dead time caused by the computer in taking an event.

There is a 'stopping muon' trigger, which consists of a fast coincidence of the top three planes and a scintillator paddle placed in the center of the array, vetoed by any of the lower three planes. Between the third and fourth plane of scintillator we have placed 4 inches of wood as a low-Z muon target. Muons which stop in the wood will subsequently decay. The stopping muon trigger allows us to see this in our detector, and provides a valuable check on the operation of our detector. Both the cosmic ray and the stopping muon triggers are prescaled down to rates of about 1 per 100 s so that they don't swamp the monopole triggers.

The monopole trigger consists of a delayed coincidence between all six planes of the detector. A pulse in the first plane generates a gate after a fixed delay. Pulses from the second plane that arrive in this gate are used to trigger a delay and gate for the third plane, etc. Currently the delays are set at 150 ns and the gate widths at 2.5 µs, which gives a sensitivity to monopoles in the range of $\beta = 4 \times 10^{-3} \rightarrow 3 \times 10^{-4}$. As mentioned previously, the monopole trigger is vetoed by the cosmic ray trigger to prevent cosmic rays from generating an excessive number of false triggers.

There is also the recent addition of a 'random' trigger, which is a pulser set at about .01 Hz. Events produced with this trigger are generally uninteresting, but contain a random sample of noise pulses from the PMT's and are useful for off-line diagnostics.

Data Acquisition System

The data acquisition system consists of 12 data acquisition (DAQ) timing modules that were modified from their original use with a drift chamber, and associated control electronics. There are two DAQ modules for each plane of scintillator. Pulses from the plane are fed in parallel to both DAQ's which record the time and the tube numbers that fired. The DAQ's keep a continuous record of the last 15 pulses from a plane scintillator, with a timing resolution of 7.1 ns. Given the known noise rates of the tubes, this results in event histories about 500 µs long.

When an event pulse comes in, it latches the type of trigger and starts a delay. The delay allows post-trigger data to be collected for a fixed length of time (approx. 10 µs). The DAQ's are then shut off, and the PDP 11/34 is interrupted so that the DAQ's may be read out. The event is then written on tape for later analysis. The 11/34 also monitors the PMT noise rates for each plane so that light leaks and bad tubes are quickly found and corrected.

DETECTOR PERFORMANCE

Table 2 gives the trigger rates observed in our detector, before prescaling of the cosmic ray and stopping muon triggers.

Table 2. Trigger Rates.

Monopole	0.056	Hz
Cosmic Ray	1.38	kHz
Stopping Muon	0.15	Hz
Random	0.03	Hz

The cosmic ray triggers are prescaled by a factor of 10^5 and the stopping muon triggers by a factor of 10 before being used to trigger events.

Afterpulsing

When one examines the event histories closely, it becomes apparent that there is some form of afterpulsing being produced. This has the effect that a 'primary' pulse will be followed some microseconds later by one or more afterpulses. The afterpulsing phenomenon is an important problem for the detection of monopoles, since the time scales associated with afterpulsing are of the same order as the monopole time scales. The afterpulsing mechanism results from the ionization of residual gas in the PMT.[1]

When a signal is multiplied in the dynode structure of a PMT, there is a possibility that one of the electrons will ionize a remnant gas atom in the tube. It is most likely that the ionization will take place between the last dynode and the anode, because the electron flux is highest in the last stage. The gas ion will then drift to the photocathode or one of the dynodes, and upon impact will liberate electrons. The electrons will then be multiplied just as they would for a normal pulse, and the afterpulse then appears at the anode of the photomultiplier tube.

The process that sets the delay from the original pulse and the afterpulse is the drift time of the ion in the PMT's electric field. Using 'typical' numbers of 1 kV tube voltage and tube dimensions of $\approx$ 10 cm, one obtains the drift times for common gases.

Table 3.   Typical Ion Drift Times.

H_2	0.65	μs
He	0.91	μs
N_2	2.4	μs
Ar	2.9	μs

The amount of gas in a PMT strongly depends upon the details of a tube's fabrication and history. Helium tends to be the worst contaminant, since it will slowly diffuse into the tube through the glass envelope. Tubes which have spent years in an environment with higher than normal helium content (such as most physics labs) will exhibit more after pulsing because of this effect.

It has also been suggested that there may be sources of afterpulsing that are related to late fluorescence in scintillator or BBQ shifter bars.[2] This may indeed explain some of the afterpulsing which has been observed, and tests are in progress to examine both of these mechanisms in detail.

Muon Lifetime

A convenient test of the performance of our detector is provided by cosmic ray muons stopping and subsequently decaying. A simple analysis of these events showed some interesting characteristics. The muon decay analysis software finds the stopping muon's track and makes a histogram of all the pulses found after the muon track. This analysis produced a decay curve with a slope $\approx$ 2 μsec in the central region (1 μsec $<$ t $<$ 3 μsec). Also, for times $<$ 1 μsec, the decay curve had a slope $\approx$ 0.5 μsec. This corresponds to afterpulsing from either the PMT's or the scintillator, which is also observed in time distributions of pulses in cosmic ray triggered events.

CONCLUSIONS

The monopole detection system described is currently in operation, collecting both 'candidate' monopole events and diagnostics. It has been designed as a prototype of a much larger detector, which would be capable of detecting monopoles unambiguously.

The problem of afterpulsing becomes serious when one addresses the problem of confident detection of monopoles, and not just setting new limits. We believe the afterpulsing problem may be minimized through proper selection of PMT's and scintillators, and we are studying the problem in detail.

A possible improvement which may be useful in our detector scheme is the addition of pulse height information, as well as timing. This would allow one to distinguish small, monopole-like pulses from the relatively large pulses of cosmic rays. In particular, it would give an indication that a pulse was produced by a cosmic ray even if it only penetrated a single plane of scintillator.

In summary, we have a detector instrumented well enough to yield the necessary information on what will be required to scale up to reach the Parker Bound with a sea level scintillator array capable of unambiguously detecting monopoles.

ACKNOWLEDGEMENTS

The Caltech Monopole Group working on this experiment consists of B. Barish, C. Lane, G. Liu and D. Wu. Work supported in part by the U.S. Department of Energy under contract DE-AC03-81-ER40050.

I would like to thank Professor Barish for his comments and suggestions, particularly with regard to the muon decay analysis. The Monopole Group thanks Professor A. Bodek for his kind loan of the scintillator which is part of the Lab-E neutrino facility at Fermilab.

REFERENCES

1. P.B. Coates, "The Origins of Afterpulses in Photomultipliers," J. Phys. D6, 1159 (1973).
2. P.M. McIntyre and R.C. Webb, "Searching for the GUT Monopole," pp. 352-392 in ANL Proton Decay Workshop Proceedings, ed. D.S. Ayres, (Feb. 1982).

DETECTORS FOR RARE EVENTS

Georges Charpak

CERN
Geneva, Switzerland

Present day particle physics has raised interest in the observation of extremely rare events whose separation from background is often critically dependent upon the energy resolution and a clean signature of some disintegration or interaction pattern. Whenever purely electronic means are used, the search for increased energy or position resolution and data redundancy is usually limited by the corresponding increase in cost.

Detectors, such as nuclear emulsions, bubble chambers and streamer chambers, where a rare event can be characterized by a photograph of a disintegration or reaction pattern, offer a clear advantage for background rejection. In this article, I wish to draw attention to the fact that it is possible to combine the advantages of photographic data retrieval with the flexibility of operation of conventional gaseous or liquid detectors operated with electronic data retrieval, and mention as an example some possible applications to such problems as nucleon decay, neutrino-electron interaction, double-β decay, and the search for magnetic monopoles.

PHOTOGRAPHY OF IONIZATION PATTERNS

Many detectors are based on the localization of electrons liberated in a medium. It is sometimes possible to detect electrons on the spot where they have been liberated: this is the case of streamer or avalanche chambers, where the electrons are exciting a large number of neighboring atoms by the applicaton of a short intense electric field.

In most detectors the electrons are transferred, by proper electric fields, to detecting wire planes where the signals induced by avalanches permit the spatial reconstruction of the event.   This is where the cost of the electronic elements and the spatial extension of the induced signals limit the quality of the data retrieval when the events are complex.   We suggest using photographic methods, with or without the help of image intensifiers, to obtain the information on the localization of avalanches, without sacrificing the quality of the information obtained by collecting the flow of charges, as a function of time, on various electrodes.   This can be obtained by exploiting two phenomena:

    i)   In multistep avalanche chambers it is possible to bring the amplification of every avalanche, even when it is initiated by a single electron, to a level where it is possible to photograph the cluster of atoms excited at the front edge of the avalanche.

   ii)   It is possible to drift electrons in intense electrical fields, where they produce VUV radiation which can be photographed, after proper wavelength shifting by means of image intensifiers.

PHOTOGRAPHY OF IONIZATION TRACKS WITH THE MULTISTEP AVALANCHE CHAMBERS

Multistep   structures[1-3] permit   the   multiplication   of electrons between parallel grids, and the transfer of a fraction of the electrons from the avalanche into drift spaces and other successive amplifying structures: multiwire chambers[4,5]   or parallel-grid amplifying gaps, or triggered or d.c. spark chambers.[6,7]

One of the main applications of this counter has been the construction of Ring Imaging Cherenkov detectors, where the capability of amplifying up to levels of 5×10^7 is exploited for the imaging of single photoelectrons produced by the absorption of VUV Cherenkov photons.[4-9]

The use of spark chambers as a final step permits indeed the easy imaging of the position of the initial electrons.   Figures 1-4 illustrate this imaging capability.   It is however subject to the well-known limitation of spark chambers concerning the rate and the number of sparks. The avalanche size which can be reached in a double-step structure is however sufficient to be photographed with the help of image intensifiers. This is demonstrated indirectly by the results of the triggered avalanche chamber coupled to image intensifiers.   In a multistep chamber the

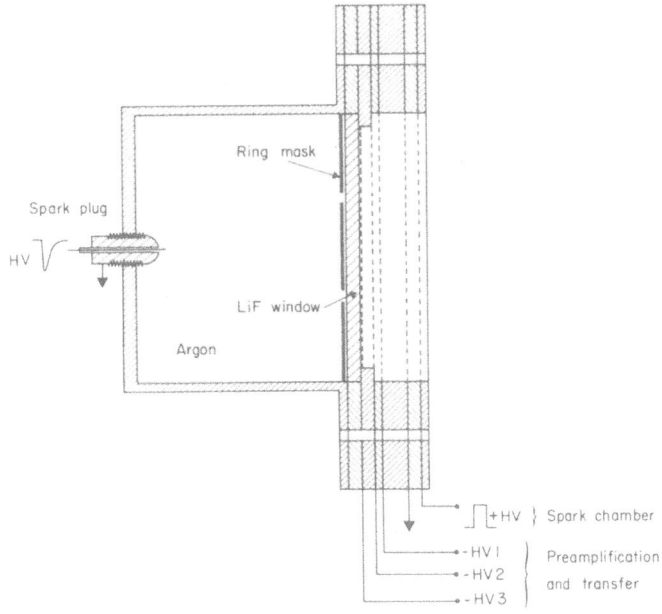

Fig. 1 Test of the imaging capability in the VUV for multiphoton
        events with the multistep chamber, where the last step is
        a triggered spark chamber.   Bursts of photons are gener-
        ated in argon by a triggered spark plug, and a mask
        simulates a ring image.   The gas fillings which give the
        best multiplication efficiency are He or He-Ne, with 2-5%
        acetone, or benzene, or triethylamine. (From Ref. 6.)

situation is more favorable since the light is produced mostly in
the last ionization mean free path of the avalanche, which is of
the order of a few hundred microns and, since the optics does not
require any depth of field, large apertures can be used. This has
been demonstrated directly by Gilmore et al.,[10] who have developed
a Ring Imaging Cherenkov detector which is capable of photographing single electron avalanches and even particle tracks, as
illustrated by Figs. 5 and 6.

    It seems to me worthwhile to consider the coupling of these
read-out methods to gaseous or liquid drift spaces where wires or
strips of wires are collecting, as a function of time, the flow of
charges, while the picture of the interaction or disintegration
pattern is obtained by a photographic method.   This has the
advantage of reducing considerably the task carried out by the
electronics, compared for instance to a conventional Time Projection Chamber (TPC), and gives a bias-free image of all the
ionizing events.

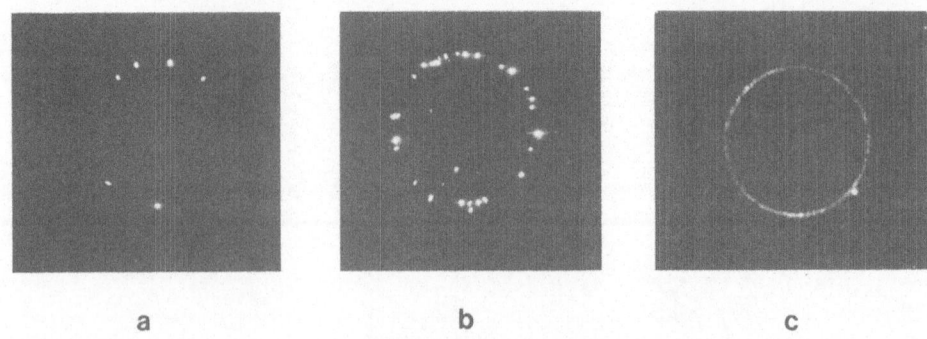

a   b   c

Fig. 2   Ring images obtained by photographic recording of single
         events with an increasing number of photons from (a) to
         (c). Ring diameter and width are 35 mm and 0.5 mm, re-
         spectively.   Film Polaroid, 3000 ASA, 1/f = 5.6.   (From
         Ref. 6.)

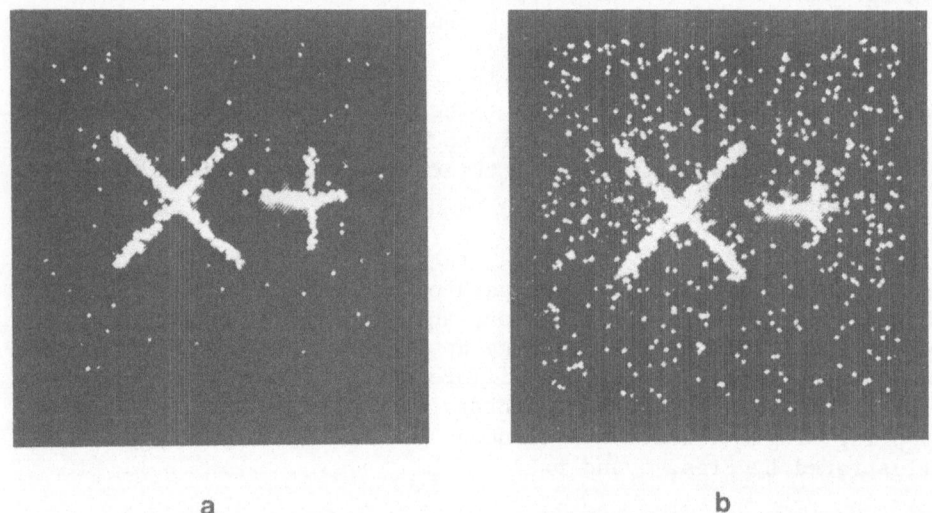

a   b

Fig. 3   Imagining of soft X-rays with the multistep spark chamber.
         In (a), the image is obtained in the self-triggered mode,
         the HV pulses to the spark chamber being initiated by the
         signals in the preamplification elements.   In (b), the
         same image, obtained by operating the chamber at constant
         voltage.   (From Ref. 6.)

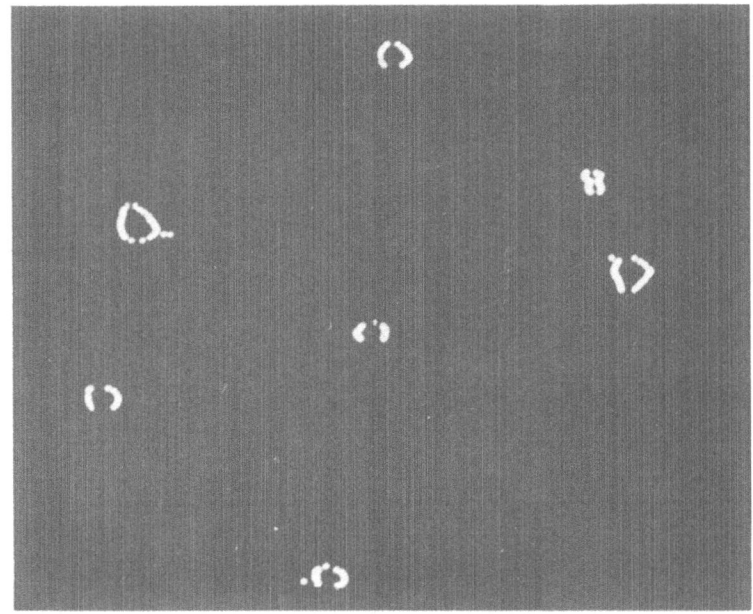

Fig. 4 Vidicon read-out. The VUV photons are produced by
        Cherenkov light. The image of a single event, viewed by a
        digitizing Vidicon camera, gives the contour of the
        spark.[7]

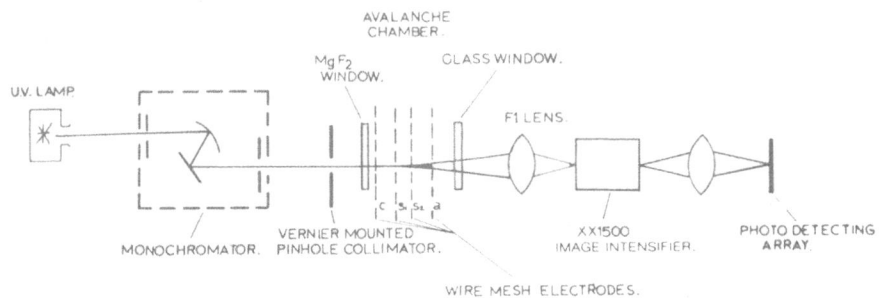

Fig. 5 Photography of an avalanche in a two-step avalance cham-
        ber. Experimental set-up described in Ref. 10. Gas
        filling: argon +2.6% acetone. Maximum gain: 10^7.

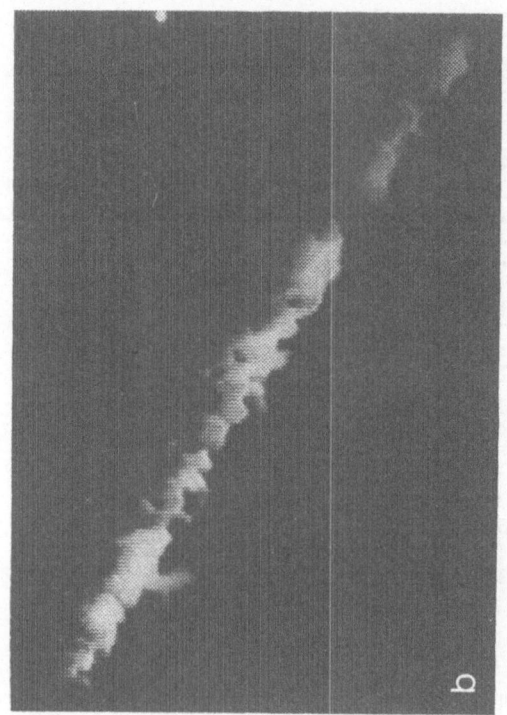

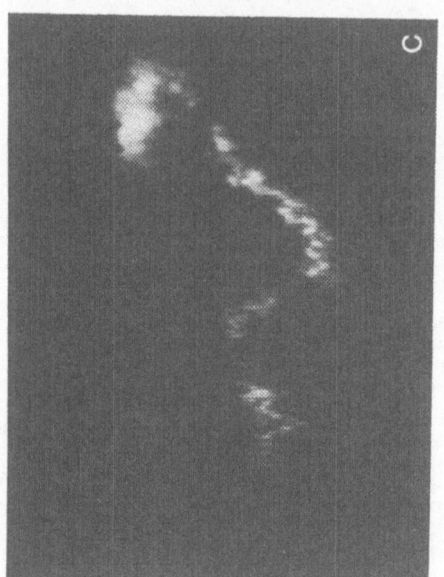

Fig. 6

Images of avalanches or tracks (from Ref. 10). Same set-up as Fig. 5. (a) Avalanche from a single photon; (b) Cosmic-ray track showing a δ-ray about 1 mm long; (c) Track of a single electron. The increase of ionization towards the end of its range is visible.

Gaseous Detectors

This is the easiest application. The electrons created in a large gaseous volume, as in a TPC, can be drifted to a multistep structure where the events are photographed while the electronic signals are detected by wire strips and appropriate electronics recording the amplitude as a function of drift time.

To take an example connected with this conference let me consider the detection of monopoles by gaseous detectors. It has been shown that massive monopoles of low velocity can be detected by the excitation and ionization of proper gas mixtures.[11] With helium as a gas carrier, the threshold for monopole detection is at velocities of about $v/c \approx 10^{-4}$. It would reach $\approx 3 \times 10^{-5}$ with argon as the main gas carrier. The crossing by a monopole of a large gaseous volume would result in a string of electrons with arrival times, at the cathode, correlated not only by the well-defined electron drift velocity, but also by the monopole velocity, which may well be of the same order of magnitude as the drift velocity of electrons in gases. The times of arrival are given by the sequence of signals on the strips, and the ionization pattern of the very rare monopole event is given by the photograph of the ionization pattern with the very powerful rejection of spurious background events.

While for very large detecting surfaces, of the order of 1000 m^2, which are necessary to reach some suggested astrophysical limits, purely electronic methods may be necessary[12]--and the interaction pattern is after all very simple--an exhaustive study of the sources of background in a model of the order of 1 m^2 may be worth undertaking by the photographic method combined with the electronic read-out of the flow of ionization in the final-stage multistep chamber.

Liquid Detectors

This is the field where the method is the most promising, if feasible. The low cost of argon together with the experience acquired with liquid-argon calorimeters in the detection of the ionization deposited by charged particles, has led to active research and the demonstration[13,14] of the possibility of drifting electrons over large distances.[13,14] Several authors[15,16] consider that liquid-argon TPCs have very promising features and do not hesitate contemplating 10^3 tonne detectors for proton decay.

Dolgoshein et al.[17] have shown that ionization electrons can be extracted efficiently from liquid argon with moderate fields. They have even succeeded in extracting electron clusters produced in liquid argon by α-particles and imaging them with a spark chamber (Fig. 7). I suggest extracting the electrons, amplifying

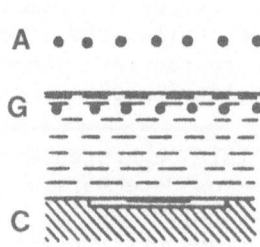

Fig. 7   Imaging of ionization electrons in liquid argon. (a) Schematic   diagram of a three-electrode chamber   with α-source
underneath.   A = anode;   G = grid;   C = cathode.
Dimensions: A - G = 1 cm; G - C = 2   cm.   The   level of
liquid argon over G is about 0.1 cm.   (b) Image of the
active surface of the α-source immersed in the liquid
argon.   Diameter: 35 mm; liquid layer thickness: 0.4 cm.
The bright ring around the active region is due to the
reflection of the light of the sparks on the rounded part
of the polished electrode.[17]

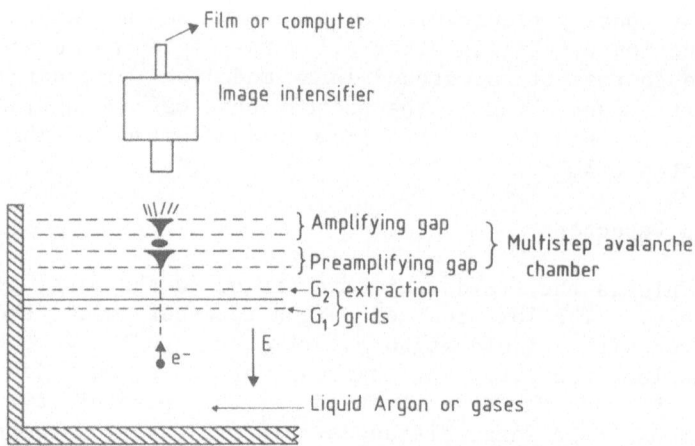

Fig. 8   Principle of the imaging detector: The electrons are
extracted from the gaseous or liquid medium   by electric
fields.   They are multiplied in a gas   in successive
stages between parallel grids.   The light produced by the
final avalanche is photographed with the help of image
intensifiers.

them in a multistep structure, photographing with an image intensifier the atoms excited in that last avalanche, and measuring as
a function of time the rate of arrival of the charges (see Fig.
8). One has to demonstrate, indeed, that a gas mixture compatible
with a liquid-argon interface can be operational in a multistep
chamber. However, the very large initial number of electrons
makes the problem easier than in a gaseous detector. One could
thus envisage, for the study of nucleon decay, 10^5 tonnes of
liquid argon with an energy resolution, around 1 GeV, at least 10
times better than that of the water Cherenkov detector, which, at
a level of 10^4 tonnes has shown no events, with a level of
background which is determined to a large extent by the energy
resolution. We should also keep in mind that liquids other than
argon can be candidates for the detector fillings, such as liquid
methane[18] or even certain liquids at room temperature which are
being actively investigated.

However, while such a gigantic detector might be utopian, a
much simiplier application seems of interest. The study of
neutrino interactions with electrons would greatly benefit from
the quality of a detector of the type envisaged. The vertex at
the production point would give one of the most convincing
signatures of the absence of recoiling hadrons. The multistep
structure permits the gating of the image intensifier in order to
reject some gross sources of background and the electronic storage
of the image can permit a high rate of computer treatment of the
images.

EXPLOITING THE STIMULATED SCINTILLATION LIGHT

Electrons drifting in gases under the influence of an intense
electrical field may produce atom excitation and photon emission
without producing any ionization. This effect is exploited in the
so-called proportional scintillation counters.[19] The photons can
be detected by photomultipliers or by counters filled with a
photo-ionizable vapor.

Gorenstein and Topka[20] have shown that it is possible, after
various stages of wavelength shifting and image intensification,
to take a picture of the light-emitting track. In their device
the electrons were produced by 5.9 keV X-rays absorbed in xenon at
2 atm. The VUV light is produced by electrons drifting in an
intense electric field between two grids over a distance of 3 or
4.8 mm. The amount of light they obtained was just at the limit
of detection. Figure 9 shows the experimental set-up.

Figure 10 displays an integrated picture of two slits
separated by 0.5 or 1 mm irradiated by 5.9 keV X-rays. There is
room left for improving the sensitivity attained by Gorenstein and

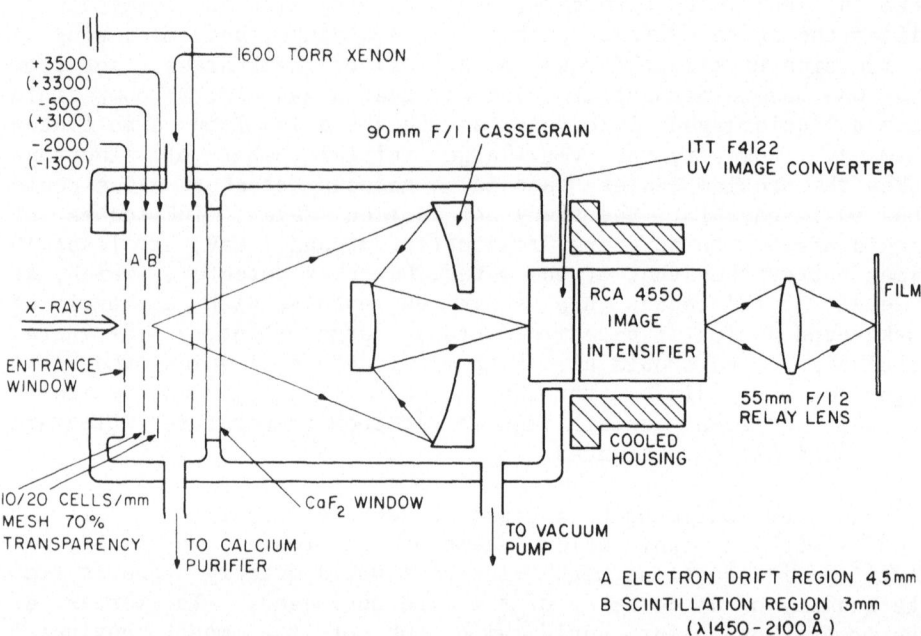

Fig. 9   Schematic diagram of the Scintillating Imaging Proportional Counter.   The upper values are typical operating
values for light production in region B.   The lower
values, in parentheses, refer to the situation where the
spatial resolution is better.   (From Ref. 20.)

Topka, and this offers an interesting possibility for a powerful
device for studying double-β decay.   It is well known that ^{136}Xe
is a good candidate for studying this process.   A TPC approach has
been proposed to give a clean signature of this very rare
process.[21]   Nguyen Ngoc[22] has reached a very good energy resolution with Xe at 10 atm: he obtained 3.2% (FWHM) at 59 keV.
The neutrinoless double-β decay would release an energy of 2.5 MeV
for the 2β's which would result in an energy resolution of better
than 1%.   It is possible at the same time to obtain a clean
picture of the tracks of 2β's giving the range of each electron
and the variation of ionization along the track, which would give
a very powerful background rejection.   Choosing the method of
Gorenstein and Topka one could imagine using one side of the
vessel for imaging and the other for retrieval of the photons for
optimum energy resolution, or a semi-transparent mirror where half
of the energy is used for imaging and the other for energy
measurement (see Fig. 11).   Electrons of 1 MeV in xenon at 10 atm

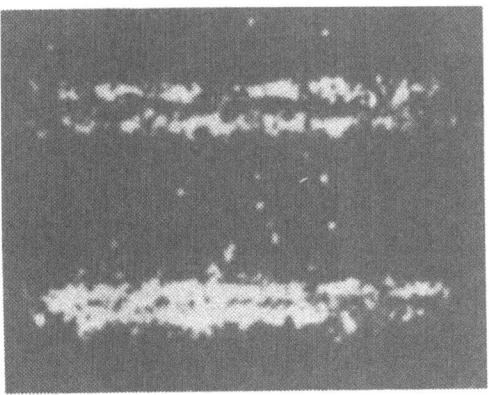

Fig. 10 Image intensifier photograph of two-dimensional image
         formed when 5.9 keV X-rays entered the detector seen on
         Fig. 9.   The slits in the resolution mask are 0.52 mm
         wide and are separated by 0.95 mm for the upper pair and
         0.58 mm for the lower one. (From Ref. 20.)

lose 7 keV per millimeter.   If one aims at a resolution of 1 mm
for the electron tracks it should be possible to have all the
tracks visible from the beginning to end.   Even if a loss of a
factor of 10 is encountered in the number of photons detectable in
the channel measuring the energy, the resolution can still be
close to 1%.   It is indeed not the only method of detecting the
VUV photons.   By having a very thin window of quartz separating
the xenon vessel from a wire chamber working at the same pressure,
with a photo-ionizing vapor, a challenging possibility of purely
electronic read-out is worth considering.   However, the possibility of obtaining a photograph of the tracks and of eventual
coincident spurious background events, with a larger redundancy
than by any purely electronic method, is an attractive feature.

CONCLUSIONS

    We have examined two processes which give rise to the
emission of light when ionizing electrons interact in gases under
the influence of an electric field.   In the first case the number
of electrons is multiplied to the highest possible value in d.c.
fields, using multistep structures to reach high gains.   In the
second case the electrons produce light by drifting in an intense
electrical field without producing ionization.

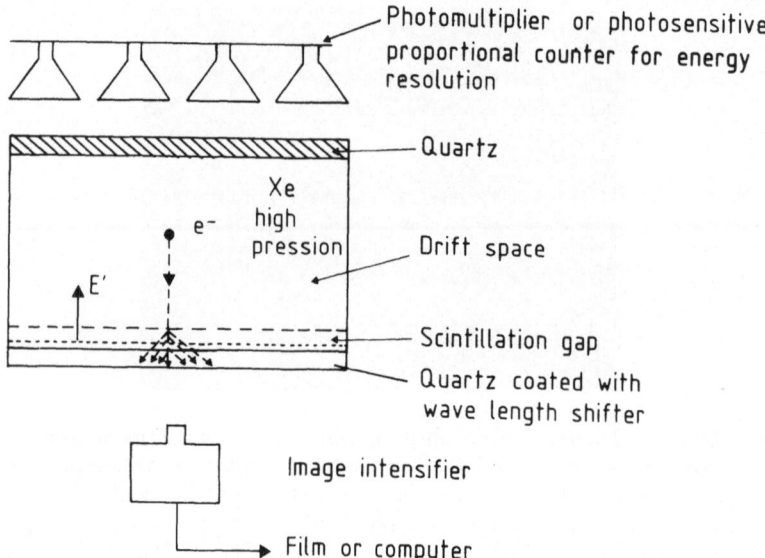

Fig. 11   Principle of the imaging detector for a xenon projection
          chamber: The electrons are drifted to a scintillation
          gap where, in an electric field between two parallel
          grids, they produce VUV photons that are converted to
          visible light by a wavelength shifter deposited on the
          near-by quartz window.   A zero-time signal can be de-
          tected from the primary excitation of the xenon, if the
          energy loss is above 100 keV.   This signal and the
          drift-time allow a three-dimensional reconstructuion of
          the event.

        In both cases the use of appropriate wavelength shifters and
image intensifiers permits the photography of the pattern of
ionization electrons arriving at an end plane. The combination of
this information with additional information concerning the energy
loss or the time of arrival at the detecting plane permits the
construction of instruments which may have powerful advantages in
various domains where very rare events have to be identified
against a large background and where the redundancy of photography
can be of essential help.

REFERENCES

1. G. Charpak,   G. Melchart,   G. Petersen, F.   Sauli, E.
    Bourdinaud, P. Blumenfeld, G. Duchazeaubeneix, A. Garin, S.
    Majewski and R. Walczack, CERN 78-05 (1978).
2. G. Charpak and F. Sauli, Phys. Lett. 78B, 523 (1978).

3. A. Breskin, G. Charpak, S. Majewski, G. Melchart, G. Petersen and F. Sauli, Nucl. Instrum. Methods 161, 19 (1979).
4. R. Bouclier, G. Charpak, A. Cattai, G. Million, A. Peisert, J. C. Santiard, F. Sauli, G. Coutrakon, J.R. Hubbard, Ph. Mangeot, J. Mullie, J. Tichit, H. Glass, J. Kirz and R. McCarthy, Nucl. Instrum. Methods 205, 403 (1983).
5. G. Coutrakon, M. Cribier, J.R. Hubbard, Ph. Mangeot, J. Mullie, J. Tichit, R. Bouclier, A. Breskin, G. Charpak, J. Million, A. Peisert, J.C. Santiard, F. Sauli, C.N. Brown, D. Finley, H. Glass, J. Kirz and R.L. McCarthy, IEEE Trans. Nucl. Sci. NS-29, 323 (1982).
6. G. Charpak, S. Majewski, G. Melchart, F. Sauli and T. Ypsilantis, Nucl. Instrum. Methods 164, 419 (1979).
7. G. Charpak, A. Peisert, F. Sauli, A. Cavestro, M. Vascon and G. Zanella, Nucl. Instrum. Methods 180, 387 (1981).
8. M. Adams, A. Bastin, G. Coutrakon, H. Glass, D. Jaffe, J. Kirz, R. McCarthy, J.R. Hubbard, Ph. Mangeot, J. Mullie, A. Peisert, J. Tichit, R. Bouclier, G. Charpak, J.C. Santiard, F. Sauli, J. Crittenden, Y. Hsiung, D. Kaplan, C. Brown, S. Childress, D. Finley, A. Ito, A. Jonckheere, H. Jostlein, L. Lederman, R. Orava, S. Smith, K. Sugano, K. Ueno, A. Maki, Y. Hemmi, K. Miyake, T. Nakamura, N. Sasao, Y. Sakai, R. Gray, R. Plaag, J. Rothberg, J. Rutherford and K. Young, π/K/p Identification with a Large-Aperture Ring Imaging Cherenkov Counter", Proc. Wire Chamber Conference, Vienna, February 1983.
9. A.S. Gilmore, W.M. Lavender, D.W.G.S. Leith and S.H. Williams, IEEE Trans. Nucl. Sci. NS-28, 453 (1981).
10. R.S. Gilmore, T.K. Gooch, W.L. Knan, I.C. McArthur, J. Malos, J.P. Melot, R.J. Tapper and R.J. Whyley, Nucl. Instrum. Methods 206, 189 (1983).
11. S. Drell, N. Kroll, M. Mueller, S. Parke and M. Ruderman, Phys. Rev. Lett. 50, 644 (1983).
12. G. Charpak and F. Sauli, preprint CERN-EP/83-128 (1983).
13. W.J. Willis and V. Radeka, Nucl. Instrum. Methods 120, 221 (1974).
14. H.H. Chen and J.F. Lathrop, paper given at FNAL Neutrino Workshop, April 1977.
15. C. Rubbia, Internal Report CERN-EP/77-8 (1977).
16. E. Gatti, G. Padovini, L. Quartapelle, N.E. Greenlaw and V. Radeka, BNL Report 23988 (1981).
17. B.A. Dolgoshein, V.N. Lebedenko and B.U. Rodionov, JETP Lett. 11, 351 (1970); B.A. Dolgoshein, A.A. Kruglov, V.N. Lebedenko, V.P. Miroshnichenko and B.U. Rodionov, Sov. J. Part. Nucl. 4, 70 (1973).
18. A.S. Barabash, A.A. Golukev, O.V. Kazachenko, V.M. Lobaster and B.-M. Ovchinnikov, Nucl. Instrum. Methods 186, 525 (1981).
19. A.J.P.L. Policarpo, Physica Scr. 23, 539 (1981).

20. P. Gorenstein and K. Topka, IEEE Trans. Nucl. Sci. NS-24, 511
     (1977).
21. E. Bellotti, O. Cremonesi, E. Fiorini, C. Lignori, S.
     Ragezzi, L. Rossi, L. Traspedini and L. Zanotti, preprint
     CERN-EP/83-144 (1983).
22. N. Nguyen Ngoc, J. Jeanjean, H. Itoh and G. Charpak, Nucl.
     Instrum. Methods 172, 603 (1980).

NEUTRINO OSCILLATIONS AND NEUTRINO ASTRONOMY

IN A LARGE FLAT DETECTOR

John C. van der Velde

Department of Physics
University of Michigan
Ann Arbor, Michigan 48109

Monte Carlo calculations of upward going muons produced by neutrino interactions in the Earth are examined to determine the sensitivity to neutrino oscillation phenomena. The detector dimensions assumed are 500 m x 10 m x 2 m (high), which would give an order of magnitude increase in rate over existing detectors. The calculations are integrated over neutrino energies in the range 3 GeV to 600 GeV. Sensitivity to possible astronomical point sources of neutrinos is also examined.

INTRODUCTION

Measurements, with good statistics, have recently been made by the Baksan[1] and IMB[2] detectors of the flux of upward going muons produced deep underground externally to the detectors. The rate, for muon energies above a few GeV, is $\sim 2 \times 10^{-4}/m^2/sr/day$. A large (5000 m^2) detector with a flat profile (to minimize edge associated inefficiencies) can thus expect to record over a thousand events per year. In this paper we focus our attention on what might be learned from data of this type.

The minimum design requirements for such a detector to be able to study such muons are: (1) angular resolution of a few degrees on straight-through muons, (2) track directionality to eliminate downward going background, and (3) a depth of $>$ 1500 mwe to keep the total background rate manageable. Further desirable properties might be good spatial resolution, the ability to distinguish electrons from muons, capability to detect monopoles, and particle energy determination; although these latter properties will not be assumed in this paper.

RATE CALCULATIONS

For neutrino energies (E) above a few GeV the spectrum falls approximately as E^{-3}. For $E \lesssim 1$ TeV the cross section rises linearly with E so that the local event rate in earth goes as E^{-2}. The muons produced in charged current interactions have energies E_μ which scale with E and, if their energy is below ~ 1 TeV where ionization loss dominates, their range will be proportional to E_μ. Hence the target volume for produced muons to be able to reach a detector grows as E^3, and with the solid angle falling as E^{-2}, one picks up another factor of E. So the total rate of externally produced upward muons goes as E^{-1}. For this reason external muons dominate the upward externally produced electron and neutral current events which must be produced within a radiation length of the detector in order to be seen. Detailed calculations were performed using a modified version of a Monte Carlo program developed by R. Bionta and D. Casper for the IMB detector. The input neutrino spectrum was that of Volkova[3] which appears to give the correct overall upward rate. The present calculations give much more precise results than ones done previously "by hand"[4].

Starting with the Volkova differential angle and energy spectrum in the E range 3 to 600 GeV the program generates muons in the earth surrounding the detector. Range-energy tables are used to determine whether a given muon will hit the detector. Only those muons which enter the bottom and exit the top are saved for further analysis. This requirement biases the parent neutrino spectrum slightly towards the vertical, as can be seen in Fig. 1.

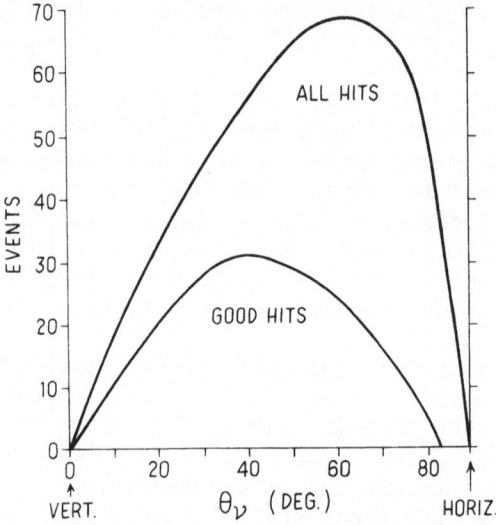

Fig. 1 Angular distribution of parent neutrinos that produce upward going muons which hit the detector.

The distribution of included angle between the neutrino and the produced "good" muons is shown in Fig. 2a. In the majority of cases the muon follows the parent neutrino to within 5°. The muon energy and angular distribution for "good" events are shown in Figs. 2b and 2c. The lower edge of the energy distribution is determined by ranging out in the detector volume, which is assigned an average density of 2.7.

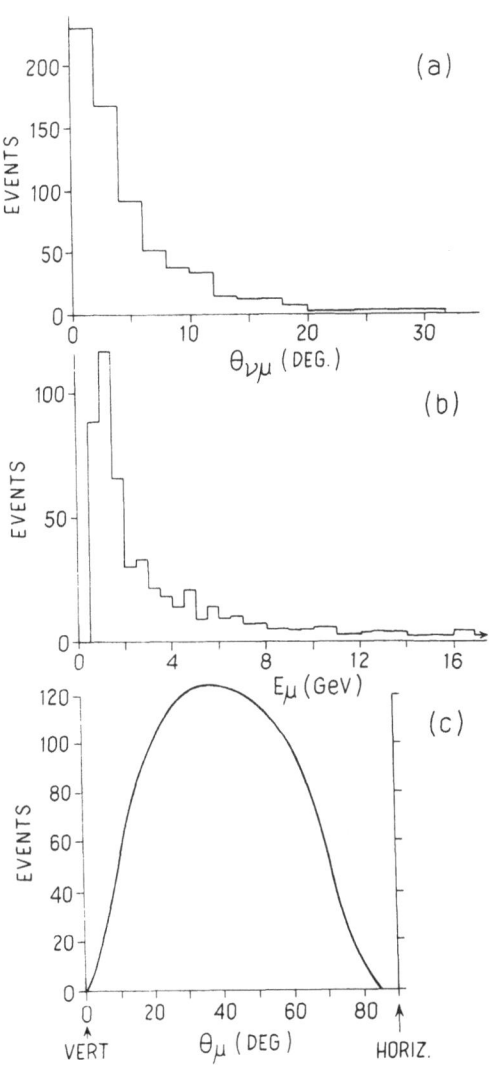

Fig. 2 (a) Distribution of angle between the parent neutrino and the produced muon. (b) Muon energy distribution. (c) Muon angular distribution. In a,b,c, we use only "good" events from Fig. 1. No neutrino oscillations are assumed.

NEUTRINO OSCILLATIONS

    Given a sample of muons of the type described above one can
then see how their properties would be modified by assumed
oscillations of their parent neutrinos[5].

    Assuming a simple two-component oscillation, the survival
probability of an initially produced muon neutrino is given by

$$P\ (\nu_\mu \rightarrow \nu_\mu) = 1 - (\sin^2 2\theta)\sin^2(1.27\ \Delta^2\ L/E)$$

where: θ = two - component mixing angle
                Δ^2 = difference of squared neutrino masses in eV^2
                L = flight path in Km
                E = neutrino energy in GeV

    For each muon in the detector we know its parent E and can
determine the L of the parent from its interaction point and
direction through the earth. We assume that the neutrino was made
in a band of atmosphere between 10 Km and 30 Km above the earth's
surface.

    We can get a rough idea of the effect of oscillations from
the following:

    Horizontal neutrinos: $L/E \stackrel{<}{\sim} 5$

    Vertical neutrinos: $50 < L/E < 5000$

    We see from this that horizontal neutrinos have the same L/E
range as accelerator and reactor experiments. Hence we don't
expect any effect. This is one reason for making a flat
horizontal detector.

    The predicted modulation factors for the upward muon angular
distribution are shown in Fig. 3 for various assumed Δ^2 values,
assuming maximal mixing. One sees that if $\Delta^2 < 2 \times 10^{-4}$ eV^2 the
effect is probably too small to be seen. If $\Delta^2 > .001$ eV^2 the
angular distribution is fairly uniformly modulated. In this case
the integrated flux would give the best measure of the effect.
For Δ^2 values in the range .001 eV^2 to $\sim$ .003 eV^2 the modulation
of the angular distribution becomes an effective means to search
for oscillations. Of course it must be emphasized that one is
looking for a deviation at small (with respect to the vertical)
angles from a calculated angular distribution. Hopefully, as
knowledge of underground neutrino interactions progresses, the
errors in the above type of search will not be dominated by
uncertainties in the expected angular distribution and flux. For
example, one can get constraints on the expected upward flux by
measuring the downward flux. In this regard, we note that the

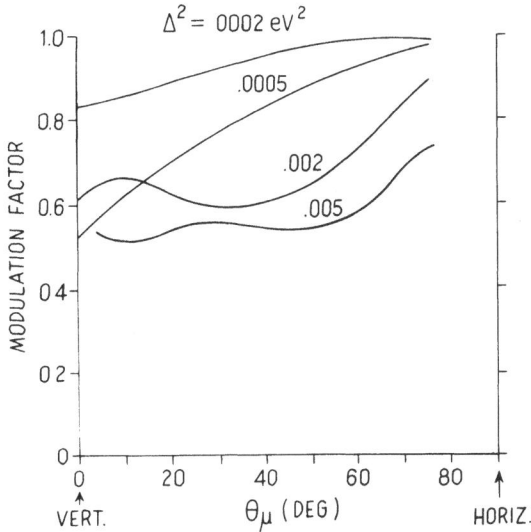

Fig. 3 Modulation factors for the muon angular distribution of
Fig. 2c due to various assumed Δ^2 values. Maximal mixing
($\sin^2 2\theta = 1.0$) is assumed. For other mixing angles one
can assume that the difference between any curve and 1.0
is proportional to $\sin^2 2\theta$.

complications arising from geomagnetic effects are less important
for externally produced muon events than for the lower energy
internally produced events.

Note that we have neglected the possible generation of muon
neutrinos by oscillating electron neutrinos from the atmosphere.
This is expected to be a small effect at the energies we are
considering. Nevertheless it should be taken into account in a
full blown calculation.

The possible contribution that an experiment of the type
described here might make to our knowledge of neutrino oscillations is depicted in Fig. 4. Some representative present and
probable future limits on oscillations from accelerators and
reactors are shown for comparison[6]. The line labeled "Baksan"
represents the present[1] limit from that experiment, based on the
integrated upward muon flux. Similar results can be expected soon
from the IMB and other large detectors. A detector with 10 times
the statistics could hope to push the Δ^2 limit down by another
factor of two or so. Such experiments will be limited in their
$\sin^2 2\theta$ sensitivity due to uncertainties in the expected flux and
by other systematics. The "finger" around $\Delta^2 = .0005$ eV^2 represents an area where oscillations would cause a detectable modulation of the upward muon angular distribution.

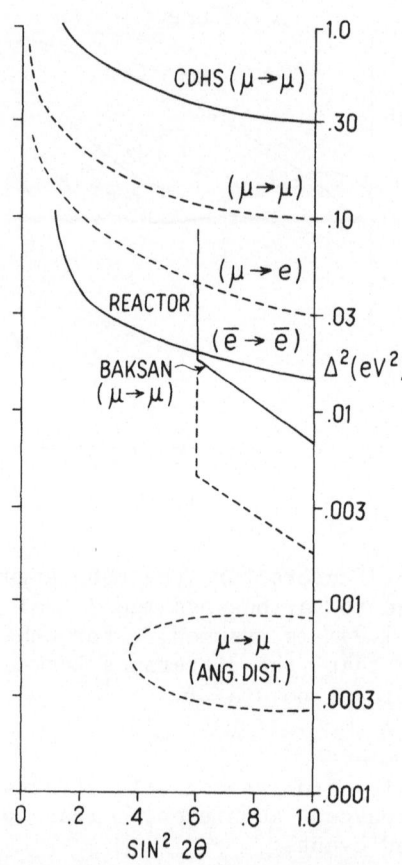

Fig. 4   Present limits on neutrino oscillations are shown as solid lines. Expected improvements in the next few years are indicated as dashed lines for accelerator experiments on $\mu \to \mu$ and $\mu \to e$. The dashed line below BAKSAN indicates the limits that experiment and other large underground devices might be able to set in the next few years using the integrated upward muon flux. The dashed "finger" represents the region that would give rise to a detectable modulation of the shape of the angular distribution of upward going externally produced muons in a detector of the type described.

NEUTRINO ASTRONOMY

    It is evident from Fig. 2a that upward external muons preserve much of the directionality of their parent neutrinos. Hence, the data gathered by a detector of the size contemplated here could be used to detect point or diffuse sources of neutrinos in the sky by looking in angular bins of about 5 square degrees.

Let us assume a data sample of ~ 1000 events (about 1 year's worth). A Mercator plot of the events (assuming the sky was approximately uniformly sampled) would have an average of 0.4 events per bin of 5 square degrees. The Poisson probability that at least one bin would have $\geqslant 5$ counts would be 16%. Hence one would have to demand $\geqslant 6$ counts in a bin in order to claim an effect at $> 90\%$ C.L.

The conclusion is that one could hope to find unknown sources which were responsible for $\geqslant 0.5\%$ of the total upward muon flux. Looking at known possible sources, such as Cygnus A, one would be about a factor of two more sensitive.

I hope this paper has shed some light on the information that can be expected in the areas of neutrino oscillations and neutrino astronomy using large underground detectors.

ACKNOWLEDGEMENTS

I would like to thank David Casper for help with the Monte Carlo program.

REFERENCES

1. M.M. Boliev et al., in Proc. 17th International Cosmic Ray Conference, (Paris), 7, 106 (1981); and M.R. Krishnaswamy et al., report at Neutrino-'82, Balaton, Hungary, 1982. See also P.V. Ramana Murthy, Rapporteur paper, Proc. 17th Int. Cosmic Ray Conference (Paris) 13, 381 (1981).

2. R.M. Bionta et al., Irvine-Michigan-Brookhaven experiment, (to be published).

3. L.V. Volkova, Sov. J. Nucl. Phys. 31, 1510 (1980); L.V. Volkova, Proc. DUMAND Workshop 1980, p. 75.

4. J.C. van der Velde, "On the Possibility of Neutrino Oscillation Information from Upward-Going Muons in the IMB Detector" (University of Michigan Research Note, May, 1980).

5. We restrict the discussion to muon events which originate outside and pass through the detector. For a discussion of what might be learned from the lower energy internal events, including the effects of electron neutrinos and three-component neutrinos, see D.S. Ayres, B. Cortez, T.K. Gaisser, A.K. Mann, R.E. Shrock, L.R. Sulak, "Neutrino Oscillation Search With Cosmic Ray Neutrinos". (Bartol Research Foundation Preprint BA-83-31)

6. M. Shaevitz, Rapporteur talk given at the International Symposium on Lepton and Photon Interactions, Cornell University, August, 1983.

REPORT ON STANFORD SUPERCONDUCTIVE MONOPOLE DETECTORS

B. Cabrera, M. Taber, R. Gardner, M. Huber and J. Bourg

Physics Department
Stanford University
Stanford, California 94305

INTRODUCTION

Since January, 1983 we have been operating a three loop superconducting detector with a signal to noise ratio of 100 for the direct detection of Dirac magnetic charges.[1] Its sensing area averaged over solid angle is nearly 500 cm^2 for double coincidence events. To date, the data contain no candidate events and set an upper flux limit of 2×10^{-11} cm^{-2} s^{-1} sr^{-1} at 90% C.L., 70 times lower than that suggested by the single candidate event[2] seen on the prototype single loop detector. If supermassive magnetic monopoles exist they are rare and the need for cost-efficient larger area superconductive detectors is evident if we are to push forward a definitive search in the cosmic rays. Superconducting loop detectors can be constructed with 1 to 10 m^2 planar areas coupled to a single SQUID while achieving a signal to noise ratio of 10 for the detection of a Dirac magnetic charge. A 100 m^2 sensing area array of such detectors, which would reach the Parker flux bound in one year of operation, can be built at a cost comparable to that of typical high energy experiments. It remains quite possible that a monopole particle flux significantly higher than the Parker bound exists within our galaxy. In this paper the most recent results from low noise operation of the three loop detector are presented. Also discussed are our plans to build a larger fixed loop detector with a sensing area of 1.5 m^2. It is designed as a prototype for a future yet larger 100 m^2 array.

THEORETICAL AND ASTROPHYSICAL CONSIDERATIONS

All aspects of monopole theory and experiment are discussed

within this volume in many detailed articles by expert authors.
In this section we summarize recent work related to the detection
of monopoles with superconductive systems.

Grand unification theories (GUTs) predict the existence of
monopoles with masses of 10^{16} GeV/c^2 or higher. Such large masses
force nonrelativistic particle velocities. The suggestion has
rekindled interest in experimental searches for magnetically
charged particles carrying the Dirac unit of charge. Cosmological
theories based on GUTs lead to predictions for monopole particle
flux limits which are impossibly high or unobservably low (inflationary universe). However, the latter results are exponentially model dependent and thus are not inconsistent with
observable levels. Astrophysical arguments provide more concrete
observational limits. In the discussion that follows we assume a
particle mass of 10^{16} GeV/c^2 to obtain representative numbers. An
absolute upper bound for the galactic monopole particle flux of
4×10^{-10} cm^{-2} s^{-1} sr^{-1} then is obtained from the limits on the
local galactic dark mass.[2]

A much smaller upper bound of 10^{-15} cm^{-2} s^{-1} sr^{-1}, the Parker
flux bound, is obtained assuming an isotropic flux from arguments
based on the existence of the 3 microgauss galactic magnetic
field.[3] However, several authors have demonstrated that models
which self-consistently couple the magnetic field energy to the
kinetic energy of a monopole distribution allow a much larger
particle flux to coexist with the observed galactic field.[4] The
entire missing mass of our galaxy then could exist as a monopole
halo and remain consistent with the observed galactic magnetic
field. Past constraints on such models based on not allowing
magnetic field changes faster than the long damping times for the
observed interstellar dust grain alignment were incorrect. Instead, the Barnett magnetic moment induced by the rotation of the
dust grains causes a precession of the angular momentum axis about
the magnetic field. Thus as the grains slowly damp into alignment
(about 10^5 years) they adiabatically follow any magnetic field
changes which only must be slow compared to the precession period
(about 10 years)[5]. All of these galactic bounds suggest particle
velocities in gravitational virial equilibrium, thus very near
10^{-3} c.

An enhanced monopole density, gravitationally bound to our
own solar system has been suggested[6] and would allow much smaller
average galactic flux levels and leads to lower particle
velocities very near 10^{-4} c. Although a mechanism for the
formation of such a cloud remains obscure, its possible existence
is not ruled out.

It has been shown theoretically that the supermassive monopoles arising from many GUTs will catalyze nucleon decay

processes.[7] If the cross section for such events is of order the hadron cross section, as has been suggested, then all attempts at direct detection of those monopoles may be doomed to failure. Arguments based on X-ray flux limits from galactic neutron stars and which assume a strong cross section lead to an upper bound for a monopole particle flux of about 10^{-22} cm^{-2} s^{-1} sr^{-1}. However, even for such GUT monopoles, there remain unanswered questions both on the catalysis cross section and on the astrophysical arguments based on our incomplete understanding of neutron stars.[8]

An important question regarding the direct experimental detection of monopoles is: can conventional ionization or scintillation devices detect the passage of single Dirac charges with velocities of order 10^{-3} or 10^{-4} c? While several phenomenological theories suggest the answer is marginally yes,[9] experimental experimental tests of these theories are not possible. More recently, calculations based on fundamental quantum mechanical arguments suggest that helium gas devices would provide such a sensitivity.[10] Experimental efforts with both conventional and superconductive detectors should be encouraged as these efforts are complementary. For the same cost, conventional detectors provide at least ten times larger sensing areas and convincing detection for any slow moving electric or magnetic particles with velocity above about 10^{-3} c whereas the superconductive detectors provide definitive identification of a magnetic charge for any velocity.

In summary, superconductive detectors allow the direct identification of a magnetic charge independent of its velocity or any other physical property.[2] Thus, such detectors are largely independent of specific theoretical monopole models and they would always provide the most definitive identification of a magnetic charge.

THE THREE LOOP SUPERCONDUCTIVE DETECTOR

Motivated by the single candidate event observed on our original one loop superconductive detector,[2] in April, 1982 we began the construction of a larger three loop detector. In January, 1983 modification to that detector proved successful in reducing the low frequency noise by a factor greater than 10. Since January 25, 1983 it has been operating continuously. Its theory of operation, design and early performance have been reported previously.[1,11] Briefly, the rms current noise levels in an effective noise bandwidth of 0.3 Hz are now 0.02 ϕ_o/L in all three loops, less than 1% of the signal expected from the passage of a Dirac magnetic charge through one of the loops (4 ϕ_o/L). The sensing area based on this low noise operation is 476 cm^2 (71 cm^2 loop area and 405 cm^2 near miss area) for events greater than 0.1 ϕ_o/L in at least two of the three loops.[12]

A signal consistent with the passage of a Dirac charge
through or near the three loops, will be a sharp step function
within the data acquisition bandwidth of 100 Hz. Figure 1
contains updated histograms of all the unexplained sharp offsets
which have occurred up to October 31, 1983. The largest is nearly
one order of magnitude smaller than the candidate event observed
on the prototype detector. The unshaded events occurred within
the first five days of the run during a settling period of
generally noisy data. Since all of these unexplained events
occurred in a single loop, with no evidence of coincident offsets
in the other two loops, they are all excluded as monopole
candidates. The probability of a monopole trajectory which
produces an offset greater than $0.1 \phi_0/L$ in one loop showing no
offset in the other two loops at the noise level of $0.02 \phi_0$ is
less than 10^{-4}. High bandwidth data for one of these unexplained
events is shown in Fig. 2 and is typical. No accelerometer or
fluxgate magnetometer disturbances are evident. These events are
thought to be due to the motion of trapped fluxons in the SQUIDs
themselves, but the trigger mechanisms are not understood. The
rate of accidental double coincidences from these events is one
every 5000 years.

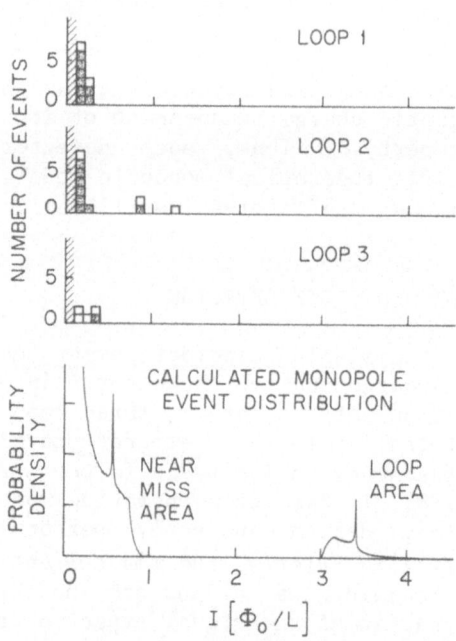

Fig. 1   Histograms of all unexplained events in the three loop de-
         tector as of Oct. 31, 1983. The signals "expected" from an
         isotropic distribution of monopole trajectories are shown.

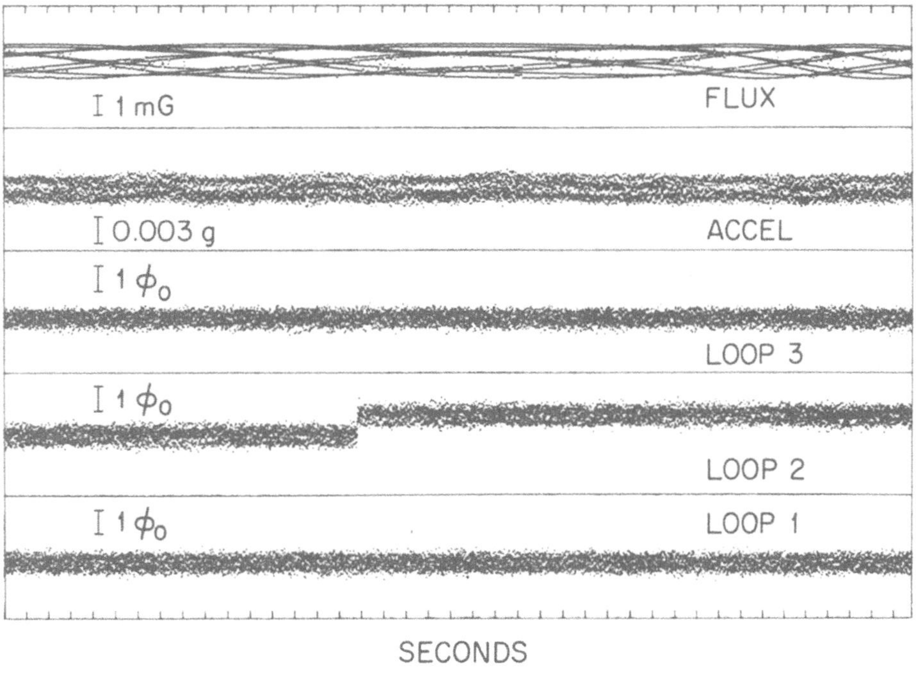

SECONDS

Fig. 2 Typical unexplained offset occuring at 3:03 PST on October 8, 1983 and shown at 100 Hz bandwidth (200 data points per second). Data are continuously taken at this bandwidth for the three loop signals, an accelerometer mounted on the probe top plate, and a fluxgate magnetometer mounted on the dewar stand. The last 15 seconds of the high bandwidth data is always stored in the computer and permanently recorded whenever an offset is detected above a threshold of 0.1 ϕ_0/L in any loop.

Since the possibility of cosmic ray showers has been mentioned many times as a possible cause for inducing fluxon motion, we have completed the installation of a simple cosmic ray shower detector. It consists of three 15 by 15 cm scintillators placed at the vertices of a 2 m on a side equilateral triangle around the monopole detector. Their thresholds are set at about three times minimum ionizing and any triple coincidences within a 50 ns window are recorded in the permanent computer data at five second intervals. About one shower event is recorded per hour on the average. Fig. 3 shows a typical eight hour computer printout of the stored data which now includes the cosmic ray shower events. The three single loop SQUID offsets which have occurred since we began operating the shower detector were not coincident with a cosmic ray shower event.

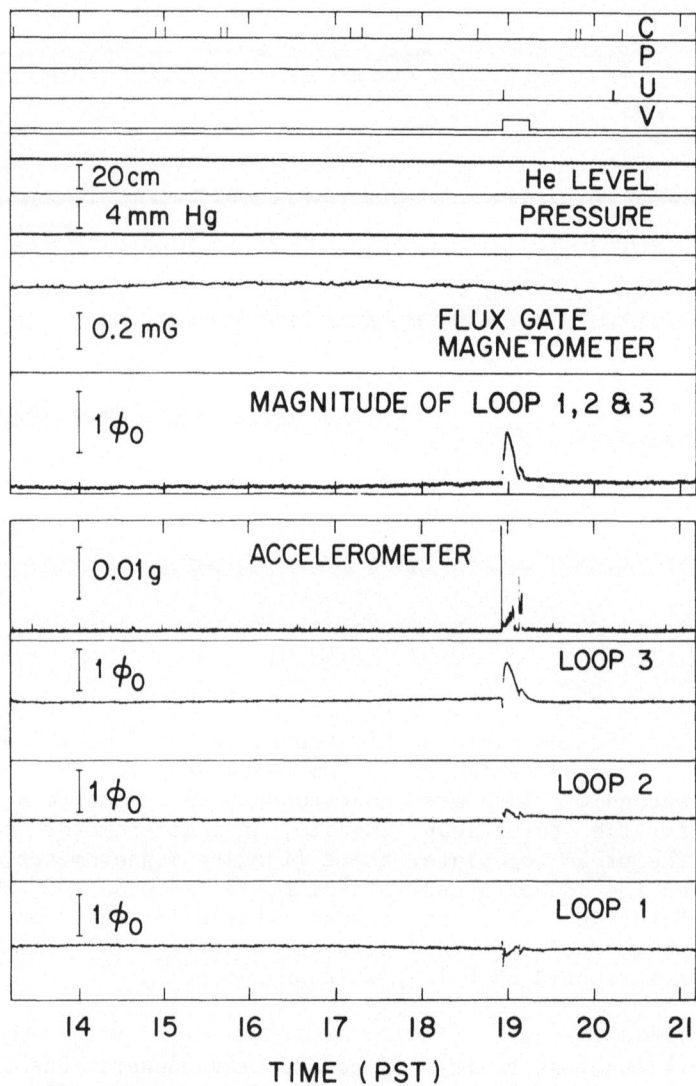

Fig. 3 Typical eight hour data summary taken on December 4, 1983.
        The continuous low bandwidth data contain a point every 5
        seconds (0.1 Hz bandwidth) for the three loop signals, the
        accelerometer output (maximum absolute value during each 5
        second interval), root mean square of loop signals,
        external fluxgate magnetometer, pressure within the dewar
        He compartment, He level within the dewar, and four on/off
        channels which monitor the high bandwidth trigger circuit
        veto (V), an ultrasonic burglar alarm within the labora-
        tory (U), power line voltage transient detector (P), and
        the cosmic ray shower detector (C).

As of October 31, 1983 we have accumulated 233 days of data (214 days of active operation). No candidate events have been seen. The data set an upper limit of 2.0×10^{-11} cm^{-2} s^{-1} sr^{-1} at 90% C.L. on any uniform flux of magnetic monopoles passing through the earth's surface at any velocity. No large spurious or real signals have been seen, casting no new light on the origin of the previously reported candidate. However, these data lower that previous flux limit by a factor of 70, increasing the probability of a spurious cause for that event.

We intend to operate the three loop detector for at least one additional year.

NEW LARGE TWISTED LOOP DETECTOR

A year ago, we had suggested a design for a scanning magnetic monopole detector.[11] A cylindrical superconducting sheet would be periodically scanned with a small loop coupled to a SQUID magnetometer. The device would have sufficient resolution to map the position of each trapped fluxon in the surface of the cylinder. The appearance of new doubly quantized fluxons would provide convincing evidence for the passage of a magnetic charge through the cylinder walls. In addition, 70% of the trajectories which intersect the cylinder wall would do so twice, providing valuable coincidence information and a precise determination of the trajectory.

Such a scanning detector would provide the most convincing evidence for the existence of magnetic charges and the technology would allow the construction of a series of post identification experiments based on precise trajectory information and aimed at measuring or setting limits on the particle mass and velocity. However, no candidate events have been seen to date by the three loop detector. Thus, if supermassive monopoles exist they are rare and we must use the least expensive way to build signif- icantly larger area detectors aimed only at identification if we are to continue a definitive search for any cosmic ray flux of monopoles. Work by C. Tesche, et al.,[13] at IBM in Yorktown Heights, New York and by H. Frisch, et al., at the University of Chicago (see papers in this volume) has convinced us that it is possible to extend the static superconducting loop detector to cover an area of 1 m^2 or more while maintaining sufficient signal to noise for observing the passage of a Dirac magnetic charge. Coincidence between nearby independent loops would provide the necessary redundancy. A superconducting array which would reach a particle flux bound of 10^{-15} cm^{-2} s^{-1} sr^{-1} (the Parker bound) in one year of operation could be built for a cost comparable to that of typical high energy physics experiments. To help demonstrate the feasibility of such an approach, we are building an eight

SQUID loop array in an existing dewar with a sensing area of
1.5 m^2 averaged over solid angle (30 times greater than that of
the three loop detector).

The detector design is based on the use of an existing liquid
helium dewar originally built to house a prototype gravitational
wave antenna.[14]   The dewar is 3 feet in diameter and 22 feet long
with an inner cryogenic compartment at 4.2 K measuring 20 inches
in diameter by 20 feet long. This compartment is surrounded by an
annular liquid helium storage vessel.   The design for the superconducting twisted loop array detector is shown schematically in
Fig. 4 and is housed inside the dewar.   It consists of eight
planar twisted loops each coupled to a SQUID.   Each loop is 16 cm
wide and 16 feet long (1 m^2) and is composed of an array of
oppositely coupled square elements each 8 cm on a side.

When compared to a simple loop, two important advantages are
derived from the twisted loop geometry.   Both are based on
relative insensitivity to external magnetic field changes.   First
the constraints on the ambient magnetic field levels are eased,
allowing use of a single mu-metal shield which will provide an
ambient field of a few milligauss rather than the 20 nanogauss
ambient field provided by the ultra low field shield used with the
three loop detector.   Second, the superconducting shield which
surrounds the loops can be brought much closer to the detection

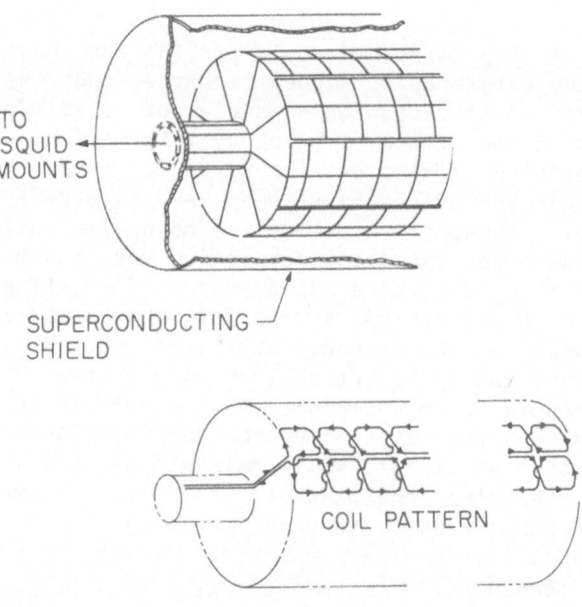

Fig. 4   Schematic of loop design for new 1.5 m^2 eight SQUID
superconductive detector.

loops without reducing the signal magnitude and increasing the width for the distribution of signals expected from the passage of Dirac charges through the detector.

We have performed a series of calculations to find the optimum design parameters for twisted loop geometry. For example, Fig. 5 shows the total coupling of a Dirac charge to the twisted loop as a function of the distance from the loop to the nearest superconducting surface. We have assumed that the coil and surface are planar and for each separation we have chosen the trajectory which most deviates from $2\phi_o/L$. Consistent with an overall signal to noise of 10, we pick a separation of 5 cm (5/8 of the loop element size of 8 cm on an edge). Then the range of all possible signal magnitudes from the passage of a Dirac charge is 15%. To achieve a comparable signal to noise at the SQUID output, the twisted loop inductance must be 35-40 microhenries. Such a low inductance can be obtained by using a 4 mm wide line pattern on a G-10 substrate (a printed circuit board!), as demonstrated by the Chicago group. We will couple these loops directly to the input terminals of S.H.E. rf SQUIDS.

We expect to complete construction and testing of the detector during 1984 and begin continuous operation in early 1985.

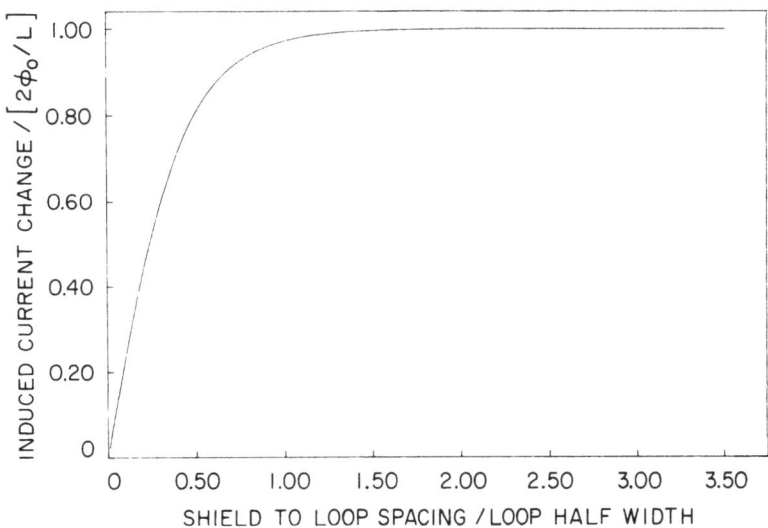

Fig. 5 Signal strength in one loop of the new detector as a function of the distance between the loop plane and the superconducting shield which is modeled as a parallel plane.

FUTURE 100 M^2 SUPERCONDUCTIVE DETECTOR ARRAY

Using existing dewar technology, a 100 m^2 array can be built. For example, the array could be housed in six dewars each 8 feet in diameter by 40 feet long. Each dewar would contain eight SQUID detectors each coupled to a 10 m^2 twisted loop planar area and positioned around the circumference as in the design of the prototype. An inductance approaching several hundred microhenries would couple to each specially designed rf or dc SQUID. A collaborative effort between several groups could provide the necessary manpower and allow timely completion of the detector.

CONCLUSIONS

It continues to be difficult to assign a spurious cause to the candidate event seen by the original single loop detector. However, the lack of confirming events on the larger three loop detector lessens the probability that it was caused by a monopole. Predictions for the existence of monopoles from grand unification theories remain as a strong motivation for continuing the search. It is now within our reach to search down to the Parker flux bound using definitive superconductive detectors, but the discovery of a much larger monopole flux remains possible. Although some theories suggest unobservably low particle flux levels, the significance of finding monopoles would be so great that the search must be continued.

ACKNOWLEDGEMENTS

The work described in this paper has been funded in part by DoE contract DE-AM03-76SF00-326.

REFERENCES

1. B. Cabrera, M. Taber, R. Gardner and J. Bourg, Phys. Rev. Lett., 51, 1933 (1983).
2. B. Cabrera, Phys. Rev. Lett., 48, 1378 (1982).
3. M.S. Turner, E.N. Parker and T.J. Bogdan, Phys. Rev. D26, 1296 (1982).
4. E.E. Salpeter, S. L. Shapiro and I. Wasserman, Phys. Rev. Lett. 49, 1114 (1982); J. Arons and R.D. Blandford, Phys. Rev. Lett. 50, 544 (1983); R. Farouki, S.L. Shapiro and I. Wasserman, Astrophys. J. (in press), contains detailed computer simulations of self-consistent galactic monopole halo models.
5. E.M. Purcell, comment at Workshop on Magnetic Monopoles, Racine, Wisconsin, October 1982; see also E.M. Purcell,

Astrophys. J. 231, 404 (1979).

6. S. Dimopoulos, S.L. Glashow, E.M. Purcell and F. Wilczek, Nature 298, 824 (1982).

7. For reference see paper by C.G. Callan in this volume.

8. For reference see paper by J.A. Harvey in this volume.

9. G. Tarle, S.P. Ahlen and T.M. Liss, Phys. Rev. Lett. 52, 90 (1984) and paper by S.P. Ahlen in this volume.

10. S.D. Drell, N.M. Kroll, M.T. Mueller, S.J. Parke and M.A. Ruderman, Phys. Rev. Lett. 50, 644 (1983).

11. B. Cabrera, Magnetic Monopoles, eds. R.A. Carrigan and W.P. Trower (Plenum Publishing Corp., 1983), p. 175.

12. B. Cabrera, R. Gardner and R. King, submitted to Phys. Rev.D.

13. C.D. Tesche, C.C. Chi, C.C. Tsuei, and P. Chaudhari, Appl. Phys. Lett. 43, 384 (1983). Also, see paper by C.C. Chi in this volume.

14. S. Boughn, Ph. D. Thesis, Stanford University, 1975 (unpublished).

MONOPOLE SEARCH AT IBM: PRESENT STATUS AND FUTURE PLANS

C.C. Chi, C.D. Tesche, C.C. Tsuei, P. Chaudhari and
S. Bermon

IBM T.J. Watson Research Center
Yorktown Heights, New York 10598

A novel magnetic monopole detection scheme using superconducting high-order planar gradiometers has been successfully designed and tested. A prototype of two parallel gradiometers monitored by two independent SQUIDs was built and operated for a period of more than 5 months. With no monopole events observed, it established an upperbound of monopole flux of 1.7×10^{-10} cm^{-2} sr^{-1} s^{-1} at a 90% confidence level. A larger detector of six planar gradiometers on the facets of a parallelopiped with a coincident detection area 40 times that of the prototype is currently running. A flux bound of 2.0×10^{-11} cm^{-2} sr^{-1} s^{-1} at a 90% confidence level has been established by this detector as of February 1984. An even larger detector with a coincident detection area about 3 m^2 is now in the planning stage.

INTRODUCTION

Due to the recent advances of grand unification theories, the existence of a magnetic monopole is deemed essential rather than heuristic. The experimental confirmation of the existence of the monopole or even an upper-bound of the flux would have a great impact on elementary-particle theories and cosmologies. Although the report of a single noncoincident event of a magnetic monopole candidate by Cabrera[1] using the superconductive induction technique cannot be used as firm proof of its existence, it has certainly stirred up interest in searching for monopoles.

A variety of techniques have been proposed and are currently being applied for detecting magnetic monopoles.[2] The superconductive induction technique[3,4] is superior in terms of its sensitivity to the monopole charge and its independence of the monopole velocity and internal structure. The challenge of using this technique is, however, to design a detector with a sufficiently large detection area so that the Parker bound[5] of monopole flux ($\sim 10^{-15}$ cm^{-2} sr^{-1} s^{-1}) can be approached or even exceeded in case of a null detection within a period of a few years. Also, it is very important for the detector to have the capability of unambiguously distinguishing monopole events from spurious ones in case of a few positive detection records.

Having these goals in mind, we have designed a novel magnetic monopole detection scheme using superconducting high-order planar gradiometers. A prototype detector with a coincidence detection area[6] of 24 cm^2 has been successfully constructed and tested. Before it was replaced by the current IBM parallelopiped detector with a coincidence detection area about 1000 cm^2, it had a continuous detection period of 165 days with no monopole events recorded. While the current detector is running, an even larger detector with a coincident detection area about 3 m^2 is under development in order to pursue the Parker bound.

MONOPOLE DETECTION WITH SUPERCONDUCTING GRADIOMETERS

The quantum of a magnetic charge is hc/4πe according to Dirac's electrical-charge quantization argument. The passage of a single monopole with a unit of magnetic charge through a superconducting loop would result in a flux change of $2\phi_0$, where $\phi_0 = hc/2e = 2.07 \times 10^{-7}$ G cm^2 is the superconducting flux quantum. The flux change in a superconducting loop can be inductively monitored by a SQUID (Superconductive Quantum Interference Device). It is clear that using a large area magnetometer as the monopole detector would require an extreme stability of ambient magnetic fields.[4] Since almost all the magnetic field fluctuations are caused by sources outside of the detector, while the monopole signal results from a monopole passing through the detector, the high-order planar gradiometer is nearly an ideal solution for large-area monopole detection.[7,8] The construction of such a gradiometer is schematically shown in Fig. 1. It has been shown that the higher the order, the less sensitive the detector is to ambient field fluctuations. But the order of the gradiometer cannot be increased without bound because the monopole signal as detected by a SQUID is reduced as the inductance of a detection coil increases with the order. When the ambient field fluctuations are no longer the limiting factor, the SQUID intrinsic noise would eventually dictate the size and the order of a gradiometer design. Another very important advantage of the gradiometer design is the reduction of the mutual inductive coupling between the detector and the superconducting shield often used to achieve the required magnetic field stability. It is

CONSTRUCTION OF HIGH ORDER GRADIOMETER

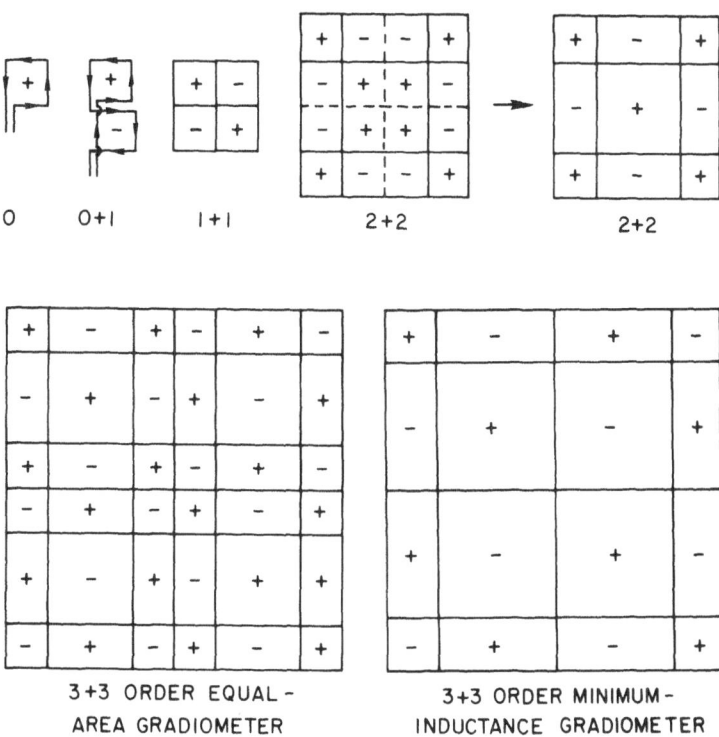

Fig. 1   The construction of high order planar gradiometers.   The
schematic winding diagrams for some lower order planar
gradiometers are shown with "+" and "-" representing the
counter-clockwise and clockwise windings respectively.
Up to the linear order of 2, there is no distinction
between the equal-area and minimum-inductance constructions.   Beyond that, the inductance of the minimum
inductance winding increases linearly with the order,
while the inductance of the equal-area winding increases
approximately 2 to the power of the order.

known that a minimum separation between the two is required for a
clean monopole detection profile.[4]   The detection profile is
defined to be the probability distribution of the flux changes in
the detector caused by an isotropical infringement of monopoles.
Such a probability distribution exists due to the coupling between
the detector and the randomly located flux created by the passage
of a monopole through the superconducting shield.   The significant
reduction of the required minimum separation between coil and
shield which results from using high-order gradiometer detectors
allows a more efficient use of a given cryogenic volume.

THE PROTOTYPE DETECTOR

The purpose of building this prototype monopole detector was to test the high-order gradiometer scheme and to gain information on the importance of the ambient magnetic field stability, the performance of various equipment and data acquisition systems, and the probability of flux jumps in the SQUID detection system. The prototype detector (Fig. 2) had two parallel phenolic plates with 5 mil-diameter Nb wires glued to the grooves on the surfaces. The Nb wires were wound according to the pattern for the (7 + 7)th-order minimum-inductance gradiometer.[8] The physical area of each gradiometer was 10 cm × 10 cm. With a 4 cm separation between them, the coincident monopole-flux detection area is reduced to about 24% of the physical area of one single detector. Each gradiometer had a measured inductance of 3.9 μH and was monitored independently by a separate SQUID with 2 μH input inductance.[9] Two concentric cylindrical superconducting Pb shields were cooled down to 4.2 K in an ambient field of a few milligauss achieved by using one μ-metal shield which was not used after the initial cooling down. Tested with a solenoidal coil wound on the outside of the cryogenic dewar, flux changes of $4.3\phi_0$ and $0.5\phi_0$ for the top and bottom detectors were observed when the external magnetic field changed by about 10 G. The ambient magnetic field

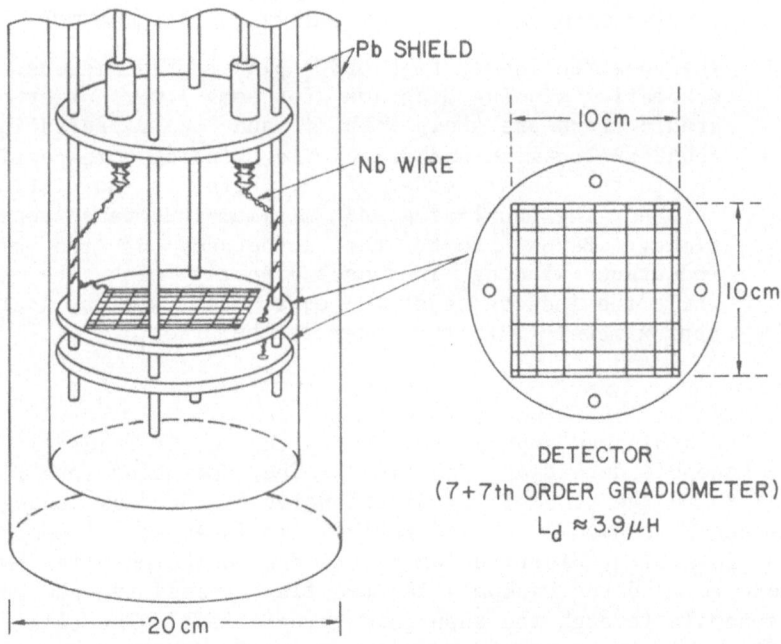

Fig. 2   The prototype monopole detector. Two parallel planar gradiometers were independently monitored by two SQUIDs.

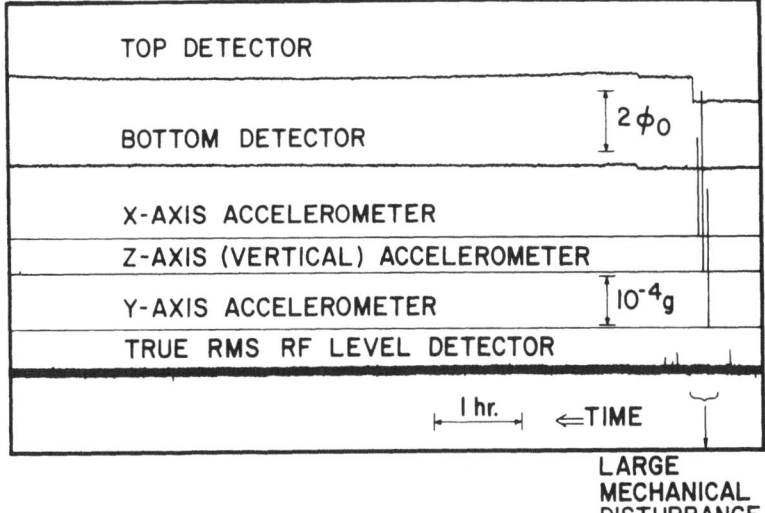

Fig. 3   A section of the typical strip-chart records of the
          prototype detector. The top two traces correspond to the
          SQUID's outputs of the flux changes in the top and bottom
          gradiometers respectively. Followed by three traces from
          the outputs of a three-axis accelerometer, and one from a
          RF level detector.   On the right-hand side of the strip
          chart, the top detector showed a flux jump induced by a
          mechanical disturbance. The widths of the detector
          traces were limited by the SQUID's intrinsic noise.

fluctuations in our laboratory are typically about 10^{-2} G or less.
The noise on the SQUID outputs, shown in Fig. 3, is limited by the
SQUID intrinsic noise rather than ambient field fluctuations. The
order of the gradiometer is so high that the mutual couplings
between the two detectors and the superconducting shields are
completely negligible. The calculated detection profile is almost
δ -function like at $\Delta\phi = 0$ (near miss events) and $2\phi_0$ (on target).
The two SQUID outputs were continuously recorded on a strip-chart
recorder and simultaneously digitized and stored using an IBM-PC
computer at 6-second intervals. To distinguish possible spurious
events due to mechanical and RF disturbances, the outputs of a
three-axis accelerometer mounted on the dewar and a near-by RF
level detector were also simultaneously recorded on the strip-
chart recorder (shown in Fig. 3) and read into the computer for
digital storage. As indicated in Fig. 3, mechanical disturbances
can sometimes induce a flux jump in one or both of the SQUIDs'
outputs. Severe RF disturbances, such as electrical sparks
generated by a Tesla coil, can also give rise to jumps in the
SQUIDs' outputs, usually with distinctive spikes. The flux

changes induced by external magnetic fields are vastly different
for the two detectors as mentioned previously. Our definition of
a monopole candidate event is that both of the SQUIDs' simultaneously register a flux change within 10% of $2\phi_0$, without
registration of any of the contingency devices. With this rather
stringent criterion for monopole events, no monopole candidates
have been observed for the 165-day period (April 8 to September
20, 1983). As a matter of fact, even without restricting the size
for the flux jumps in the SQUID outputs, there was only one
coincident event which was not associated with outside disturbances. This coincident event corresponded to about $0.4\phi_0$ and
$5.2\phi_0$ flux changes for the top and bottom detectors, respectively.
This event might have been due to the rearrangement of trapped
flux somewhere close to the bottom of the superconducting shield.

Figure 4 shows the histogram of the flux-jump distributions
of the two detectors as a function of the magnitude of the flux
jump in units of ϕ_0 (a change of $2\phi_0$ corresponded to a 13.6 mV
change in the SQUID output voltage). About 2/3 of those flux
jumps can be correlated with the external disturbances. As a
matter of fact, all the five noncoincident jumps near $2\phi_0$ were
clearly induced by some mechanical disturbances. The remaining
1/3 of the flux jumps were probably due to the motion of the
trapped flux in the superconducting shields and the SQUID Nb
housing, because most of those occurred with in a few hours after
some large mechanical or thermal disturbances, such as refilling
liquid He or moving the cryogenic dewar.

In addition to the 80 and 136 flux jumps shown in Fig. 4,
there were 5 and 81 very large noncoincident flux jumps recorded
for the top and bottom detectors, respectively. It is interesting
to note that they were all about the same size for each detector,
i.e. 91 ϕ_0 change for the top detector and 69 ϕ_0 change for the
bottom detector. It is unlikely that such jumps were due to a
large flux change in the gradiometer loops because of their
noncoincident character. On the other hand, a single trapped
vortex jumping between two stable sites near an individual SQUID
loop could have induced such large changes in the detector
outputs.

The monopole flux limit can be calculated from the null
detection result of this prototype detector. If the more stringent requirement of coincidence detection is used, then the upper
bound of monopole flux is set to be 5.2×10^{-10} cm^{-2} sr^{-2} s^{-1}. On
the other hand, excluding those flux jumps associated with
external disturbances, there were no monopole candidate events
observed for either planar detector. Therefore, the noncoincident
detection area can be included to give a lower upperbound of
1.7×10^{-10} cm^{-2} sr^{-1} s^{-1}. Both figures are calculated for a 90%
confidence level.

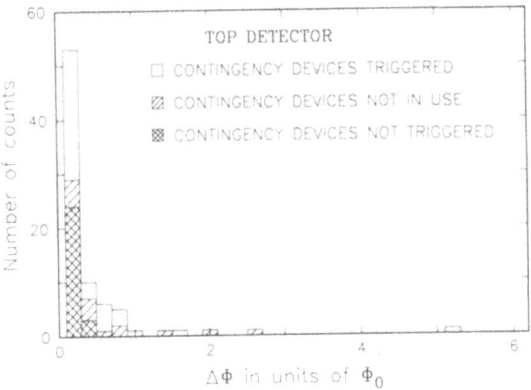

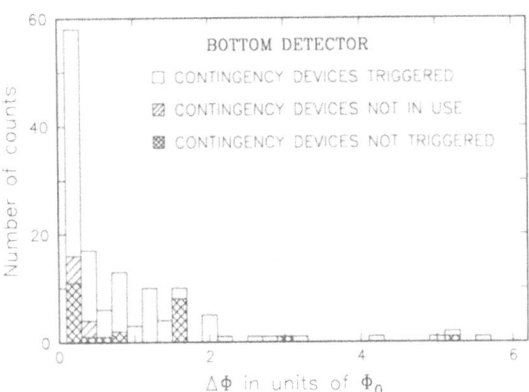

Fig. 4   The histograms of the counts of the flux jumps as a
         function of the size of the flux jump for the top and
         bottom gradiometers of the prototype detector.   The flux
         jumps were recorded during a 165-day continuous running
         period (April 8 to September 20, 1983).

THE CURRENT DETECTOR

     The monopole detector which is currently running at the IBM
T.J. Watson Research Center, was designed and built based on our
experience with the prototype detector.   In addition to the
nontrivial size extension, the current detector, shown in Fig. 5,
has the following improved features: (1) 6 independent gradiometers were placed on the six faces of a parallelopiped so that
nearly 100% of the monopole detection would be coincident;
(2) four nested concentric μ-metal cylinders were used to provide
a low field environment of about 10 μG for the cooling of the Pb
superconducting shield and detector; (3) A three-axis flux-gate
magnetometer and an electrical power line monitor have been added

to enhance our capability to discriminate against spurious events;
(4) A proportional counter is to be added to analyze any possible
correlation between cosmic ray activity and the monopole candidate
events if there is any.

The current detector is expected to be running for a year.
The detailed results will be published at a later date. In case
of no monopole events observed in one year, this detector would
give us a monopole flux bound of 5.8×10^{-12} cm^{-2} sr^{-1} s^{-1} at the
90% confidence level. As of February 1984, this detector has
established a flux bound of 2.0×10^{-11} cm^{-2} sr^{-1} s^{-1} at the 90%
confidence level, with no monopole candidate events observed.

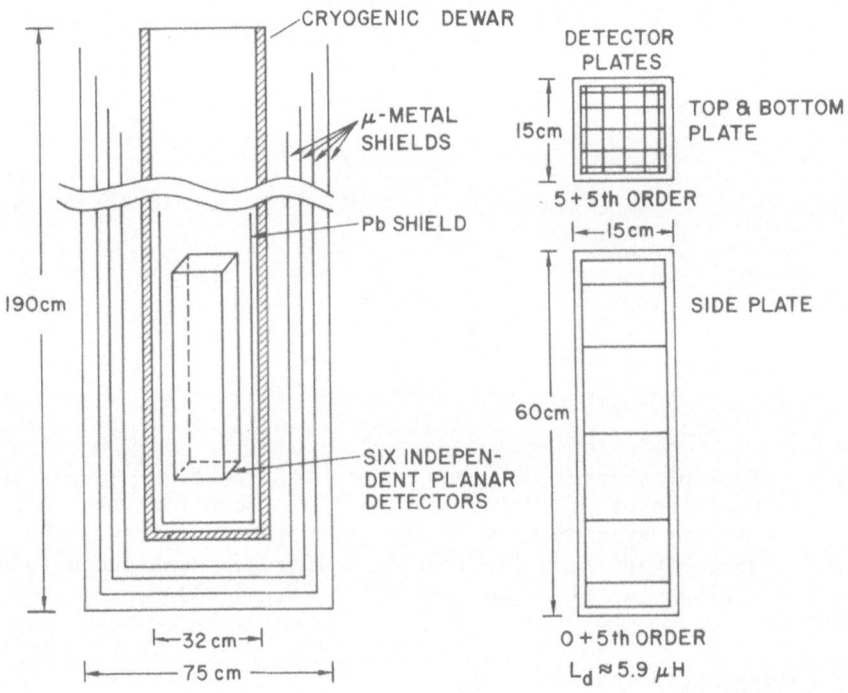

Fig. 5   The current monopole detector.   Six planar gradiometers
on the facets of a parallelopiped are independently
monitored by six SQUIDs' (not shown).

FUTURE PLANS

The coincident detection area of our current detector is a
factor of 40 larger than the prototype.   Admittedly, it is still
too small to reach the Parker bound within any reasonable time
frame. We do not think however, that we have reached the limit of
the superconductive induction technique yet.   It is possible to

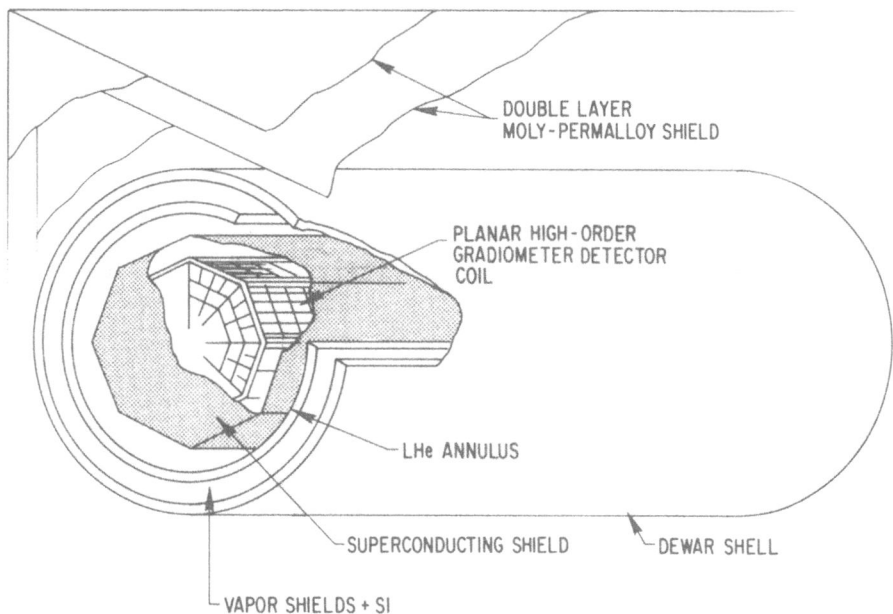

Fig. 6.   The large area detector under design. The detector con-
          sists of 8 rectangular and 2 octagonal planar gradiometer
          coils monitored by 10 independent thin film dc SQUIDs.
          The dewar is roughly 7 ft. in diameter by 14 ft. long.
          Two closed ferromagnetic shields surround the entire
          dewar.   The coincident detection area is 3.2 m^2 (for 4π
          steradians).

increase the detection area in a single larger cryogenic dewar by
at least a factor of 30. The system under design is shown in Fig.
6.   The detection coils are located on the faces of a retangular
octahedron.   The side plates are 2.8 m long and 0.45 m wide. With
detection coils also wound on the end plates, this detector has a
nearly 100% coincident detection as our current detector.   Ten
high resolution thin film dc SQUIDs fabricated at IBM will be used
to monitor the detector coils. Thus, the area of the detector has
been increased by a factor of 32 with less than a factor 2
increase in the readout system.   The detector and the SQUIDs are
located in exchange gas surrounded by a liquid helium annulus.
The system is cooled within a closed double layered molypermalloy
shield. The room temperature shield is used to reduce the ambient
field for the initial cool down, thus reducing the probability of
flux trapping in the detector coils and superconducting shields.
The detector is expected to be operational in early 1985.   We are
hopeful that this system will serve as a prototype for an array of
large area detectors capable of reaching the Parker bound in a few
years.

ACKNOWLEDGEMENTS

    We would like to acknowledge helpful discussions concerning
the design of high-order planar gradiometers with S. Kirkpatrick
and M. Gutzwiller.

REFERENCES

1. B. Cabrera, Phys. Rev. Lett. 48, 1378 (1982).
2. G. Giacomelli, Magnetic Monopoles, edited by R.C. Carrigan Jr.
    and W.P. Trower (Plenum Press, 1983) p. 41.
3. B. Cabrera, ibid., p. 175.
4. C.C. Tsuei, ibid., p. 201.
5. E.N. Parker, Astrophys. J. 160, 383 (1970); M.S. Turner, E.N.
    Parker and T.J. Bogdan, Phys. Rev. D26, 1296 (1982); R.A.
    Carrigan Jr. and W.P. Trower, Nature 305, 673 (1983).
6. The coincident detection area of a coincident detector is
    defined to be the sensing area of a single planar detector
    which has the same detection rate as the coincident detector
    for an isotropic flux of monopoles. The sensing area of a
    planar detector, averaged over 4π solid angle, is one half of
    its physical area.
7. C.D. Tesche, C.C. Chi, C.C. Tsuei and P. Chaudhari, Appl.
    Phys. Lett. 43, 384 (1983).
8. C.D. Tesche, Proceedings of the Fourth Workshop on Grand
    Unification, edited by H.A. Weldon, P. Langacker, and P.J.
    Steinhardt (Birkhäuser, 1983) p. 121.
9. The SQUIDs used in this experiment are rf SQUIDs manufactured
    by SHE Inc.

FIRST RESULTS FROM THE CHICAGO-FERMILAB-MICHIGAN

COSMIC RAY MAGNETIC MONOPOLE DETECTOR

J.R. Incandela[a], M. Campbell[a], H. Frisch[a],
S. Somalwar[a], M. Kuchnir[b], and H.R. Gustafson[c]

[a]Enrico Fermi Institute
The University of Chicago
Chicago, Illinois 60637

[b]Fermi National Accelerator Laboratory
P.O. Box 500
Batavia, Illinois 60510

[c]University of Michigan
Randall Physics Lab
Ann Arbor, Michigan 48109

The design and performance of a prototype detector with two 60 cm diameter superconducting loops is presented. During one month of data-taking, no candidate events are observed and an upper limit on the flux of cosmic ray magnetic monopoles $F_m \leqslant 6.9 \times 10^{-11}$ cm^{-2} s^{-1} sr^{-1} is set. This exposure corresponds to 21 times the original exposure of Cabrera.[1] The detector demonstrates the possibility of large, high-ambient-field inductive detectors.

We have sought to answer the following question: Can a superconducting induction detector be built sufficiently large to have sensitivity to monopole fluxes at the Parker bound[2] of $F_m \leqslant 10^{-15}$ cm^{-2} s^{-1} sr^{-1}? (Such a flux translates into roughly one monopole per year per 1,000 m^2 of detection area.) This presentation will set forth the necessary design considerations for such a detector, a description of our detector prototype, and a report of its performance during a one month data-taking period.

For monetary reasons alone, a large detector requires certain limitations. First, to minimize cryogenic volume, the detector should fill accessible space inside the superconducting shielding as efficiently as possible. Second, for a 1000 m² array we believe that the detector should be essentially planar, having small vertical separation between coincidence planes as compared with overall horizontal dimensions. Third, to circumvent the major cost of μ-metal shielding for such a large system, the detector should operate in fields of 1-10 mgauss--obtainable using relatively inexpensive degaussed steel.

The main problem encountered in trying to design a large area detector is the inductive coupling of the superconducting ("SC") detector loop and the surrounding SC shield. A monopole passing through both the shield and the loop directly induces supercurrents in each. The supercurrents in the shield in turn produce a magnetic flux, a portion of which impinges upon the detector loop. The supercurrent induced in the loop as a result of the shield flux is opposed to the current directly induced in the loop by the passing monopole. The net result is a reduction of signal. In fact, when a circular loop of maximum possible diameter is placed inside a cylindrical shield in an attempt to maximize detector area, no signal can be obtained.

For the case of a monopole traveling along the axis of a closed cylindrical shield, there is an expression[3] for the fraction F of the full signal one would see using a coaxial circular detection loop (Fig. 1). If the loop has radius b and the shield has radius R and half-height h, then the expression for F is

$$F(R,b,h,Z) = 1 - [\frac{R^2}{b^2} + \frac{2R}{b} \sum_{n=1}^{\infty} (\frac{\cosh(K_nZ)}{\cosh(K_nh)}) \frac{J_1(K_nR)}{X_{1_n}[J_2(X_{1_n})]^2}]$$

where X_{1_n} is the nth root of the first order Bessel function $J_1(X)$ and $K_n = X_{1_n}/b$. Figure 2 illustrates F for various values of R/b and h/b.

We tested this expression by building a small prototype detector with h = 1", b = 3", and R = 1", for which we predict F = 0.875. We simulated the monopole trajectory by passing a current through a solenoid fixed along the axis; we now use these very slender solenoids, "pseudopoles", for calibrating our detectors. (The current is determined by requiring that the flux in the solenoid equal 2 ϕ_o). We obtained a measured value F = 0.87.

The expression for F predicts that for a broad and flat detector, e.g. h = 3.5", b = 12", the loop size must be restricted to R ≤ 1.2" in order to see 90% of the full signal. Thus, a circular loop cannot be used in any efficient large area scheme and one is forced to seek alternatives.

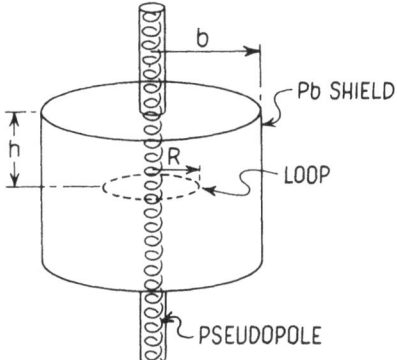

Fig. 1   Geometry for calculation and test of F(R,b,h,Z).   In this case Z = 0 (loop midway between cylinder endcaps).

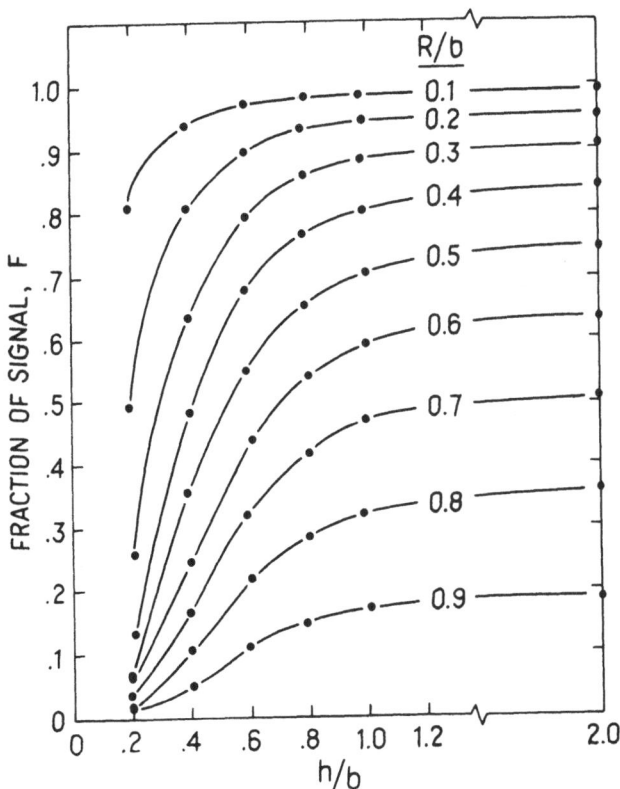

Fig. 2   Fraction of signal F.

Our solution[4] is to take a single SC filament and "weave" it into a square cell grid pattern so that adjacent cells will circulate current in opposite senses (Fig. 3). We call this a "macrame" since early models were actually "woven" by hand. Macrames now used successfully are printed circuit boards--solder which is plated onto copper-clad G-10. (With the aim of making prototypes for large detectors, we chose this relatively inexpensive and accessible technique.)

With the macrame in place of the circular loop, the flux created by currents in the shield will try to induce opposing currents in adjacent cells. These currents tend to cancel one another--with greater degree of cancellation for finer grid patterns. The monopole, however, does not induce opposing currents because it passes through a single cell. Consequently, its direct interaction with the macrame yields a net induced current.

Fig. 3   Photograph of the macrame with 8 cells along the diameter. A half-row of cells is begun on one side of the G-10 and completed on the other with continuation through the board being accomplished by plated through holes.

Therefore, in a simple model one need only consider the single cell through which the monopole passes. One then expects the fraction, F, of signal to be limited by the relation of cell size--rather than macrame size--to the dimensions of the shield. (It should be kept in mind that although one obtains a larger fraction F using smaller cells, the maximum signal itself is decreased with use of a finer grid because the inductance of the macrame increases.)

Our prototype detector operates two macrames, each of diameter 60 cm (b ~ 12") and cell size 5 cm (R ~ 1"), located inside cylindrical shields of half-height h ~ 3.5" radius b ~ 12". These dimensions are familiar from the above example and together with the simple model just discussed predict F > 0.9. This is in fact observed for cells near the center of the macrame. F falls off to 0.6 for cells located along the edge of the macrame and against the cylinder walls. This latter effect is the result of the fact that the actual shield tapers to a smaller half-height near the edge. (This should be avoided in later models.)

The actual prototype system that we have been operating has the following characteristic features (Fig. 4):

a) macrame - the loops are 60 cm in diameter with 5 cm equal size cells.

b) coincidence - the two 60 cm diameter macrames are set 21.6 cm apart for a coincident detection area averaged over 4π of 698 cm^2.

c) vacuum - the detectors operate in a vacuum rather than in a liquid helium (LHe) bath. They are cooled by conduction through flexible copper braids which do not rigidly couple the detectors to the LHe reservoir. The vacuum eliminates all pressure changes or bubbling effects which we have found to cause noise when operating in liquid in a high ambient field. It also decouples the detector from most external mechanical perturbations.

d) isolation - each detector is suspended by its own set of support rods extending to vibration isolation devices located on the exterior shell of the cryostat. The external positioning of the mounting devices allows us direct mechanical access to the detectors for vibration sensitivity tests. The entire set-up mechanically decouples the two detectors. Each detector also has its own SC shielding. This prevents the possibility of spurious signals appearing coincidentally in the outputs of both detectors as a result of shield flux migration. Finally, by operating in a vacuum, cooling by conduction,

and providing radiation shielding, the thermal time
constants are large enough to assure that no abrupt or
localized thermal changes (which would cause noise and
drift of signal) can occur.

e) pseudopoles - each detector has permanently fixed slender
    solenoids used for calibration at a flux of $2\phi_o$.

f) impedance matching - niobium wire SC transformers are
    used to interface the 25 μH macrame to the 2 μH SQUID
    loop in order to maximize signal.

g) Fe shield - the entire apparatus is surrounded by a steel
    pipe which we use to reduce ambient fields to
    ~1-10 mgauss.

h) high field - we operate with a "pinned" magnetic field
    which was trapped in a residual external field of
    ~1-10 mgauss.

For additional assurance in discriminating spurious signals
caused by some sort of macroscopic disturbance from true monopoles, we have installed a variety of monitoring devices. These
monitor orthogonal components of ambient magnetic field, sound,
vibration, atomspheric pressure and pressure of the liquid
nitrogen fill system, as well as the temperature and pressure of
the cryogenic system.

Data from detectors and monitors is written simultaneously
onto a chart recorder and a magnetic tape using a 32 channel ADC:
we operated at 1 Hz bandwidth on the SQUIDS (although steps would
be clearly visible at 100 Hz).

So far, detector performance indicates that the bottom
detector is very well shielded while the top detector is sensitive
to external field changes. An external field change of 1 mgauss
elicits little or no response from the bottom detector, while the
top detector responds with a 5 to 35 mV signal depending on which
component of ambient field changes. (A monopole would give
approximately a 6 mV step). For this reason, we have chosen not
to trust the top detector for singles information in the data run
described below. The top detector was sufficiently quiet to be
used for coincident information.

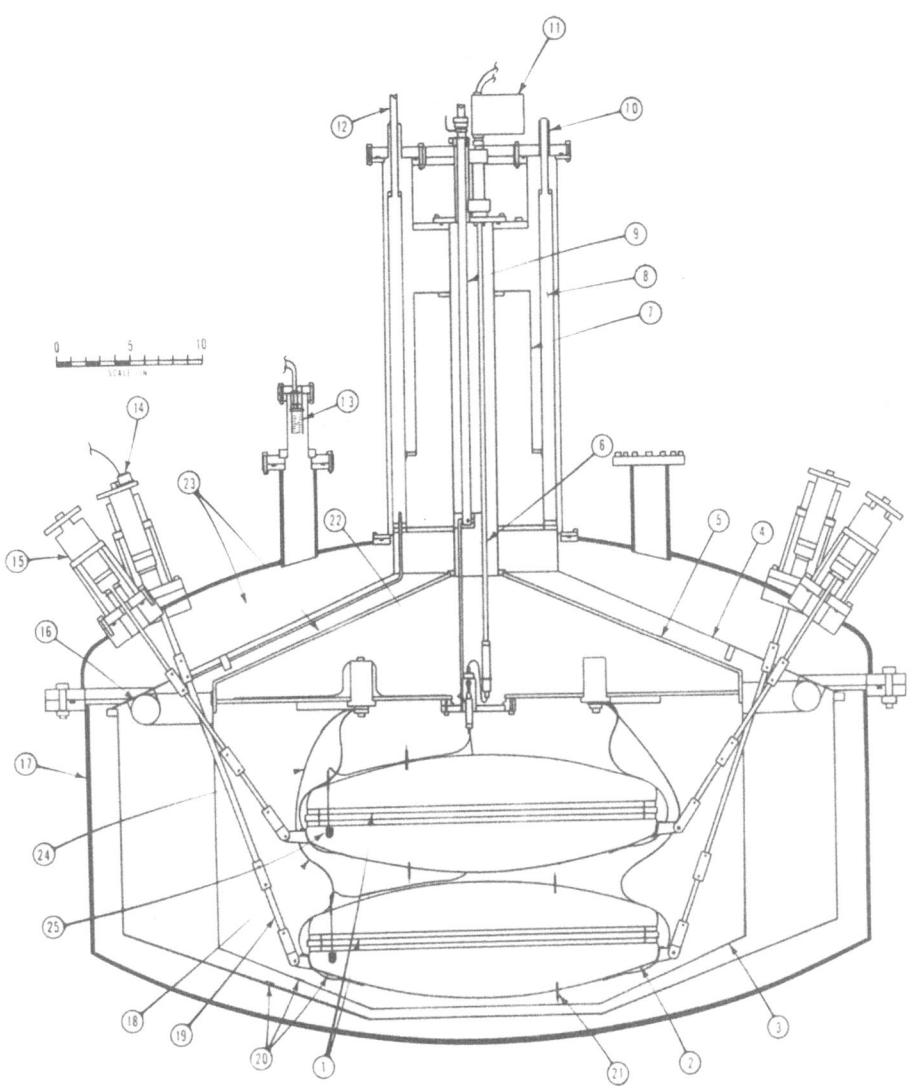

Fig. 4 Monopole Detector Assembly: 1. detector plane, 2. lead lined copper shield, 3. helium temperature shield, 4. nitrogen temperature shield, 5. helium reservoir, 6. SQUID tube, 7. heat pick-up, 8. nitrogen jacket, 9. helium fill tube, 10. nitrogen vent, 11. SQUID electronics, 12. nitrogen fill line, 13. vacuum gauge, 14. accelerometer, 15. vibration isolator assembly, 16. nitrogen cooling ring, 17. vacuum vessel, 18. cooling straps, 19. suspension rod, 20. thermometers, 21. pseudopole, 22. vacuum feedthrough, 23. super insulation, 24. cooling straps, 25. SQUID matching transformer.

Figure 5 illustrates a daytime SQUID output sample. The double mesas located in the center of the SQUID outputs are caused by remote stimulation of calibration pseudopoles. The computer stimulates one pseudopole in each detector for a period of one minute at the beginning and end of two hour data runs to verify sensitivity. The size of the steps corresponds to a single Dirac charge monopole passage and indicates the visibility of the signal in both detectors. The pseudopoles are stimulated simultaneously.

During the entire data-taking period (Aug. 29 - Oct. 4, 1983) no coincident steps were observed. Figure 6 is a histogram of all steps observed in our bottom detector for this period. No steps consistent with a single Dirac charge were observed. Figure 7 indicates the output for one of the largest two steps observed. This histogram depicts only those steps not vetoed by correlation with external monitor indicators.

From August 29 to October 4, we were sensitive with our single bottom detector for 22 days, 6 hours. The effective area of this detector is 1385 cm^2 averaged over 4π. Thus, having seen no candidate events we set an upper limit (90% C.L.) for the cosmic monopole flux at $F_m \lesssim 6.9 \times 10^{-11}$ cm^{-2} s^{-1} sr^{-1}. (Our coincidence live time[5] was ~ 13.5 days for 698 cm^2, allowing a coincident limit of $F_m \lesssim 2.2 \times 10^{-10}$ cm^{-2} s^{-1} sr^{-1}.)

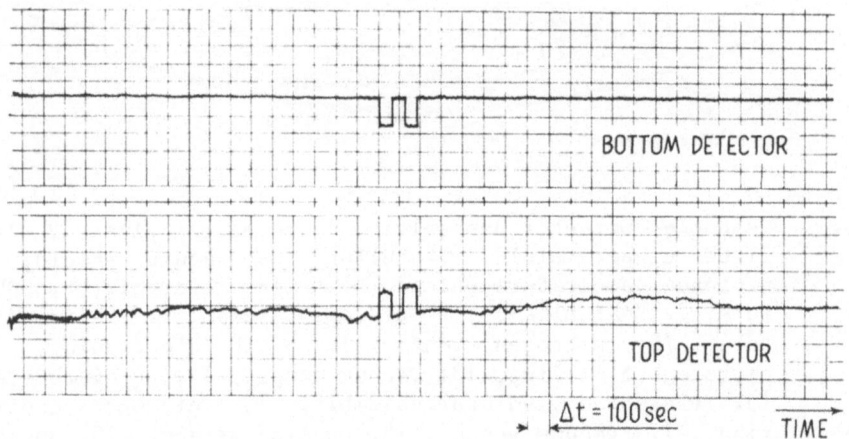

Fig. 5 One hour of daytime output from each detector with calibration solenoid data.

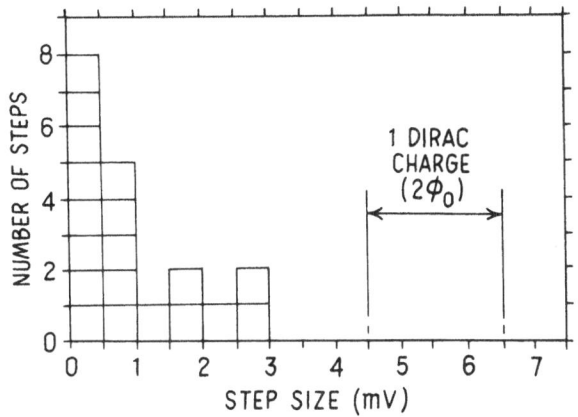

Fig. 6 Histogram of uncorrelated steps in the bottom detector.

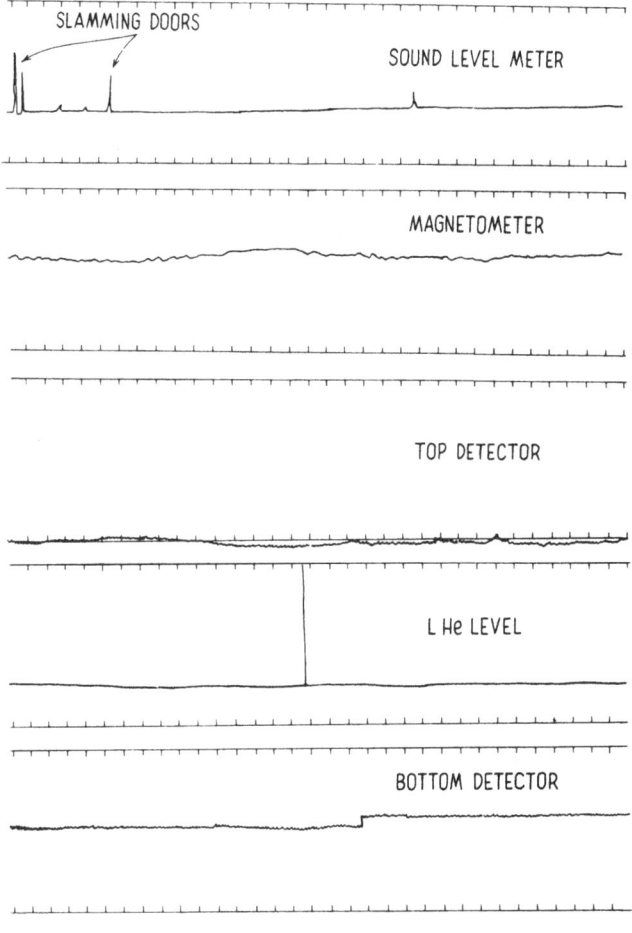

Fig. 7 Stripchart output during the time of a 3 mV step.

CONCLUSIONS

    We have collected 21 times the sample of Caberea's first
experiment and observe no candidate monopole events. The probability that these two results are compatible is less than 5%. In
addition, we have demonstrated that large area cheaply constructed
superconducting monopole detectors can be stably operated in large
ambient fields, which can be achieved with inexpensive iron
magnetic shielding.

ACKNOWLEDGEMENTS

    We would also like to thank: R. Carrigan and J. Lach
(Fermilab); R. Smith (Argonne/Fermilab); A. Guthke, C. Kendzione,
and K. Stanfield (Fermilab); B. Brown, F. Juravic, H. Jostlein and
J. Tague (Fermilab); S. Lucero and R. Szara (University of
Chicago); R. Armstrong, L. Fiscelli, R. Gabriel, P. McGolf and R.
Northrop (University of Chicago); S. Rice (University of Chicago);
D. Berley and A. Dzierba (NSF); J. Vrba (CTF Systems); D. Crum and
R. Fagaly (SHE); and S. Dickinson for editorial help.

REFERENCES

1. B. Cabrera, "First Results From a Superconducting Detector for
    Moving Magnetic Monopoles", Phys. Rev. Lett. 48, 1378 (1982).
2. M.S. Turner, E.N. Parker, T.J. Bogdan, "Magnetic Monopoles and
    the Survival of Galactic Magnetic Fields." Phys. Rev. D26,
    1296 (1982). Also see E.N. Parker in this volume.
3. H. Frisch as discussed by C.C. Tsuei in "Magnetic Monopoles,"
    R.A. Carrigan, Jr. and W.P. Trower, eds., (Plenum Publishing
    Corp.,1983) pp. 209-11.
4. This solution is similar to those found independently by the
    IBM group and the Imperial College group. For IBM see: a)
    C.D. Tesche, C.C. Chi, C.C. Tsuei and P. Chaudhari, Appl.
    Phys. Lett. 43, 384 (1983). b) C.C. Tsuei, "Magnetic Mono-
    poles", R.A. Carrigan, Jr. and W.P. Trower, eds., (Plenum
    Publishing Corp., '83) p. 201. c) C.D. Tesche, to be
    published in the Proceedings of the Fourth Workshop on Grand
    Unification, University of Pennsylvania, April 1983. d) C.C.
    Chi in this volume. For Imperial College see: a) J.G. Park,
    C.N. Guy, "Inductive Monopole Detectors of Coil-Shield Area
    Ratio Approximately 1: This Signal Size Distribution" Black-
    ett Lab., UK-preprint (1982). b) See discussion by C.C. Tsuei
    in "Magnetic Monopoles" R.A. Carrigan, Jr. and W.P. Trower,
    eds., (Plenum Publishing Corp., 1983) p. 210-211. c) C.N.
    Guy and J. Schouten in this volume.
5. The smaller time when both detectors were quiet and taking
    data is due to the fact that we were instrumenting the
    detector during much of this running time.

THE IMPERIAL COLLEGE MONOPOLE DETECTOR

J.C. Schouten, A.D. Caplin, C.N. Guy, M. Hardiman[a] and J.G. Park[b]

Blackett Laboratory
Imperial College
London, England

Shortly after the publication of a candidate event for the passage of a magnetic monopole by Cabrera[1] our group decided to set up a similar experiment. Because of the far reaching consequences of the existence of monopoles, if confirmed, an independent and carefully prepared experiment seemed to be necessary. The main new features of our experiment would be first, a larger effective area, an order of magnitude greater than Cabrera's original; second, provisions for coincidence/anti-coincidence measurements; and third, a stable low field magnetic environment around the superconducting shield, so as to reduce the risk of flux motion in this shield. The experiment needs to be isolated from external influences as well as possible mechanical and electrical influences, etc. It is necessary also to monitor all relevant parameters such as magnetic field, mechanical stability, r.f.-level, electrical interference.

The heart of the detector is formed by a simple superconducting coil arrangement consisting of one circular coil A, and two semi-circular D-shaped coils, B and C (see Fig. 1). These extra coils are placed immediately under the circular one. Each of the three coils is coupled to a different SQUID. The passage of a magnetic monopole would give a signal, quite sharply defined in magnitude, and consisting of a step in the output of either A and B and not C, or A and C and not B. Any spurious flux movement would most likely give a signal in all three coils.

The detector coils have radius R_{coil} of 11 cm, giving an effective area of ca. 200 cm^2 sr^{-1}, and are placed within a one meter long, 25 cm diameter superconducting lead shield. In order

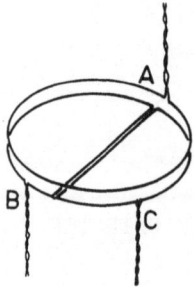

Fig. 1.   Coincidence/anti-coincidence triple coil arrangement.

to avoid the large reduction in signal height caused by this large ratio R_{coil}/R_{shield}, we place a compensating coil in each of the detector circuits. These additional coils have the same area-turn product as the large pick-up coils but 1/10 of the area, so forming an Asymmetric Astatic Pair (AAP). Such arrangements will cancel almost all of the mutual inductance between detector coil and shield, while enabling us to use as large an area as possible. The small coupling between the AAP and the shield means that monopole tracks of different angle and impact parameters all yield much the same signal size. The different distributions are calculated[2] and plotted in Fig. 2 for several ratios of R_{coil}/R_{shield}.

The lead shield is closed at the top and bottom except for three long "chimneys" on the top to give access to the rf-lines. This shield design has been tested extensively on a small scale and we are confident that the full scale shield will provide an extremely stable field. The lead shield will be placed in the lower half of a 2.5 m tall cryostat. This vapor-cooled liquid helium cryostat is constructed of aluminum and fiberglass and is designed to have a boil-off of less than 5 ℓ/day. Extensive tests have confirmed this performance and with the 40 liter reservoir capacity of helium above the lead shield it will give us a run time of more than one week between helium transfers.

The cryostat hangs within a set of 5 nested mu-metal shields. These shields have been designed to give a maximum shielding factor for a given weight (i.e., cost of mu-metal) using established procedures.[3] After demagnetization in situ the residual field is about 1 nT, and the dynamic shielding factor is better than 5×10^4. The cryostat and mu-metal shields are suspended on air-springs to give maximum mechanical isolation. The resonance frequency is less than 2 Hz, which gives an attenuation of 10-30 dB for frequencies between 10 and 250 Hz. Mechanical vibration is monitored by an accelerometer which has been designed and tested. A schematic drawing of the detector is given in Fig. 3.

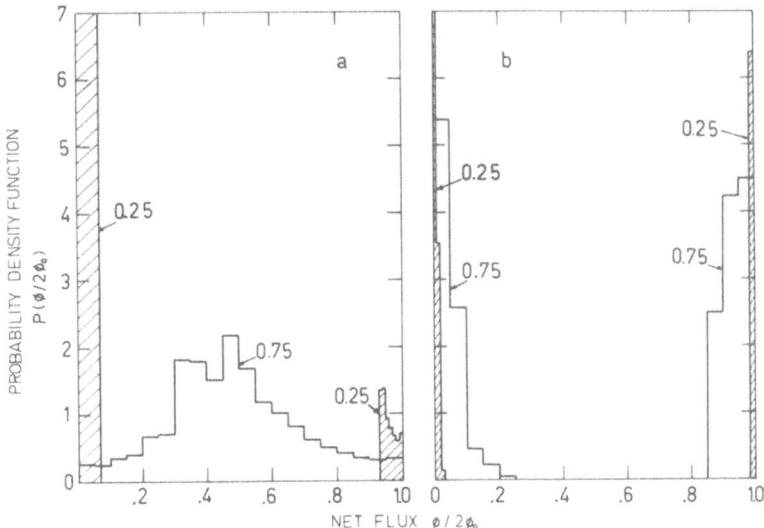

Fig. 2. Probability distribution $P(\phi)$ for single coil detectors (a) and AAP-detectors with $\Delta = R_{coil}/R_{shield} = 0.25$ and 0.75 (b).

A small computer will analyze the outputs of the three SQUIDs in real time using a bandwidth of up to 100 Hz, looking for the steps that are candidate monopole events. Simultaneously it will monitor the outputs of the accelerometer, an ultra-sensitive low-noise magnetometer, rf-level indicator, temperature and pressure sensors.

We have now (October 1983) had one test run of 300 hours duration with a single small (10 cm^2) detector coil. Except for the first few hours after cool down when significant flux jumps do occur, the system was exceedingly quiet, even though our shielding arrangements were still somewhat primative.

During the last 18 months other inductive detectors have been operational, but no more candidate events have been seen (Cabrera, Frisch et al., Chi et al.,).[4] Thus the accumulated area-time of observation has increased by a factor of 100 since February 1982, and the estimated upper bound on monopole flux has diminished by the same factor. Furthermore, the absence of large spurious flux movements in the superconducting shields of these experiments and our own experience with small shields gives us a high degree of confidence that our large shield will be free of such spurious events. Because it is now important to reduce the likely upper-bound on monopole flux as quickly as possible, we are relaxing our constraint on the coincidence by using only two instead of three horizontal circular coils. The third coil is being replaced by a

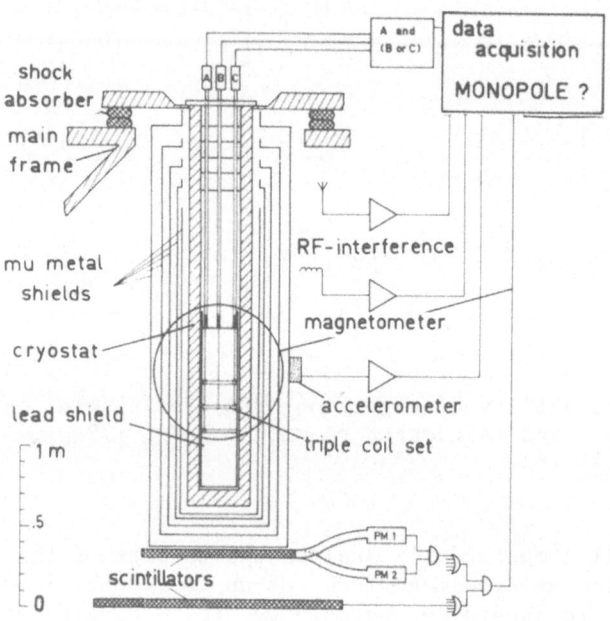

Fig. 3.   Schematic view of the inductive monopole detector.

vertical window frame detector which will give an effective area
of more than 3000 cm^2 sr^{-1}.   The physics of this detector will be
discussed in the article by C.N. Guy, also in these proceedings.
The detector should be running before the end of 1983 and will be
the largest one available.   If it too fails to see any candidate
monopole events the upper bound on monopole flux will be reduced
significantly in a few months of run time.

REFERENCES

a. Now at University of Sussex, Physics Division, Falmer, Brighton,
    BN1 9QH, Sussex, England.
b. Died June 1983.
1. B. Cabrera, Phys. Rev. Lett $\underline{48}$, 1378 (1982).
2. C.N. Guy and J.G. Park, accepted for publication in J. Phys. D.
3. D.U. Gubser, S.A. Wolf and J.E. Cox, Rev. Sci. Instr. $\underline{50}$, 751
    (1979).
4. See their contribution in this volume.

MONOPOLE DETECTOR STUDIES AT NBS

F. R. Fickett, M. Cromar, and A.F. Clark

Electromagnetic Technology Division
National Bureau of Standards
Boulder, Colorado 80303

The work at the National Bureau of Standards has had three major goals. First, to investigate sources of noise in SQUID-based detector systems and to develop techniques to minimize their disruptive effects. Second, to investigate and identify sources of signals similar in size and signature to those expected from a monopole passage and, again, to eliminate them. Third, to participate in the search for the monopole. To these ends, we have constructed and operated a two-coil coincidence system in several configurations for well over 1000 hours. Because our efforts have been concentrated on the investigation of anomalous effects, not many of these hours can be considered as true detector time.

Figure 1 shows the detector configuration, presents data on the basic system, and describes the three major coil configurations used in our testing to date. The system has several features that are unusual. The structure is totally rigid; there is relatively little shielding of the earth's field; and the system is operated in an environment that is extremely noisy both electrically and mechanically. A copper coil, placed between the detector coils, is used to calibrate the system. The lead-plated can is suspended from the top plate by a structure of stainless steel tubing that is not shown in the figure. The clamps that connect the SQUID tubes to this structure are a recent addition to the system and are needed to eliminate the mechanical effects that we have determined to be the cause of most large offsets. The location of the clamps is critical. They must be near the top of the can where they will be in a nearly isothermal environment for most of the run time. In the most recent runs, a low-temperature strain gauge is attached to one of the SQUID cans to monitor mechanical shock and motion.

SYSTEM DATA:
  Two coaxial coils (coinc.)
  Typ. run time (3 fills)= 100 hrs
  Depth to SQUIDs = 68 cm
  Can (shield) diameter = 13 cm
  Can (shield) length = 40 cm
  Field at coils= 56 mA/m (0.7 mOe)

COIL CONFIGURATIONS:
  A) turns/coil = 5
    coil diameter = 3.1 cm
    coil separation = 1 cm
    can/coil area ratio =18
    run time= 522 hrs

  B) turns/coil = 3
    coil diameter = 6.4 cm
    coil separation = 1 cm
    can/coil area ratio = 4
    run time = 502 hrs

  C) turns/coil = 3
    coil diameter = 6.4 cm
    coil separation = 13 cm
    can/coil area ratio = 4
    run time = 145 hrs

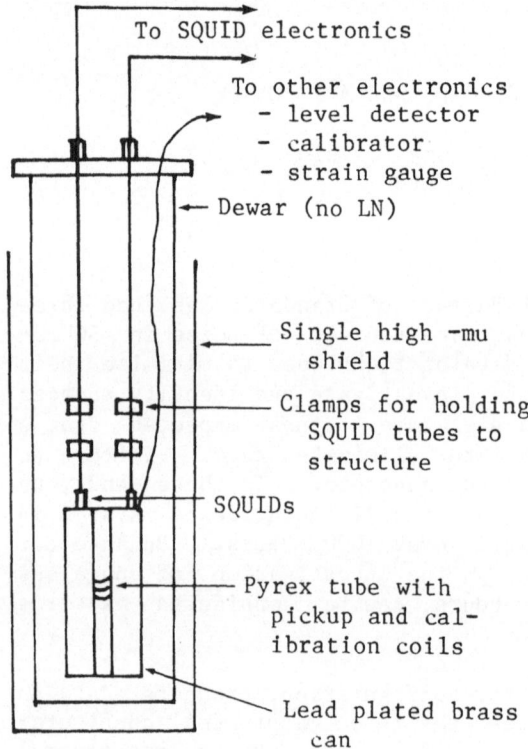

To SQUID electronics

To other electronics
  - level detector
  - calibrator
  - strain gauge

Dewar (no LN)

Single high -mu
  shield

Clamps for holding
SQUID tubes to
structure

SQUIDs

Pyrex tube with
  pickup and cal-
  ibration coils

Lead plated brass
can

Fig. 1 System data and detector configurations used in the
      experiments.

    With this system we have performed a large number of tests
using disturbances created by mechanical shocks, pressue changes
in the dewar, tensile and bending stress in the SQUID tubes,
extraneous magnetic fields, and rf signals and other electrical
transients. We have also looked at the effect of operation
without external magnetic shielding (the superconducting shield
was still used) and have evaluated many of these effects when
applied to the SQUID along with its input either shorted or open.
Table 1 summarizes our observations regarding the disturbance
tests. In general, it is difficult to induce offsets of any size,

Table 1. Summary of observations on the effect of various
          disturbances on the detector and unattached SQUIDS.

Disturbance	Effect	How to minimize
Mechanical shock	Very large shock may produce an offset.	Rigid mounting.
	Moderate shocks produce significant noise.	Clamp SQUID probes.
Stress in SQUID probes	May produce offsets of monopole size.	Clamp SQUID probes to structure in cold region.
		Monitor SQUID can with strain gauge.
Magnetic signals	Very large disturbance can create offsets.	Keep large, moving magnets away.
Flux leakage into can	Negligible in our system.	Keep access holes relatively small.
Pressure changes	Increased noise and offsets occur if can plating is loose.	Good plating.
	Large changes affect SQUID operation.	Thermally stable dewar.
No magnetic shields	Sensitivity to noise and shock increased.	Don't run unshielded.
	No offsets observed.	
Electrical transients	Noise, but not offsets even at gross interference levels with proper filtering and shielding.	Filtered electrical feedthroughs.

SQUID alone, input open - Above disturbances have little effect.
SQUID alone, input shorted - Small deformation of SQUID probes
   gives increased noise and has also been observed to give offsets.
Large magnetic disturbance gives offsets.
Large shock gives offsets.

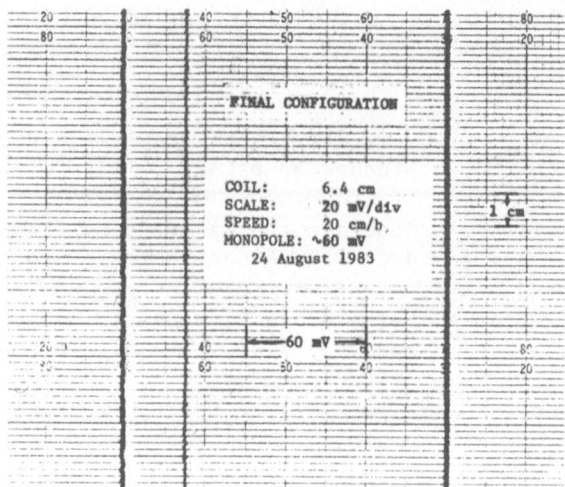

Fig. 2   Typical trace from the detector in its final configura-
         tion. The center trace is from the strain-gauge shock and
         motion detector.

but noise is easily created.   We also list in this table
techniques that we have used to minimize the various effects.
They are more than adequate, at least for our system.   Figure 2
shows a typical trace from a recent run of the detector.   The
center trace is from the strain gauge shock monitor. The noise is
actually somewhat smaller than it appears; the ink tends to
saturate the paper at this low chart speed.   We now have several
hundred hours of run time in this configuration with no offsets.

     Our investigation thus far has led to several conclusions.
With respect to noise, it is quite possible with a little care to
operate a system essentially at SQUID noise level; under any
conditions, it is possible to operate with adequate signal-to-noise
ratio for detection of a direct hit event and most near miss
events.   With respect to disturbances, monopole-like signals can
be induced even in a coincidence system both magnetically and
thermo-mechanically; it is possible to eliminate all sources of
such signals that we have observed to date; it is possible to run
a detector in an electrically noisy environment without spurious
events.   With respect to monopoles, we seen no candidate
events and no near misses.

     We plan to continue the project with the construction of a
larger system (15 cm loops) using three coils (à la Cabrera), but
maintaining the rigid structure concept.   The external magnetic
shielding will be increased somewhat in this new system.

ZERO-QUANTUM SUPERCONDUCTING MAGNETIC SHIELD

John R. Clem

Ames Laboratory-USDOE and Department of Physics
Iowa State University
Ames, Iowa 50011

ABSTRACT

The design of a new superconducting magnetic shield suitable for use with magnetic monopole detectors is described, together with a method for achieving a final state in which no magnetic flux quanta are trapped in the shield's central region.   The magnetic shield consists basically of a specially designed super-conducting cylindrical tube, equipped with electrical and magnetic devices to remove flux quanta trapped during cooldown.   An electrical current is applied along the length of the tube to cause mutual annihilation of trapped vortices and antivortices in the tube's central region and thus to remove flux quanta trans-verse to the cylinder axis.   A parallel applied magnetic field then moves an appropriate number of vortices and antivortices from tube-end reservoirs to opposite ends until no longitudinal flux quanta thread the tube.   Also, a procedure is described by which the shield itself could be used as a monopole detector.

INTRODUCTION

Good magnetic shielding is desirable for use with supercon-ducting monopole detectors.   With poor shielding, vortices can be trapped in the superconducting lines of a SQUID or a high-order gradiometer detector.   Mechanical, thermal, or electrical distur-bances can cause these vortices to move to new positions, thereby producing signals mimicking the transit of magnetic monopoles through the detector.

When a SQUID or detector made of superconducting lines with

481

width w is cooled through the superconducting transition temperature in an ambient magnetic field B, thermodynamic considerations
dictate that vortices with spacing $a \approx (\phi_0/B)^{1/2}$ will be trapped.[1]
When $w \gg a$, flux trapping is practically certain, but when
$w \ll a$, flux trapping is highly unlikely. Another way of looking
at this is to compare w with a critical superconducting linewidth,
$w_c \approx (\phi_0/B)^{1/2}$. If $w > w_c$, flux trapping is likely, and when
$w < w_c$, flux trapping is unlikely. To minimize spurious signals
arising from the motion of trapped flux, it is thus important to
use an appropriate combination of small linewidth w and small
magnetic field B.

The behavior of superconducting magnetic shields is governed
by Ampere's law, London's equation, and the equation of fluxoid
quantization. These equations, when applied to an infinitely long
superconducting cylindrical shell of thickness d and radius R,
result in the following expressions for the magnetic flux density
B_i inside the shield when it is subjected to an applied field B_a
(penetration depth $\lambda \ll R$ and $d \ll R$):

$$\frac{B_i}{B_a} \approx \left[(R/2\lambda) \sinh(d/\lambda) + \cosh(d/\lambda) \right]^{-1} , \tag{1a}$$

$$\approx (4\lambda/R) \exp(-d/\lambda) , \quad \lambda \lesssim d , \tag{1b}$$

$$\approx \left[(R/\lambda_\perp) + 1 \right]^{-1} , \quad d \ll \lambda , \tag{1c}$$

where $\lambda_\perp = 2\lambda^2/d$. These equations apply to a zero-quantum magnetic shield containing no trapped flux quanta (vortices or
antivortices) either in the hole or in the walls of the shield.
Equation 1 may be thought of as dictating the minimum thickness
d_{min} of superconductor required to produce a desired level of
internal field B_i when a long cylindrical shell of radius R is
subjected to an ambient field B_a. When $\lambda \ll R$, we obtain from
Eqs. (1b) and (1c)

$$d_{min} \approx \lambda \ln \left(\frac{4\lambda}{R} \frac{B_a}{B_i} \right) , \quad d_{min} \gtrsim \lambda , \tag{2a}$$

$$\approx \lambda \left(\frac{2\lambda}{R} \frac{B_a}{B_i} \right) , \quad d_{min} \ll \lambda . \tag{2b}$$

Once the dimensions d and R are chosen to produce a desired
field level B_i in applied field B_a, one may now use a magnetic
shield of finite length L. The transverse magnetic field $B_{i\perp}$ at
the middle of a cylindrical shield subjected to an applied
transverse yield $B_{a\perp}$ is given by

$$B_{i\perp}/B_{a\perp} \approx 1.324 \exp (-0.921 \ L/R) . \tag{3}$$

This equation yields the residual transverse field leaking into the middle of the shield from the open ends. The corresponding longitudinal field leaking in from the ends is typically much smaller than $B_{i\perp}$.

Equation 3 yields the minimum length L_{min} required to produce a desired field level $B_{i\perp}$ at the center of a shield subjected to an ambient transverse field $B_{a\perp}$:

$$L_{min} \approx 1.086 \ R \ \ln \ (1.324 \ B_{a\perp}/B_{i\perp}). \tag{4}$$

For example, a magnetic shield of radius R = 10 cm and penetration depth λ = 1000 Å in an ambient field of B_a = 1 mG, with dimensions chosen to produce a magnetic flux of less than 0.1 ϕ_0 through a circle of radius R at the center of the working volume ($\phi_0 = 2.07 \times 10^{-7}$ G-cm^2), should be able to achieve an internal field $B_i \approx 10^{-10}$ G with superconductor thickness $d_{min} \approx 3,960$ Å and length $L_{min} \approx 178$ cm.

Despite the many useful properties of superconductors, an unfortunate fact of life is that, when a type-II superconductor is cooled through its transition temperature in an ambient magnetic field, vortices (each of which carries one flux quantum ϕ_0 out of the shell wall) and antivortices (each of which carries one flux quantum ϕ_0 in through the wall) are trapped in the superconducting shell. The characteristic spacing of vortices and antivortices is $a \approx (\phi_0/B)^{1/2}$. Below T_c, the observed net microscopic magnetic field $\vec{b}$ is the sum of the zero-quantum response field $\vec{b}_0$ and a dipole-like field distribution $\vec{b}_{v-av}$ generated by the trapped vortices and antivortices, as sketched in Fig. 1. The magnitude of $\vec{b}_0$ inside the cylinder is given by Eq. (1).

How might it be possible to get rid of the transverse magnetic field $\vec{b}_{v-av}$ generated by the trapped vortices and antivortices? Numerous experiments have shown that it is possible to set vortices in motion by applying a current density greater than the critical current density j_c.[2] It is also known that vortices and antivortices at close range attract each other and mutually annihilate.[3,4] This suggests that the vortices and antivortices sketched in Fig. 1 can be eliminated by passing a current along the length of the cylinder. A current of density $j > j_c$ directed into the paper should drive the vortices in the clockwise direction and the antivortices in the counterclockwise direction, thereby causing them to mutually annihilate where the vortices and antivortices meet each other. To avoid nucleating new vortices, however, the current required to depin the existing vortices and antivortices must be sufficiently low that the self magnetic field is less than the lower critical field $H_{c\ell}$ of the superconductor. Otherwise, new vortices and antivortices would be generated at a rate that would replenish those being annihilated. These two

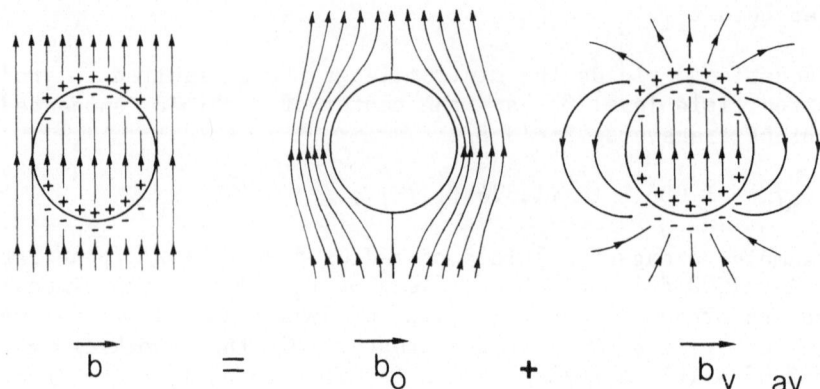

$$\vec{b} \quad = \quad \vec{b}_0 \quad + \quad \vec{b}_v \;_{av}$$

Fig. 1 Sketch of transverse magnetic field distribution $\vec{b}$ trapped in a thin superconducting cylindrical shell: $\vec{b} = \vec{b}_0 + \vec{b}_{v\text{-}av}$, where $\vec{b}_0$ is the zero-quantum response field and $\vec{b}_{v\text{-}av}$ is the field generated by the trapped vortices (top) and antivortices (bottom).

conditions thus require that the current density j be in the range

$$j_c < j < j_{c\ell} \; , \tag{5}$$

where $j_{c\ell} = cH_{c\ell}/4\pi d$, in Gaussian units. To have the widest possible range of operation, it is thus desired to have $j_c \ll j_{c\ell}$ or, equivalently, $d \ll d_{max}$, where

$$d_{max} = cH_{c\ell}/4\pi j_c \tag{6}$$

is the maximum thickness for proper operation of the annihilation process. It should be possible to achieve the condition $d \ll d_{max}$ easily in many superconducting materials. For example, for an amorphous or glassy superconductor with $H_{c\ell} = 10$ G and $j_c = 10^4 A/cm^2$, Eq. (6) yields $d_{max} = 80,000$Å. Typically the minimum thickness of superconductor to achieve a high degree of shielding is only about 5000Å, which is indeed much less than d_{max}.

MAGNETIC SHIELDING APPARATUS

Sketched in Fig. 2 are the main elements of a proposed magnetic shield assembly[5] that should be capable of achieving

extremely low magnetic fields in its interior. The key element in
this assembly (part B) is a superconducting shell of radius R and
total length L_2. The superconducting shell has: (i) a central
low-pinning region of length L_1 and thickness d; (ii) high-pinning
regions at both ends; and (iii) transition regions of length L_t
between the central region and the end regions.   The central
region is made of material with a relatively low critical
depinning current, the end regions are made with relatively high
values of the critical depinning current, and the transition
regions are made such that the critical depinning current varies
monotonically from the low central value to the higher values on
the ends.

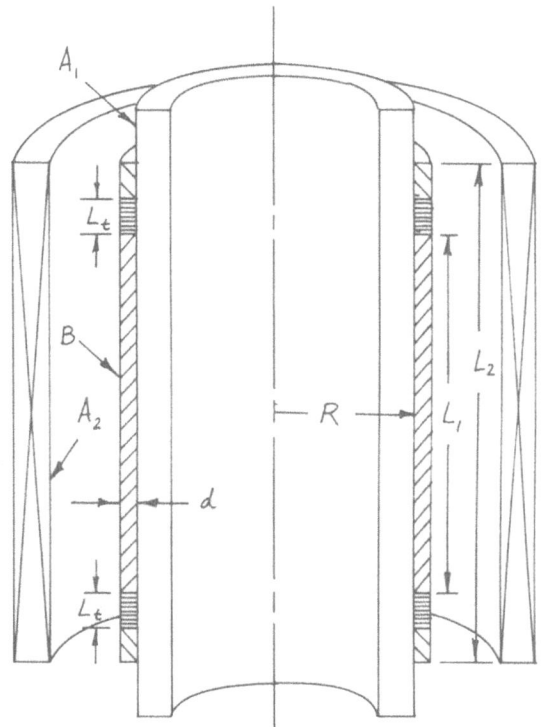

Fig. 2 Sketch of zero-quantum superconducting magnetic shielding
        apparatus. Part B shows the design of the superconducting
        cylindrical shell with a central low-pinning region, high-
        pinning regions at the ends, and transition regions in
        between.   Part A_1 shows the mechanical support structure,
        and part A_2 shows the coaxial solenoid magnet.   The
        dimensions in the radial direction are greatly exaggerated
        for clarity.   The actual dimensions should be chosen
        according to Eqs. (1)-(6) and normally should obey
        $\lambda < d \ll R \ll L_1 < L_2$.

Part A_1 in Fig. 2 is the mechanical support structure, here shown inside the superconducting shell. An alternative arrangement would be to have the superconducting shell inside the support structure. Not shown in Fig. 2 are cylindrically symmetric electrical conductors connected to both ends of the superconducting shell, to which an electrical direct current can be applied. In order that all of the current might flow through the superconducting shell, it is desirable to have the support structure made of a nonmagnetic insulating material, such as fused quartz. Special electronic equipment may be needed to detect quickly any inadvertent superconducting-to-normal transition in the superconducting shell when it carries a current, so that the current could be turned off rapidly and burnout avoided.

Part A_2 in Fig. 2 is a coaxial solenoid magnet, which can apply a magnetic field of either polarity parallel to the superconducting shell and thereby can induce azimuthal currents in the shell wall. Having described the main elements of the superconducting magnetic shielding of the apparatus, I now disucss the various steps required to achieve a zero-quantum state of the central region of length L_1.

COOLDOWN STEP

The superconducting cylindrical shell is cooled down through its transition temperature in the presence of an ambient magnetic field $\vec{H}$, the primary source of which is the earth's magnetic field, whose magnitude is of the order of 1 G. It may be convenient, however, to use the superconducting shell in conjunction with a conventional mu-metal shield, in which case the magnitude of the magnetic field $\vec{H}$ is of the order of 1 mG. The magnitudes of the components of the magnetic field $\vec{H}$ parallel and perpendicular to the shell's longitudinal axis are denoted by $H_\parallel$ and $H_\perp$, respectively. Following cooldown, the magnetic field $\vec{H}$ is trapped in the form of a distribution of both vortices (outwardly directed flux quanta) and antivortices (inwardly directed flux quanta) in the superconducting shell wall, as well as a number of longitudinal magnetic flux quanta (each carrying magnetic flux $\phi_o = 2.07 \times 10^{-7}$ G-cm^2) threading the interior of the central region. During cooldown, the values of $H_\parallel$ averaged over the top and bottom of the central region ($H_\parallel$(top) and $H_\perp$(bot), respectively) should be smaller than the corresponding values of $H_\perp$ ($H_\perp$(top) and $H_\perp$(bot), respectively). In particular, the ratios $H_\parallel$(top)/$H_\perp$(top) and $H_\parallel$(bot)/$H_\perp$(bot) must each obey the inequality

$$H_\parallel/H_\perp < 0.64 \ fL_t/R, \tag{7}$$

where f is the fraction of vortices or antivortices, trapped in

either transition region, that can be conveniently depinned and
moved to the opposite end of the central region during the
solenoid field manipulation step (which will be described later).
A typical value of f is 0.1. The symbol L_t denotes the length of
either transition region, and R denotes the radius of the
superconducting shell. The above inequality, Eq.(7), guarantees
that, after the current-induced annihilation step, there will be a
sufficiently large reservoir of readily movable vortices or
anti-vortices in the transition regions to cancel the trapped
longitudinal magnetic flux quanta during the solenoid field
manipulation step. In order to make both $H_\parallel$(top) and $H_\parallel$(bot)
sufficiently small so that Eq. 7 will be satisfied, it may be
necessary to use two additional coaxial solenoid field coils (not
shown in Fig. 2), arranged to surround the top and bottom of the
central region.

CURRENT-INDUCED ANNIHILATION STEP

An electrical direct current is next applied to the superconducting shell shown in Fig. 2. The magnitude of the applied
current is adjusted to be slightly larger than the critical
depinning current in the central region. On the other hand, this
current should not exceed the critical depinning current of either
the end regions or any appreciable portion of the transition
regions. All magnetic flux quanta trapped in the central region
are driven into circumferential motion around the central region
at speeds typically exceeding 1 cm/s. The depinned flux quanta
will move in circular paths around the perimeter of the central
region, with vortices moving in one direction and antivortices
moving in the opposite direction. During this step nearly all the
vortices will annihilate with antivortices (which are present in
equal number), thereby leaving the central region nearly free of
trapped magnetic flux quanta. After about a 10-second duration,
the current is turned off.

SOLENOID FIELD MANIPULATION STEP

At the end of the current-induced annihilation step, it is
possible that there remain a few vortices and antivortices whose
paths do not intersect with each other and which consequently do
not mutually annihilate. Further reduction of the magnetic field
remaining within the central region is achieved through use of the
coaxial solenoid A_2 which surrounds the superconducting shell B.
At this point, the coaxial solenoid is energized, causing a
longitudinally directed magnetic field to be set up external and
parallel to the cylindrical shell. Circumferentially directed
(azimuthal) currents will be induced in the superconducting cylindrical shell. The magnetic field magnitude is adjusted such that

the induced current in the superconducting shell is slightly
larger than the critical depinning current in the central region.
On the other hand, the induced current should not exceed the
critical depinning current of either the end regions or any
appreciable portion of the transition regions. The longitudinally
directed magnetic field, while leaving largely unaffected the
vortices and anitvortices trapped in end regions and transition
regions, will cause any vortices and antivortices remaining in the
central region to move to oppostite ends of that region. After
about a 60-second duration, the solenoid current is turned off.

In order to assess the need for further iterations of the
above-described procedure, the magnitude and direction of the
longitudinally oriented remnant magnetic field inside the central
region is measured with a magnetic flux detector which is
sensitive down to the level of a fraction of a single flux
quantum, ϕ_0. For example, a SQUID detector, connected to a
superconducting coil capable of being rotated 180 degrees, is
suitable for this measurement. After a single application of the
solenoid field, it is probable that the magnetic field measured
will exceed the theoretical minimum field calculated from Eqs. (1)
and (3) using the assumption that no longitudinal flux quanta are
trapped within the central region. Because the combination of the
current-induced annihilation step and the first application of the
solenoid field removes all vortices and antivortices from the
shell wall in the central region, the magnetic field at the middle
of the central region has a negligibly small component perpendicular to the cylinder axis. The value of the component of the
magnetic flux directed upward along the cylinder axis, however, is
equal to the product of ϕ_0 and the net number of upwardly directed
flux quanta that thread the volume enclosed by the central
region.

The number of flux quanta threading this volume now can be
reduced to zero by moving an appropriate number of vortices or
antivortices out of storage in a transition region at one end of
the central region to the transition region at the opposite end of
the central region. The net number of upwardly directed flux
quanta threading the volume enclosed by the central region
increases by one each time a single vortex moves from the lower to
the upper transition region, or when a single antivortex moves
from the upper to the lower transition region. Such motions of
vortices or antivortices can be caused by applying an upwardly
directed longitudinal solenoid field of such a magnitude that the
induced azimuthal current exceeds the depinning critical current
over parts of the transition regions. Similarly, the net number
of upwardly directed flux quanta threading the volume enclosed by
the central region can be decreased by applying a downwardly
directed longitudinal solenoid field of the appropriate magnitude.

An iteration of the solenoid field manipulation step is now carried out, with the magnitude and direction of the applied solenoid field so chosen that the number of depinned vortices or antivortices, and their direction of motion, will lead to a reduction in the number of flux quanta threading the hole. At the end of this iteration, the magnitude and direction of the remaining magnetic flux is measured again, and the need for an additional iteration is assessed. Such iterations are repeated until the number of longitudinal flux quanta threading the interior of the central region is reduced to zero.

The above steps lead to a zero-quantum state of the central region (length L_1 and radius R) of the superconducting shell. That is, all vortices and antivortices are removed from the superconducting shell wall in this region, and no longitudinal flux quanta thread the space surrounded by the central region. The resulting magnetic field at the middle of the central region is thus given by the larger of the values from Eqns. (1) and (3).

USE OF THE SHIELD AS A MONOPOLE DETECTOR

The proposed shield could be used to provide an extremely low field environment for a magnetic monopole detector placed inside the shield's central region. In addition, the shield itself could be used as a monopole detector using the following procedure. The shield shown in Fig. 2 is then also equipped with pairs of vortex detectors placed on the top and bottom ends of the central low-pinning region. Each vortex-detector pair consists of both a loop surrounding the cylinder to detect the motion of vortices along the length of the cylinder and a contact pair attached at points parallel to the cylinder axis to detect azimuthal motion of vortices. The superconducting cylindrical shell is cooled down through its transition temperature, and the current-induced annihilation and solenoid field manipulation steps are carried out so as to depin and annihilate all vortices and antivortices in the shield's central region. When a current and a parallel magnetic field are simultaneously applied to the shield, no vortex motion occurs and the SQUID vortex detectors give null readings. The current and the parallel applied field are then turned off.

If a Dirac magnetic monopole passes through the central region of a magnetic shield so prepared, it will leave behind a doubly-quantized vortex[6] (or two vortices) at one intersection of the monopole trajectory with the shell wall and a doubly-quantized antivortex (or two antivortices) at the other intersection of the trajectory with the shell wall. The vortices and the antivortices will be pinned near where they were produced.

Periodically a current I exceeding the depinning current of the central region and a parallel magnetic field are simultaneously applied to the superconducting shield. Any monopole-generated vortices and antivortices will follow spiral trajectories around the central region of the superconducting shield and will pass through the SQUID vortex and antivortex detectors at opposite ends of the central region. The double vortex and the double antivortex signatures will be read easily by the SQUID detectors. The initial positions of the vortices and antivortices can be determined from their times of flight. This is possible because the vortices and antivortices, in response to the applied current and field, quickly reach a terminal velocity and travel with constant speed until they pass through the SQUID vortex and antivortex detectors.

ACKNOWLEDGEMENTS

Ames Laboratory is operated for the USDOE by Iowa State University under contract No. W-7405-Eng-82. This work was supported by the Director for Energy Research, Office of Basic Energy Sciences.

I thank S. Bermon, D.K. Finnemore, H. Frisch, C. Heiden, M. Kuchnir, J.E. Ostenson, and C.Tseui for a number of stimulating discussions. I also thank B. Mansfield for his contributions to this paper.

REFERENCES

1. S. Bermon and T. Gheewala, IEEE Trans. Magn. MAG-19, 1160 (1983).
2. Y.B. Kim and M.J. Stephen, in Superconductivity, Vol. 2., R.D. Parks, Ed., (Marcel Dekker, New York, 1969), p. 1107.
3. R. P. Huebener, Magnetic Flux Structures in Superconductors, (Springer-Verlag, Berlin, 1979), p. 65.
4. J.R. Clem, Phys. Lett. 22, 125 (1966).
5. J.R. Clem, IEEE Trans. Magn. MAG-19, 1278 (1983).
6. B. Cabrera, Phys. Rev. Lett. 48, 1378 (1982).

THE STRAIGHT WIRE MONOPOLE DETECTOR

Chris N. Guy

Physics Department
Imperial College
London, England

In the eighteen months that have elapsed since the original report of a candidate monopole event by Cabrera[1] three second generation induction detectors have, between them, accumulated nearly one hundred times the time area product of the original exposure without detecting any further candidate events.[2-4] Clearly there is now an urgent need to run detectors of very much larger area. In this paper we describe a novel coil arrangement that maximizes the collection area without concern for coincidence that can be obtained using cylindrical geometries of large aspect ratio.

Figure 1 shows the geometry that we are discussing here. The pick-up coil is mounted to conform to the path ABCDE inside a long superconducting shield of radius a and length H closed at each end by flat circular superconducting discs. The loop portion AB is on the axis of the shield and portions BC, CD, DE are arranged to be very close to the interior surface of the shield. We show that this arrangement repsonds to the passage of a monopole through the detector along a track such as IO in Fig. 1 as if the return wires CD and BC, DE were absent. This makes the detector axially symmetric and hence uses the full area, $2\pi aH$ for collection. All superconducting induction detectors have to be surrounded by a superconducting shield in order to provide a sufficiently stable magnetic environment, otherwise background field fluctuations far exceed the anticipated signal from a magnetic monopole. Generally this shield makes the induced signal produced by the passage of a monopole dependent on the precise monopole trajectory. Calculations of the expected distribution of signals arising from a random distribution of tracks are complicated; for each possible trajectory the distributed screening currents set up on the shield

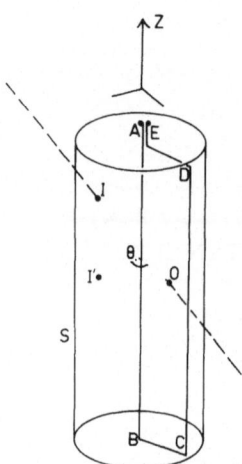

Fig. 1   The   geometry   of   the   straight wire   detector.   S   is   a
         superconducting hollow cylinder closed at the ends.   The
         detection loop is wound along the path ABCDE.   A typical
         monopole track enters at I and leaves at O.   The projec-
         tion of this track in the plane perpendicular to the z
         axis I'O subtends an angle θ at the axis.

must be allowed for, because they too contribute to the signal
induced in the detector loop. For our purpose it is sufficient to
calculate the integrated effect that the monopole has on the
detector; the current induced by a monopole can be calculated by
replacing it by a fine solenoid of turn area N, carrying current
i_s stretched along the monopole track.[5]   To simulate the passage
of a Dirac monopole we arrange $N_{i_s} = 2\phi_0$, the total flux emerging
from such a magnetic charge. This replacement allows us to retain
the familiar form of Maxwell's equations in which there is no
magnetic charge and hence treat the magnetic flux as a conserved
quantity inside superconducting rings or cylinders.   It can be
shown that the current induced in a detector loop such as ABCDE
shown in Fig. 1 when a Dirac monopole traverses the path IO is
given by[6]

$$i_c = \frac{-1}{L_c} \frac{2\phi_0}{\mu_0} \int_I^O \left(B_c(\vec{r}) + B_s(\vec{r}) \right) \cdot d\vec{\ell} \qquad (1)$$

where L_c is the self inductance of the loop. The line integral is
performed along the path IO and the integrand $B_c(r) + B_s(r)$ represents the field that would be produced at a point r on the line IO
if the loop ABCDE were to carry unit current.   The part $B_c(r)$
represents the field arising directly from the loop current in the
absence   of   the   shield   and $B_s(r)$ represents   the   field

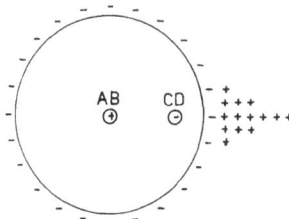

Fig. 2 A schematic cross section of the distribution of screen-
        ing currents, flowing parallel to the z axis, set up in
        response to unit current flowing in ABCDE. The negative
        signs represent screening current flowing in response to
        the current along AB. The positive signs represent the
        screening current flowing in response to the negative
        current flowing along CD.

arising from the distributed screening currents set up on the
interior surface of the shield in order to keep the flux through
the shield constant when unit current is turned on in the detector
loop. For symmetrical geometries $B_s(r)$ can be calculated in a
straightforward manner by numerical means.[6]

The loop geometry shown in Fig. 1 is far from symmetrical and
hence apparently very difficult to handle. Fortunately the limiting case that we are interested in, where the portions BC, CD, DE
are very close to the shield turns out to be remarkably simple.
To see this we consider a very long detector, H >> a, ignoring the
effects of the ends, and consider separately the screening current
distributions set up in response to the current lines AB, CD.
Figure 2 shows schematically the expected distribution of the Z
component of the screening current distributions. It can be seen
that there is a uniform negative part associated with the positive
current in AB and a positive part associated with the negative
current in CD that becomes increasingly localized as CD approaches
the shield. By symmetry the uniform part always contributes no
field on the interior of the cylinder. As the limit is approached, the current line CD together with its image screening
current, form a dipole pair whose associated far field is
vanishingly small. This leaves a single contribution that arises
from the central portion of the loop, AB. Away from the ends of
the detector this is given by

$$\vec{B} = \frac{\mu_0 \vec{\phi}}{2\pi r} .$$

Since the field has only an azimuthal component it is only the
projection of the track in plane perpendicular to Z that is

needed, and Eq. (1) becomes

$$i_c = \frac{1}{L_c} \frac{2\phi_0}{2\pi} \int_{I'}^{0} \frac{\mu_0 \vec{\phi} \cdot \vec{d\ell}}{|\vec{r}|} = \frac{2\phi_0 \theta}{L_c 2\pi} \quad , \qquad (2)$$

where θ is the angle subtended at the detector axis by the projected entry and exit points I', O shown in Fig. 1. All tracks with same angle θ, give the same current signal i_c independently of the azimuthal orientation of the track with respect to the loop plane ABCDE. Thus the detector is axially symmetric and all tracks lying in the plane $II'O$ produce the same signal. Equation 2 does, however, predict a signal variation with θ and it is thus necessary to calculate the expected distribution of signals $i_c(\theta)$ obtained from an isotropic flux, I m^{-2} sr^{-1} s^{-1} intersecting the detector. This is accomplished by considering the number of tracks entering across an infinitesimal patch of area $adzd\phi$ centered on I that leave across a patch $adz_0 d\phi_0$ central around O. Integration over the strips dz_I and dz_0 separated by the angle θ gives[7]

$$dN(\theta)d\theta = I \int_0^H \int_0^H \frac{a^4 \sin^4(\theta/2)dz_I dz_0 d\theta}{[(z_I - z_0)^2 + a^2 \sin^2 \frac{\theta}{2}]^2}$$

hence

$$dN(\theta)d\theta = 4\pi a \, HI \, \sin \frac{\theta}{2} \, \tan^{-1} \left(\frac{H}{2a \sin \theta/2}\right) d\theta \quad ,$$

and for $H \gg a$

$$dN(i_c)di_c \sim \sin\left(\frac{i_c}{k}\right)di_c$$

where $k = 2\phi_0/2\pi L_c$. This shows that the majority of tracks produce signals, i_c, that are large fractions of $2\phi_0/L_c$, the signal expected from an isolated loop. The total number of particles intersecting the detector is obtained by a final integration over θ to give

$$N = I\left(4\pi a^2 H - 2\pi a^2 + O(a^2/H^2)\right)$$

where the second and third terms reflect the neglect of those tracks entering or leaving across the end caps of the detector.

One very important consideration for large detectors is the self inductance of the coil that is connected to the SQUID. In the case of the straight wire, neglecting the end effects,

$$L_c = \frac{\mu_0}{2\pi} \log \left(\frac{a}{\rho}\right) , \quad \text{where } \rho$$

is the radius of the wire used for the loop ABCDE. Again the contribution of the wire running along the inside of the shield is very small. In the Imperial College detector H = 1 m, a = 0.12m. The collection area/sr is 0.36 m^2 and $L_c \sim 1.6$ μH.

The straight wire detector has the twin advantage of large collection area and small self inductance, but the disadvantage of producing a fairly wide distribution of signals from an isotropic flux of monopoles. Fortunately, it appears from the experience of the past eighteen months that superconducting lead shields are extremely stable[2-4] so that we can expect to operate such a detector, accepting those signals with $i_c > 0.2\phi_0/L_c$ without encountering a large number of spurious events produced by flux motion in the shield itself.

ACKNOWLEDGEMENTS

The author thanks all his colleagues at Imperial College for the many discussions that we have had. In particular, he thanks Dr. A.D. Caplin and Prof. J. Cowen.

REFERENCES

1. B. Cabrera, Phys. Rev. Lett. <u>48</u>, 1378 (1982).
2. B. Cabrera, Status report at Monopole '83 Conference (3 axis Stanford Detector), in this volume.
3. H. Frisch, Status report at Monopole '83 Conference (Chicago-Fermilab-Michigan Detector), in this volume.
4. C.C. Chi, Status report at Monopole '83 Conference (IBM detector), in this volume.
5. J.G. Park, Report at Monopole '82 Conference.
6. C.N. Guy, J.G. Park, to be published in J. Phys. D. (1984).
7. J.G. Park, C.N. Guy, in preparation.

NOISE CONSIDERATIONS IN A "ROOM TEMPERATURE" INDUCTION

TECHNIQUE MAGNETIC MONOPOLE DETECTOR

Michael J. Price

E.F. Division
CERN
Geneva, Switzerland

The prediction of magnetic monopoles in Grand Unified Theories and the candidate monopole found by Cabrera[1] have given an impetus to recent searches for these particles. Arguments involving the draining of the galactic magnetic field[2] lead to the conclusion that the flux of magnetic monopoles should be low and that, consequently, large area detectors (100's of square meters) would be needed in order to obtain a reasonable event rate. Although detectors employing conventional ionization and scintillation techniques offer reasonably inexpensive means of achieving the required surface area (and are being used and proposed for this purpose) they suffer from the disadvantage of being unable to detect very low velocity monopoles ($\beta < 6 \times 10^{-4}$) - a velocity range in which it might be expected to find a comparatively large monopole flux. By contrast, induction techniques are, in principle, sensitive to all velocities and, in addition, offer a means of measuring the monopole's magnetic charge, a quantity of fundamental theoretical importance.

Cabrera's superconducting technique presents a very elegant method of detecting monopoles and variants of it are being constructed to scale up the dimensions from 20 cm^2 to several square meters. The next step up, to detectors having areas of several hundreds of square meters, presents rather formidable logistic and financial problems.

The possibility of using coils at room temperature to detect magnetic monopoles has been considered by Rubbia[3] and by the present author[4]. In both these reports the predominant role of amplifier noise and its interaction with the input impedance was underestimated. In this note the two sources of noise - coil and amplifier - are analyzed and it is shown that it should be possible to construct a system for inductively detecting magnetic monopoles without having recourse to superconducting techniques.

Dirac has predicted the magnetic monopole charge to be

$$g = nhc/2e \tag{1}$$

where n is an integer. The instantaneous e.m.f. generated in an M turn coil during the passage of a monopole will be, by Faraday's law, equal to the product of the rate of change of flux through the coil and the number of turns. The time integral of the induced e.m.f. will be

$$4.12 \times 10^{-15} \text{ Mn volt·sec} = V_m Mn.$$

Above a frequency of a few kHz the predominant source of noise will be Johnson (or thermal) noise where the r.m.s. noise voltage is given by

$$V = (4kTR(f_2 - f_1))^{1/2} \tag{2}$$

(k is Boltzmann's constant, $1.38 \times 10^{-23} \text{ JK}^{-1}$, T is the absolute temperature, R the coil resistance and $f_1 - f_2$ the bandwidth). If the coil is toroidal with torus radius a and cross-sectional radius pa then

$$R = 2\rho M^2 / p^2 a \tag{3}$$

showing that the r.m.s. noise voltage, like the signal, is proportional to M so that the signal to noise ratio (for the coil) is independent of M.

To extract the signal from the coil it is convenient to use an operational amplifier as shown in Fig. 1. Since such devices

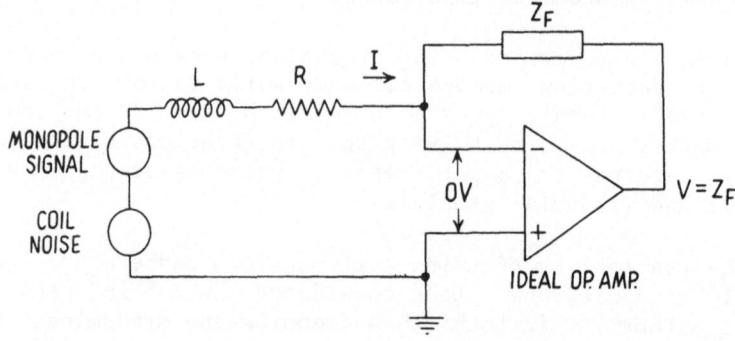

Fig. 1.   Signal extraction using an ideal operational amplifier.

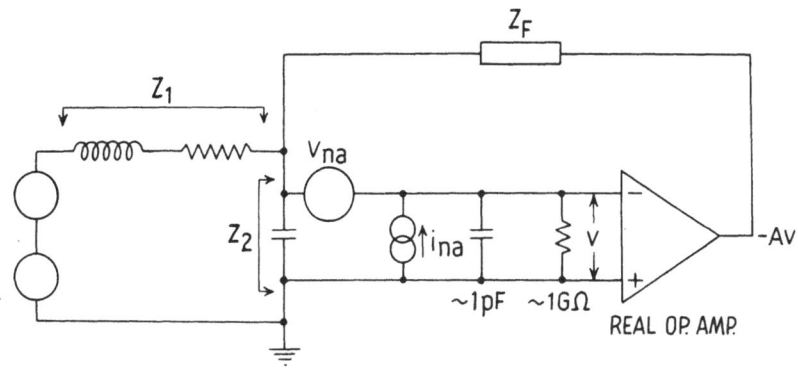

Fig. 2.   Signal extraction using a real operational amplifier.

have very large gain, the voltage across the input terminals is very small so that the coil, in a first approximation, is short circuited. If, in addition, the operational amplifier has high input impedance (as is the case for operational amplifiers having FETs in their first stages) all the current induced in the coil will pass through the feedback impedance. In practice the operational amplifier also has associated noise sources and a certain unavoidable input capacitance; the coil will also have a winding capacitance, however, by suitable winding techniques, this can be made small. The equivalent circuit of the detector and amplifier is as shown in Fig. 2, where it can be seen that the amplifier has both voltage and current noise sources; in the present application the effect of the latter can be shown to be negligible. The input resistance, typically of the order of 10^{12} Ω for a FET front stage, can also be neglected.

The frequency components of the amplified signal and noise can be shown to be

$$V_s(\omega) = -v_s(\omega) \, Z_F/Z_1 \tag{4}$$

$$V_{nc}(\omega) = -v_{nc}(\omega) \, Z_F/Z_1 \tag{5}$$

$$V_{na}(\omega) = v_{na}(\omega)(1 + Z_F/Z_1 + Z_F/Z_2) \quad . \tag{6}$$

If L is the inductance of the coil and R its resistance then $Z_1(\omega) = j\omega L + R$; and $Z_2 = 1/j\omega C$ where C is the capacitance across the operational amplifier input terminals. Z_F is the feedback impedance, as yet undefined. Clearly, the net signal at the output will be the Fourier sum of the three quantities defined in Eqs. (4), (5) and (6).

To reduce the amplifier noise contribution to a minimum, Z_2 should obviously be maximized, however, optimum values of the other components clearly exist since, were the coil inductance (which is proportional to M^2) to be reduced the signal would become smaller. If the number of coil turns were increased, Z_2 would be comparatively small and the amplifier noise would predominate. In addition Z_F also has an influence on the ultimate signal to noise ratio obtained. A good choice for Z_F is to use a purely resistive element so that both the signal and the noise are integrated (note that the familiar op. amp. integrator with a resistance as the input impedance and a capacitance as the feedback impedance is mathematically equivalent to a system with an inductor and resistor as input and feedback elements). In practice, a resistive element would itself contribute noise. This inconvenience can be avoided by using, for example, a feedback element with impedance mZ_1 (where m is a real constant) and connecting the output from this amplifier to an integrator.

To evaluate the signal and noise contributions at the output, the frequency contributions in Eqs. (4), (5) and (6) must be Fourier summed and integrated with respect to time. Since we know that the time integral of the e.m.f. in the coil induced by a through-going monopole is $V_m M_n$ (Vs) the final peak output signal will be

$$mMnV_m/R_1C_1$$

(R_1 and C_1 are the values of the input and feedback components of the integrator). This signal will decay exponentially with time constant L/R. To find the r.m.s. values of the two noise contributions the instantaneous white noise voltages can be considered as Fourier series of the form

$$v_n = \frac{e_n}{\sqrt{\pi}} \, \delta\omega^{1/2} \sum_{\omega_1}^{\omega_2} \cos(\omega_i t + \phi_i) \quad .$$

In this expression the cosine term contains the random phase term ϕ_i. The summation from ω_1 to ω_2 is made in steps of $\delta\omega$ and e_n has dimensions $VHz^{-1/2}$. The following expressions are obtained for the r.m.s. output voltages due to the amplifier and coil noise contributions

$$Vn_a = \frac{e_{na}}{(2\pi)^{1/2} R_1 C_1} \left[(1+m)^2 \left(\frac{1}{\omega_1} - \frac{1}{\omega_2}\right) - 2m(m+1)\, LC(\omega_2 - \omega_1) \right.$$

$$\left. + \frac{m^2(LC)^2}{3} (\omega_2^{\ 3} - \omega_1^{\ 3}) \right]^{1/2} \tag{7}$$

$$Vnc = \frac{e_{nc}mM}{(2\pi)^{1/2} R_1 C_1} (\frac{1}{\omega_1} - \frac{1}{\omega_2})^{1/2} \tag{8}$$

where $e_{nc} = (4kTR_0)^{1/2}$ $VHz^{-1/2}$.

The inductance of a coil can be approximated by $L = aM^2L_0$ where L_0 has dimensions Hm^{-1} and is a slowly varying function of the coil shape. For a reasonably large value of p; $L_0 \simeq 10^{-6}$ Hm^{-1}. Since the coil and amplifier noise voltages are uncorrelated they will add quadratically so that the signal to noise ratio (in the limit of large m) can be expressed as

$$\frac{S}{N} = [(2\pi)^{1/2}nV_m)] \times [\frac{e_{na}^2}{M^2}(\frac{1}{\omega_1} - \frac{1}{\omega_2})+(-2e_{na}^2(aL_0C)(\omega_2 - \omega_1)+$$

$$e_{nc}^2 \times (\frac{1}{\omega_1} - \frac{1}{\omega_2}))+e_{na}^2 \times M^2\frac{(aL_0C)^2}{3}(\omega_2^3 - \omega_1^3)]^{-1/2} . \tag{9}$$

This expression can be differentiated to obtain a value of M which maximizes S/N

$$M^4 = (\frac{1}{\omega_1} - \frac{1}{\omega_2}) \frac{3}{(aL_0C)^2(\omega_2^3 - \omega_1^3)} \tag{10}$$

The value for M^2 can now be substituted back into Eq. (9). If the lower radian frequency ω_1 is defined in terms of ω_2 by $\omega_1 = \omega_2/\alpha$ the signal to noise ratio can be written as a function of ω_2. A value of ω_2 can thus be found which maximizes S/N. From the previous definitions of e_{nc} the following expressions can finally be derived for ω_2 and the signal to noise ratio.

$$\omega_2 = \frac{2}{e_{na}ap} [\frac{kT\rho}{L_0C}]^{1/2} g_1^{-1/2} (\alpha)$$

$$\frac{S}{N} = \frac{\pi^{1/2} n V_m p^{1/2} g_2^{-1/4} (\alpha)}{2(e_{na})^{1/2} (kT\rho)^{1/4} (L_0C)^{1/4}}$$

where

$$g_1(\alpha) = 3^{-1/2}(1 - \frac{1}{\alpha^3})^{1/2} (\alpha - 1)^{-1/2} - 1/\alpha$$

and

$$g_2(\alpha) = (\alpha - 1)^2 g_1(\alpha).$$

Note that the behavior of S/N as a function of the various parameters is much as one might expect; e.g. low temperature and resistivity increase S/N and a low input capacitance and low amplifier noise are desirable. Note also that the maximum attainable signal to noise ratio does not depend on the coil radius, however, both ω_2 and M do vary with this dimension. Taking the example of a Dirac monopole (i.e. n = 1) with p = 0.2, a = 0.1 m, T = 300 K, $\rho = 1.6 \times 10^{-8}$ Ωm, $L_0 = 10^{-6}$ Hm, $C = 10^{-12}$ F, $\alpha = 10$ and $e_{na} = 10^{-9}$ VHz$^{-1/2}$ (corresponding to a very low noise, but technically feasible, operational amplifier) we find

$$\omega_2 = 2.7 \text{ Mhz (i.e. } f_2 = 430 \text{ kHz)}$$

$$M = 2700$$

$$S/N = 11.$$

A signal to noise ratio of 11 is, in fact, very comfortable, for the above conditions, since it means that noise will only simulate the signal once in 10^{15} years! It should also be noted that the figure for signal to noise ratio represents a lower limit – the signal takes the form of a slowly decaying exponential whereas the calculation of the noise voltages takes into consideration all possible output forms of these voltages (e.g. short pulses which do not have the same time behavior as the signal).

Since the proposed detector works at very high frequencies the problems of shielding are greatly alleviated. Aluminum sheet shielding of the order of 1 cm thickness should be adequate to protect the device from external disturbances.

REFERENCES

1. B. Cabrera, Phys. Rev. Lett. 48 1378 (1982).
2. M.S. Turner, E.N. Parker and T.J. Bogdan, Phys. Rev. D26 1296 (1982).
3. C. Rubbia, E.P. Internal Report, 82-01, CERN, Geneva.
4. M.J. Price, E.F. Internal Report, 83-02, CERN, Geneva.

SQUID FLUXMETER SEARCH FOR MONOPOLES IN 400 kg OF OLD IRON ORE

Tadashi Watanabe and Takeo Ebisu

Department of Physics, Faculty of Science
Kobe University
Nada-ku, Kobe 657, Japan

ABSTRACT

With a superconducting fluxmeter a search was performed for magnetically trapped monopoles in ~ 400 kg of old iron ore, by heating the sample above its Curie temperature. An upper limit of 2.3×10^{-6} monopoles/g is obtained for the density of magnetic poles heavier than 6×10^{14} GeV/c^2 in the iron ore of ~ 10^8 years old. We can estimate the flux of cosmic monopoles as less than 6.8×10^{-13} cm^{-2} s^{-1} sr^{-1} according to the age and the density of the sample examined.

SUPERCONDUCTING DETECTOR

The structure of our detector is depicted in Fig. 1. A three-turn niobium wire is wound on an 8 cm diameter glass tube.[1] The search coil, connected to a SQUID, is inside an 11 cm diameter cylindrical superconducting shield made of lead foil and closed at the top and bottom with inlet pipes. The ratio of the search ring to shield cross-sectional areas slightly exceeds 0.5, in which case the flux response is largely modified through the interaction between the coil and the shield.

A one-turn coil and seven trajectories of monopoles are shown as an illustration in Fig. 2a. The flux responses (Fig. 2b) are known to be A through D and a through c when non-shielded. If the coil is shielded by placing a cylindrical superconducting shield co-axially around it, then for vertical trajectories such as A, D and a, the coupling of the coil to the shield is the strongest. Induced magnetic flux would be shared between the coil and shield,

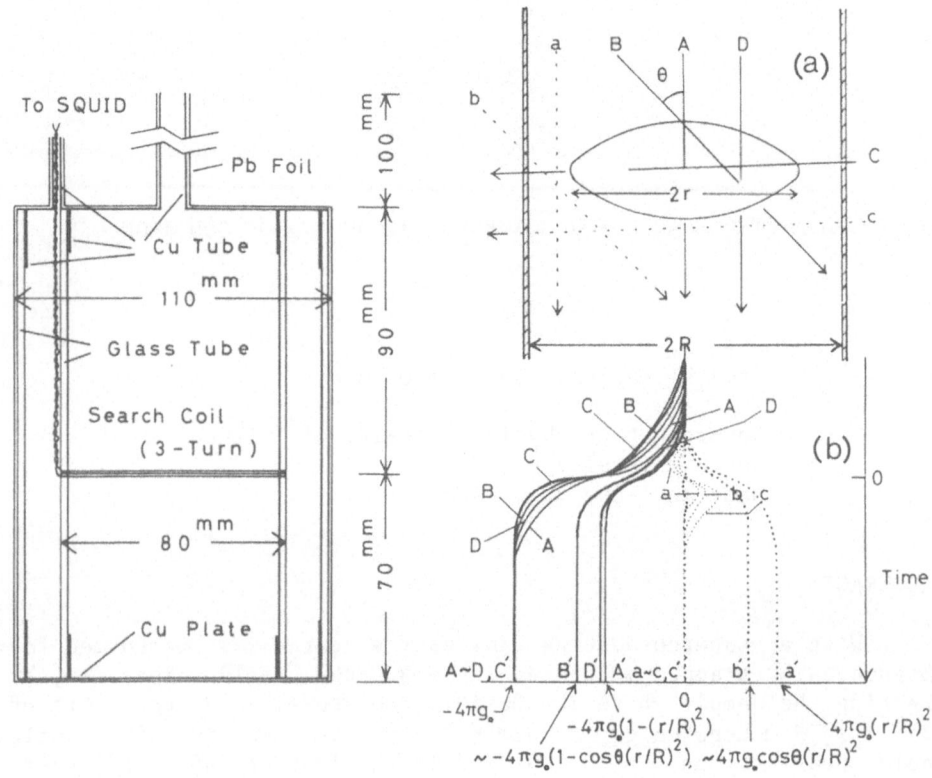

Fig. 1 (Left) Schematic drawing of the superconducting search
coil and shield case.

Fig. 2 (Right) (a) A one-turn coil shielded with a cylindrical
superconducting case with seven typical trajectories of
monopoles. (b) Flux responses A-D and a-c when non-
shielded, and A'-D' and a'-c' when shielded.

so that the current due to the penetrating pole is diminished,
e.g. from A to A'. However, the current due to a non-penetrating
pole grows to a finite and observable magnitude from a to a'. On
the other hand, the coupling is the weakest for horizontal
trajectories so that flux responses C and c are barely modified to
C' and c', respectively. By making use of the shield effect, we
can increase the effective area of detection.

We have calibrated the sensitivity of our detector by
generating a magnetic flux 4.77×10^{-6} G $\times$ cm^2 with the end of a
long solenoid at the bottom of the shield case. Results are given
in Fig. 3. The most left and right points show that persistent

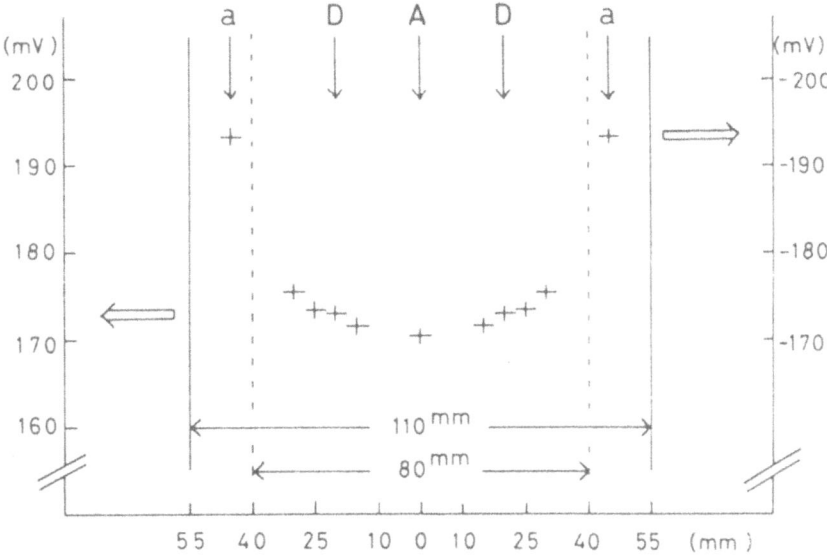

Fig. 3 Results of calibration by a solenoid with 4.77×10^{-6} $G \times cm^2$.

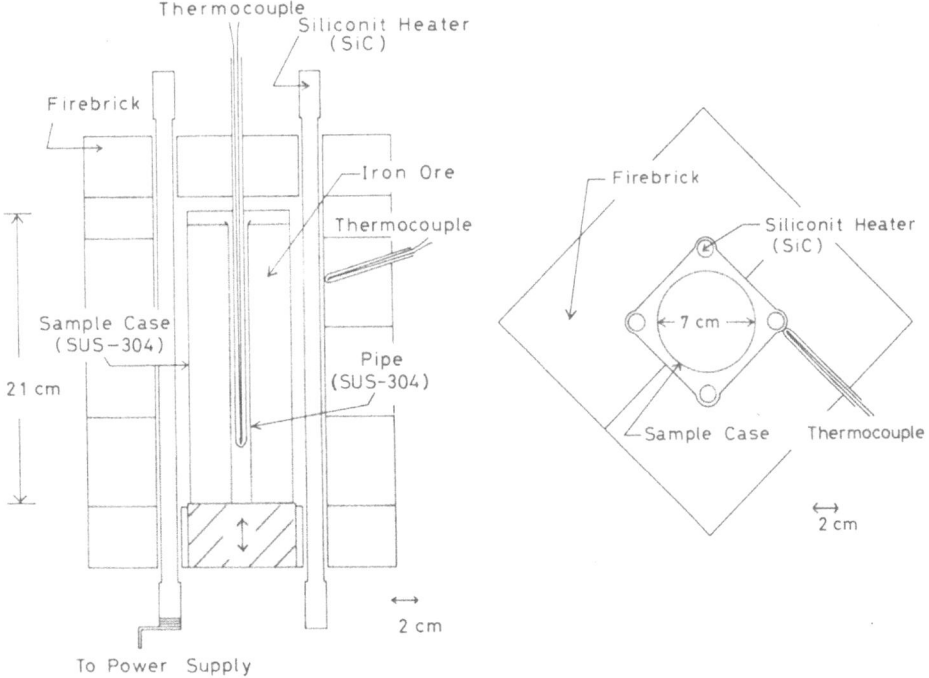

Fig. 4 Cross sectional view of the furnace.

currents are caused with negative signs in the search coil by near-miss monopoles. On the basis of these results, an output of 14.9 mV is obtained for the trajectory A. This value is consistent with the one calculated. Hence, we have the S/N ratio of our fluxmeter between 7:1 and 8:1 for the axial passage of a single Dirac charge. For the magnetic pole of trajectory C, the expected output signal is about 30 mV.

SEARCH IN OLD IRON ORE

In our monopole search experiment, iron ore is heated above the Curie temperature in an electric furnace (Fig. 4) placed 180 cm above the coil, and kept there for more than 20 minutes. Monopoles released from the demagnetized samples will reach the detector with at most 5 degrees of deflection angle. Magnetic

Table 1. Properties of the measured samples.

	Magnetic Sand	Maghemite	Grate-Bars
Molecular Formula	Fe_3O_4	$\gamma-Fe_2O_3$	(Fe-24% Cr)
Curie Point Treatment at	585 °C > 875 °C	675 °C > 925 °C	Fe:770 °C > 1150 °C
Saturation Magnetization	471 G	417 G	1735 G
Age (years)	$(6\sim25) \times 10^6$	$(105\pm10) \times 10^6$	
Place of Collection	Waipipi point (New Zealand)	Santa Barbara (Chile)	(Kobe Steel Ltd.)
Total Weight (kg)	214.5	213.9	46 kg($\approx$1410 tons of iron ore)
Density (g/cm^3)	5.08	4.9	

poles with mass greater than 6×10^{14} GeV/c^2 can pass through the detector's effective area (11 cm diameter).

Magnetite and maghemite were examined (Table 1), both are ferromagnetic. The total weight of two samples is 428.4 kg. The 3rd sample is called grate-bar, which has significantly different properties from the first two samples. Our grate-bars are worn-out after 2 years of use at a sintering plant in the steel mills of Kobe Steel Ltd. in Japan. If the retrapping of released monopoles were possible, in spite of the gravitational force ~ 1 GeV/m, 46 kg of grate-bars would contain as many monopoles as 1400 tons of iron ore.

The total running time is 1010 hours which includes 270 hrs for the heat treatment of the sample. Not only trapped monopoles but also free inflight monopoles have been searched for.

EXPERIMENTAL RESULTS AND ANALYSES

We have obtained 20 events exceeding $0.2\phi_0$ during the whole running time with and without samples (Fig. 5). The event recorded with the largest magnitude (7.2 mV) corresponds to $1.4\phi_0$. The maximum flux change of our 3-turn coil is $6\phi_0$. The event is shown in the strip chart (Fig.6) with the stable level offset from the base line. The event was recorded on July 28, 1983 when a grate-bar was being heated.

Due to the small deflection angle of a released monopole, as explained before, we have only to wait for events within the hatched region in Fig. 5. However, no event found corresponds to a trajectory within the region. In conclusion, we have not found any magnetic poles in 428.4 kg of our sample. An upper limit of 2.3×10^{-6} monopoles/g is obtained for the concentration in the iron ore. This is approximately 2 orders of magnitude less than the limit obtained by Ross et al.[2] in lunar materials 10 years ago.

Furthermore, it is possible to estimate the flux of cosmic monopoles by applying the age times area law. Taking 100 million years as the age and about 5g/cm^3 as the density of our sample, we obtain a flux less than 6.8×10^{-13} cm^2 s^{-1} sr^{-1}.

Can we assign the 20 events to cosmic monopoles incident on our detector? There are 5 events (Fig. 5), for example, which belong to the group of the greatest magnitude. If they were caused by true monopoles, they had to "near-miss" the search coil in certain directions. Due to the ratio of the ring to shield cross-sectional areas, it is almost the same probability to have penetrating and near-miss poles, respectively. But we have no such events penetrating the coil region. Therefore we conclude that the

20 events are not due to cosmic monopoles but originate in the detector itself. With the effective area, we can derive an upper limit of free monopoles incident on the earth as 4.6×10^{-10} cm^{-2} s^{-1} sr^{-1}, independent of the monopole mass.

FUTURE PLANS

We have two plans for future experiments. One is to search for monopoles in very old iron ore, say, of 10^9 years old by the heat treatment method. Our second plan is to search for free monopoles with a larger scale coil and superconducting cylinder, diameters of 14 and 20 cm respectively. By these improvements we could achieve good statistics for the distribution of events plotted versus the magnetic flux change in the search coil.

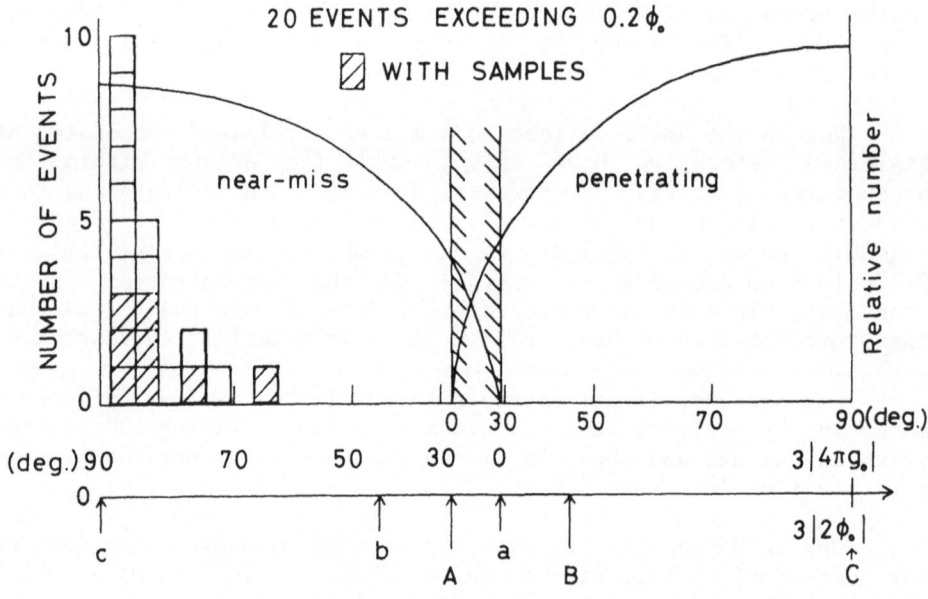

Fig. 5 Experimental results. Two curves represent the relative number of monopoles versus incident polar angles of trajectories. The abscissa is the magnitude of induced flux in a three-turn coil. Events due to released monopoles are expected in the hatched overlapping region.

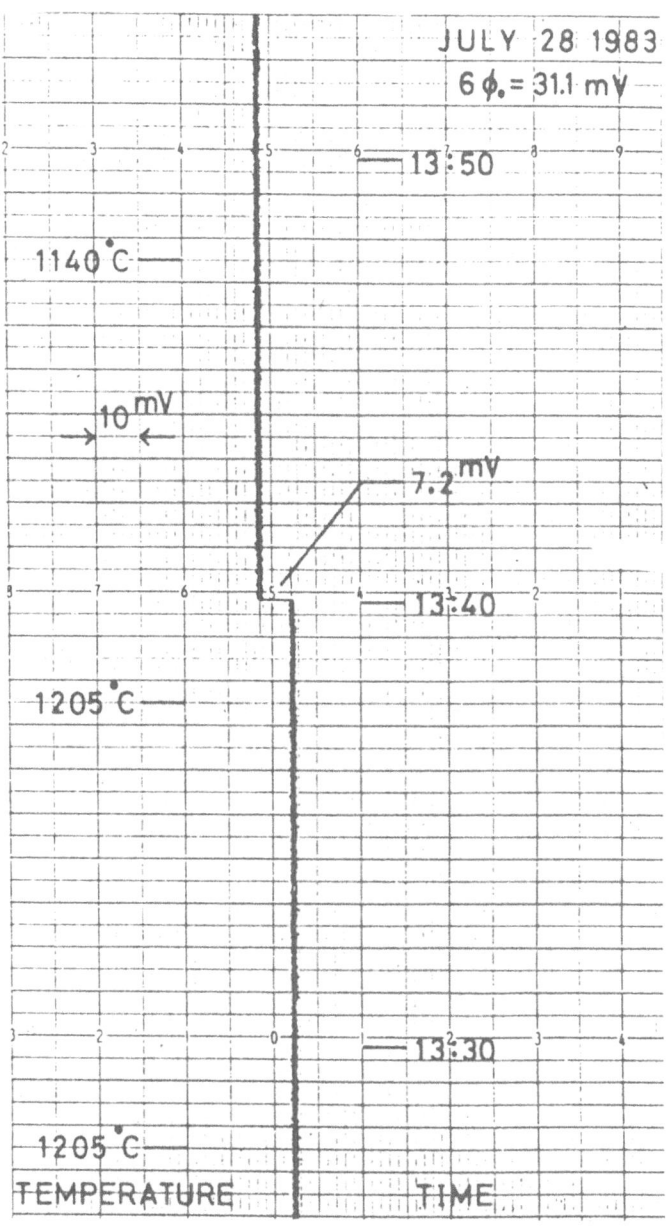

Fig. 6 A part of the strip chart showing the event recorded with
        the greatest magnitude.

REFERENCES

1. T. Ebisu and T. Watanabe, J. Phys. Soc. Japan 52, 2617 (1983).
2. R.R. Ross, et al. Phys. Rev. D8, 698 (1973).

A DETECTOR FOR RELIC MONOPOLES TRAPPED IN MAGNETIC ORES

Robert March, David Cline and David Joutras

Department of Physics
University of Wisconsin
Madison, Wisconsin 53706

ABSTRACT

We propose the construction of a four-coil superconducting SQUID magnetometer to search for relic monopoles trapped in sedimentary iron ores of age 2×10^9 years. Such monopoles may be released when the ore is heated through its Curie point as part of the industrial processing of the ores, permitting industrial-scale batches to be sampled. Engineering design of the device has been completed.

MOTIVATION AND EXPERIMENTAL PROCEDURES

Efforts to detect a flux of cosmic monopoles in ionization detectors of substantial area have not been successful to date, although interpretation of these results is clouded by the difficulty of estimating the ionization produced by a slow-moving monopole. Aside from one possibly spurious event observed by Cabrera[1], the same can be said for magnetic induction detectors. These results are consistent with various upper limits set by astrophysical constraints, which fall in the range 10^{-1} to 10^{-5} per square meter per year.

It is unlikely that sensitive superconducting magnetometers with areas on the order of square meters can be built at reasonable cost. Thus the detection of low fluxes of slow-moving cosmic monopoles may not be possible with existing techniques. Because of the importance of these objects as one of the few experimental handles for Grand Unified Field Theories (GUTs), it may be worthwhile to explore other means of establishing their existence. One promising route is to search for monopoles bound in solid matter.

The strongest binding force experienced by a monopole in matter would occur at a domain boundary in a ferromagnet, where it would experience an attraction to its image pole analogous to that experienced by an electric charge at the surface of a conductor. A detailed calculation by Kittel and Manoliu[2] predicts a binding energy of about 35 eV in magnetite. For a 10^{17} GeV monopole mass, the gravitational potential at the Earth's surface is about one eV per angstrom. Hence such monopoles could be immobilized in the Earth's gravitational field by this binding mechanism.

When the ferromagnetic material is heated to its Curie point, the monopole is no longer bound, and the strength of its inter-action with the lattice binding forces is comparable to the gravi-tational potential drop across a lattice spacing. A monopole at the high end of the GUTs mass range will essentially fall freely. At the low end, it will diffuse rapidly out of the lattice.

The north-central US contains extensive deposits of sedimen-tary magnetite in beds of age roughly two billion years. Economic exploitation of these ores requires that they be formed into pellets and heated to around 1400 C, well above the Curie point, before shipment. The ore passes through a compact kiln on a conveyor, so that the point at which it reaches the Curie temperature is well-defined. Annual production at a single kiln ranges from one million to ten million tons. We have located two plants in which there is free space suitable for a modest-size apparatus directly below the kiln.

One of these is an Inland Steel Corporation plant at Black River Falls, Wisconsin, a two hour drive from the Madison campus. Its annual throughput is about a million tons. We have made arrangements with the mine management to locate our apparatus there. The other is a US Steel Corporation facility at Mount Iron, Minnesota, on the north shore of Lake Superior. We have contacted the management, but no definite arrangements have been made. This plant has a ten million ton throughput. Figure 1 illustrates how our apparatus would be deployed at the Black River Falls facility.

At present, Black River Falls is not in service, due to reduced demand for steel. We have no reliable estimate of when it will resume operations, so we are pursuing negotiations with US Steel.

Our proposed detector would have a sensitive aperture of about 80 cm^2, sufficient to effectively sample roughly 4% of the Inland throughput, and somewhat less than 1% in the US Steel facility.

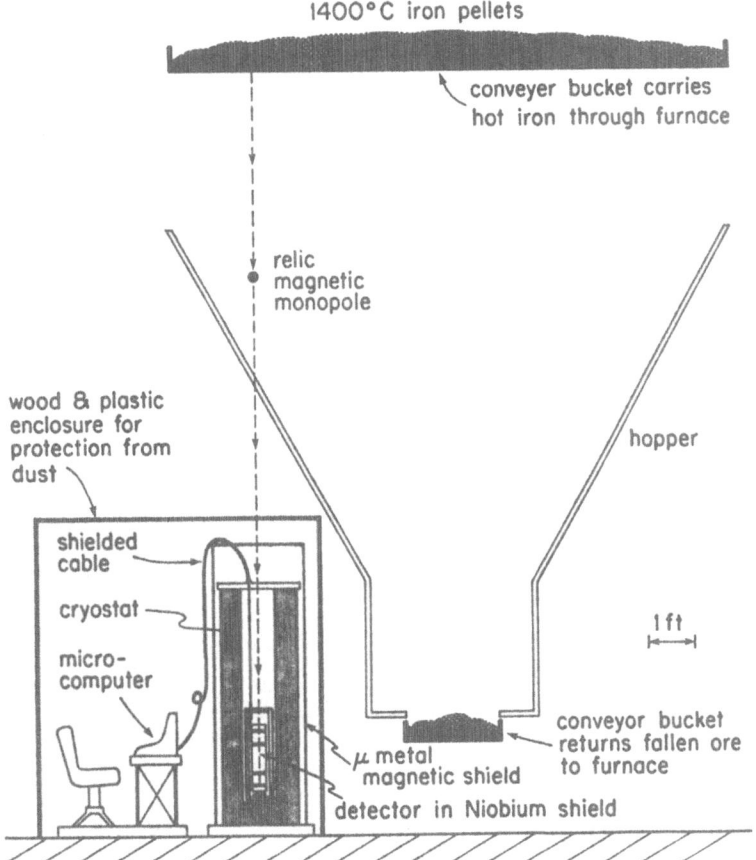

Fig. 1. Proposed monopole detector as it would be deployed at the
Black River Falls facility.

It is difficult to estimate the probability that a monopole
will actually be captured when it passes through a magnetic
mineral. But given an integration time on the order of billions
of years, the probability need not be high to produce a substantial amplification of the cosmic flux, especially when
industrial-scale batches of the mineral can be sampled.

In 6 months of data-taking at Black River Falls, we would
have an effective sample of about 20,000 tons, or 5000 m^3, for a
total of 10^{13} m^3-years. Thus if the capture probability is in
excess of 10^{-13} per meter of monopole path length, our detector
would be more sensitive than a 1 m^2 cosmic monopole detector.

DESIGN OF THE APPARATUS

The conceptual design of the apparatus was developed by the authors of this proposal, with the assistance of the Cryogenic Engineering Research group of the University of Wisconsin College of Engineering. The detailed mechanical design and the design of the readout system were carried out by Quantum Design, Inc. on a grant provided by the Wisconsin Alumni Research Foundation.

The apparatus consists of four single-loop superconducting coils 10 cm in diameter, aligned on a common axis in a single cryostat.

The middle two coils are read out as a gradiometer; they are counterwound in a single readout circuit. Thus their response to the passage of a monopole would be a pulse of duration determined by the velocity of the monopole. The two outer coils are read separately as single-loop magnetometers, and would experience a step in flux in response to a monopole. Thus a monopole falling through all four loops would have a highly redundant signature; all three readout systems must give a signal of consistent polarity, with the time lag from top to bottom consistent with the duration of the gradiometer pulse and with the expected velocity of a monopole in free fall from the kiln. It seems unlikely that any of the usual sources of spurious signals in a device of this type (e.g. "flux jumping" in the magnetic shield) could mimic this signature.

The readout devices will be commercial SQUID units, coupled to the coils through wideband resonant circuits, with a bandwidth of 0 to 100 Hz. The monopole signal amplitude at the SQUID would be 0.17 ϕ_0, the unit of quantized flux. Typical noise values of such systems are less than 10^{-4} ϕ_0 per Hz bandwidth, so the device will have a signal to noise ratio of about 20/1.

The readout devices will feed an 8-channel waveform digitizer, which will record not only the signal itself but an interval of roughly 10 seconds surrounding the event. The other 5 channels will digitize environmental sensors (accelerometers, acoustic sensors, electromagnetic sensors) to give evidence for any unusual external factor coincident with the event. The data will be logged to floppy disk via a DEC LSI 11/23 computer.

REFERENCES

1. B. Cabrera, Phys. Rev. Lett. 48, 1378 (1982).
2. C. Kittel and A. Manoliu, Phys. Rev. B15, 333 (1977).

SUMMARY OF INDUCTION DETECTORS AND TECHNIQUES

Henry J. Frisch

Enrico Fermi Institute and Physics Department
University of Chicago
Chicago, Illinois

ABSTRACT

I summarize the present experimental limits from induction searches for both free and trapped monopoles. Other topics which were discussed in the induction topical sessions at MONOPOLE '83 are the sources of spurious steps in SQUID outputs, the feasibility of room temperature induction detectors, and superconducting shielding. A discussion of the possibility of large area superconducting induction detectors ends the summary.

INTRODUCTION

Faraday induction[1] is the theoretically simplest method of searching for monopoles. It is remarkable in that it relies only on a single Maxwell equation for the directly detected effect, and in principle needs no assumptions from more modern physics.[2] In contrast, measurements involving the ionization or excitation of matter are quite complex in principle, and depend on acquired properties of the monopole such as its velocity, and whether or not in its most recent past it has picked up a companion nucleus.[3] To an experimentalist, it would seem clear that the method of choice for searching for a monopole would be by induction, the technique which is sensitive to the only property of a magnetic monopole which is assured, its magnetic charge.

The principle of the technique dates from 1831; the first[4] detection of a magnetic pole (though not a monopole) traversing a loop is shown in Fig. 1.

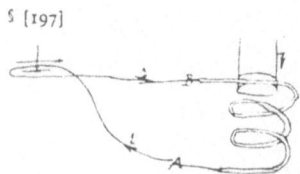

Fig. 1   The introduction of a magnetic pole into a loop, complete
         with readout. This picture is from Michael Faraday's Diary
         Dec. 8, 1831. (See References 1 and 4).

     The experimental problem of detecting magnetic monopoles,
however, is one of scale.   A monopole is a very prominent object
on an atomic scale -- at a typical atomic radius from the pole
there exists a magnetic field on the order of 10^8 gauss.
Detectors which depend on these short-range effects of the
monopole as it passes through and near atoms, produce signals
which are independent of the size of the detector, and whose
properties (especially the signal/background ratio) are determined
on the atomic rather than the macroscopic scale.

     Induction detectors, on the other hand, are macroscopic
detectors; their signals depend on the area of the whole loop,
which is on the order of 1 meter in diameter rather than the
10^{-10} m of atomic dimensions. The signal detected due to the
passage of a Dirac monopole (flux = $4\pi g$ = 4.14 × 10^{-7} gauss-cm^2)
is the current induced in the loop due to the monopole flux

     δI = $4\pi g/L$ = 4.14 × $10^{-15}/L$ amperes

where L is the inductance of the loop in mks units (Henrys).

     The inductance of a simple planar loop of circular cross
section wire is given by[5]

     L = $0.004\pi D$ $\left[\ln(8D/d) - 1.75\right]$ μH

where D is the diameter of the loop in cm, d is the diameter of
the wire in cm, and L is in μH.   For example, a one-meter loop of
250 μ diameter wire has an inductance of about 10 μH, and a Dirac
monopole traversing such a loop would induce a current in the loop
of ~4 × 10^{-10} amperes.

     The change in flux due to the monopole, however, corresponds
to a very small change in field compared to the scale of
macroscopic fields.   For a one meter diameter loop, a change of a
more-or-less uniform field of ~10^{-11} gauss would mimic the signal.
This, for example, corresponds to the field from one ampere in a
long straight wire a distance of 100,000 miles away.[6] More to the

point, changes in the earth's field and typical electromagnetic disturbances in a laboratory are orders of magnitude larger than this. The major technical problem involving induction detectors, then, is not so much in sensitivity to signal as in shielding the loops from background signals.

The present experimental situation divides itself into two cases for which the background problems are quite different: superconducting shielding, and non-superconducting (i.e. transient eddy current) shielding. The superconducting case has been the subject of intense work in the past year. Largely motivated by the step seen by Cabrera,[7] several groups are now building bigger detectors. It is now clear, I feel, that very large superconducting induction detectors can be built: however, much still has to be learned in the way of engineering detail. With respect to the other case, to my knowledge no one has yet succeeded in building even a very small non-superconducting detector.

The strengths of the superconducting induction technique are threefold:

(a) The monopole signal is a step, not a pulse. The signal then is DC and one can work at low bandwidths to eliminate noise.

(b) One can use superconducting shielding to exclude external field lines. Superconducting shielding also traps and holds rigid external field lines to a very high degree. Again, these are DC properties.

(c) One has available highly sensitive amplifiers in the form of SQUID's. These are commercially available[8] and easy to use. DC SQUIDs can be several orders of magnitude better than the RF SQUIDs.

This paper summarizes and compares the induction searches for free heavy monopoles; discusses the searches for monopoles trapped in matter, including some history; and reports other topics which were presented at MONOPOLE '83, such as calculations of signal-to-noise for room temperature detectors, calcuations on shielding, and new low field shield designs. Also, I give a personal prognosis for building bigger detectors.

SEARCHES FOR FREE MONOPOLES

The status of monopole searches as of the Racine Conference[9] of October 1982 is shown in Table 1. At that time several groups had small prototypes, but only the Stanford group of Cabrera was actually running a sophisticated detector.[10]

Table 1.   The Status of Monopole Searches as of the Racine
           Conference of October 1982.

Group	Coincidence?	Independent Shields	Loop Diameter	Comment
Stanford	Yes	No	4 in.	Running (3 orthogonal loops)
Wisconsin	Yes	Yes	4 in.	--
Texas A & M	No	--	6 in.	Frozen in Argon
IBM	Yes	No	4 in.	Ordered new dewar for 7 in. loop
Imperial College	No	--	undecided	Ordered dewar (1/83)
Chicago/Fermilab/Michigan	Yes	??	24 in.	--
Virginia	undecided	undecided	undecided	undecided
Minnesota	undecided	undecided	undecided	undecided
LBL	not yet		6 in.	Not S. C.; Uses LN temp FET's, eddy current shielding.

A year later, at our conference here in Ann Arbor, results
have been reported by four groups. Detectors with loops over
thirty times in area of the loops of the second generation
Stanford detector are running.   It has been shown that ultra low
magnetic fields are not necessary for quiet operation, and that
large detectors are possible.   I will give a brief summary and
comparison of the technique and results of each of the groups.
For more details one should consult the individual papers contributed to the Conference.

The Stanford Three-Loop Superconductive Detector:

(B. Cabrera, M. Taber, R. Gardner, S. Felch, J. Bourg, R. King and M. Huber)

The detector consists of three orthogonal 4" diameter loops mounted on a sphere in a vacuum. The three loop detector is shown in Fig 2. In Fig. 3, the original single loop detector with a monopole trajectory is shown.

The fundamental characteristics of the detector are listed in Table 2. Of particular note are the very low value of the ambient field, 20 ngauss, which corresponds to about $8\phi_0$ linked with each loop, and the geometry of three orthogonal detectors in a common superconducting shield.

These two properties allow the use of the shield itself as a detector of trajectories which miss the loops themselves. The low field makes for very quiet detector operation, as shown in the sample data in Fig. 4. The fact that several orthogonal loops look at a common volume gives redundancy in the sense that even monopole trajectories which miss the loops themselves have predictable signals in each of the three loops. This allows a large increase in effective area for the detector; the loops themselves have an effective area of 71 cm^2 averaged over all solid angle (i.e. assuming an isotropic monopole flux), but the sensitivity to the currents induced in the shield allows an extra effective area of 405 cm^2 after the cuts described below.

The results of the detector are plotted in Fig. 5. To place a limit on monopole flux, a signal greater than 0.1 ϕ_0/L in two out of the three loops is required. No such coincidences are seen (and no large single offsets are seen).

The detector has been run a total of 214 days as of Oct. 4, 1983. Using a solid-angle averaged total area, including the near misses, of 485 cm^2, one obtains a 90% C.L. limit on the monopole flux of

$$F_M \leqslant 2.1 \times 10^{-11} \text{ cm}^{-2} \text{ sr}^{-1} \text{ s}^{-1} \quad .$$

This corresponds to 67 times the exposure of the first experiment, which had an angle averaged area of 10 cm^2 and was run for 151 days. A discussion of the compatibility of the results, and possible explanations of the original step, will be given later.

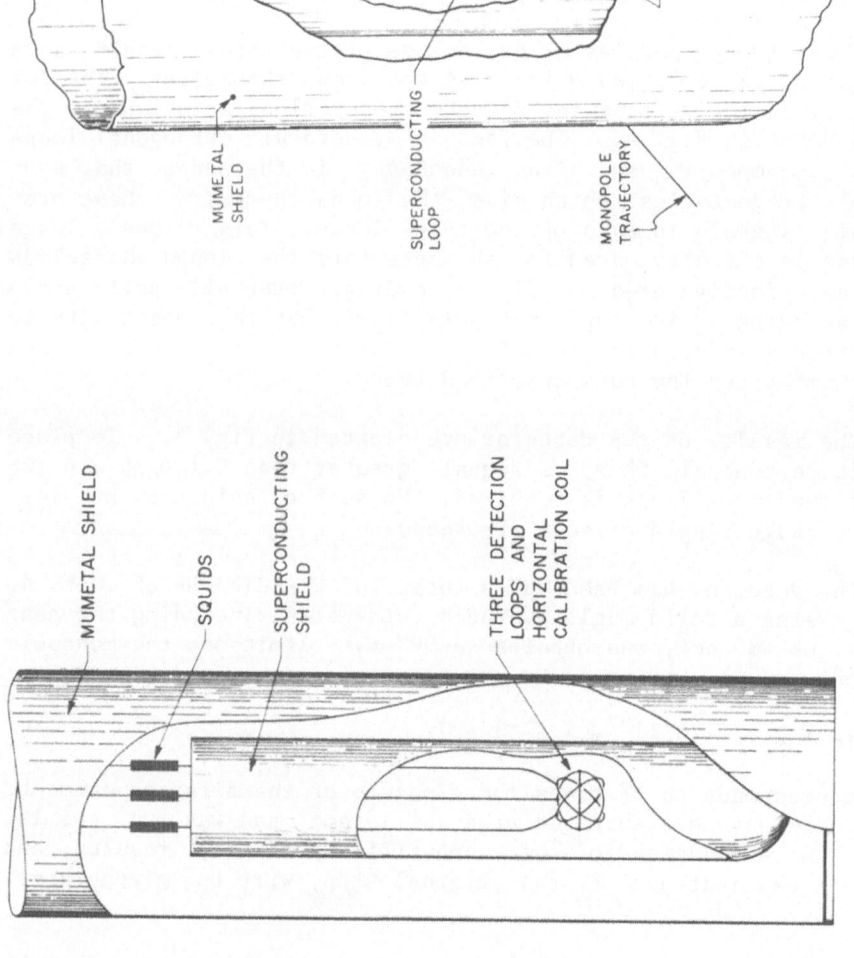

CALIBRATION COIL

SUPERCONDUCTING SHIELD

MUMETAL SHIELD

SUPERCONDUCTING LOOP

MONOPOLE TRAJECTORY

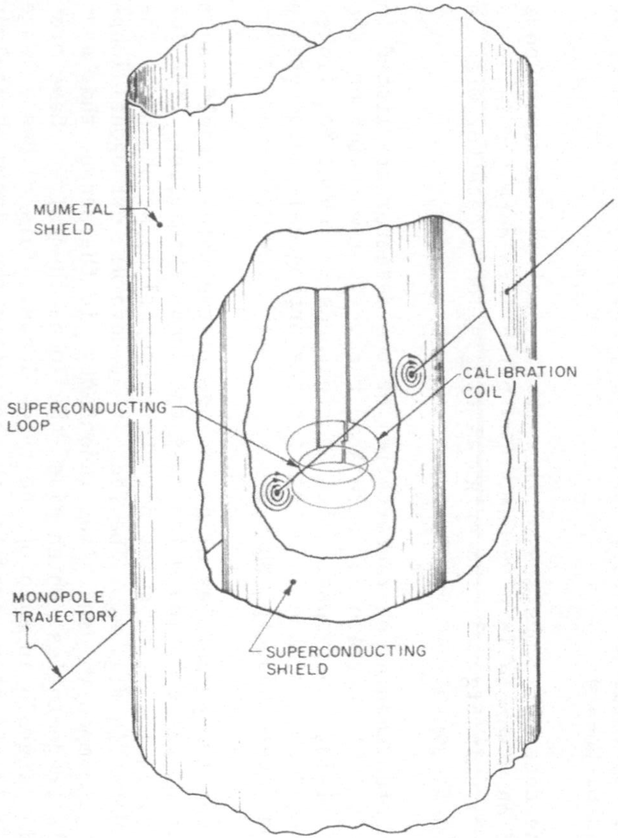

Fig. 3 A monopole trajectory in the original single loop detector showing vortices left in the superconducting shield.

MUMETAL SHIELD

SQUIDS

SUPERCONDUCTING SHIELD

THREE DETECTION LOOPS AND HORIZONTAL CALIBRATION COIL

Fig. 2 The Stanford Three-Loop Superconductive Detector and Cryostat.

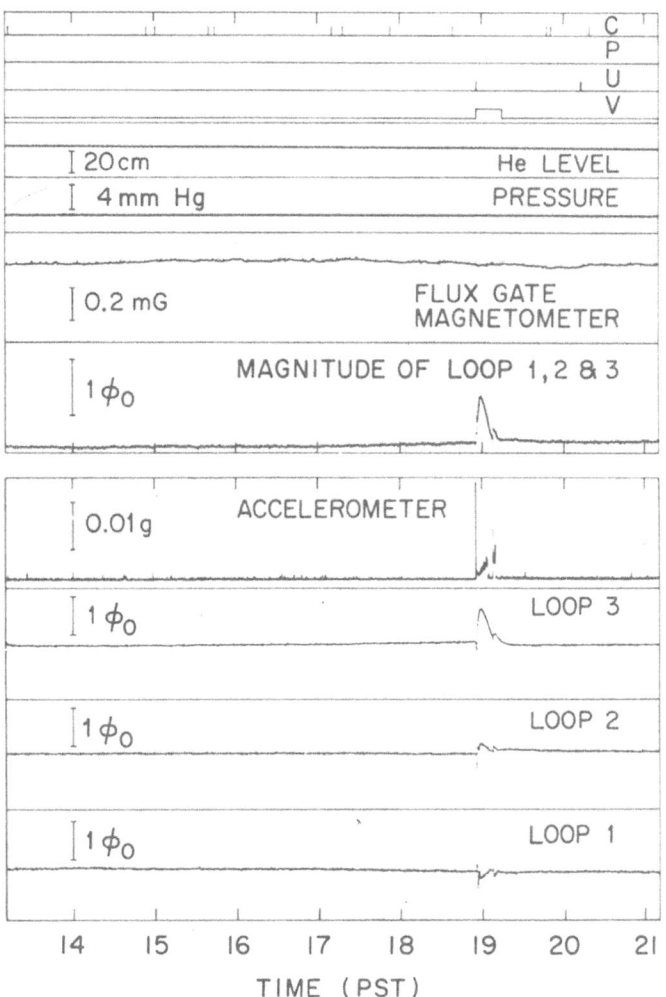

Fig. 4 Typical 8 hour data summary from the Stanford detector. The expected monopole signal would give a step of $\sim 4\phi_0$.

Table 2. Highlights of the New Stanford Three-Loop Detector.

---

1. Coincidence expt.:   three 10 cm diameter orthogonal loops – 2
    turns each.
2. $\langle B \rangle$ = 10 nG (about 8 ϕ_0/loop).
3. Loops are in vacuum and are mechanically precise and rigid.
4. Precise calibration for monopole signal via calibration coil.
5. Well instrumented with other monitors – accelerometer, mag-
    netometer, AC monitor, burglar alarm, smoke detector, etc.
6. 'High' bandwidth data acquision: computer samples at 200/s-
    events are stored at this rate; a slow average is recorded at
    (0.1) 1 Hz.
7. Extremely quiet loop environment allows the use of the near
    misses:

$$\begin{array}{lcl} \text{Coincidence loop area} & = & 70.5 \ \text{cm}^2 \\ \text{Coincidence shield area} & = & \underline{405 \quad} \ \text{cm}^2 \\ \text{Total} & = & \overline{475} \ \text{cm}^2 \end{array}$$

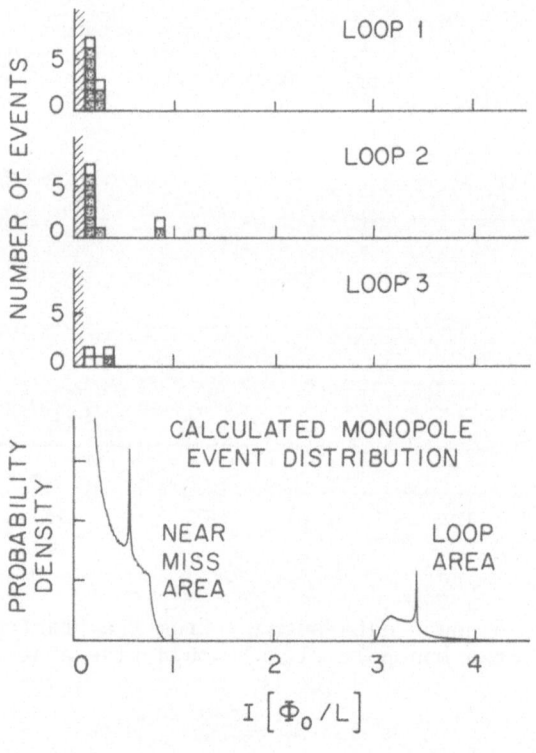

Fig. 5 The results of the Stanford detector as of June 30.   No
coincidences are seen.

Monopole Searches at IBM

(C.C. Chi, C.D. Tesche, C.C. Tsuei, P. Chaudhari, and S. Berman)

The IBM group's prototype detector is shown in Figure 6. It consists of two planar gradiometers, 10 cm × 10 cm, 4 cm apart.

The prototype detector has been run for 165 days. A typical chart recorder trace is shown in Fig. 7. The step in one SQUID trace, and the later small coincidence 'event', are due to known mechanical disturbances.

The area of each of the loops of the prototype detector is 100 cm^2. The coincidence area, averaged over all directions, is 25 cm^2. The 165 day exposure gives a limit on the (isotropic) monopole flux of F_M < 5.1 × 10^{-10} cm^{-2} sr^{-1} s^{-1} (90% C.L.). This is 2.1 "Cabreras" (one Cabrera = the first Stanford experiment exposure of 10 cm^2 direction - averaged area × 151 days).

A larger detector, based on the same principles, consisting of a six-sided rectangular box of gradiometers is being brought on the air now. It is shown in Fig. 8. The side plates are linear gradiometers of fifth order, the top and bottom are 5th × 5th order. The inductances are small (~6 μH) and so are reasonably well matched to the 2 μH inductance of the SQUID input. The detector is cooled inside a carefully tuned set of μ metal shields which reduce the ambient field at cooldown to the microgauss region. Table 3 lists the highlights of this new detector.

A future IBM detector is shown in Fig. 9. It utilizes the same design, a box of gradiometers, but would be appreciably larger with a coincidence area of ~5 m^2. This design would then be used as a basis to 'step-and-repeat' to get to a very large area.[12]

Table 3. Highlights of the New IBM Detector.

1. Coincidence experiment - has six loops arranged in a rectangular box.
2. Uses high order gradiometer to reduce sensitivity to magnetic field.
3. Largest coincidence area so far ~1000 cm^2.
4. Good magnetic shielding - get shielding factor of 10^8.
5. Cools down in fields much less than the earth's field (B_T = 3 ± 2 μg).
6. Lots of expertise on SQUIDs - can build their own.
7. Very narrow range of predicted signal (due to gradiometer).
8. Technique is applicable to much larger scale detectors.

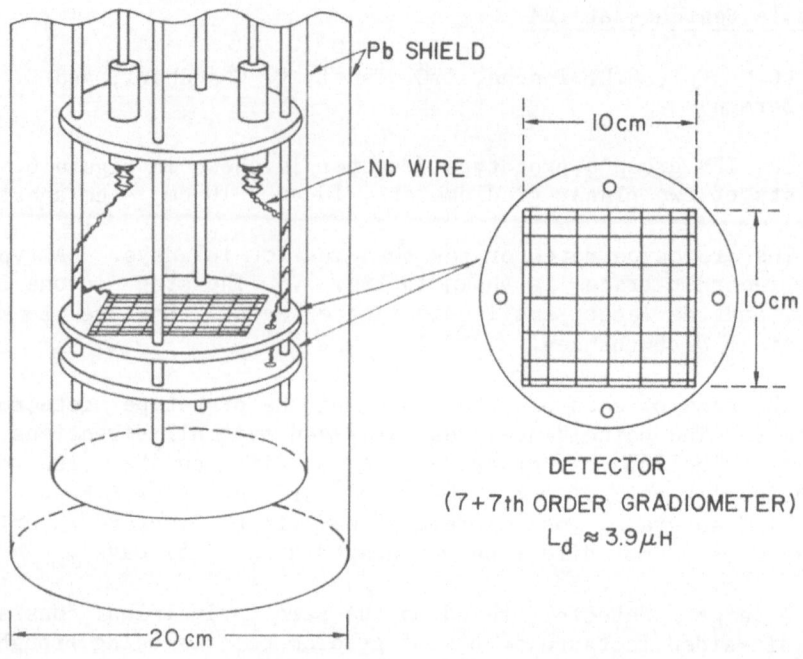

Fig. 6   The IBM prototype detector for which results were reported
         at this conference.

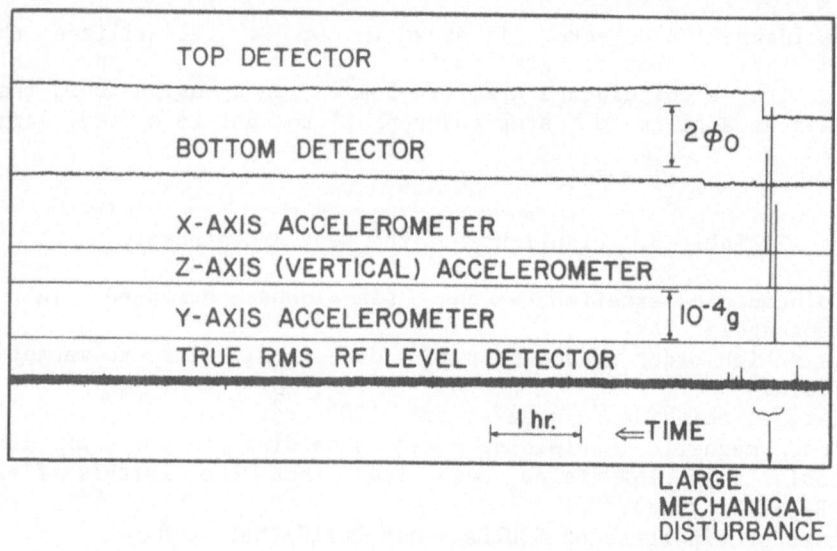

Fig. 7   Typical data from the IBM prototype.

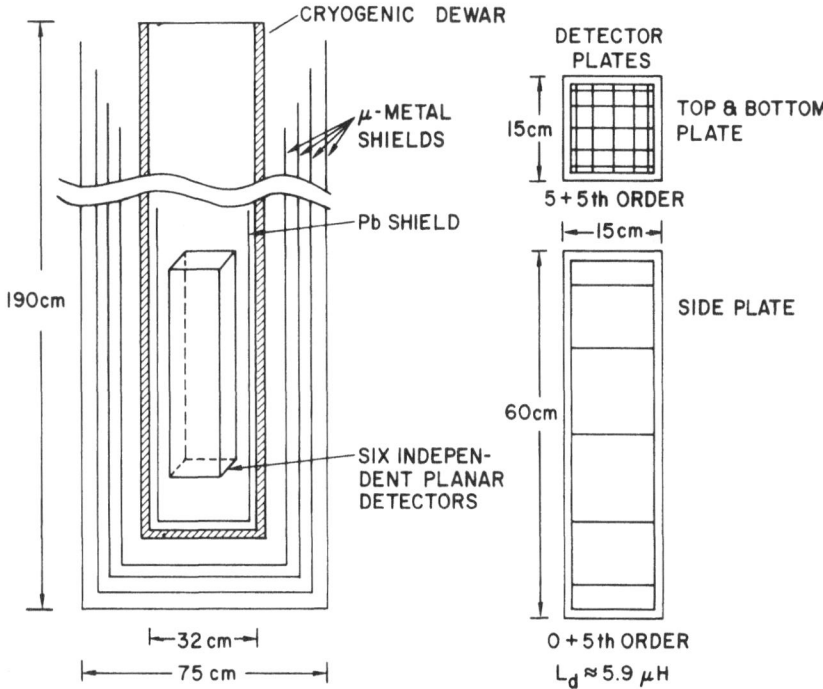

Fig. 8   The   IBM detector currently in operation.

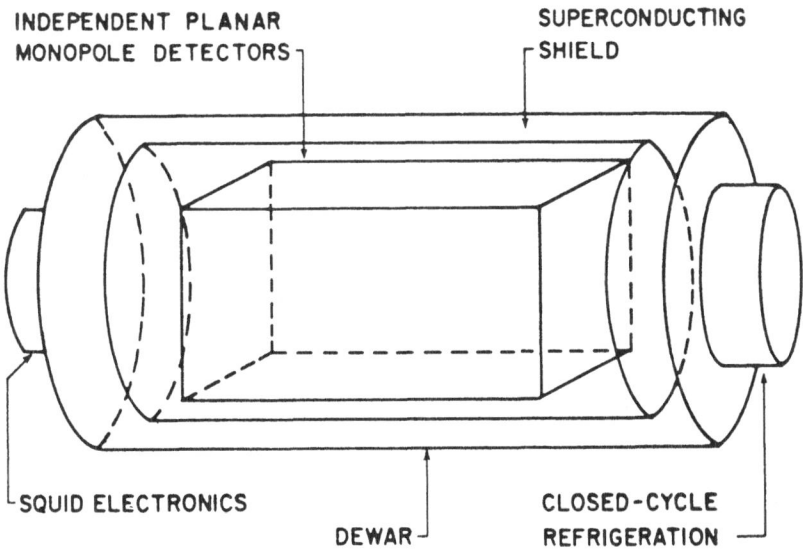

Fig. 9   The proposed large area IBM detector.

The Chicago-Fermilab-Michigan Prototype

(M. Campbell, H. Frisch, J. Incandela, S. Somalwar (UC),
M. Kuchnir (Fermilab), and H.R. Gustafson (Michigan))

The prototype detector this group has been running is shown in Fig. 10. Built to study the possibility of big induction detectors, it depends on good mechanical and thermal stability rather than low ambient magnetic field. Two 60 cm diameter loops, also of gradiometer design (the local (independent) invention was called the macramé, as the first ones were made by hand), but with equal size cells[13] rather than the cosine expansion of IBM, sit in superconducting shields (see Fig. 11). The detector arrays are in vacuum with cooling by conduction to the helium resevoir. This arrangement makes for much better mechanical and thermal stability as the detectors are in thermal contact through a long-time-constant connection with a point which is always covered by liquid helium. The detectors are free from the mechanical motions associated with bubbling or convection of the liquid. Table 4 gives the highlights of the detector.

A typical chart recorder trace is shown in Fig. 10. The left-most trace is the SQUID output from the bottom detector. The two small mesas shown are calibrations induced by exciting small diameter solenoids to a flux of $2\phi_o$ (one Dirac monopole). The computer does this remotely at the start and finish of each data run - what is shown here is the end of one run and the start of the next. The inductance of the 3000 cm^2 loops is large (25 μH); the signal-to-noise, however, is excellent.

The second trace from the left in Fig. 10 is the output of the upper detector. At this time, it had a break in the shielding of its leads to the SQUID, and so was sensitive to external magnetic field changes. The calibration steps, however, are distinct enough to be used in coincidence with the lower detector.

The detector was first run from August 29 - October 4, 1983. During this time the detector was being instrumented with other monitors (sound level meter, magnetometers, etc.). The live time for the bottom detector alone, i.e. not requiring a coincidence with the top detector, was 22 days and 6 hours. The direction-averaged-area of the one detector is 1385 cm^2. The spectrum of the steps seen by this single detector is shown in Fig. 13. No monopole candidates were seen, leading to an upper bound on the monopole flux of $F_M < 6.9 \times 10^{-11}$ cm^{-2} sr^{-1} s^{-1} (90% C.L.). This corresponds to 21 times the original Stanford exposure of 151 days. We note that this was done in 22 days of running.

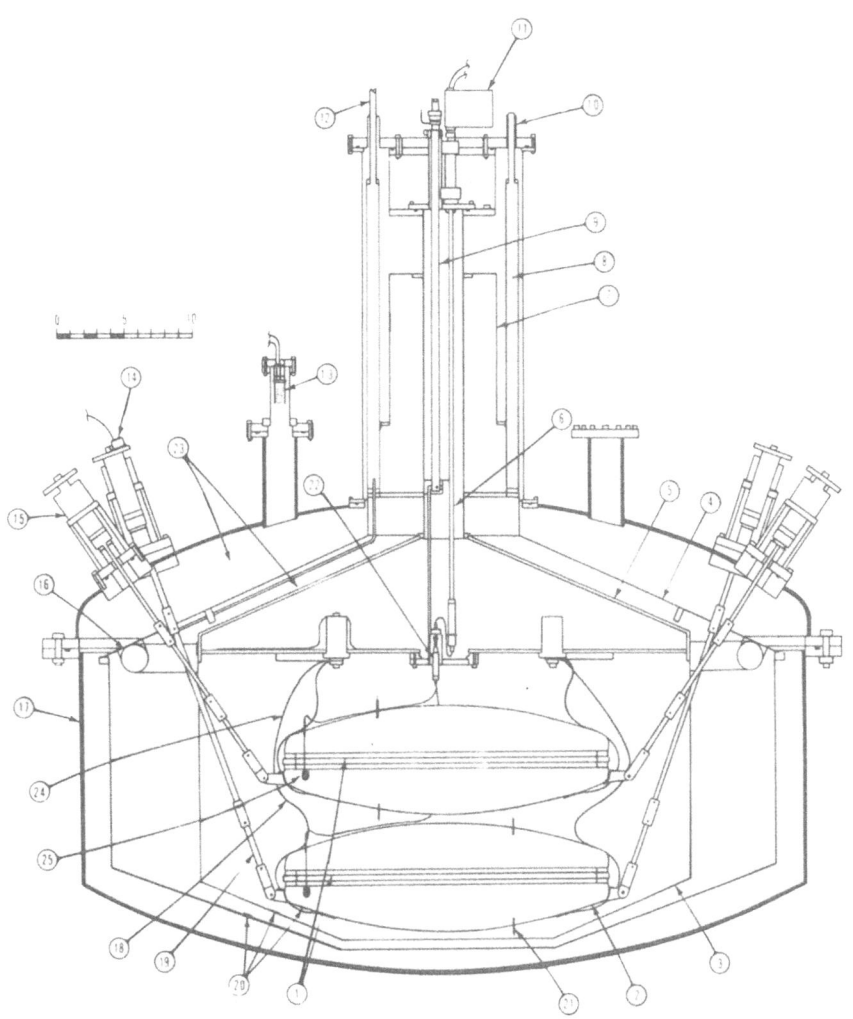

Fig. 10 The Chicago-Fermilab-Michigan prototype detector. The
scale is set by the diameter of the loops, which is two
feet. Key: (1) Detector plane, (2) Lead lined copper
shield, (3) Helium temperature shield, (4) Nitrogen
temperature shield, (5) Helium reservoir, (6) SQUID tube,
(7) Heat pick-up, (8) Nitrogen jacket, (9) Helium fill
tube, (10) Nitrogen vent, (11) SQUID electronics,
(12) Nitrogen fill line, (13) Vacuum gauge, (14) Accelerometer, (15) Vibration isolator assembly,
(16) Nitrogen cooling ring, (17) Vacuum vessel, (18)
Cooling straps, (19) Suspension rod, (20) Thermometers,
(21) Pseudopole, (22) Vacuum feedthrough, (23) Super insulation, (24) Cooling straps, (25) SQUID matching transformer.

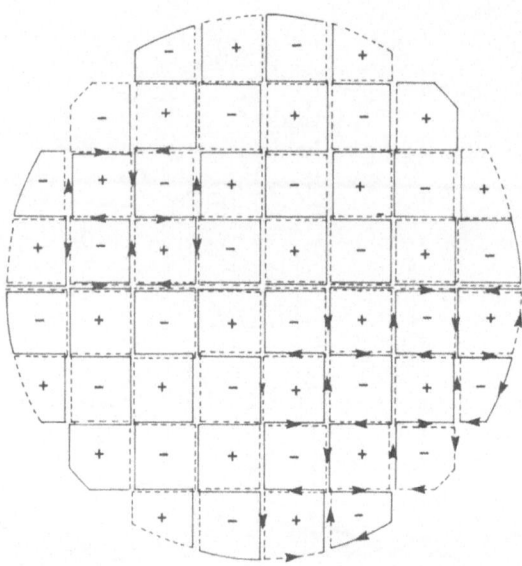

Fig. 11   The folded loop (macrame) of the CFM group. Note the
         equal size cells. The loop is made with standard printed
         circuit board technology as a two-sided board, with
         plated through holes.

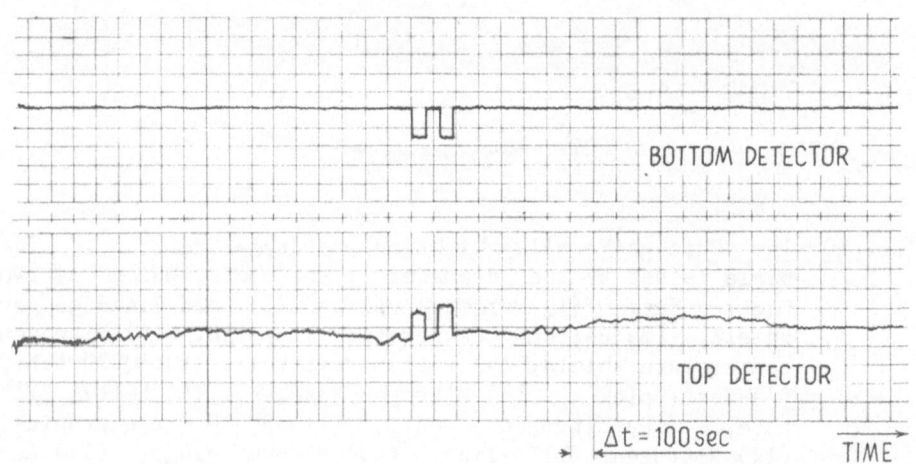

Fig. 12   A typical output of the CFM detector.   The two small
         mesas are $2\phi_0$ calibrations, induced by exciting small
         diameter solenoid to a flux of $2\phi_0$. The 'top' detector
         was noisy due to a broken shielding tube, which has since
         been fixed.

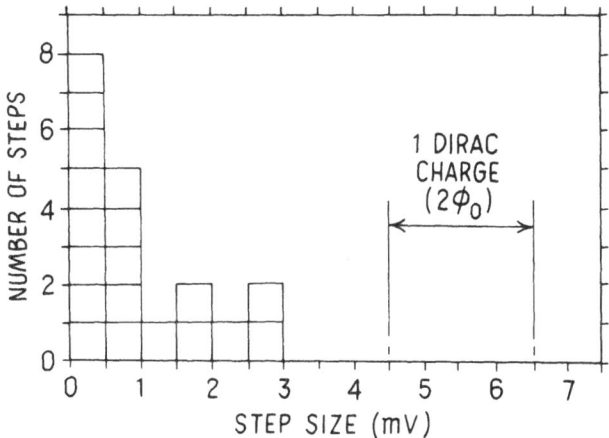

Fig. 13 The spectrum of steps seen by the CFM detector.

Table 4. CFM Detector - Highlights.

1. Coincidence experiment-has 2 60cm loops. Biggest loops so far.
2. Uses a 'macrame' winding to reduce sensitivity to currents in the superconducting shield.
3. Loops are in individual shields, mechanically isolated.
4. Detectors are in a vacuum to reduce mechanical coupling, bubbling, etc. Cooling is by conduction.
5. Good calibration - each detector has 5 small solenoids threading the loop and shield to mimic a monopole trajectory automatically fired by computer at $2\phi_0$ every two hours during data taking.
6. Mechanically very rigid - printed circuit coils and lots of mechanical support.
7. Has large inductance coils and impedance matching superconductance transformer.
8. Well instrumented - 2 magnetometers, sound level meter, accelerometer, 23 thermometers, 2 pressure sensors, etc.
9. Technique is designed for being replicated into a big area detector.

The Imperial College Monopole Detector

(Chris Guy, David Caplin, Mike Hardiman, Gavin Park[+], Josh
Schouten, Dave Websdale (HENP), and Jerry Cowen (MSU))

This group has done thoughtful calculations on the shielding
problem while building a large dewar surrounded by a fine set of
μ-metal shields. The dewar, shields, and a possible detector
layout are shown in Fig. 14. They solve the problem of coupling
to the shield by having a small loop in series with the main loop,
but with multiple turns in the opposite direction. A three-loop
layout, two D's and a full loop, is shown in Fig. 15. The
calculated spectra for monopoles with loop/shield radii values of
0.25 and 0.75 are shown in Fig. 16. The detector is being
designed at present.

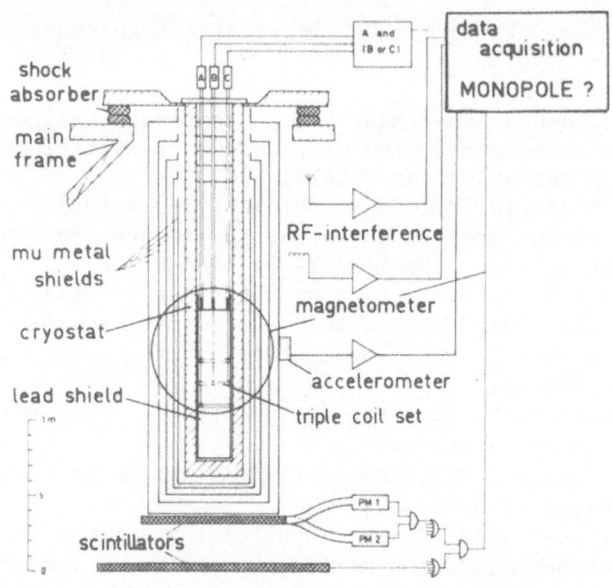

Fig. 14 The Imperial College Cryostat and proposed detector.

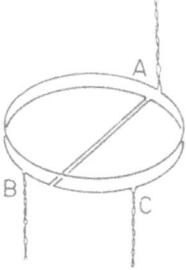

Fig. 15 A proposed three-loop layout for the Imperial College detector.

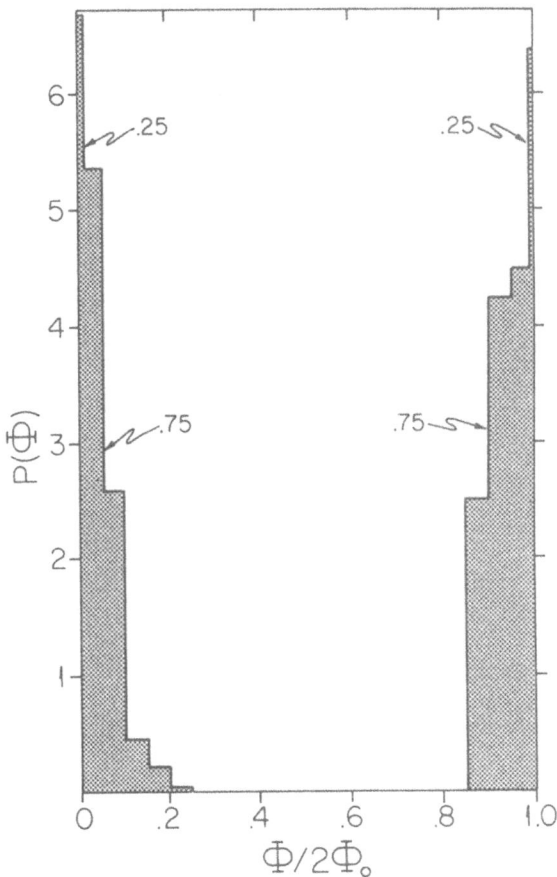

Fig. 16 The calculated spectra for the Imperial College asymmetric astatic pair for two different values of the ratio of coil diameter to cylindrical shield diameter.

The Kobe University Detector

(T. Ebisu and T. Watanabe, Kobe University)

This detector has as its primary focus the search for monopoles trapped in old iron ore. In the process, however, they have set a limit on free monopoles. As they also have plans for larger free monopole detectors, I include the detector description here. The technique of their search for trapped monopoles will be discussed below.

The detector is shown in Fig. 17. A three-turn-8 cm diameter coil is wound on a glass tube, which is inside an 11 cm diameter lead-covered glass tube. They also use a calibration solenoid (of about $20\phi_0$).

The experimentally observed spectrum as well as the calculated response curve for the detector are shown in Fig. 18. No events penetrating the loop are seen, leading to a limit of

$$F_M \lesssim 2.0 \times 10^{-9} \text{ cm}^{-2} \text{ sr}^{-1} \text{ s}^{-1} \text{ (90\% C.L.)}$$

This exposure is 0.7 "Cabrera's."

The Kobe group has plans for a larger area coil to do free monopole searches - 14 cm diameter, inside a 20 cm shield.

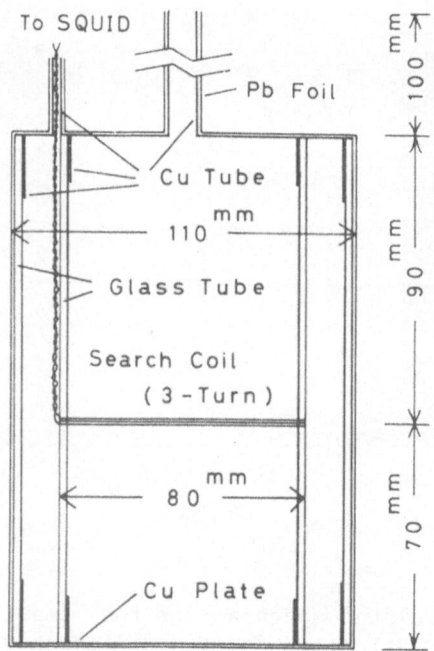

Fig. 17 The Kobe University single-coil detector.

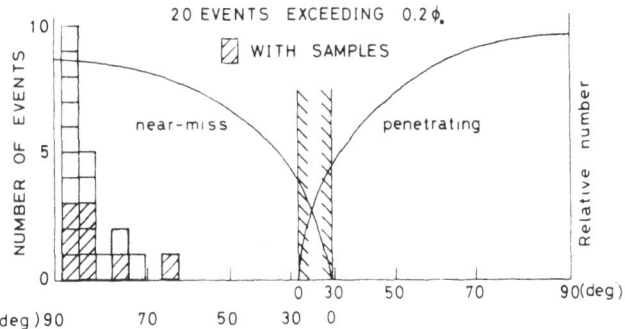

Fig. 18 The results from calibrating the Kobe detector and the experimentally observed spectrum. Note the event at $1.4\phi_o$!

The NBS Studies on Sources of Spurious Steps in RF SQUIDS

(Alan Clark, Michael Cromar, and Frederick Fickett, NBS, Boulder, Colorado)

The NBS group up to now has concentrated their attention on the question as to what explanations other than that of magnetic monopoles there are for Cabrera's step. They have taken a pair of SHE Model 30 SQUIDS (the same model used by Cabrera), and have deliberately tried to force steps. A major difference from Cabrera's detector is that the ambient field is much larger in the NBS setup. This presumably magnifies the sensitivity to mechanical motion by many orders of magnitude.

The best search for a possible spurious source for the original Stanford step would have been with that self-same detector, but installed in an ambient field many orders of magnitude larger.[14] A measurement of background as a function of ambient field would possibly allow an extrapolation to the very low field of the actual experiment.

In lieu of that controlled experiment, one has only tests on similar but not identical setups. The NBS tests, and the experience of others (for example, the IBM coincidence step presented at the Racine Conference in 1982) show how easy it is to get spurious steps with RF SQUIDs if care is not taken.

The NBS experimental layout is shown in Fig. 19. A mechanically induced pair of steps is shown in Fig. 20. According to the NBS group, they are able to manufacture these steps at will by stressing the SQUID probes. A table of sources of spurious steps and their antidotes is given in Table 5. The conclusion of the NBS group is that with the proper care in clamping the SQUID probes, these spurious steps can be completely eliminated.

While this work does not yet answer the question as to exactly what are the sources of the steps, I think it is important in the context of the Cabrera candidate monopole.   One knows that larger, more redundant, and more rigid detectors (Stanford II, and CFM) have put a limit 100 times lower than the 'flux' associated with the one event (see the next section of this talk).   The NBS work as well as the combined experiences of the other groups gives an alternative explanation, though at some unknown probability level (in the Stanford case the probability of the step is presumably proportional to a very low field times a high degree of possible mechanical motion).

The NBS group is in the process of constructing a triplet of larger area coils - 15 cm diameter, inside a 20 cm shield.   The loop geometry and shield is similar to the Stanford three loop detector.

SYSTEM DATA:
    Two coaxial coils (coinc.)
    Typ. run time (3 fills)= 100 hrs
    Depth to SQUIDs = 68 cm
    Can (shield) diameter = 13 cm
    Can (shield) length = 40 cm
    Field at coils= 56 mA/m (0.7 mOe)

COIL CONFIGURATIONS:
  A) turns/coil = 5
     coil diameter = 3.1 cm
     coil separation = 1 cm
     can/coil area ratio =18
     run time= 522 hrs

  B) turns/coil = 3
     coil diameter = 6.4 cm
     coil separation = 1 cm
     can/coil area ratio = 4
     run time = 502 hrs

  C) turns/coil = 3
     coil diameter = 6.4 cm
     coil separation = 13 cm
     can/coil area ratio = 4
     run time = 145 hrs

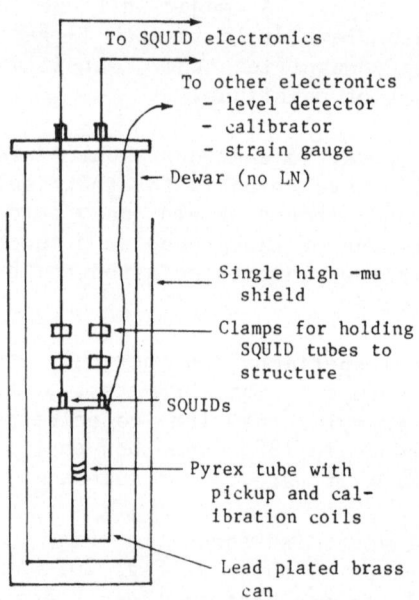

To SQUID electronics

To other electronics
  - level detector
  - calibrator
  - strain gauge

Dewar (no LN)

Single high -mu
  shield

Clamps for holding
  SQUID tubes to
  structure

SQUIDs

Pyrex tube with
  pickup and cal-
  ibration coils

Lead plated brass
  can

Fig. 19 The NBS prototype detector layout.

Table 5. A Summary of the NBS Observations on Sources of Spurious Steps in RF SQUIDs.

Disturbance	Effect	How to minimize
Mechanical shock	Very large shock to give offset Relatively large for significant noise	Rigid mount system Clamp SQUID probes
Stress in SQUID probes	Offsets of monopole size	Clamp SQUID probes to structure in cold region Monitor SQUID can with strain gauge
Magnetic signals	Very large disturbance can create offset	Keep large, moving magnets away
Flux leakage into can	None in our system	Keep access holes small
Pressure changes	Increased noise and offsets if can plating is loose	Good plating Thermally stable dewar
No magnetic shields	Increased sensitivity to noise and shock No offsets	Don't run unshielded
Electrical transients	Noise, but no offsets even at gross interference levels	Filtered feedthrough Cap all leads into dewar

SQUID alone, input open - Above disturbances have little effect
SQUID alone, input shorted - Small deformation of SQUID probe gives increased noise and has
    also been observed to give offsets; Large magnetic disturbance gives offsets; Large shock
    gives offsets.

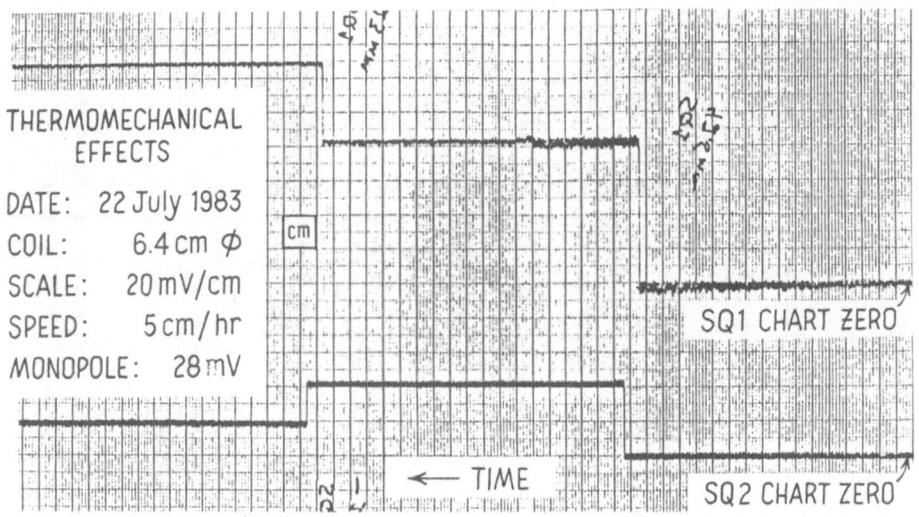

THERMOMECHANICAL
EFFECTS

DATE: 22 July 1983
COIL: 6.4 cm ϕ
SCALE: 20 mV/cm
SPEED: 5 cm/hr
MONOPOLE: 28 mV

SQ1 CHART ZERO

← TIME   SQ2 CHART ZERO

Fig. 20   Mechanically induced steps in the SQUID output of the NBS
setup.

A Summary of the Results of Operating Induction Detectors

Table 6 gives a summary of the results presented at the
conference.   Each detector has listed its area averaged over all
directions (Column 5), and the limit achieved as of October 4,
1983 or so in units of the Stanford "one-event" flux.

The two most important facts are:

(1)   In almost 100 times the exposure of the first
experiment, no events are seen.
(2)   Detectors ~ 100 times the area of the original experiment area are now running, with bigger ones on the
horizon.

The probability that one event occurs in the first of 100
identical time periods is 1%.   The first experiment, however, was
not identical to the others.   In particular it was only a single
loop and it was clearly less mechanically rigid than the newer
versions. Although the detector was also in a very low field, the
step could well have been induced by some mechanical effect.   It
has proved impossible to show it was not a monopole - it would
take a detailed experimental program with the same apparatus, say
in a field 1000X the original, to explore backgrounds.   But it has
been shown (see Fig. 20 for example) that it is possible to get

Table 6. A Summary of the Results of Operating Induction Detectors Searching for Free Heavy Magnetic Monopoles.

Group	# of loops	loop diam. (cm)	arrangement	Physical loop area (cm²)	Area av. over 4π (cm²)	Calendar Exposure (days)	Sensitive time (days)	Limit (90% C.L.) (cm⁻² s⁻¹ sr⁻¹)	Exposure Stan. I
Stanford I	1	5.0	----	20	10		151	1.4×10^{-9}* (6.1×10^{-10}*)	$\equiv 1$ (+1)
Stanford II	3	10	3-axis	79	71 hit 405 near miss 476 total coinc.	≈250	214	2.1×10^{-11}	67
IBM I	2	10	parallel 4 cm apart	100	25 coinc.		165	5.1×10^{-10}	2.7
IBM II	6	2@15 x 15 4@15 x 60	rect. box	225 900	1000 coinc.	1 week	---	----	---
CFM	2	60	parallel 21.6 cm apart	2800	1400 single 700 coinc.	30	22	6.9×10^{-11}	21
Imperial College	3	25	Circle + 2 D's	625	~300 coinc.	----	----	----	---
Kobe U⁺	1	8	----	50	25		42	2.0×10^{-9}	0.7

TOTAL 93.4

$F_m < 1.5 \times 10^{-11}$ (90% C.L.)

*6.1×10^{-10} is the 1 candidate flux
+main thrust is for trapped monopoles

clean steps out of SQUID systems. In the spirit of the rules of
physics of Luis Alvarez in his 'Analysis of a Reported Magnetic
Monopole',[15] the mechanical relaxation is a reasonable alternative. It is worth saying that Blas Cabrera has been very careful
not to claim the candidate as an event, and that I think the
publication of the step was the correct move, and has served as a
major stimulus to both experimental and theoretical work.

The limit from the total of all the induction experiments is
shown in Fig. 21. One sees that there are still three orders-
of-magnitude to go until one is even barely into the astrophysically interesting region.

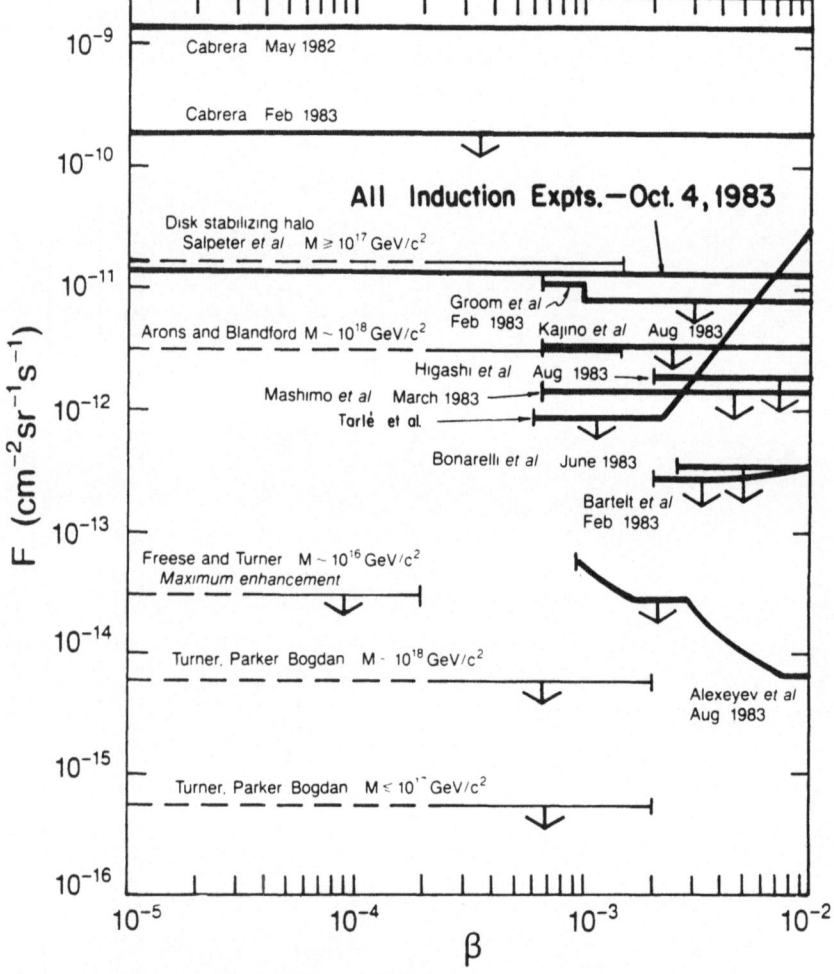

Fig. 21 The limits on monopole fluxes from induction detectors,
         superimposed on limits from ionization detectors. (From
         the graph of Tarlé et al.; I have modified their 60% C.L.
         limits into 90% C.L. limits).

SEARCHES FOR MONOPOLES TRAPPED IN MATTER

Some Modern History

The use of magnetic induction in a superconducting loop as a detector for the passage of a magnetic monopole was first suggested by L.W. Alvarez[16] in 1963, and independently by L.J. Tassie[17] in 1965. The size of the expected signal was also worked out by P.H. Eberhard[18] in 1964.

The first proposal for such a detector using this principle was that of Alvarez, Eberhard, et al.[19] in which it was proposed to pass lunar material through a superconducting loop multiple times. This detector[20-21] was used to search 8.3 kg of lunar material, and about 20 kg of non-lunar material, including 2.4 kg of ocean sediment and an emulsion containing a peculiar stopping track from an emulsion search by Kolm et al.[22] No monopole candidates were seen.

The first use of a SQUID as the detector of the current induced by a magnetic monopole was in 1968 by L.L. Vant-Hull.[23] (This in fact precedes the Alvarez - Eberhard et al. measurements, but not the proposal. Experiments go quicker if they do not require lunar matieral). Samples were passed directly through the SQUID interferometer loop. The samples were 0.043 kg of CU, and a total of 0.049 kg of W and Au. A beam of 7×10^{21} electrons was also passed through the loop.

This latter experiment set a limit on the magnetic charge of electrons of $g_E < 4 \times 10^{-24}$ Dirac charges per electric charge. From the other samples g(proton) - g(electron) $< 10^{-26}$ Dirac charges. I believe that these limits could be easily improved by orders of magnitude if there were any theoretical incentive (the second number was already considerably improved by the Alvarez-Eberhard experiment).

The Kobe University Serach in 430 kg of Magnetic Ore

The Kobe University contribution (T. Ebisu and T. Watanabe) to the conference is a search for monopoles bound to ferromagnetic materials, where they could be expected to accumulate over geologic times as suggested by Goto[24] in 1958. The experiment heats iron ore in a furnace which is located above the SQUID coil described above. At the Curie point, monopoles of mass larger than $\sim 3 \times 10^{15}$ GeV/c^2 are expected to fall through the loop. This is the same technique as proposed by the Wisconsin group (see below).

The results presented in this conference were on heating 215 kg of old magnetic sand (estimated age 20 million years),

214 kg of Maghemite, and 46 kg of blast furnace grate-bars over the coil. The furnace arrangement is shown in Fig. 22. Several kilograms of the sample at a time were heated above the Curie point, and the SQUID output was recorded to look for monopoles falling (gravitationally) through the loop. No monopoles were seen, although a step of $1.4\phi_0$ was seen while a sample was hot. The step is shown in Fig. 23. (Notice how nice and clean it is. It seems clear that much more redundancy is required to believe any of these single events. A single loop will never convince.)

The Kobe result corresponds to an upper limit of 6.9×10^{-6} monopoles/gram (95% C.L.) for iron ore.[25] This result is to be compared to the result of Alvarez et al.[22] of 1.7×10^{-4} monopoles/gm (also 95% C.L.) While the authors translate this into an upper limit on monopole fluxes of 6.8×10^{-13} cm^{-2} s^{-1} sr^{-1} (60% C.L.), I feel rather strongly that such calculations are too uncertain to take as serious experimental limits on monopole fluxes. The future plans of the Kobe group are to increase the size of their apparatus to a 14 cm diameter loop inside a 20 cm shield, and to process 10^9 year-old ore.

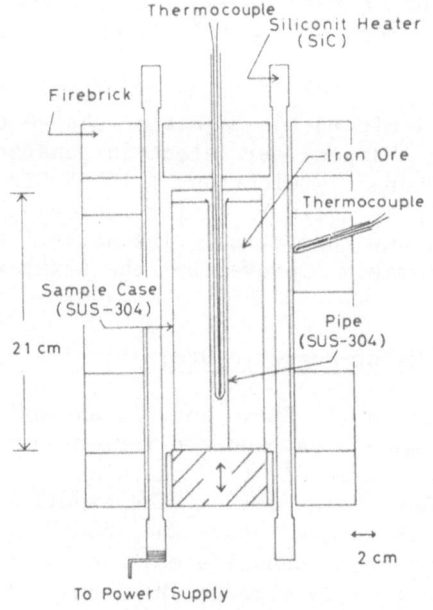
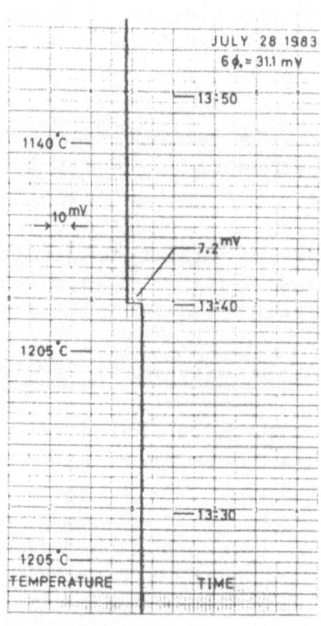

Fig. 22 (Left) The furnace arrangement of the Kobe University search for monopoles trapped in old iron ore.

Fig. 23 (Right) The largest step seen in the Kobe experiment. It corresponds to $1.4\phi_0$.

The Wisconsin Proposal to Search for Relic Monopoles in Iron Ore

(R. Boom, D. Cline, D. Jantras, and R. March – U. W., Madison)

The Wisconsin proposal is to search truly large amounts of iron ore for captured monopoles by putting a multi-loop superconducting detector below the furnace of one of the large iron ore smelters of Minnesota or Wisconsin. These mines heat $10^6 - 10^7$ tons of iron ore per year through the Curie point. They estimate they could search on the order of $10^4 - 10^5$ tons/year this way.

The proposed experimental setup is shown in Fig. 24. The detector itself has the required redundancy in that there are four coils hooked up to three SQUIDs in an arrangement that gives a very nice unique signal: two steps of opposite sign, and perhaps[26] a pulse of a width determined by the monopole gravitational velocity. While a negative result would be impossible to translate into a flux, it is an imaginative experiment.

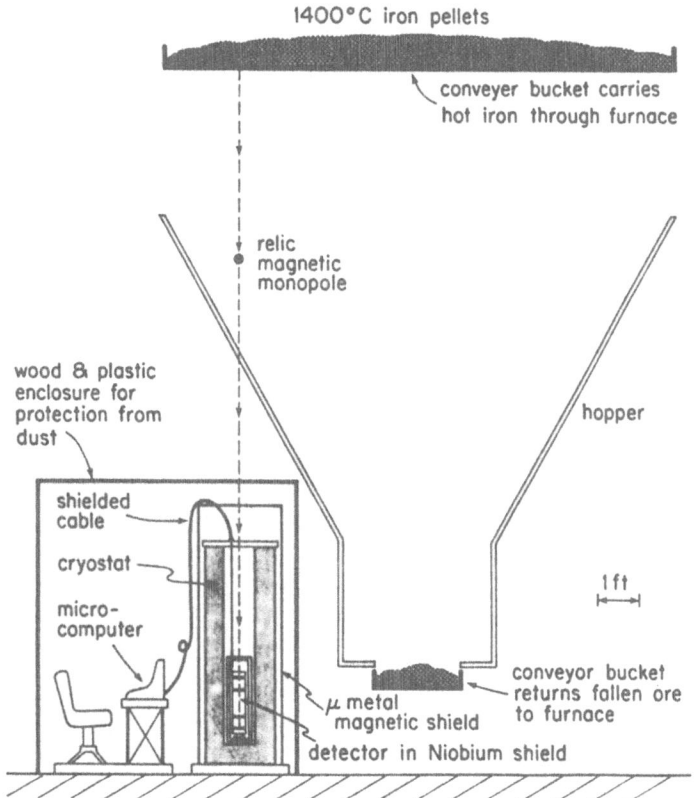

Fig. 24 The Wisconsin proposal apparatus to search large values of iron ore, showing the commercial oven.

OTHER RELATED TOPICS

Room Temperature Induction Detectors

(M. Price, CERN)

Faraday induction is the most direct way of searching for monopoles. The superconducting techniques described above suffer both complication and expense from the necessity of vacuum and cryogens. Room temperature (or even liquid nitrogen temperature) induction coils would possibly allow much larger area induction detectors.

Calculations of signal-to-noise of room temperature induction detectors have been made by D.E. Morris,[27] C. Rubbia,[28] and M. Price.[29] In his contribution to this conference, Price has addressed the Johnson noise in the coils as well as amplifier noise.

Price concludes that in principle for a very bulky coil (i.e. a conductor thick compared to the radius of the loop) the Johnson noise can be reduced to the point where a signal-to-noise ratio of 10 is achievable for monopole velocities greater than 1.4×10^{-4}.

Price has also done a detailed analysis of an amplifier and integrator for the loop. Again he finds that a signal-to-noise ratio of about 10 is achievable in principle.

Zero Quantum Superconducting Magnetic Shield

(J. Clem, Iowa State)

J. Clem has proposed a device to allow very low fields over substantial volumes. The device consists of a hollow cylinder of superconductor with a central region in which flux quanta are lightly pinned, and end regions in which they are strongly pinned. The cylinder is cooled in some ambient field, and vortices are frozen in as shown[30] in Fig. 25. When a current is flowed longitudinally down the cylinder, the vortices in the central region will circulate around the tube, opposite signs going in opposite directions. A small solenoidal field drifts the vortices in the central region longitudinally as well so that the vortices of opposite sign intersect and annihilate.

Once the shield is dequantized it could be used to provide a zero-field region (e.g. for Josephson junction computers), or used as a monopole detector as proposed in Fig. 26.

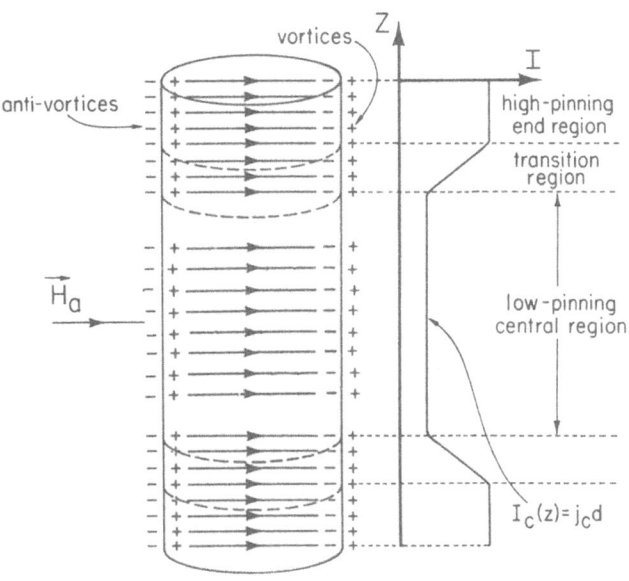

Fig. 25 John Clem's zero flux shield with vortices from a transverse field frozen in.

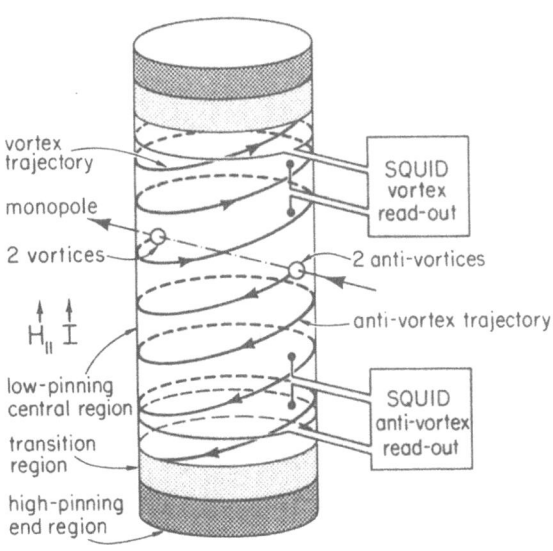

Fig. 26 The use of the zero flux shield as a monopole detector.

Large Area Detectors - Optical Pumping and Large Loop Arrays

(J.D. Silver, Oxford)

J. Silver presented ideas for going to 1000 m² arrays. In particular he elucidated use of zero-field crossing resonances in ⁸⁷Rb as measured by Dupont-Roc et al.[30] He also presented schemes very similar to the macrame and gradiometer array ideas of CFM and IBM. The Oxford group has a very large cryostat (2m diameter) and is seriously considering construction of one module for a very large array.

PROGNOSIS FOR VERY LARGE INDUCTION DETECTORS (100 m² - 1000 m²)

The size of superconducting induction detectors has grown by two orders of magnitude (from .0020 m² to 0.3 m²) in less than two years. What are the limits on how big a detector one can build? Detectors on the order of 1-5 m² are now on the drawing boards (e.g. IBM - see Fig. 9). To make a detector of ~10 m² one could clearly just duplicate the ~1 m² detectors by brute force. But what of going to a detector sensitive at the Parker bound level? (500m² × 1 year gives a sensitivity of about 10^{-15} cm^{-2} s^{-1} sr^{-1}). Experience with the next size will be necessary, but here's my present guess:

How Large Can a Submodule Be?

The Chicago-Fermilab-Michigan idea of a large detector is shown in Figs. 27 and 28, and has the following features:

* Modular subunits for ease of construction and mechanical stability
* The subunits have SQUIDs mounted directly on them
* The subunits and SQUIDS live in a vacuum, and are cooled by conduction only
* The LN and LHe systems consist only of pipes which cool both cylindrical shields and the submodules

For a very large array, the optimum geometry is fundamentally flat. It is not yet clear whether or not having arrays of flat independent modules (à la CFM) or having three-dimensional coils (Stanford and IBM) is better for redundancy. (See below.)

How Large Can a Submodule Be?

We can scale from the present detectors. The largest area submodule at present is that of the CFM; it is 0.3 m² in area. The rather small cell size of the macrame makes for a large inductance of 25 μH. The signal size is 7 mV (see Fig. 12), and

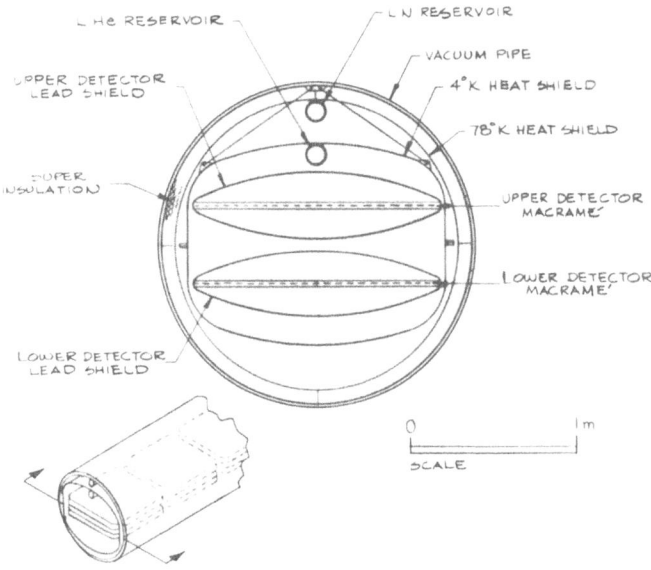

Fig. 27 A possible design for one unit of a large array. Key features are the use of vacuum with conduction cooling for the detectors, cheap iron cryostats, and a fundamentally planar geometry.

Fig. 28 The large array as it would be installed in the Fermi Institute.

the signal-to-noise is ≥ 15:1. If no matching transformer is used, the signal would fall inversely with the inductance, which in turn is proportional to the diameter (to within log terms). If we ask for a signal no smaller than half the present CFM signal, a loop of ≥ 4' diameter is possible. With a matching transformer, however, the signal falls as the square root of the diameter. The maximum diameter is then ≥ 6'.

Finally, one can buy commercially DC SQUIDs which have a signal-to-noise ratio 10X better than the RF SQUIDs. The maximum diameter then could be 800 feet (!). I think the ultimate limitations are more likely to be mechanical stability and the need for redundancy rather than signal-to-noise. Experience with the next generation of large. detectors will be necessary to optimize the submodule size.

A Wild-Eyed-Guess at the Cost of a 100 m^2 Array

There have been some very high estimates (40M$/100 m^2) of the cost of a large inductive array. These have probably been based on the presumptive need for ultra low field regions, quartz support structures, etc. As a counterbalance to these, for the fun of it let's make an estimate using the new techniques presented at the conference. The estimates are possibly wrong in the other direction, but have good precedent in the art of building large new objects.[33]

Assume:

* brute force - 1 SQUID/subunit
* SQUIDs are bought from SHE at list unit price
* assume LHe and vacuum pumps are free (may have to be at Fermilab)
* make 16 m^2 modules à la Fig. 28

Costs per 16 m^2 module

Vacuum cryostat - 8' diameter pipe @ 250$/foot...... 10K
Flanges... 20K
Misc. plumbing...................................... 20K
Heat and field shields.............................. 32K
Macrame... 32K
SQUIDs.....................................32 × 7K = 224K

TOTAL per 16 m^2 = $338K

which equals ~$21K/$m^2$, of which 14K$ is SQUIDS.

The SQUID cost is dominant, and could go way down: several institutions build their own (e.g. IBM and NBS). DC SQUIDs can be fabricated 32/wafer, and the electronics if produced in quantity would be cheap. In conclusion, I believe a cost of 5K$ - 10K$/m^2 to be achievable for superconducting monopole detectors. This is 5-10 times the cost per square meter of ionization detectors.

CONCLUSIONS

* Superconducting induction detectors have grown in size by more than two orders of magnitude in less than two years. There is still lots of room for clever ideas leading to arrays large enough to probe the Parker bound. There are no obvious limitations in size yet. (CFM, IBM)

* The present generation sets a combined limit on the monpole flux of $F_M < 1.5 \times 10^{-11}$ cm^{-2} s^{-1} sr^{-1} (90% C.L.).

* This is 93 times the original Cabrera exposure in which one candidate was seen. The compatibility of the new limit with the one previous candidate is on the order of 1%. (Stanford, CFM)

* RF SQUID systems do give steps, primarily due to mechanical stress in the SQUID region. Although good technique can avoid this, single-loop systems should be suspect and, a fair degree of redundancy is required for any claim that a step is a monopole. (NBS, IBM, CFM)

* New searches for monopoles trapped in matter have searched 400 kg of material, with no candidates. The limit on the abundance of monopoles with mass $> 10^{15}$ GeV in magnetic iron is

$$F_m < 2.3 \times 10^{-6} \text{ monopoles/gm (Kobe)}$$

* Warm detectors may be possible (M. Price).

This is a field rich in ideas and techniques; would that it were also rich in events.

The induction technique has the marvelous virtue that it can be calibrated directly, in contrast to ionization techniques. It has the pitfall, however, that it is a new technique, and one doesn't yet know what the backgrounds are. I would like to emphasize that any large detector must have enough information about an event so that even one event is highly convincing. Whether this is achieved by measuring multiple points on a trajectory, time-of-flight, or just having many loops in a common volume (or some combination of the above) is not yet clear. But no fate could be worse for an unwary experimenter than to have one event and not be able to 'see' enough about it to really know whether or not it is genuine. In preparing this talk late Thursday night after two days of sessions (and the banquet!) I turned to the Gideon Bible in my room. Just reading for a short while I found the following:

> 'The wise man's eyes are in his head
> but the fool walketh in darkness
> and I myself perceived also
> that one event happeneth to them all.'

Ecclesiastes 2:14

REFERENCES

1. M. Faraday, Faraday's Diary, Vol. I, October 17, 1831. Ed. by Thomas Martin; G. Bell & Sons, LTD (1932).
2. It should be noted that the effect does not depend on superconductivity.
3. See, for example, the nice review at this conference by Steven Ahlen.
4. Actually the first observation was on October 17. Figure 1 shows the apparatus of the measurement of Dec. 8 (see Ref. 1).
5. F.W. Grover, Inductance Calculations, D. Van Nostrand and Co. (1946).
6. Sorry - this is a little silly as the wire is also very long. But it is a small field.
7. B. Cabrera, Phys. Rev. Lett. 48, 1378 (1982).
8. SHE Corporation, 4174 Sorrento Valley Blvd., San Diego, CA 92121.
9. The proceedings are: Magnetic Monopoles, edited by Richard Carrigan, Jr. and W. Peter Trower (Plenum Press, 1983).
10. B. Cabrera, in Magnetic Monopoles, op, cit., p. 175.
11. C.D. Tesche et al., Appl. Phys. Lett. 43, 384 (1983); C.C. Tsuei, in Magnetic Monopoles, edited by Richard Carrigan, Jr. and W. Peter Trower (Plenum Press, 1983), p. 201.
12. C.D. Tesche, private communication.
13. As the purpose of the alternating cells is to decouple the loop from the shield, the equal size cells are the optimum solution. (The source due to a vortex in the shield has an infinite number of Fourier components.) The loops are called macrames by the group as early ones were woven out of wire.
14. H.J. Frisch, Monopole '82, Conference at Wingspread, Wisconsin.
15. L.W. Alvarez, in Proceedings of the 1975 Symposium on Lepton-Photon Interactions, Stanford 1975, p. 967 (T.W. Kirk, ed.).

16. L.W. Alvarez, LRL Physics Note 470, 1963 (unpublished).

17. L.J. Tassie, Nuovo Cimento 38, 1935 (1965).

18. P.H. Eberhard, LRL Physics Note 506, 1964 (unpublished).

19. L.W. Alvarez, Proposal to NASA to Search for Magnetic Monopoles in Returned Samples of Moon Surface Material, Jan. 30, 1966.

20. L.W. Alvarez et al., Review of Sci. Inst. 42, 326 (1971).

21. H. Kolm et al., Phys. Rev. D4, 3260 (1971).

22. P. Eberhard et al., Phys. Rev. D4, 3260 (1971).

23. L.L. Vant-Hull, Phys. Rev. 173, 1412 (1968).

24. E. Goto, J. Phys. Soc. Japan, 10, 1413 (1958), E. Goto et al., Phys. Rev. 132, 307 (1963), See also R.L. Fleischer et al., Phys. Rev. 184, 1393 (1969).

25. I have converted their number of 95% C.L. to compare to the published result of Eberhard et al. (Ref. 22).

26. Bandwidth?! (Both SQUID and readout).

27. D.E. Morris (LBL), talk at the Monopole '82 conference, Wingspread, Wisconsin, pp. 203-204, R. Carrigan, Jr. and W. Peter Trower, editors.

28. C. Rubbia, CERN preprint EP 82-01.

29. M. Price, CERN preprint EF 83-02, and paper submitted to this conference.

30. A purely transverse field is shown. For the more general case see the paper by J. Clem submitted to this conference.

31. J. Dupont-Roc et al., Phys. Lett. 28A, 638 (1969).

32. e.g. the SHE Corporation DC SQUID. See Ref. 9.

33. R.R. Wilson is a wonderful example.

A SEARCH FOR GUT MONOPOLES WITH A SINGLE THICK SCINTILLATOR

AT SEA LEVEL

G. Tarle[a,b], S.P. Ahlen[c] and T.M. Liss[b]

[a]Physics Department
University of Michigan
Ann Arbor, MI   48109

[b]Physics Department
University of California
Berkeley, CA   94720

[c]Physics Department
Indiana University
Bloomington, IN   47405

INTRODUCTION

Experimental searches for magnetic monopoles have prolif-erated in recent years due to the prediction of certain Grand Unified Theories (GUTs) that monopoles should exist and be extremely massive ($M > 10^{16}$ GeV/c^2). In contrast to monopoles of much lower mass, GUT monopoles are expected to be moving at very low speeds[1] ($V \sim 10^{-3}c$) and to have a greatly reduced rate of energy loss.[2] The possibility that a large reservoir of low velocity monopoles may have gone undetected has provided a challenge to experimentalists to devise new methods of detecting slow particles. Many experimenters have employed direct methods that rely only on the electromagnetic interactions of monopoles. Superconducting loop detectors, ionization detectors and various detectors based on excitation, such as scintillators, fall into this category. Of these, superconducting loop detectors are unique in that they respond to monopoles of arbitrary velocity. Several indirect techniques have been used to set impressive limits on the monopole flux. These techniques, however, often require additional assumptions regarding the velocity, charge state and the interactions of monopoles with nuclei that compli-cate the problem of setting hard limits on the monopole flux. For example, limits[3-4] based on the Callan-Rubakov effect[5] require

551

that baryon number is not strictly conserved, an assumption that has yet to be experimentally verified.[6] If monopoles do not catalyze baryon decay, then stable bound states of monopoles with nuclei may exist[7,8] and monopoles traversing the earth may pick up abundant heavy nuclei such as Al or Mn which have large magnetic moments. Such composites can be detected by tracks left in ancient mica[9] or may have sufficient energy loss to stop in the earth and be detected in stable matter searches. It is possible, however, that monopoles may capture a proton in interstellar space via an Auger mechanism[8] or by ordinary radiative capture in the early Universe.[10] If monopoles are accompanied by a positive electric charge then Coulomb repulsion may prevent the capture of heavy nuclei and searches which depend on the enhanced stopping power of such composites may be nullified. The problem with all the indirect searches is that so many loopholes exist that it is unlikely that any will be strong enough to provide compelling evidence against monopoles. In view of the above difficulties, it seems preferable to search for monopoles with a detector which relies on as few assumptions as possible and which is sensitive to monopoles having velocities and magnetic and electric charges within a reasonable band of expected values.

Ahlen and Tarle[11] have argued that scintillators offer an ideal compromise between costly superconducting systems, which have the widest dynamic range in velocity, and low cost gaseous detectors which may well have a velocity threshold above the astrophysically interesting region, near $10^{-3}c$. They conclude that polyvinyltoluene (PVT) based organic scintillators should exhibit a sharp threshold near $V \approx 6.5 \times 10^{-4}c$ whereas argon proportional detectors should have a threshold near $V = 2 \times 10^{-3}c$. Drell et al.[12] have suggested that enhanced excitation of hydrogen and helium at velocities as low as $V \approx 10^{-4}c$ may occur as a result of monopole induced level mixing. This has encouraged some workers to experiment with helium filled proportional detectors that can be constructed in very large areas and at very low cost. It should be noted that if monopoles are accompanied by a positive electric charge, then electrons will be readily captured and the combined system may be excited rather than the helium detection medium.[13] In this case, helium filled proportional chambers would offer no advantages over other types of gas proportional counters for the detection of monopoles.

For the experiment described here we have chosen to use PS-10 acrylic scintillator (10% naphthalene, 1% PPO, 0.01% POPOP). We have calculated the response of PS-10 to monopoles, dyons and monopole-Al composites as a function of velocity according to the technique of Ahlen and Tarle.[11] In Fig. 1 we present the results of these calculations along with the expected light yield for a minimum ionizing singly charged particle. There are several differences between PS-10 and NE110 which result in slightly

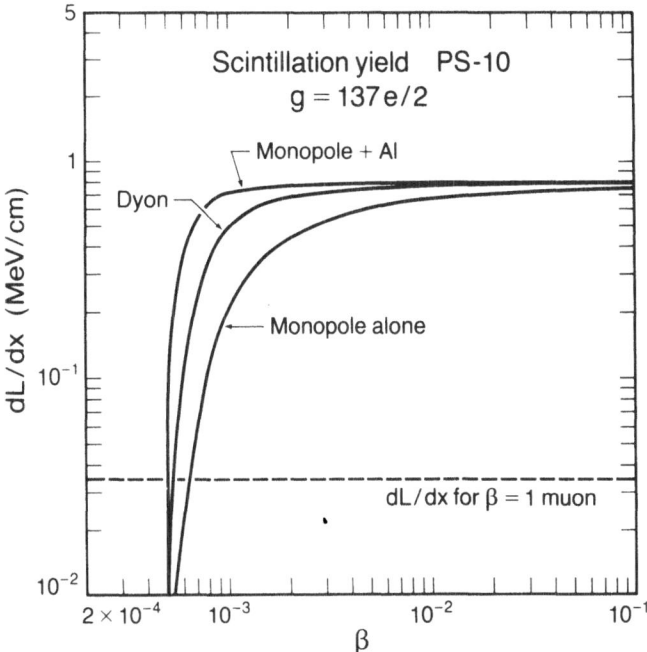

Fig. 1. Scintillation yield for monopoles, dyons and monopoles
with a bound Al nucleus in PS-10 acrylic scintillator as
a function of velocity V = βc.   The dashed line is the
scintillation yield for a minimum ionizing singly charged
particle.

different response curves than those presented in Ref. 11.   In
PS-10, naphthalene and perhaps acrylic are the primary energy
absorbing molecules.   Both have a minimum excitation energy, ε_m,
near 4.0 eV resulting in slightly lower threshold velocity,
$V_{th} \approx 5.1 \times 10^{-4}$ c, compared to PVT scintillators with $\varepsilon_m \approx 5.0$ eV
and $V_{th} \approx 6.5 \times 10^{-4}$c. We have measured and incorporated into our
calculation the α/β ratio[14] (a measure of saturation) and the
conversion efficiency of PS-10.   The α/β ratio is 6% for PS-10
compared to 7.5% for NE110 and the conversion efficiency for PS-10
is 50% of that for NE110.

THE THICK SCINTILLATOR DETECTION TECHNIQUE

From April through October 1983 we have operated a novel
monopole detector in the physics building at the University of
California at Berkeley.   The detection technique involved the use
of a single thick (7.6 cm) slab of PS-10 acrylic scintillator of
surface area 2.8 m^2 and geometry factor a full 17.5 m^2sr.   A

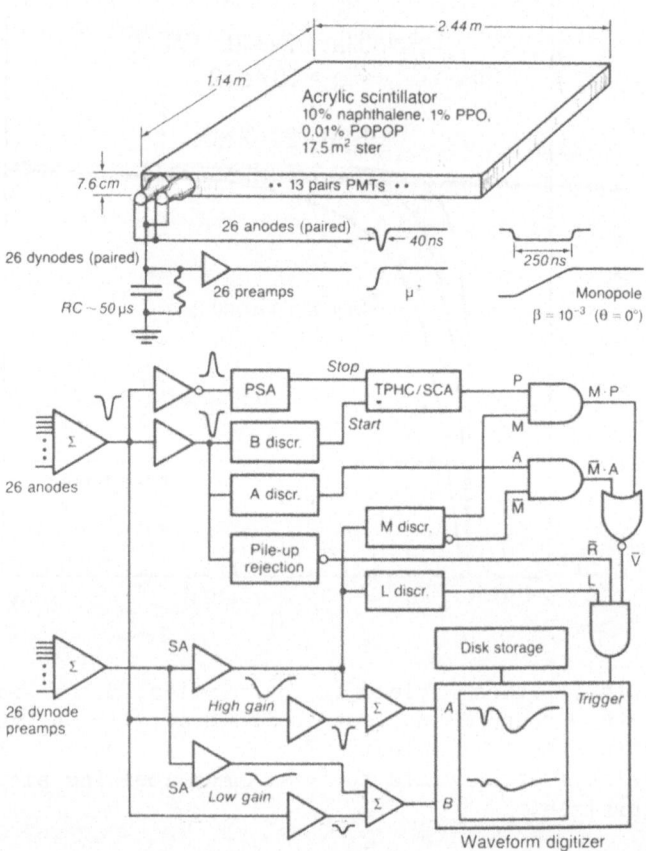

Fig. 2. Schematic   outline   of   the thick   scintillator detector
         system and block diagram of the   trigger and recording
         electronics.

schematic drawing of the detector and associated trigger electronics is shown in Fig. 2.   The scintillator was viewed on two
edges by 52 RCA 4900 photomultiplier tubes (PMTs) which collected
approximately 5000 photoelectrons from a minimum ionizing muon.
Pairs of adjacent tubes were selected for matching gains and a
common anode signal and common dynode signal were obtained from
each pair.   The anode signals were separately summed with appropriate weighting to compensate for PMT gain variations.   The
dynode signals from each pair of PMTs were sent to charge
integrating preamps with 50 μs decay times.   These signals were
summed with appropriate weighting to obtain a common integrated
dynode signal used in the monopole trigger.   A map of the response
of the detector was obtained using atmospheric muons and was found
to be uniform to ±10% over the entire area.

Slow monopoles would be distinguished from the enormous flux of fast penetrating electrically charged particles by examining the pulse shape of each fast anode signal. Relativistic particles always traverse the detector in a time much shorter than the response time of the overall detector system. The full width half maximum (FWHM) of the common anode pulse from such fast particles is 40 ns. A massive GUT monopole with velocity $V = 10^{-3}$ c will take a minimum of 250 ns to traverse the thick slab and the width of the anode signal is a measure of this time (see Fig. 2). The signature of a slow GUT monopole in our detector is an anomalously wide anode pulse with a pulse height constant in time. The thick slab technique is equivalent to a large number of redundant thinner detectors where continuity of pulse height and timing information are used to eliminate backgrounds.

The trigger electronics shown schematically in Fig. 2 were used to reject over 10^6 fast particles which passed through the detector each hour. To detect very slow particles it is important that integrated signals be used to initiate the trigger logic. In addition to reducing noise, this guarantees that events are not discarded which have large integrated signals, but by virtue of their extended duration, have instantaneous signals that fall below the trigger threshold. The discriminator labeled L in Fig. 2 was used to identify events with an integrated light output L_T greater than 0.6 I_{min} (I_{min} is the most probable light output for atmospheric muons). Note that due to the sharp cutoff in scintillator response at low velocities (see Fig. 1), much lower thresholds were unnecessary. The remainder of the trigger electronics was used to veto events which had a normal anode pulse width. Over most of the dynamic range of the instrument ($L_T > I_{min}$, identified by the discriminator labeled M) the anode pulse width was measured with a pulse shape analyzer (PSA) and a time to pulse height converter/single channel analyzer (TPHC/SCA). The discriminator labeled B initiated this process. If the pulse width was less than 120 ns, a veto signal resulted and the event was not recorded. At low light levels ($L_T < I_{min}$) a second method was used to veto fast particles. Normal events have a fixed ratio of anode pulse height to integrated dynode pulse height. Anomalously long pulses have a ratio considerably less than the nominal value. By properly choosing the threshold for the discriminator labeled A, a veto for normal events with $L_T < I_{min}$ was provided by the logic signal $\overline{M} \cdot A$. A pileup inspector supplied an additional veto for events consisting of multiple pulses (μ decays, accidentals, etc.) which appeared wide to the remainder of the trigger electronics. When an event was not accompanied by a veto signal, a shaped, integrated pulse, preceeded by the fast anode signal was recorded by a waveform digitizer which digitized the pulse height to 8 bits every 10 ns. These events were stored on a floppy disk and were examined each day. Two gain modes and digitization channels were used to cover the range $0.6\ I_{min} \lessgtr L_T \lessgtr 70\ I_{min}$.

Only about two of the more than 10^6 events per hour were recorded by the waveform digitizer. Approximately 90% of these were multiple pulses, which, because of overlap between pulses appeared wide to the electronics and were not rejected by the pileup inspector. Because of the enormous number of photoelectrons per event, multiple pulses were readily identified by visual inspection of the waveform. The remaining 10% of the triggers were muons with energy depositions far out on the Landau tail or showers with $L_T > 70$ I_{min}, which saturated the amplifiers and were again easily identified.

RESULTS

As of October 1983, the experiment had been operated for a total of 3852 hours. Over 4×10^9 fast particles had penetrated the detector and were rejected whereas 7137 events resulted in a trigger. Of the latter, 6150 were double pulses, 107 were triples, 7 were quadruples, one was a quintuplet, 52 were single pulses with normal widths that "leaked" through the trigger and 820 were events that saturated the amplifiers. Nearly all of these events could be eliminated as monopole candidates by visual inspection of the anode waveform. Two multiple pulse events had such small amplitudes that noise and photoelectron fluctuations made it difficult to resolve the individual peaks. Least squares fits of these events were performed using a sum of individual anode pulses where each pulse had an adjustable amplitude and starting time. In addition, both events were fitted to a simulated monopole parent function of variable width and amplitude. Fluctuations in pulse height were determined by adding photoelectron, noise and digitization errors in quadrature. For both events there was a reduced χ^2 for a multiple pulse fit which was considerably smaller than that for a monopole fit. Because of uncertainties in the calculated values for the pulse height fluctuations, F-tests were performed to determine which of the parent functions gave the best fit. In both cases there was more than one multiple pulse function that provided a significantly better fit than a monopole pulse. Further evidence that these events were not monopole induced comes from comparison of their integrated light levels with those in Fig. 1. The durations of the two pulses were 140 ns and 100 ns and both had $L_T = 0.7$ I_{min}. If these events were monopoles, their velocities would have to be at least 1.7×10^{-3} c and 2.3×10^{-3} respectively (normal incidence). At these velocities the measured light levels fall more than an order of magnitude below the curve in Fig. 1. We consider this to be well outside the uncertainties of the calculation. A more detailed discussion of these events and sources of multiple pulse events will appear in a later publication.[15]

DISCUSSION

Having seen no event with a signal characteristic of a slow monopole, we can set a limit of $F < 4.1 \times 10^{-13}$ cm^{-2}sr^{-1}s^{-1} (1 event significance) on the flux of monopoles with 6×10^{-4}c $< V < 2.1 \times 10^{-3}$c. The lower velocity limit corresponds to that velocity where the calculated light output for bare monopoles (Fig. 1) falls below the threshold of 0.6 I_{min} set by discriminator L. For velocities in excess of 2.1×10^{-3} c, the geometry factor for particles with transit times greater than the minimum detectable transit times (120 ns), gradually decreases from the full 17.5 m^2 sr.

In Fig. 3 we compare our limit with other experiments[16-24] (bold lines) which require no assumptions such as binding of nuclei or monopole catalysis of nucleon decay. We have taken the liberty of truncating the lower velocity limits for several of the experimental results at the velocity that we calculate they would become insensitive to monopoles. Experiments which rely on PVT based scintillators have been truncated at $V = 6.5 \times 10^{-4}$c and those that rely on ionization of Ar have been truncated at $V = 2 \times 10^{-3}$c, unless trigger threshold considerations demanded otherwise. For example, the experiment of Alexeyev et al. at Baksan uses a trigger set at 0.25 I_{min} for a signal integrated for only 50 ns. The signal for a slow monopole traversing the 30 cm thick Baksan scintillation modules will be stretched out in time and will have a reduced amplitude. Using the results of Ref. 11, we calculate that the velocity threshold for bare monopoles in the Baksan instrument should be raised to $V_{th} = 9.8 \times 10^{-4}$c. Also shown in Fig. 3 are theoretical limits[1,25-27] which purport to constrain the flux of GUT monopoles. The so-called Parker limit,[1] $F < 5 \times 10^{-16}$ cm^{-2} sr^{-1} s^{-1} $M < 10^{17}$ GeV/c^2 and $F < 5 \times 10^{-15}$ cm^{-2} sr^{-1} s^{-1} for $M \approx 10^{18}$ GeV/c^2, results from the dissipation of the large scale Galactic magnetic field (GMF) in the presence of a flux of monopoles. Dimopoulos et al.[28] have suggested that the sun may trap GUT monopoles into semistable orbits resulting in an enhanced flux at meteoritic velocities, $V \sim 10^{-4}$c. Freese and Turner[25] have concluded, however, that the orbits are too short lived to produce sizeable enhancements. If monopoles are given a radial "kick" by large magnetic fields within the sun, then enhancements by at most a factor of 50 are possible. Several authors have argued that if monopoles transfer energy to the GMF the Parker limit may be circumvented and a much larger flux of monopoles can be accommodated. An intriguing possibility is that monopoles might provide the mass needed to stabilize the Galactic disk. Salpeter et al.[25] have concluded that for $M > 10^{17}$ GeV/c^2, a disk stabilizing halo of monopoles and antimonopoles with $F < 1.7 \times 10^{-11}$ cm^{-2} sr^{-1} s^{-1} could exist and produce the observed GMF via magnetic charge density fluctuations. A problem with this model is that the GMF appears to have a non-zero curl[29] making it

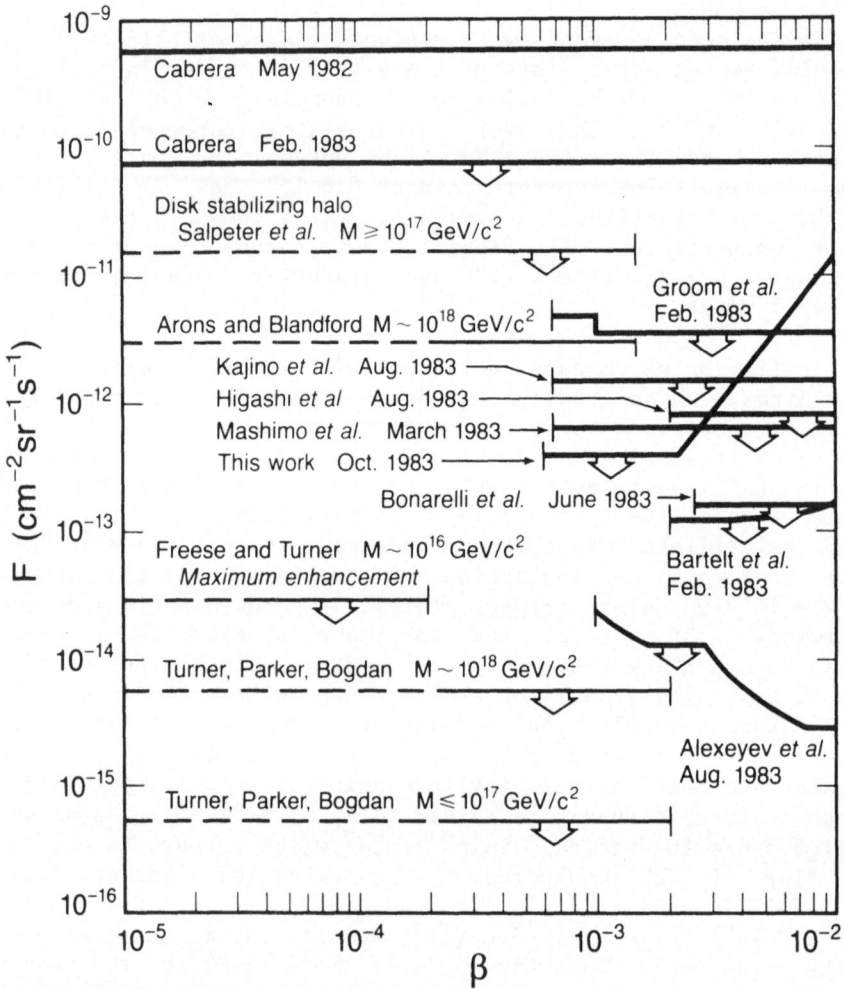

Fig. 3. Experimental (bold lines) and theoretical (light lines) limits on monopole flux as a function of velocity V = βc. For the theoretical limits the region of expected monopole velocity is indicated by solid lines.

impossible that monopoles are solely responsible for the field. Arons and Blandford[27] have concluded that the observed GMF can be maintained in a flux of monopoles, $F = 3 \times 10^{-12}$ cm^{-2} sr^{-1} s^{-1} with $M = 10^{18}$ GeV/c^2 which would provide sufficient matter to stabilize the Galactic disk. Rather than produce the GMF directly, the monopoles would resonantly transfer energy to and from a GMF produced by a standard Galactic dynamo. In their model monopoles form a halo about the Galaxy with a Maxwellian velocity distribution with dispersion speed ~ 200 km/s in the rest frame of the Galaxy. The motion of our solar system about the Galactic center (V ~ 250 km/s) would result in an anisotropic flux centered about the apex of the solar motion (ℓ_A). The velocity distribution would fall off for V > 1.5×10^{-3} c for monopoles arriving from ℓ_A and at a smaller velocity in any other direction.

The limit set by this experiment, although well above the Parker limit, is sufficiently low to address the disk stabilizing halo models. Our limit of 4.1×10^{-13} cm^{-2} sr^{-1} s^{-1} is about a factor of 40 below that of Salpeter et al. Since this model presents only an upper limit to the flux, we are unable to draw any conclusions regarding their model based on our results. Arons and Blandford predict a monopole flux of 3.2×10^{-12} cm^{-2} sr^{-1} s^{-1} so that it would appear that our limit, being almost a factor of eight below this, is in serious conflict with their model. It should be noted, however, that Arons and Blandford have not properly taken into account the directionality or the distribution in velocity of monopoles associated with their model so that the correct flux prediction will be smaller. Of the experiments shown in Fig. 3, ours provides the best combination of large area and low velocity sensitivity needed to address models in which monopoles cluster with the Galaxy. In addition, our flat detector is more suitable than telescopic detectors for detecting a directional flux which will exist in any such model.

ACKNOWLEDGEMENTS

We gratefully acknowledge M. Solarz for assisting in the construction of the detector and P. Buford Price for numerous discussions and for providing the excellent environment for research that exists within his group. This work was supported in part by a California Space Institute Award No. CS72-83 and by National Science Foundation Grant No. PHY-8024128.

REFERENCES

1. M.S. Turner, E.N. Parker and T.J. Bogdan, Phys. Rev. D 26, 1296 (1982).
2. S.P. Ahlen and K. Kinoshita, Phys. Rev. D 26, 2347 (1982).
3. E.W. Kolb, S.A. Colgate and J.A. Harvey, Phys. Rev. Lett. 49, 1373 (1982); S. Dimopoulos, J. Preskill and F. Wilczek, Phys. Lett. 119B, 320 (1982).
4. S. Errede et al., Phys. Rev. Lett. 51, 245 (1983).
5. V. Rubakov, JETP Lett. 33, 644 (1981); C.G. Callan, Jr., Phys. Rev. D 26, 2058 (1982).
6. R.M. Bionta et al., Phys. Rev. Lett. 51, 27 (1983).
7. W.V.R. Malkus, Phys. Rev. 83, 899 (1951); D. Sivers, Phys. Rev. 2D, 2048 (1970); C. Goebel, Monopole Seminars at Univ.

of Wisconsin, p. 51 (1981).

8. L. Bracci and G. Fiorentini, Phys. Lett. 124B, 493 (1983).
9. P.B. Price, Shi-lun Guo, S.P. Ahlen and R.L. Fleischer, submitted to Phys. Rev. Lett. (1983).
10. G. Fiorentini, private communication (1983).
11. S.P. Ahlen and G. Tarlé, Phys. Rev. D 27, 688 (1983).
12. S.D. Drell et al., Phys. Rev. Lett. 50, 644 (1983).
13. S. Parke, in proceedings of this conference.
14. J.B. Birks, The Theory and Practice of Scintillation Counting, Pergamon Press, Oxford (1964).
15. T.M. Liss, G. Tarlé and S.P. Ahlen, submitted to Phys. Rev. D (1984).
16. B. Cabrera, Phys. Rev. Lett. 48, 1378 (1982).
17. B. Cabrera, Stanford preprint, March 1983.
18. E.N. Alexeyev et al., Proc. 18th Inter. Cosmic Ray Conf., Bangalore, India 5, 52 (1983).
19. D.E. Groom, E.L. Loh, H.N. Nelson and D.M. Ritson, Phys. Rev. Lett. 50, 573 (1983).
20. T. Mashimo et al., Univ. of Tokyo preprint UTLICEPP-83-03 (1983).
21. F. Kajino et al., Proc. 18th Inter. Cosmic Ray Conf., Bangalore, India 5, 56 (1983).
22. S. Higashi, S. Ozaki, T. Takahashi and K. Tsuji, ibid. 5, 69 (1983).
23. R. Bonarelli et al., Phys. Lett. 126B, 137 (1983).
24. J. Bartelt et al., Phys. Rev. Lett. 50, 655 (1983).
25. K. Freese and M.S. Turner, Chicago EFI Preprint EFI 82/56 (1982).
26. E.E. Salpeter, S.L. Shapiro and I. Wasserman, Phys. Rev. Lett. 49, 1114 (1982).
27. J. Arons and R.D. Blandford, Phys. Rev. Lett. 50, 544 (1983).
28. S. Dimopoulos, S.L. Glashow, E.M. Purcell and F. Wilczek, Nature 298, 824 (1982).
29. C. Heiles, Ann. Rev. Astron. Astrophys. 14, 1 (1976).

USE OF LARGE VOLUME LIQUID SCINTILLATOR FOR MONOPOLE DETECTION

P. Galeotti, G. Badino, C. Castagnoli, W. Fulgione,
and O. Saavedra

Istituto di Cosmo-geofisica del CNR
Istituto di Fisica Generale dell' Universita di Torino
Torino, Italy

INTRODUCTION

Experimental searches for monopoles have been steadily increasing during the last few years, and many results on monopole flux limits obtained from several experiments using different techniques have been reported at this Conference. Among the ionization/excitation detection method, there are experiments, running or planned, in which either plastic or liquid scintillator is used. The main difference between these experiments is that plastics may be used for very large "area" array while the liquids may be used for very large "volume" detectors. Both techniques have advantages and disadvantages with respect to the critical parameters for monopole detection, viz., sensitivity in energy loss, in velocity range, and achievable flux limit, i.e. the detector's acceptance.

However, in our opinion, there is another important parameter which should be taken into account, namely the versatility of a huge (and costly) detector to perform multipurpose physics. This quality has been achieved in all of the large volume underground detectors. Indeed, even though the main object of all the existing liquid scintillation detectors (Artyomovsk, Baksan, Homestake, Mont Blanc) is neutrino astrophysics, they can be used as well for other physics: proton lifetime measurements, cosmic ray physics, high energy neutrino, electromagnetic and hadronic interactions, and searches for magnetic monopoles. In fact, this latter topic is a by-product of these detectors and not the principal aim for which they have been built. However, since all the liquid scintillation underground experiments have a large mass

of sensitive material, it seems natural to use them for doing
physics in as many ways as possible.

One could, however, question if the sensitivity of these
"wide spectrum" experiments to monopole detection is comparable to
that of single purpose experiments, built mainly to look for
monopoles with plastic scintillators in a telescope array.   The
aim of our communication is to describe a method for monopole
detection with large volume liquid scintillation experiments that
can reach the sensitivity of the best plastic experiments.

DETECTION METHOD

The sensitivity of any experiment for monopole detection
depends strictly on the detector's geometry. However, most of the
plastic telescopes work at the single photoelectron level, and the
sensitivity in energy loss is hardly better than 10^{-2} times the
ionization of a minimum ionizing particle.   Their sensitivity in
velocity, obtained by measuring the time of flight among the
layers in a telescope array, is of the order of $\beta \sim 10^{-3}$.   We use
these values as a reference in comparing the sensitivity of a
liquid scintillation counter with that of plastic telescopes.

Here we describe the general method of monopole detection
that later on will be applied to our Mont Blanc experiment, since
the detector's geometry should also be considered in this case.
The parameters that should be taken into account are, obviously,
the same as for plastics: energy loss sensitivity and velocity
sensitivity.   In fact the energy loss gives the total number of
photoelectrons produced during the transit time, and the velocity
determines how monopole signals are spread in time.

Regarding the sensitivity to energy loss, both in plastic and
liquid scintillators, the energy loss of a minimum ionizing
particle is ~2 MeV g^{-1} cm^{-2}.   Thus the number of photons produced
per unit path length is comparable, since the density of these
materials is comparable. However, the efficiency of light collection is generally much higher in plastics, in which light-guides
are used, than in liquids, in which internal reflections and
diffusion of light help the photons reach the photomultipliers.
For this reason, the number of photoelectrons produced per unit
path length in plastics is higher than in liquid scintillation
detectors.

Regarding the velocity sensitivity, one should start by
discussing the response of a single module, i.e. a single layer of
plastic or a single counter of liquid scintillator, and afterwards
discuss the response of the whole detector.   Also, in this case
the response time of the photomultipliers is comparable (generally

one integrates over a time of 50 ns). Hence, the time resolution is very similar in both cases. However, the total signal is spread over a very short time in plastics and over a longer time in liquids: the transit time of a monopole with $\beta = 10^{-3}$ is 200 ns through a plastic layer a few centimeters thick, and 3.3 µs through a liquid scintillation counter 1 m thick. Indeed, this difference is important, since photoelectrons originate randomly in time and their time distribution may be used for a better signature of monopole events in liquid scintillation detectors.

In our calculation we assumed that N photoelectrons are produced at random during the monopole's transit time, and that Poisson statistics can be used to represent their time distribution. We define a signal, s, whenever the number of photoelectrons, NP, recorded during the photomultiplier's integration time δt ns is larger than s:

$$NP = N\left(1 - \sum_{k=0}^{j-1} \frac{(\bar{n})^k}{k!} e^{-\bar{n}}\right) \quad ,$$

where $\bar{n} = N(\beta\delta t)/3.3$ is the average number of photoelectrons produced in δt in a 1 m path length. In this way, a value s = 1 corresponds to detecting a single photoelectron, while s > 1 can be used for detection at other energy levels. Furthermore, by using Poisson statistics, one takes into account the statistical fluctuations in the time delays among photoelectrons, which is a very important effect when working at the level of several photoelectrons. In fact, photoelectrons may bunch together during the transit time, with a probability depending on their density. This is very likely not uniform and may be better analyzed according to a Poisson law.

APPLICATION TO THE MONT BLANC DETECTOR

A liquid scintillation detector (LSD), built in collaboration between our Institute and the Institute for Nuclear Research of the Academy of Sciences of USSR, Moscow[1], will be running in the next few months in the Mont Blanc laboratory. The laboratory is located at a vertical depth of ~5,200 m.w.e. underground. The LSD consists of 72 counters ($1.0 \times 1.5 \times 1.0$ m³ each) for a total mass of ~ 90 tons of liquid scintillator arranged on 3 layers. The scintillator (C_nH_{2n+2}, density ~0.8 gr cm⁻³, attenuation length > 20 m) is watched by 3 photomultipliers, 15 cm diameter. The PMT's, mounted on top of each counter, are in a 3-fold coincidence within 100 ns.

For the LSD, we apply the method described above for monopole detection and discuss: (i) the range of detectable energy loss

and velocity; and (ii) the monopole flux limit that can be
obtained.   From experimental measurements made in the Mont Blanc
laboratory, we know that the background counting rate above
0.7 MeV is 0.02 to 0.04 counts during 50 μs, which is quite
negligible even for very slowly moving monopoles.   Also, a cosmic
ray muon loses ~160 MeV while crossing vertically one counter of
scintillator (corresponding to the expected value of
2 MeV g^{-1} cm^{-2}) and produces about 1,000 photoelectrons in the 3
photomultipliers during their coincidence time.   A low ionizing
particle, such as a GUT monopole, with ionization losses ε times
that of a cosmic ray muon, i.e.,

$$\left(\frac{dE}{dx}\right)_M = \varepsilon \left(\frac{dE}{dx}\right)_\mu \quad,$$

gives an average of ε × 1,000 photoelectrons in the 3 photomultipliers of one counter.   Therefore, on the average, the number of
photoelectrons from each photomultiplier is ~330 × ε.   We use this
value and the detector geometry to evaluate the limit of detectability of low ionizing particles in the LSD.

The second parameter that we need to consider is that the
signal induced by a monopole during its transit time is spread
over a time t = 3.3/β ns, while the integration time of a single
photomultiplier is 50 ns.   Thus, for a velocity β ⩾ 6.6 × 10^{-2},
the transit time of the monopole is shorter than the integration
time of the photomultiplier, and all the photoelectrons are
recorded as a single pulse.   For a velocity β < 6.6 × 10^{-2}, the
number of photoelectrons recorded in 50 ns depends on their time
structure, which is a function of the monopole's energy loss and
velocity.   Assuming that the time distribution of photoelectrons
is Poisson, we calculated at 90% confidence level the probability
of detecting at least 1 photoelectron in 1 photomultiplier during
the integration time as a function of ε and β.   The result is
shown in Fig. 1 by the dashed line "a".

Afterwards, we calculated the probability, at 90% confidence
level, that all 3 photomultipliers of 1 counter detect at least 1
photoelectron each during their coincidence time (100 ns).   This
result is shown in Fig. 1 as the solid line "b".   Finally, we made
the same calculations at the level of 5 photoelectrons, corresponding to > 1 MeV energy loss in our counters.   In Fig. 1, the
circled and dotted line "c" shows the sensitivity of one counter
to 5 photoelectrons in a single photomultiplier.   The dotted line
"d" in Fig. 1 shows the sensitivity to 5 photoelectrons in the
3-fold coincidence of the 3 photomultipliers.   Figure 2 displays,
at the 90% confidence level, the energy loss of a monopole with a
given velocity β as a function of the number of pulses detected
during its transit time in one counter (with 3 PMTs in 3-fold
coincidence within 100 ns).   The dashed lines are computed at the
single p.e. and solid lines at the 5 p.e. level.

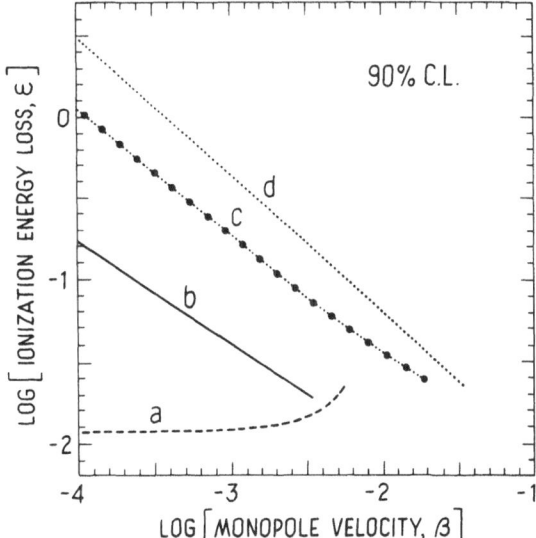

Fig. 1 Probability of detecting a monopole in liquid scintillator versus the monopole's energy loss and velocity. (a) single photoelectron (pe) in 1 PMT; (b) 1 pe in each of 3 PMT's of a single counter module; (c) 5 pe's in a single PMT; and (d) 5 pe's in each of 3 PMTs.

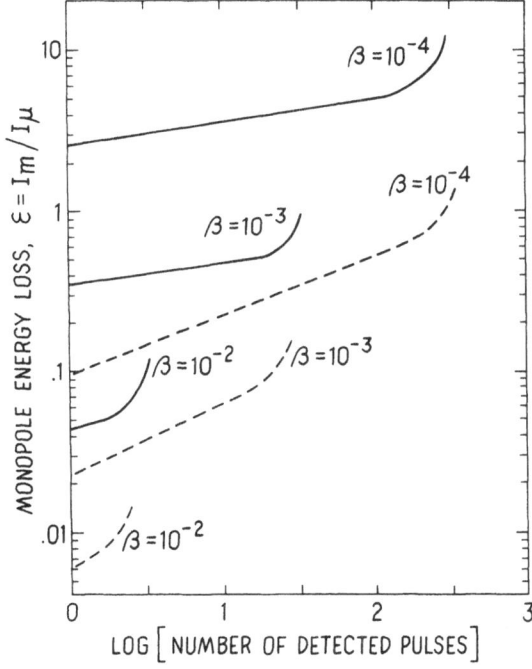

Fig. 2 Energy loss of a monopole vs. number of detected pulses. Solid lines = 5 p.e. level and dashed lines = 1 p.e. level.

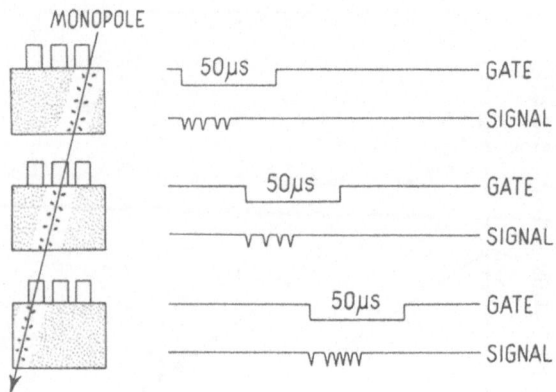

Fig. 3 Signature of a monopole in the LSD.

CONCLUSIONS

The signature for monopole detection in the LSD, as shown in Fig. 3, is given by a train of small pulses into a counter and by the delayed coincidence between contiguous counters (at least 3 counters are involved by a monopole crossing our detector). The time structure of pulses into 1 counter and the time of flight between counters give the velocity information, while the energy loss information is given by the number of pulses recorded during the transit time (Fig. 2). Since our apparatus works with 3 photomultipliers in a 3-fold coincidence and measures the pulse height during the resolution time of 100 ns, we are also able to record every single pulse ($\geqslant$ 1 photoelectron per photomultiplier) produced during the transit time of the monopole. On the contrary, most other experiments usually integrate the signal over the entire transit time of the monopole. The monopole flux limit (in the velocity and energy loss ranges given by the solid line "b" of Fig. 1) that can be reached in 1 year lifetime is $\sim 5 \times 10^{-14}$ cm^{-2} sr^{-1} s^{-1} at the 90% confidence level.

REFERENCES

1. G. Badino, G. Bologna, C. Castagnoli, B. D'Ettorre Piazzoli, W. Fulgione, P. Galeotti, G.P. Mannocchi, P. Picchi, O. Saavedra, V.L. Dadykin, P.V. Korchagin, V.B. Korchagin, A.S. Mal'gin, O.G. Ryazhskaya, V.P. Talochkin, A.L. Tsyabuk, V.F. Yakushev, G.T. Zatsepin, Proc. 8th Int. Workshop on Weak Interactions and Neutrinos, Javea, Spain, 1982.

MONOPOLE SEARCH AND NEUTRINO ASTROPHYSICS WITH LIQUID

SCINTILLATION DETECTORS

R. I. Steinberg, K. Brown, M.L. Cherry, S. Corbato,
I. Davidson, D. Keida, K. Lande and C.K. Lee

Department of Physics
University of Pennsylvania
Philadelphia, Pennsylvania   19104

ABSTRACT

We describe the 140 ton Large Area Scintillation Detector now nearing completion in the Homestake Mine and our plans for a one kiloton liquid scintillation solar neutrino detector.

INTRODUCTION

The Large Area Liquid Scintillation Detector (LASD) we are now building in the Homestake Mine at a depth of 4200 mwe consists of a hollow box 8 m high × 8 m wide × 16 m long, constructed from 200 liquid scintillator modules (see Fig. 1).

The 1500 m^2-sr aperture of the LASD makes it an exquisitely sensitive detector for magnetic monopoles in the velocity range above 1.5×10^{-4} c. The LASD will also be used for studying neutrino transmission through the earth, measuring properties of very high energy cosmic rays, and searching for neutrino bursts from collapsing stars.

In addition to these physics goals, this large area (but hollow) detector also serves as a prototype for a proposed solidly stacked 1000 ton detector using the LASD module design. This instrument (the LMSD-Large Mass Scintillation Detector) will be optimized for the detection of low energy neutrinos in the presence of various backgrounds. The primary physics goal for the LMSD will be the detection via neutrino-electron elastic scattering of solar neutrinos produced by the decay of ^8B.

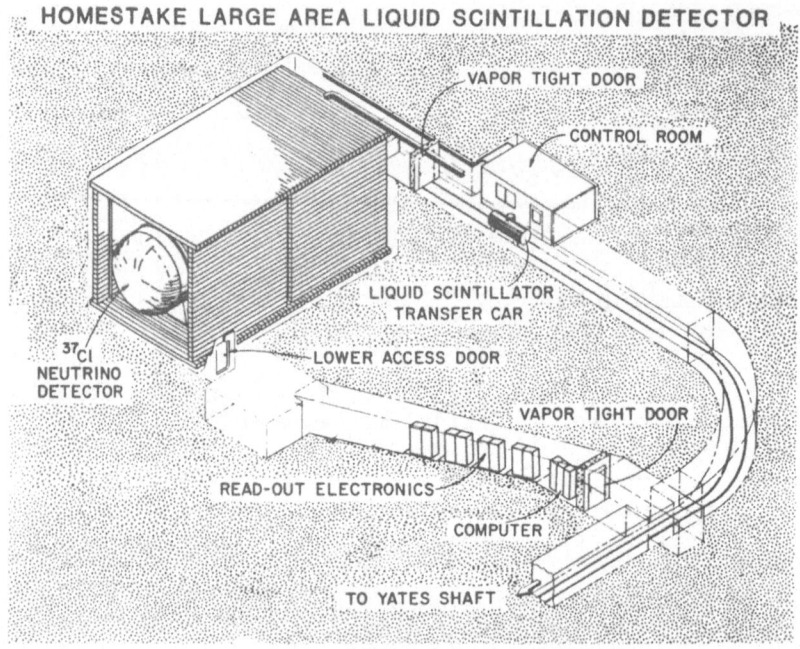

Fig. 1   Isometric view of experimental room on the 4850 ft. level
         of the Homestake mine showing the Large Area Scintillation
         Detector (LASD).

In this paper we discuss the following physics topics: (1)
the search for massive magnetic monopoles; (2) neutrino bursts
from the final gravitational collapse of massive stars; (3)
detection of solar neutrinos with a solidly stacked array of
liquid scintillator modules. We conclude with the discussion of the
LASD electronics and construction status.

MAGNETIC MONOPOLES

The Grand Unified Theories that predict nucleon decay also
provide for the creation of large numbers of magnetic monopoles
early in the history of the universe.   Although there are no
direct estimates of the density of monopoles remaining today,
upper limits have been established from consideration of the
closure of the universe and the stability of galactic magnetic
fields.   These considerations imply that, on galactic scales, the
mean flux of monopoles can be no greater than 10^{-15} cm^{-2} s^{-1} sr^{-1}
(the Parker limit).   This limit does not apply to monopoles

trapped locally in the Galaxy—for example, monopoles orbiting like micrometeoroids in the Sun's gravitational field. The expected velocities of such local monopoles would be of order 10^{-4} c, comparable to the earth's orbital velocity.

Cabrera has reported the possible detection of a magnetic monopole with magnetic charge $(137/2)e$ in a small superconducting loop. The corresponding flux would be much higher than the maximum fluxes allowed by the galactic field and closure constraints. Since no additional monopole-like events have been detected in exposures totalling almost 100 times that which yielded the original event, it seems imprudent at this time to consider the original Cabrera event as genuine. Future searches must be measured by the length of time required to reach the Parker limit.

Fast monopoles are expected to ionize several thousand times as intensely as muons and thus be readily recognizable. At a velocity of 10^{-3} c, the monopole energy loss rate is about 10 times that of a muon. At a velocity between 1.5×10^{-4} c and 6×10^{-4} c, it is probable that monopoles would no longer produce molecular excitations and thus would be undetectable in a scintillation detector. There is considerable controversy about the behavior of slow monopoles and both the exact value of the low velocity cutoff and the energy loss rate of monopoles near the cutoff. Quantitative estimates of slow monopole behavior in scintillator may improve as the problem is more thoroughly studied.

It is therefore prudent to insist that any monopole detector be sensitive to slow particles ionizing at rates as low as 10% of minimum ionizing or 0.2 MeV cm^2/g. The Homestake LASD has been designed to fulfill this stringent requirement. With a scintillator thickness of 25 g/cm^2, a monopole threshold at 10% of minimum ionizing yields a pulse height of 5 MeV, well above the level of typical energy deposits from background radioactivity. In addition, the 30 cm scintillator thickness yields a pulse width of $1/\beta$ ns, thus providing a clear broadening of the detector pulse for slow monopoles and therefore another constraint on the monopole signal.

In all, the Homestake LASD provides four constraints, allowing unique characterization of a trasversing monopole, together with its velocity and ionization: (1) and (2) the pulse widths in the entering and exiting counters should each be 1 ns/β; (3) the delay between the pulses in the entering and exiting counters should be 25 ns/β; and (4) the two pulse heights should be consistent with each other and with expectations for a monopole moving at the velocity determined by (3). Such multiple redundancy is particularly important in searching for a particle as elusive as the monopole.

For the velocities of interest, the delay time between the two counter pulses will be 25-250 μs. Such long delays can best be correlated in a very low background environment, such as that available in a deep mine. The most severe background will be due to two independent, traversing muons, each of which is detected in only one of the two counters through which they pass. The accidental coincidence rate at our depth for independent muons is 4×10^{-5}/yr. This already small accidental rate will be even further reduced by consideration of the pulse heights and widths in the two counters and by necessity of missing the outgoing pulse from each of the two accidentally correlated muons.

With the LASD aperture of 1500 m^2 sr, one event per year would correspond to a flux of 2×10^{-15} cm^{-2} s^{-1} sr^{-1}, about 3×10^5 times lower than the flux corresponding to the original Cabrera event. The Parker limit would be attained in a two year run.

NEUTRINO BURSTS FROM COLLAPSING STARS

In the final gravitational collapse of sufficiently massive stars, a sudden drop in the electron degeneracy pressure supporting the stellar core can lead to the release of a large pulse of electron capture neutrinos, thermal $\nu-\bar{\nu}$ pairs, and (in the case of an asymmetric collapse), gravitational radiation. The total energy radiated may be 10^{53} ergs, corresponding to nearly 10^{58} neutrinos with energies of 10 to 100 MeV. The rate of such events per galaxy is highly uncertain, but rates of once every 10 to 40 years have been suggested based on optical and radio surveys of supernovae in external galaxies and supernova remnants in our own galaxy. Pulsar birth rates have indicated collapse rates as high as once every 4 to 6 years.

In order to detect neutrino bursts, a lengthy search is required with massive, well-shielded detectors located deep underground or underwater, having a low background counting rate and a high efficiency for detecting low-energy ν_e and $\bar{\nu}_e$. For a typical stellar collapse model, a burst of 10^{58} 10-20 MeV ν_e and $\bar{\nu}_e$ at the galactic center would result in 10 to 20 interactions in the detector in less than a second, with roughly equal contributions from each neutrino species.

The main detector background is due to Compton scattering of gamma rays from local radioactivity. In order to separate clusters of counts due to real neutrino bursts from the steady radioactivity-induced background, we study the distribution of events in time-i.e., we look for bursts of N events occurring within a time interval t_N shorter than would be expected from random statistics. The published results of a one year neutrino burst search with our old 300 ton water Cerenkov detector clearly demonstrate that the background rate is sufficiently low to permit detection of events consisting of 5 or more neutrino counts within 0.1 s or less, or more than 10 counts in several seconds, corresponding to a burst of 10^{58} neutrinos with $E_\nu \gtrsim 10$ MeV at the Galactic Center.

SOLAR NEUTRINOS

Status of the Search for Neutrinos from the Sun

One of the most pressing questions in astrophysics is to demonstrate experimentally that hydrogen fusion is the source of energy generation in the Sun. The impressive radiochemical experiment of Davis and collaborators utilizing a perchlorethylene detector has made enormous strides in searching for the neutrinos emitted in the fusion process. Their signal of 2.2 SNU is about four times the background due to conventional cosmic ray processes induced by muons and neutrinos. It is also within a factor of three of the prediction of the standard solar model.

Unfortunately, the radiochemical technique integrates the neutrino signal over time, source direction, and neutrino energy. These integrations not only make it difficult to identify the reactions that give rise to the neutrinos, but also prevent clear identification of the Sun as the source of the detected neutrinos.

Solar Neutrino Detection with the LMSD

We can overcome most of the shortcomings of radiochemical detection and verify Davis' chlorine detector results by searching for solar neutrinos with our proposed Large Mass Scintillation Detector. This detector will permit determination of the time of neutrino arrival as well as measurement of the shape of the neutrino energy spectrum.

Of the neutrino emitting processes, only the decay of ^{8}B produces sufficiently energetic neutrinos to be easily detected in a scintillation detector. This decay is also the primary assumed source of the neutrinos detected by the chlorine radiochemical detector. Since the beta spectrum of ^{8}B extends to 14 MeV, a large fraction of the emitted neutrinos have energies above the level associated with radioactivity background.

We intend to look for two signals, that produced by the elastic scattering of solar neutrinos on the electrons of our detector, and that produced by the interaction of antineutrinos with the free protons in our scintillator. We are sensitive to a very small admixture of antineutrinos in the solar emission because the antineutrino interaction probability in our detector is 20 times that for neutrinos, and because of the additional label of delayed neutron capture in either the scintillator or the ^{35}Cl of the detector housing. Although there is no particular theoretical motivation to expect that neutrinos oscillate into antineutrinos during their passage to Earth, the great sensitivity of this detector to antineutrinos makes this search quite interesting.

The rate of solar neutrino events above a threshold of 5 MeV in a one kiloton fiducial volume detector would be 6 per day assuming the ^{8}B neutrino flux at the earth is $6 \times 10^6/cm^2$ s. Since the chlorine detection rate sets an upper limit of 1/3 of that flux, it would be prudent to anticipate a considerably lower detection rate.

The major problem in detection of solar neutrinos has always been the reduction of background to a level sufficiently below that of the signal. The background can originate from cosmic ray muons and neutrinos, from radioactivity (beta, gamma, or neutron emission) in the surrounding medium, and from radioactive contaminants within the detector.

Our proposal is to build a detector with large linear dimensions, so that the outer regions of the detector can serve as an active shield for the inner fiducial volume. It is also necessary to construct the detector from materials that are low in uranium and other radioactive impurities and to subdivide the detector into modules with linear dimensions at most equal to the mean free path for Compton scattering.

Previous attempts to build a solar neutrino detector using scintillator (by Reines and collaborators in South Africa) failed because the linear dimensions of their detector were too small. Their main background resulted from gammas produced by spontaneous fission of ^{238}U undergoing Compton scattering, with the recoil electrons simulating neutrino scattering. The Compton scattering length at 10 MeV in scintillator is about 50 cm. The South African detector was only 165 cm in diameter or about 3 scattering lengths. There was, therefore, not enough material to absorb the gammas nor could gammas be recognized by multiple scattering within the detector since more than 5% would undergo only one scattering in traversing the detector.

We avoid such problems by designing the linear dimensions of the LMSD to be 8 or 10 m, or about 20 Compton scattering lengths. The chance that a gamma will undergo one and only one scattering is thus reduced to about 10^{-9}. Building such a detector requires liquid scintillator with a long attenuation length such as the scintillator we have developed for the present 8 m detector modules. These modules, furthermore, have cross sections of 30 cm, considerably smaller than the Compton scattering length, thus insuring that multiple scatterings occur in different modules and can thus be recognized.

Liquid scintillator and PVC housings also make ideal materials for the LMSD since both are highly purified and are made from low radioactivity raw materials. The long module length insures that the photomultiplier tube and its radioactive noisy glass housing are far from the fiducial volume of the detector. Indeed, the major material concern for the interior region of the LMSD is the support structure. In the present LASD we use steel, a material we are proposing to avoid in the interior of the solar neutrino detector.

The cosmic ray muon background consists of the direct passage of muons through the detector, the stopping and decay of muons, the conversion by muons of carbon into short-lived nuclei with high beta end points, and the production of neutrons by muon interactions either within the detector or in the materials surrounding the detector. The muon flux at our depth is 4 muons/m^2/day. Of these less than 1% will stop in the detector. The muon entry and passage through the detector will fire 20 or more modules. Since multiple module firing is inconsistent with solar neutrino elastic scattering, throughgoing or stopping and decaying muons can be easily vetoed. The decay of an unstable nucleus within a few seconds after the passage of a muon can also be recognized and vetoed since the path of each muon is well known and the number of muons per day is so small. The final problem is that of neutrons made in the surrounding rock. Their attenuation length in the detector is comparable to that of 10 MeV gammas so that we expect them to be confined to the outer layers of the detector.

An important demonstration that external background is being properly attenuated by the outer layers of the detector will be provided by observing the decrease of background count rate as a function of distance from the outside. By having a detector thickness of over 20 modules it should be straightforward to carry out such a demonstration.

In order to assess the background problems at Homestake, we plan to build a small stack, consisting of 25 counters, at one edge of the LASD. With this ministack we should be able to measure the background rate from the local radioactivity and determine the extent to which we can exclude it. We are particularly interested in the spatial and energy distribution of the background signals and plan to try various strategies for dealing with the background.

DETECTOR ELECTRONICS FOR THE LASD

The electronics required for the detection of monopoles, especially slow ones, differs considerably from that required for most particle physics experiments. For example, a monopole with $v/c = 10^{-3}$ would spend about a microsecond in one of the detector modules and deposit about the same amount of energy as a minimum ionizing particle. If we set a threshold of 1/10 minimum ionizing (about 35 photoelectrons at each photomultiplier), it would require about 100 ns to accumulate enough charge at each discriminator to satisfy the trigger condition. We thus require charge integration at the discriminator input and a slow discriminator that does not fire repeatedly during the microsecond pulse duration. Since the detector must also be able to respond to fast muons or monopoles, we have also included a fast discriminator section in the trigger and readout system. Figure 2 is a schematic diagram of the LASD front end electronics.

The distinguishing features of a slow monopole would be its long flight time across the detector and long passage through each detector module (Fig. 2). The latter feature, a pair of multi-microsecond pulses, is particularly striking and would uniquely label the monopole. The transit time across the LASD can vary from 25 ns for a muon to 250 μs for a slow monopole. In order to cover this range, we have designed the readout electronics to have both fast and slow clocks, as shown in Fig. 3. The fast clock covers the first 500 ns, while the slow UTC clock covers the time

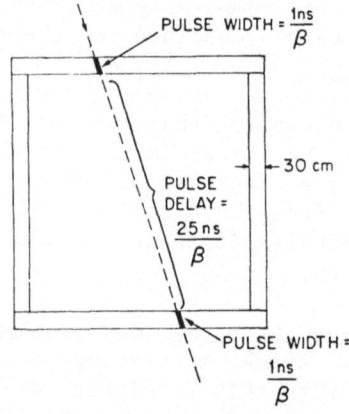

Fig. 2   Timing signature for a monopole traversing the Homestake Large Area Scintillator Detector.

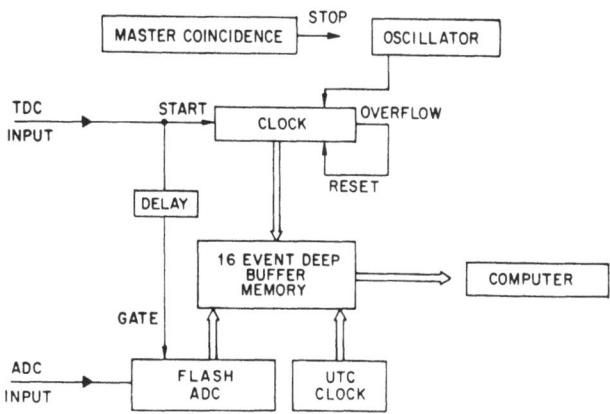

Fig. 3   System for digitizing time and pulse height information
          from each phototube. The fast clock runs at 400 MHz,
          giving 2.5 ns timing bins, while the slow UTC clock
          provides a 10 MHz signal with corresponding bins of
          100 ns. The flash ADC provides 6 bits of pulse height
          information. The computer is an LSI-11/23 with 30 Mb of
          on-line storage.

span thereafter.   Each phototube has a 16 word deep memory buffer
so that multiple pulses for each event can be recorded.

     The electronics for the LASD consists of a front end amplifi-
er, fast and slow discriminators, a trigger logic system that
establishes a two fold coincidence between the two ends of each
module, a remotely controlled high voltage board, a time and pulse
height digitizing board and a readout system.

     The amplifier, discriminators and the high voltage board are
mounted on the outside of the module end cover plate adjacent to
the phototube. This provides an extremely short path, <10 cm,
between the phototube output and the electronics.   Not only does
this arrangement preserve the output pulse, but it also avoids
long runs of large numbers of high voltage cables.   The phototube
assembly is merely supplied with low voltage (±15V DC) power and
the control signals for the HV supply.   Each detector sends a
digital timing pulse and an analog amplitude pulse back to the
central electronics together with a signal proportional to the
phototube high voltage.

     The high voltage supply, which is fed by a 15 volt source,
produces 2.5 kV at 2 ma.   An external control line permits
measurement and adjustment of the high voltage from the computer
at the main electronics control area.   For each trigger, the
relative firing time of each tube, the pulse amplitude of that

tube and the absolute UTC (Universal Coordinated Time) of the trigger are transferred to the computer memory and subsequently to the output tape.

Cosmic ray muons and neutrino-induced muons will usually produce two pairs of pulses, one pair from each of the modules traversed. From the time difference between the pulses at each end of a given module we can locate the position of a muon to ±15 cm. We can also recognize multiple muons passing through a given module by the increased pulse height associated with the signal and the mismatch between time differences and pulse height ratios from the two ends. Since each module is an independent muon detector, we are able to recognize and count very high muon multiplicities associated with a given cosmic ray interaction.

All throughgoing muons must pass through two walls of the detector, one on entering and the other on exiting. From the location of the entering and exiting points (± 15 cm), the muon direction can be determined to ± 3 degrees.

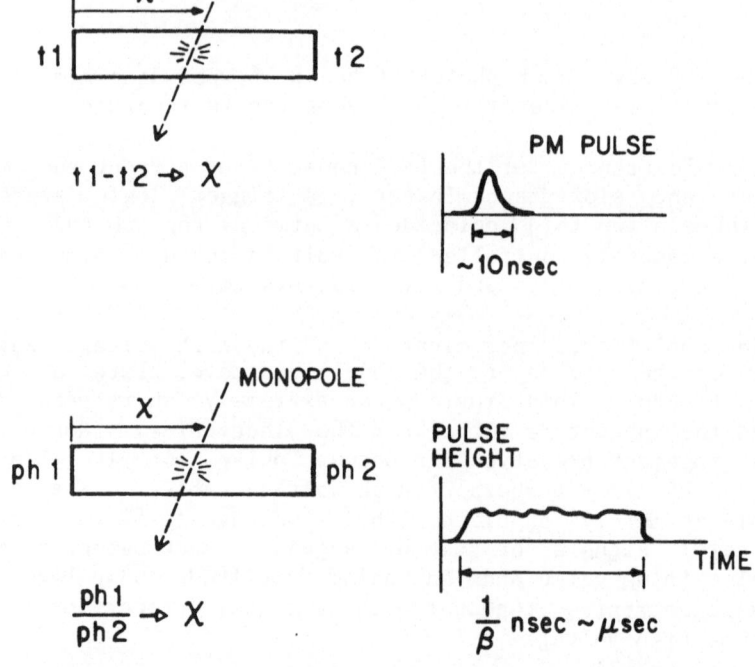

Fig. 4 Muon and monopole signals as seen in one of the scintillator modules. Both timing and pulse height information are used to determine the position of muons; for monopoles, pulse height information alone is used. Observed spatial resolution with the detector is ± 15 cm.

For a throughgoing monopole, we expect a slow coincidence between the phototubes at the two ends of a given module. The position of the monopole within that detector element will be determined by comparison of the pulse heights observed by the two PMT's, as shown in Fig. 4. The monopole transit time across the detector can be so long that its passage through the other detector wall must be treated as a separate event. The transit time of the monopole will be measured by the difference in UTC time between the two triggers.

CONSTRUCTION STATUS

We have essentially completed all mechanical and electronic work on the detector. Twenty tons of liquid scintillator (out of a total requirement of 140 tons) are currently being pumped into the modules, with forty more tons on order. Full operation of the instrument will commence during the spring of 1984 after delivery of the remaining liquid scintillator.

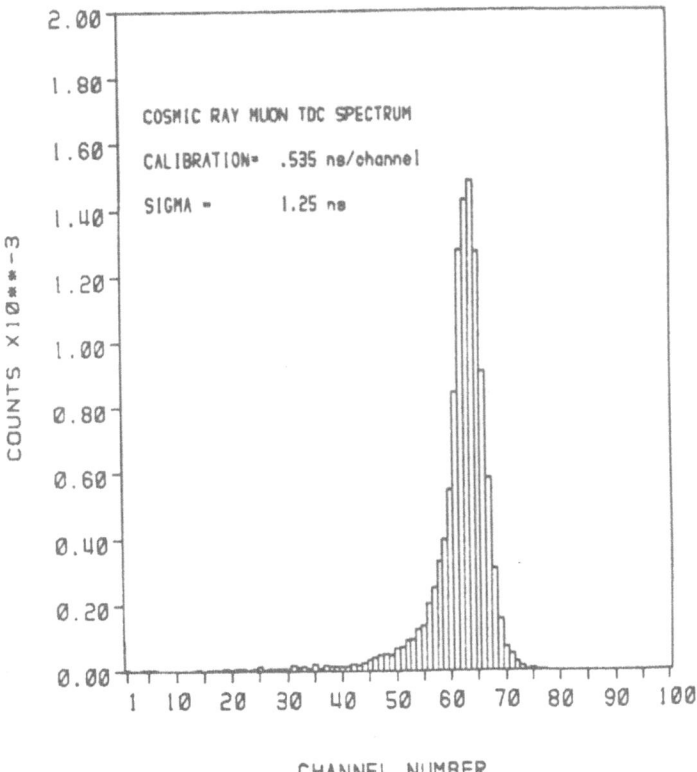

Fig. 5 Experimentally-observed pulse height spectrum for cosmic ray muons.

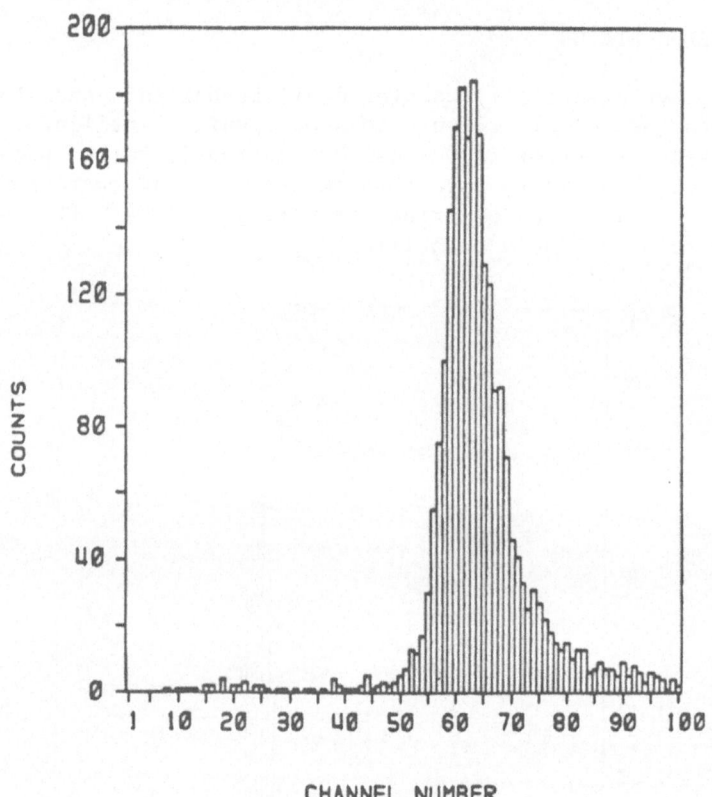

Fig. 6 Observed TDC spectrum for cosmic ray muons showing time
resolution of ±1.25 nanoseconds.

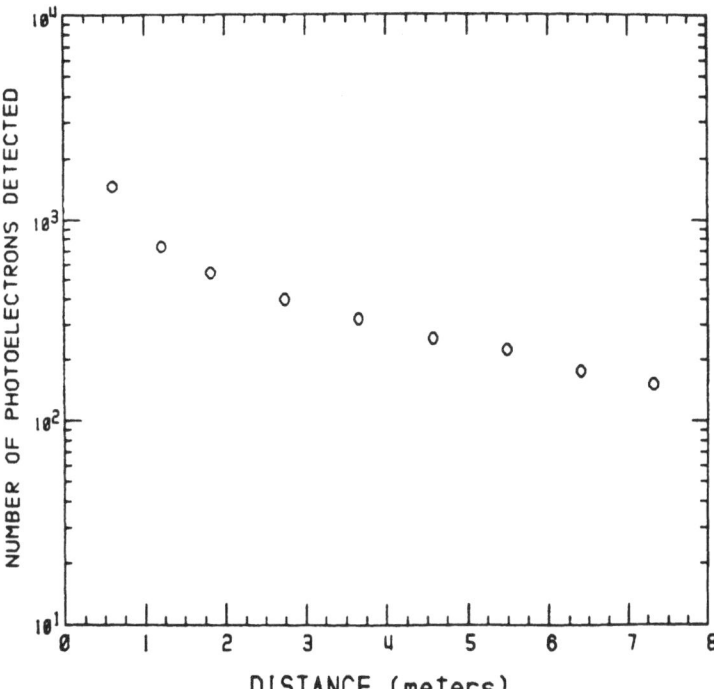

DETECTED PHOTOELECTRONS/MUON VS DISTANCE

MUON ENERGY LOSS = 54 MEV
8 m x 30 cm x 30 cm DETECTOR MODULE

Fig. 7 Measured photoelectron yield for cosmic ray muons for a
typical detector element as a function of distance from
phototube.

ACKNOWLEDGEMENTS

It is a pleasure to thank the Homestake Mining Company for
its continuing generous cooperation on these experiments.   Work
supported in part by the U.S. Department of Energy under contract
DE-ACO2-81-ER-40012.

STATUS OF THE TEXAS A&M GUT MONOPOLE SEARCH

Robert C. Webb

Physics Department
Texas A&M University
College Station, Texas   77843

ABSTRACT

We present here a report on the status of the GUT monopole search being staged by the Texas A&M University group. This experiment is located underground at a depth of 1500′ in a nearby salt mine and utilizes a large area ($\sim$ 53 m^2) scintillation counter telescope to detect the passage of a slow moving super-heavy, magnetic monopole. A description of our detector and the expected signature for a passing magnetic monopole will be presented.

INTRODUCTION

Searches for the Dirac magnetic monopole, since its incep-tion[1] have been quite numerous[2], however, none of these save perhaps the recent experiment of B. Cabrera[3] have yielded positive results.   In addition, with the advent of Grand Unified Theories (GUT's)[4], we find that the properties of the magnetic monopole may be quite strange indeed.   So unusual, in fact, that until very recently most of the searches conducted may have excluded the classes of particles with the properties these GUT monopoles might possess.[5] We have designed an experiment to search specifically for these objects in cosmic rays.   The properties of the GUT monopoles of interest are: i) a heavily penetrating cosmic ray particle; ii) a slowly moving particle with velocity in the range $10^{-4} < \beta < 10^{-1}$; iii) a moderately ionizing particle between 1-100 x minimum ionizing; iv) a rare object with fluxes of 10^{-13} cm^{-2} s^{-1} sr^{-1} or less in cosmic rays.   We will briefly describe this detector and its status in this report.

DETECTING THE SLOW MONOPOLE

As has been pointed out in recent literature on the subject[6], the detection of very slow moving magnetic poles pose some definite experimental difficulties. While this topic is still being debated[7], there appears to be a consensus emerging, that for velocities in the range of $\beta > 5 \times 10^{-4}$, plastic scintillation detectors have a significant efficiency for detecting the moderate ionization being deposited by the slowly moving monopoles. This result is based solely on the electronic ionization loss for these heavy projectiles in the scintillation material. Furthermore, in an early preprint from our group[8] on the subject of a monopole's energy loss in scintillator, we discussed the possibility that the pole's strong magnetic field could severely disturb the electron orbitals of the naphthalene molecules giving rise to a level crossing phenomenon between singlet and triplet states similar to that calculated by Drell et al.[9] for helium and atomic hydrogen. The relaxation of these crossed levels should also produce scintillation light or delayed fluorescence with a characteristic lifetime of triplet-triplet interactions in the scintillation material, which is approximately 1 µs. Figure 1 shows the expected light yields in plastic scintillator for each of these forms of energy loss versus the monopole's velocity.

The electronics being employed to search for these rare monopole events has been designed to be sensitive to both the prompt light given off as a result of the conventional electronic energy loss and this possible delayed fluorescence coming from the level crossing interaction. The special features of this aspect of the detector will be addressed in the next section.

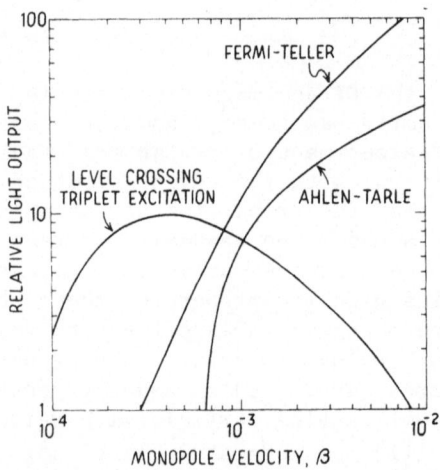

Fig. 1   Expected light output from plastic scintillator versus monopole velocity taken from References 6 and 8.

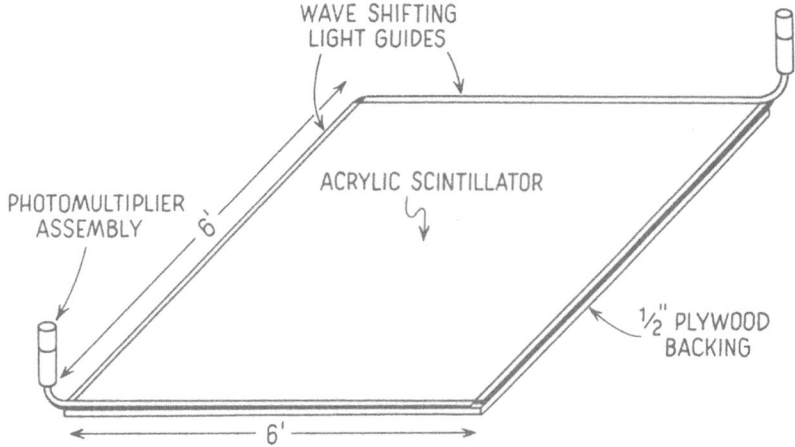

Fig. 2   Detail of single acrylic scintillation detector.

THE DETECTOR

This detector utilizes acrylic based scintillation material as the detection medium. The primary elements of the detector are 6' × 6' × 3/8" sheets of PS-10^{10} plastic scintillator, which are readout along the perimeter using BBQ wave guides. The details of this arrangement are shown in Fig. 2. This acrylic scintillator is composed of 10% naphthalene, 1% PPO, 0.1% POPOP and the balance is ultra pure PMMA. These detector elements are viewed by two 2" photomultiplier tubes with bialkali photo cathodes (EMI 9839B), which have been selected to have high gain first dynodes and low noise rates above the 1/2 photoelectron threshold. We have measured the uniformity of response of this detector geometry using cosmic ray muons and intense radioactive sources. These tests show that on the average we observe a total of 2.2 photoelectrons/minimum ionizing particle passing through the center of a detector panel. Thus, at an ionization threshold of 3 times minimum ionizing, these panels have detection efficiency for coincidences between phototubes of a single panel of greater than 90%. The complete detector is composed of 64 single counter elements arrayed on the surfaces of a rectangular parallelopiped of dimensions 24' × 24' × 6'. Figure 3 shows the layout of the spectrometer.

The electronics being utilized in this monopole search were specially selected and designed to be sensitive to the properties of the GUT monopoles described earlier. In order to identify these rare events the recording system was required to measure both pulse height and relative timings of the counters struck in

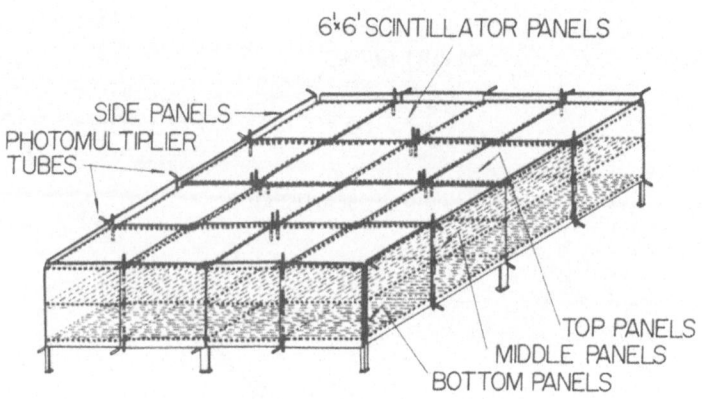

Fig. 3   Assembly drawing of monopole spectrometer.

any particular event. This requirement was further complicated by
the fact that the average transit time of a monopole through the
detector at $\beta \sim 10^{-3}$ is of order 10 µs. The signals from the
photomultipliers on each panel are therefore processed separately
at two levels in order to accomplish both the triggering of the
system and the data recording.

     To generate the trigger for this detector the output of each
phototube on a given detector element is sent to a discriminator
system with the following properties.   First, this discriminator
is sensitive to the prompt bursts of scintillation light (time
constant of ~ 100 ns) at the single photoelectron level corresponding to the conventional energy loss in the plastic.   Then,
secondly this system has an additional discriminator designed to
trigger on the delayed fluorescence of triplet de-excitation (time
constant typically ~ 1 µs) at the several photoelectron level.
Thus on each phototube output we placed two separate discriminator
channels with input integration time constants matched to these
two scintillation processes and with remotely programmable thresholds.   The outputs of these two discriminators are then logically
ORed for each phototube and a panel "struck" signal is formed by
the coincidence of the ORed outputs for the two phototubes of a
particular panel.   These panel signals are then further ORed
together in groups of 16 to form signals for the three horizontal
layers and the sides of the detector box.   The resulting four
signals are finally processed in a majority logic unit requiring
any three of the four signals being in "delayed" coincidence with
a resolving time of 25 µs. This "delayed" coincidence removes the
prompt cosmic ray muon background from the trigger by requiring a
minimum delay between any two of the three signals used in this
coincidence of 50 ns.   This 3-fold majority coincidence is the
basic monopole trigger for the system.

The data recording path for the individual detector elements is a separate parallel track from the trigger processing. The data from the two photomultipliers on each panel are linearly added and sent to a wave form digitizing circuit based on a 450 element analog shift register CCD. The details of this wave form recording circuitry can be found in Ref. 11. This unit works as a continuously sampling wave form storage device until a good trigger is received. The CCD's are being clocked in this sampling mode at 20 MHz, hence they have the capability of storing 22.5 μs of phototube pulse height information in 50 ns slices. Following the receipt of a trigger the system asynchronously digitizes, to 8 bit accuracy, all 450 samples of each detector panel and stores this data in a CAMAC memory module for future recording.

The on-line data acquisition system upon receipt of a trigger, takes control of the system and reads out the relevant counter information and records it onto disk for further on-line and off-line processing. In addition, the on-line system accepts and generates other triggers: i) prescaled cosmic ray muon events for efficiency monitoring, ii) LED flasher events for monitoring PMT gains and iii) random triggers for monitoring the pedestals in the pulse height recording system during the runs.

Figure 4 shows some sample data taken with the full data acquisition system in the laboratory using a small cosmic ray test set up. The details of this cosmic ray telescope geometry are shown in Fig. 5. These data show the typical response of the system to a cosmic ray muon passing through the detector, both in time and in pulse height development. The pulses appearing in Fig. 4 from the detector panel photomultiplier tubes, correspond to single photoelectron pulses. Figure 6 shows the response of the pulse height recording system for a cosmic ray passing through a thick slab (1") of acrylic scintillator. If one looks closely at the falling edge of this recorded wave form it is possible to see the presence of delayed single photoelectron bursts which may arise from the delayed scintillation light component of this cosmic ray event.

The detector is approximately 80% completed at this time. 60 of the scintillation panels have been installed at the mine and are presently being equipped with their necessary photomultiplier tubes and cabling. The electronics for the trigger and the data storage are completed and have been under laboratory tests on campus for the past several weeks. At the present time we are debugging and testing the on-line data acquisition programs before moving this system to the mine site. We expect to begin early data taking in January, 1984.

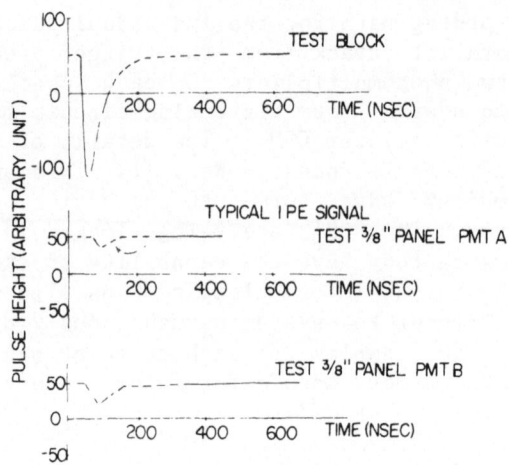

Fig. 4 Recorded wave forms from three photomultiplier tubes viewing samples of acrylic scintillator as the result of the passage of a cosmic ray muon through the system.

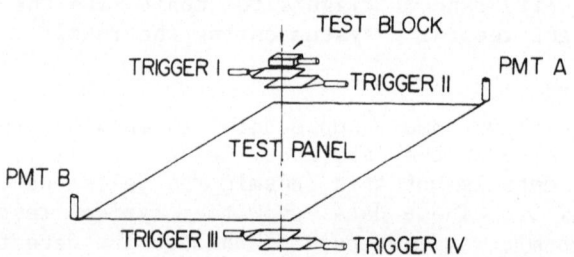

Fig. 5 Layout of counters used in the cosmic ray test set up.

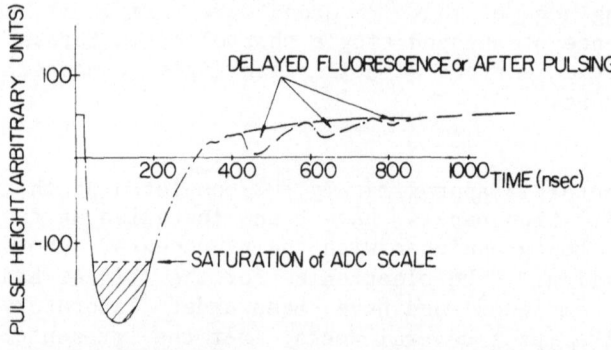

Fig. 6 Recorded wave form from a photomultiplier when a very large scintillation pulse due to a cosmic ray muon is processed through the wave form digitizing electronics.

BACKGROUND AND EXPECTED SENSITIVITY

In our original proposal we had carried out several detailed calculations to assess the effects of background events in the raw trigger and in the processed off-line data. These calculations considered several possible sources for these background triggers: (1) tube noise, (2) radioactivity, and (3) cosmic ray μ and ν interactions and μ decays. From all of these sources we expect a background trigger rate of $\sim$ 1 event/hour. From this raw data we expect to be able to further suppress these backgrounds by our off-line analysis. By imposing very conservative temporal and pulse heights cuts on this raw data we will be able to reduce this background rate to $\sim$ 0.1 event/year. Thus in one year of running, a single monopole event with no background will correspond to a monopole flux of 10^{-14} cm^{-2} s^{-1} sr^{-1}.

ACKNOWLEDGEMENTS

The experiment reported on is a collaboration of the following persons: C. Gagliardi, P. Green, P. McIntyre, T. Meyer, M. Shepko, R. Tribble and R. Webb, all at Texas A&M University. It is our pleasure to acknowledge the support which our group has received from the Warren Ranch Estate and the United Salt Corporation in allowing our group access to this underground site. This work was supported in part by the U.S. Department of Energy, contract DE-AS05-81ER40039, and by grants from the Robert A. Welch Foundation and the Texas Center for Energy and Mineral Resources.

REFERENCES

1. P.A.M. Dirac, Proc. R. Soc. A133, 60 (1931).
2. For an extensive bibliography of the literature on magnetic monopoles in the areas of both experiment and theory see: R.E. Craven, W.P. Trower and R.A. Carrigan, "Magnetic Monopole Bibliography", Fermilab Report 81/37, unpublished.
3. B. Cabrera, Phys. Rev. Lett. 48, 1378 (1982).
4. H. Georgi and S.L. Glashow, Phys. Rev. Lett. 32, 438 (1974). H. Georgi, H.R. Quinn, S. Weinberg, Phys. Rev. Lett. 33, 451 (1974).
5. Pre-GUT monopole searchers supposed they were looking for a relativistic and very heavily ionizing object instead of the slower moving more moderately ionizing super heavy GUT monopole.
6. S.P. Ahlen and K. Kinoshita, Phys. Rev. D26, 2347 (1982). S.P. Ahlen and G. Tarle, Phys. Rev. D27, 688 (1983). D. Ritson, "Fermi-Teller Theory of Low Velocity Ionization Losses Applied to Monopoles", SLAC-PUB-2950 (unpulished).

7. See S.P. Ahlen's paper in this volume.

8. P.M. McIntyre and R.C. Webb, "Searching for the GUT Monopole", Texas A&M University, Preprint DOE-ER40039-4 (1982).

9. S.D. Drell, N.M. Kroll, M.T. Mueller, S.J. Parke and M.A. Ruderman, Phys. Rev. Lett. 50, 644 (1983).

10. PS-10 is the brand name for Naphthalene Doped Arcylic Scintillator marketed by Polycast Technologies, Stamford, Connecticut, USA.

11. R.J. Ducar and P.M. McIntyre, IEEE Trans. Nucl. Sci., NS-26, 301 (1979).

A SCINTILLATOR-PROPORTIONAL COUNTER SEARCH FOR MONOPOLES

F. Kajino[a], S. Matsuno[b], Y.K. Yuan[c], T. Kitamura[a]

[a]Institute for Cosmic Ray Research
University of Tokyo
Tanashi, Tokyo 188, Japan

[b]College of Liberal Arts
Kobe University
Nadu-ku, Kobe 657 Japan

[c]Institute of High Energy Physics
Chinese Academy of Sciences
Beijing, Peoples Republic of China

ABSTRACT

A search for slowly moving superheavy magnetic monopoles has been performed using a combined detector of scintillation counters and proportional chambers. At the first stage, the minimum threshold of the energy loss was set at 1/20 minimum ionization for both detectors. PR gas was used for proportional chambers. An upper flux limit of 1.8×10^{-12} cm^{-2} sr^{-1} s^{-1} for monopoles was obtained over a velocity range from 2.5×10^{-4} c to 1.0×10^{-1} c at 90% confidence level. At the second state, mixed gas of helium + 10% methane was used for proportional chambers, because a methane molecule is ionized through the Penning effect by a helium metastable state excited by the monopole. For this condition, an upper flux limit of 1.6×10^{-12} cm^{-2} sr^{-1} s^{-1} for monopoles was obtained over the velocity range from $\sim 3 \times 10^{-4}$ c to 1c at 90% confidence level.

INTRODUCTION

Grand Unified Theories (GUT) predict the existence of superheavy magnetic monopoles with a mass of about 10^{16} GeV. Such heavy monopoles are accelerated to velocities around 10^{-3} c by galactic magnetic fields. Supersymmetric GUT and the Kaluza-Klein theory point out monopoles heavier than 10^{16} GeV, so the velocities for such monopoles might be rather smaller than 10^{-3} c. It is very important to search for the monopoles with velocities around 10^{-3} c and smaller than that. The expected flux for the monopoles is fairly small, so it is significant to make a large area detector by using scintillators and proportional counters. The energy loss of the monopoles was not clear for velocities less than 10^{-3} c. Recently, Drell et al.[1] presented an excellent theory about the energy loss of monopoles in a simple material such as helium at the velocity range from 10^{-4} c to 10^{-3} c. We report a search for monopoles based on their theory about the energy loss together with the search by scintillation counters.

EXPERIMENTAL APPARATUS

Figure 1 shows the experimental apparatus used to search for monopoles. There are 6 layers of scintillation counters(SC), 9 layers of proportional counters(PRC) and 14 iron layers. Each size of SC is 240cm × 20cm × 1cm. Effective volume of each PRC is 246cm × 92cm × 2cm. The wire spacing between two adjacent wires in the PRC is 2cm. The time-of-flight was measured by fast time-to-digital converters(TDC) with a time resolution of 0.5 ns and by slow TDCs with the resolution of 50 ns for SCs, and by slow TDCs with the resolution of 50 ns for PRCs. Ionization losses were measured by analog-to-digital converters for SCs and PRCs. Details of a data acquisition system are reported elsewhere.[2]

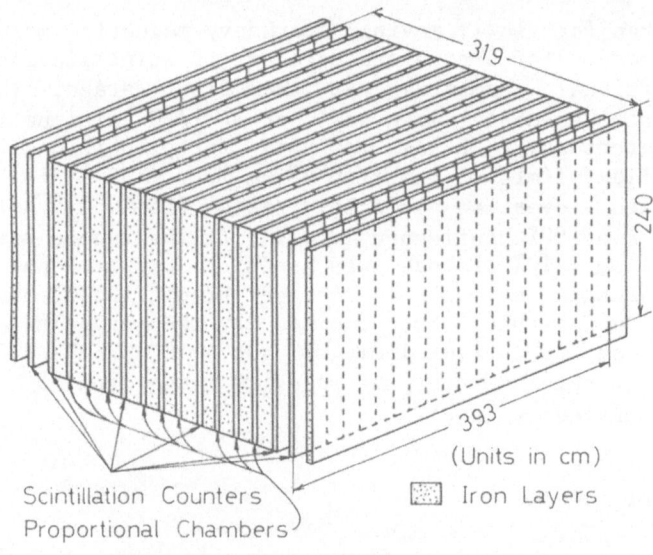

(Units in cm)

Scintillation Counters
Proportional Chambers   ▦ Iron Layers

Fig. 1 Schematic view of the monopole detector.

1ST STAGE

At the first stage, a trigger signal was generated by a successive delayed six-fold coincidence using the logical OR signals of respective layers of SCs. Muons which go through the detector from the first to the sixth layer of the SCs were rejected from this trigger. Hence, the velocity region of the trigger was between 10^{-4} c and 0.1 c. The area-solid-angle product for this trigger was 11.0 m^2 sr. Threshold levels for SCs and PRCs were set at 1/20 minimum ionization (I_{min}). PR gas (Ar + 10% CH_4) was circulated through the PRCs.

The energy losses in materials were calculated by many authors. We can measure the monopoles with velocities from 2.5×10^{-4} c to 0.1 c if Ritson's calculation[3] is applied. However, Ahlen and Tarle[4] pointed out that the lower limit for the velocities to detect monopoles with SC was about 6×10^{-4} c. Details of the result of the first stage are reported in Ref. 2.

2ND STAGE

At the second stage, the gas in PRCs was changed to He + 10% CH_4. Drell et al. calculated the energy loss of the monopole in the helium gas. Large energy losses occur in this gas for the low velocity monopole. If the monopole goes through the helium gas, helium atoms are excited as

$$He \rightarrow He^* \quad ,$$

where He^* is a metastable state of the helium atom. The metastable state of the helium atom collides with a methane molecule, which leads to an ionization of the methane through the Penning effect as follows,

$$He^* + CH_4 \rightarrow He + CH_4^+ + e^- \quad ,$$

because an ionization potential of the methane molecule is 13 eV which is smaller than the energy level of the metastable state of the helium atom of 20 eV. The life time of the metastable state is larger than 10^{-3} s and the mean collision time between the metastable state of the helium atom and the methane molecule is about 30 ns. Thus, the Penning effect effectively occurs before the metastable state radiatively decays.

The ionization loss of monopoles in He + 10% CH_4 is shown in Fig. 2, for which an ionization efficiency for the Penning effect is assumed as 83%[5] and a new calculation for the energy loss in the helium is used, which is smaller than an old one by a factor of 2.[6]

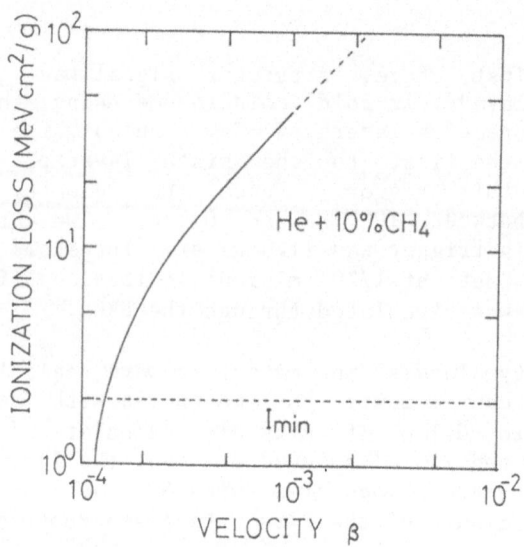

Fig. 2   Ionization loss as a function of velocity in helium and
10% methane.

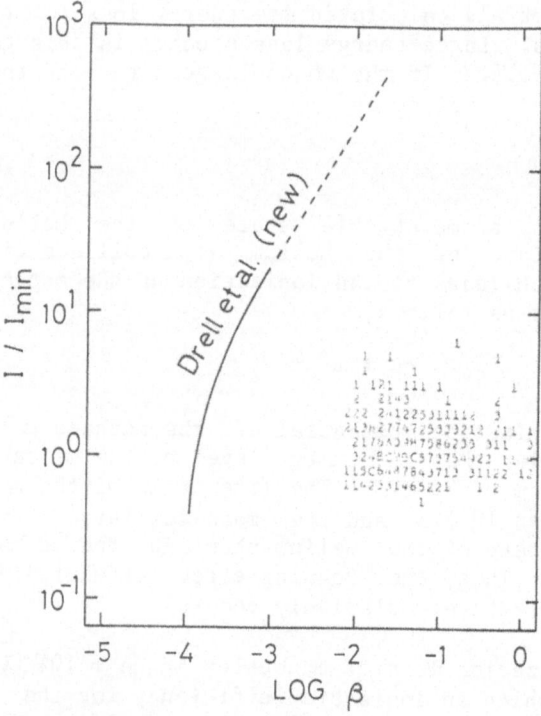

Fig. 3   Ionization losses as a function of muon velocities.

The trigger signals were generated by the successive delayed six-fold coincidence of respective layers of PRCs. Threshold level for the ionization loss in PRCs was set at $3 \times I_{min}$. An estimated efficiency for six-fold coincidence is 1.0 for the energy losses larger than about $7 \times I_{min}$. The area–solid–angle product for this trigger was 24.7 m^2 sr.

Figure 3 shows the ionization losses as a function of the velocities for muons. The events by the muons distribute around the velocities of about $10^{-1.5}$ c. This fact depends on the maximum drift velocity in the PRC being about 1 μs. It can appear again by a Monte Carlo calculation using the above conjecture.

Main conditions of event selections for the second stage are as follows:

(1) Number of hit SCs in each layer is less than 5.
(2) Number of hit PRCs in each layer is less than 3.
(3) There is at least one straight track in the PRCs.
(4) $\chi_t^2 < 4$ (for the time-of-flight).
(5) $\chi_i^2 < 3$ (for the ionization loss).

The terms χ_t^2 and χ_i^2 are defined as follows

$$\chi_t^2 = \frac{1}{n-2} \sum_{j=1}^{n} \frac{(t_j - t_{fit})^2}{\sigma_t^2} \quad ,$$

where t_j is a time in layer j, t_{fit} is a fitted value and σ_t^2 = (drift time error)^2 + (systematic error)^2. And

$$\chi_i^2 = \frac{1}{m-1} \sum_{j=1}^{m} \frac{(I_j - \bar{I})^2}{\sigma_i^2} \quad ,$$

where I_j is ionization loss in layer j, I is the average value of ionization losses and σ_i^2 = (ionization error)^2 + (systematic error)^2.

RESULTS

Figure 4 shows ionization losses as a function of velocities for a typical run after previous selections were carried out. The energy loss calculated by Drell et al. is also shown. We find no candidate for the magnetic monopole at both the first and the second stages.

The magnetic monopole flux upper limit obtained in the first stage was 1.8×10^{-12} cm^{-2} sr^{-1} s^{-1} at 90% confidence level with velocities between 2.5×10^{-4} c and 0.1 c. If we apply a conjecture of energy loss by Ahlen et al., the lower limit of the velocity range is about 6×10^{-4} c. The live time was 3.3×10^3 hours.

The upper limit of the monopole flux at the second stage was 1.6×10^{-12} cm^{-2} sr^{-1} s^{-1} at 90% confidence level over a wide velocity range from $\sim 3 \times 10^{-4}$ c to 1 c. The live time was 1.6×10^3 hours. The electrical trigger range covered the velocities between 10^{-4} c and 1 c.

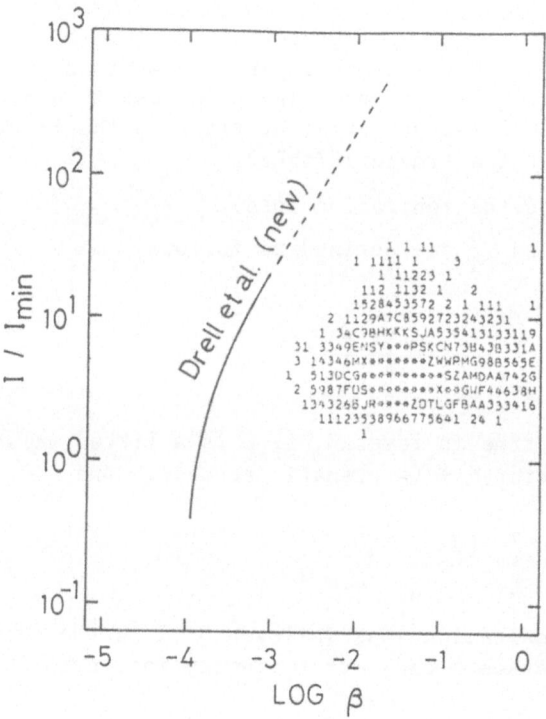

Fig. 4   Ionization losses as a function of velocities for a typical run.

ACKNOWLEDGEMENTS

We would like to acknowledge Y. Ohashi, K. Mitui, A. Okada and T. Aoki for their help and support. S. Ozaki, T. Takahashi, J. Arafune and K. Hayashi are also acknowledged for useful suggestions.

REFERENCES

1. S.D. Drell, N.M. Kroll, M.T. Mueller, S.J. Parke and M.A. Ruderman, Phys. Rev. Lett. 50, 644 (1983).
2. F. Kajino, T. Kitamura, K. Mitsui, Y. Ohashi, A. Okada, Y.K. Yuan, T. Aoki, S. Matsuno, 18th Int. Cosmic Ray Conf., Bangalore 5, 56 (1983); J. Phys. G (to be published).
3. D.M. Ritson, SLAC-PUB-2950, 1982 (unpublished).
4. S.P. Ahlen and G. Tarle, Phys. Rev. D27, 688 (1983).
5. W.P. Jesse, J. Chem. Phys. 41, 2060 (1964).
6. N.M. Kroll, in this volume.
7. J.D. Ullman, Phys. Rev. Lett. 47, 289 (1981).
8. T. Mashimo, K. Kawagoe and M. Koshiba, J. Phys. Soc. Jpn. 51, 3067 (1982). T. Mashimo, S. Orito, K. Kawagoe, S. Nakamura and M. Nozaki, Phys. Lett. 128B, 327 (1983).
9. E.N. Alexeyev, M.M. Boliev, A.E. Chudakov, B.A. Makoev, S.P. Mikheyev and Y.V. Sten'kin, Lett. Nuovo Cimento 35, 413 (1982).
10. R. Bonarelli, P. Capiluppi, I. D'antone, G. Giacomelli, G. Mandrioli, C. Merli and A.M. Rossi, Phys. Lett. 112B, 100 (1982), and 126B, 137 (1983).
11. D.E. Groom E.C. Loh, H.N. Nelson and D.M. Ritson, Phys. Rev. Lett. 50, 573 (1983).
12. J. Bartelt, H. Courant, K. Heller, T. Joyce, M. Marshak, E. Peterson, D.S. Ayres, J.W. Dawson, T.H. Fields, E.N. May and L.E. Price, Phys. Rev. Lett. 50, 655 (1983).

MONOPOLE FLUX LIMITS FROM THE SOUDAN I EXPERIMENT

John E. Bartelt

SLAC
Stanford, California 94305

We have searched for magnetic monopoles using the Soudan I proton decay detector, located 600 meters underground on the 23rd level of the Soudan Mine[1]. The detector consists of 3,456 proportional tubes arranged in 48 layers of 72 tubes each, embedded in heavy, iron-rich concrete. The proportional tubes use a gas mixture of 91% argon and 9% carbon dioxide. Alternate layers are oriented orthogonally, to provide complimentary two-dimensional views (see Fig. 1). Additionally, the individual tubes are staggered vertically to provide better coverage. The tubes are 2.8 cm in diameter and spaced by 4 cm in each dimension. They are 2.9 meters long. Overall, the detector is 2.9 m by 2.9 m by 2.0 m tall, and has a total mass of 31.4 tons. By weight, its composition is 57% iron, 29% oxygen, 13% calcium, and 1.2% hydrogen. It is surrounded on the top and four sides by scintillation counters which serve as an active shield.

The proportional wires are maintained at a voltage of about 2200 V. They are read-out through a three-stage amplifier, whose signals are then sent to time-over-threshold circuits. The time over threshold provides a logarithmic measurement of the ionization deposited in the proportional tubes (the average exponential decay constant for our tubes is 1.05 microseconds). This information is stored in 16-bit shift registers, which are initially clocked every 185 nanoseconds. After a trigger occurs, the shift registers are clocked 14 more times; the two bins before and the 14 bins after the trigger are saved. A second shift register is then clocked 16 times, every 225 nanoseconds; these data are also stored, providing a total of 6.5 microseconds of history for each tube. (For the trigger, all the signals from the 72 tubes

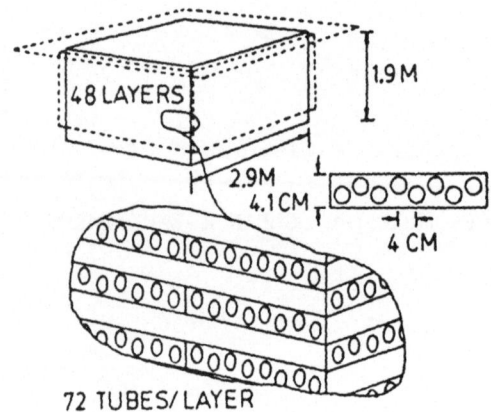

Fig. 1 A schematic view of the Soudan I detector.   Dotted lines
        represent the active shield.

within each layer are ORed; the trigger requirement is then any
three out of any four adjacent layers in coincidence.) This
information can also be used to measure the velocity of slow
particles. Our timing measurement is limited by the drift time of
about 400 nanoseconds from the edge of a tube to the center wire.

Since we have both timing and ionization information, we have
searched for monopoles three ways: (1) very high ionization
tracks, as should be produced by high velocity monopoles; (2)
measurably slow particles that are able to penetrate to the
detector's depth and through the detector itself; and (3) monopole
catalysis of nucleon decay events, as might be produced through
the Rubakov-Callan effect. This last category was defined by the
requirement that the event have at least 12 tubes hit, of which at
least 7 were "late". "Late" means that the pulse began in the
sixth time bin or later (more than 555 nanoseconds after the
trigger). These events (like those in the other categories) were
then hand-scanned by physicists. The critieria were deliberately
kept loose in order to accept as wide a variety of topologies as
possible. For our acceptance calculations, we parameterized the
cross-section for nucleon decay catalysis as cross-section =
$(1/\text{proton-mass})^2$ (c/v) (X) where the first term is just a typical
strong-interaction cross-section (441 microbarns), the second is
the velocity dependence (neglecting the attraction of the proton's
dipole moment for the monopole), and the third term is an
arbitrary factor, presumably of order 10^{-3} to 10^3. It should be
noted that for free protons (i.e., hydrogen nuclei), v is just the
magnitude of the monopole's velocity. For bound nucleons, one
must use the magnitude of the vector sum of the monopole's
velocity and the nucleon's Fermi motion. Since this velocity is

about 0.2 c, the 1.2% hydrogen totally dominates the catalysis probability for monopole velocities of 0.001 c or less. Any further suppression in heavy nuclei[2] is irrelevant. (Additionally, our limits on single nucleon decays can be used to set limits on non-ionizing monopoles with small catalysis cross-sections.)

THE SEARCH FOR HIGHLY-IONIZING PARTICLES

We determined that vertical, non-showering, non-interacting muons produce a pulse length of 1.3 microseconds on average. This corresponds to 1.58 times minimum-ionizing[3]. In searching for highly-ionizing particles, we set our criterion at an average pulse length of 3.7 microseconds, which corresponds to 16 times minimum-ionizing, or 10 times that of a muon. We further required that no more than 30% of the total tubes in an event be outside of the fit track; this helped eliminate muon-induced electromagnetic showers, which do have a core of high ionization. A minimum of eight tubes in the event was also required. All events meeting these criteria were then hand-scanned by physicists. The events were invariably very short tracks with a few anomalously long pulses, or muons with narrow showers which passed our cuts. We found no events with consistently high ionization along a non-showering track.

Monopoles with velocities greater than 0.01 c have a relatively small probability of catalyzing a nucleon decay in the calorimeter or the surrounding rock. If they do, the decays would be "on time" (i.e., not contain late tubes), and so would not make the cuts for catalysis events. If the cross-section is large enough ($Xc/v=1$), about 10% of the events are lost from the high-ionization category due to catalyzed nucleon decays pulling down the average pulse length. At $Xc/v=100$, nearly all events are lost. If the number of decays is large enough, however, and they contain muons, the events can get "accidentally" caught in the catalysis event category because of the late tubes from the muon decays. The detector's timing resolution is insufficient to reliably distinguish velocities in this range from c, except near 0.01 c.

For our flux limit, we have conservatively estimated that the ionization produced by a monopole would be above our 16x minimum threshold only for velocities equal to or greater than 0.01 c. Ritson[4], however, calculates ionization of this level for velocities above 0.005 c; and that at 0.01 c, the ionization would be 40x minimum. If there is no nucleon decay catalysis, our acceptance is a purely geometrical quantity. A Monte Carlo procedure, which took into account the detector's trigger requirement, and the selection and scanning efficiency showed that the experiment's acceptance for such tracks is about 73 m^2 sr.

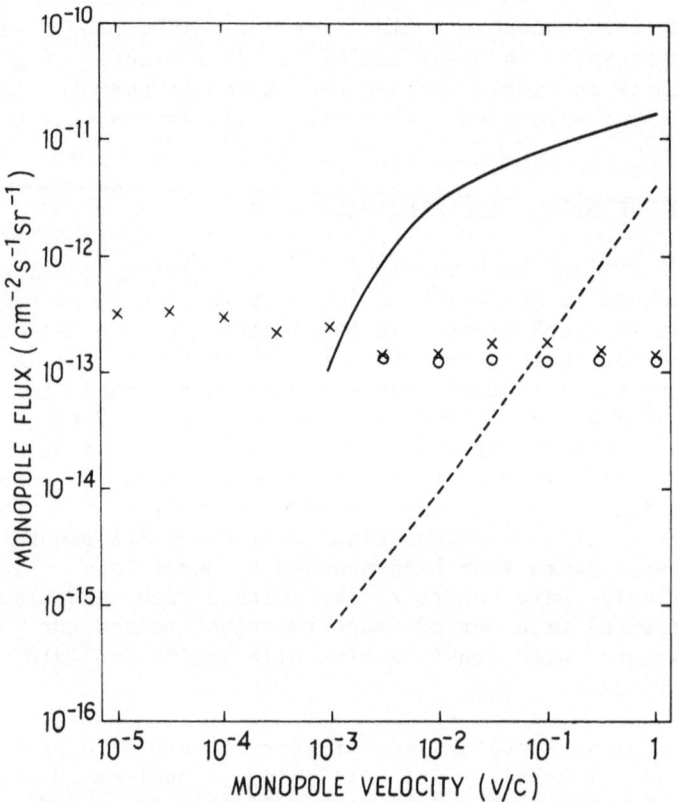

Fig. 2 A comparison of monopole flux limits. The solid and
        dashed curves are the Parker bound for monopoles of mass
        10^{19} and 10^{16} GeV respectively. The crosses are data from
        this experiment assuming a catalysis cross section of 441
        μbarns. The circles are limits based on ionization in the
        proportional tubes.

Together with our live-time of 0.78 years, this yields a flux
limit of about 1.3×10^{-13} monopoles/cm^2/s/sr (90% confidence
level), if nucleon decay catalysis is insignificant (see Fig. 2).
For flux limits when catalysis is important, see Figs. 2 and 3.

THE SEARCH FOR SLOWLY MOVING PARTICLES

     The detector's timing resolution, given sufficient track
length, is capable of measuring velocities equal to or less than
0.01 c. Following Ahlen, Liss and Tarle[5], we have assumed that
monopoles with velocities of 0.002 c or greater will deposit
sufficient ionization in the proportional tubes to be detected.

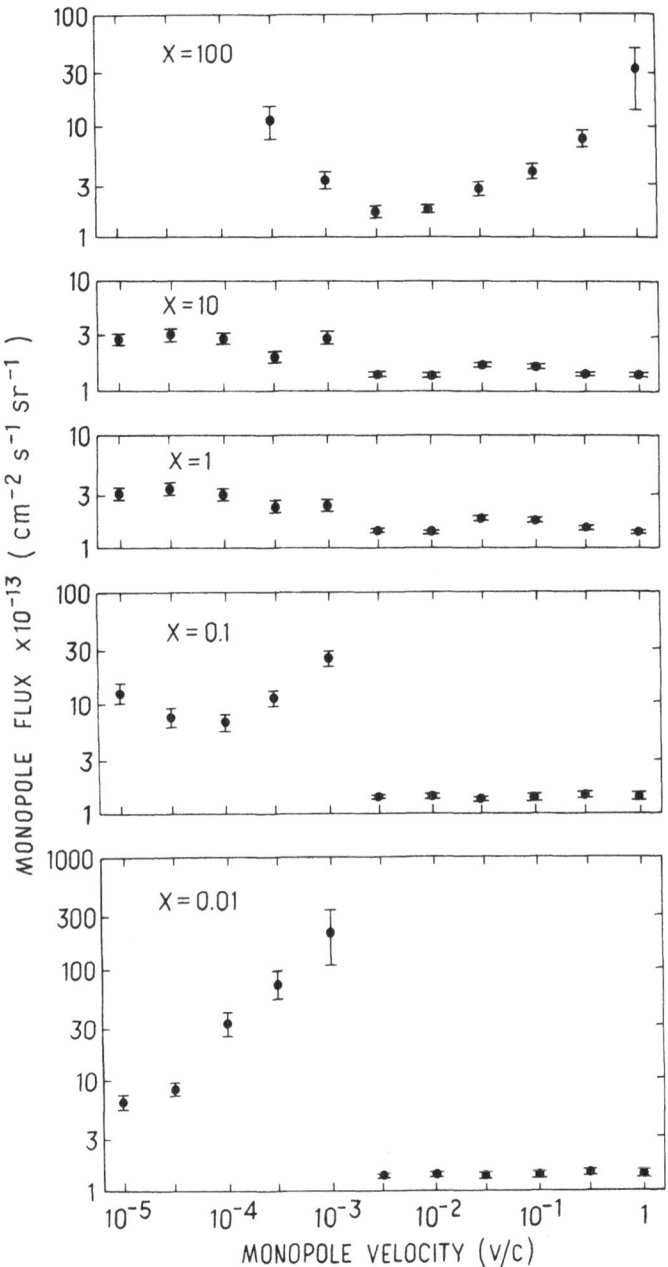

Fig. 3 90% C.L. limits on magnetic monopole flux, from Soudan I for different values of the nucleon decay catalysis cross-section. Error bars are 1 standard deviation uncertainties in the acceptance calculations.

Therefore in this velocity range, we have searched for monopoles by making a velocity fit to the starting times of pulses in the various tubes. A straight-line fit with constant velocity is required. Statistical cuts on the quality of the fit were used to eliminate muons. It is possible for a muon track to mimic that of a non-relativistic particle because the drift times in the proportional tubes can vary from 0 to 2 bins (of 185 nanoseconds each), or more, in some cases, depending on how close the muon passes to the collection wire. We also find that electromagnetic showers often contain late hits; we believe these to be caused by the excitation of nuclear states with lifetimes on the order of a microsecond. These, along with the late pulses from a muon decay, can also contribute to an apparent non-relativistic velocity. Thus we can not reliably measure a velocity greater than 0.01 c, and have set the following requirements on the fit: (1) chi-squared per degree of freedom between 0 and 6 for the straight-line velocity fit; (2) velocity fit more than 4 standard deviations away from c; and (3) more than 4 degrees of freedom in the fit.

Only one event has met these requirements (with an apparent velocity of 0.008 c), and we believe it to be a muon which has, by chance, simulated a non-relativistic velocity well enough to pass our cuts. This belief is substantiated by the fact that its pulses are of normal length (indicating normal ionization), in contrast to our expectations of ionization several times greater than that of a muon; Ritson[4], in fact, calculates the ionization at this velocity to be 30x minimum (19 times as much as a muon), corresponding to pulses 4.4 microseconds long.

It should be noted that monopole catalysis of nucleon decay would decrease our acceptance for this type of event, as the decays would degrade the fit (raising the chi-squared, for example). Nearly all the events lost from this category, however, would be caught by the catalysis event category, since the monopole is moving slowly enough in that part of its track and/or some of the decays would be "late".

THE SEARCH FOR MONOPOLE CATALYSIS OF NUCLEON DECAY EVENTS

Our third category for monopole events is intended to catch a wide variety of events that might be produced by a monopole catalyzing multiple nucleon decays in the calorimeter, with or without a monopole track, or a track and a single decay. This is our only method for detecting monopoles with a velocity less than 0.002 c, since they probably do not produce detectable ionization in argon. The criteria are so loose, however, that even a slow track that does not catalyze might easily be caught. The requirement of a minimum of 7 "late" tubes (more than 555 nanoseconds after the trigger) was set arbitrarily to cut the number of events requiring hand-scanning down to a reasonable number. Monopole catalysis events produced by monopoles of velocities greater than about 0.02c probably won't contain sufficient late hits to make this cut; there is, however, no real lower bound on the velocity. Subsequent decays would have to be nearly coincident in space in order to fit within our 6.5 microsecond time window; but Monte Carlo events have shown us that these are very distinct and recognizable types of events. In practice, we have cut off our acceptance calculation at 0.00001 c, which seems reasonable from physical and astrophysical considerations. The same physicists who scanned the Monte Carlo events (see below) scanned the real data. No events passed the scanning.

ACCEPTANCE

Our acceptance for monopole events is extremely complicated, for two reasons: (1) we search for 3 distinct types of monopole events with different velocity dependences (besides single, contained nucleon decays); and (2) catalysis of nucleon decay presents the possibility of nucleon decay in the surrounding rock, producing relativistic particles which can trigger the detector long before the monopole arrives; and which is followed by about 1 second of dead time. Hence there are competing effects as the cross-section increases. For slow monopoles, a larger number of decays within the detector improves our acceptance; but the increasing catalysis of decays in the rock hurts. And, as stated before, catalysis hurts our acceptance at high velocities.

For this reason, we have made an extensive Monte Carlo study for different values of the catalysis cross-section and for velocities from 0.00001 c up to c. In each case, an isotropic flux of monopoles was generated; each monopole was propagated through a meter of rock before it entered the detector. If the monopole catalyzed nucleon decays in the rock, the daughter particles were allowed to enter the calorimeter (and possibly trigger it), if they had sufficient range and were travelling in the proper direction. The velocity dependence of the ionization was included (in a rough way). To take into account the Fermi motion of bound nucleons, we took an average of the reciprocal of the vector sum of the monopole's velocity and a distribution of velocities as found in an iron nuclei. Those events which triggered the detector were output to a file, which was then run through the same analysis program as the real data, and which selected candidate events on the basis of the same criteria as the data (for all categories of events). These events were then scanned by physicists. This enabled us to determine our acceptance with a high degree of accuracy.

RESULTS

Using the results of our Monte Carlo study, and the lack of any events, we calculated 90% confidence level flux limits. Figure 2 shows our results for two cases: no nucleon decay catalysis, and when X (in our parameterization) equals 1 [recall that cross-section = (441 mb) (c/v) (X).] We found that when X = 1000, we have essentially no acceptance at any velocity. There are so many decays that at high velocities the events look like (rather peculiar) electromagnetic showers. At lower velocities, the detector is always triggered by the daughter particles of decays in the rock. These events are again rather distinctive in their time structure and topology, but we do not feel that we could reliably identify them with magnetic monopoles. At X = 100 we are still somewhat limited by the number of decays in the detector (at high v), and in the rock (at low v). Our acceptance is best for X = 10 and X = 1. For smaller cross-sections (X = 0.1 and X = 0.01), we are more handicapped by the loss of the monopole track at low velocities. Nevertheless, we are able to set a flux limit lower than any other dE/dx experiment, except Baksan[6], which has a considerably more restricted velocity range.

ACKNOWLEDGEMENTS

The Soudan I collaboration includes H. Courant, K. Heller, T. Joyce, M. Marshak, E. Peterson, and K. Ruddick of the Univesity of Minnesota and D. Ayres, J. Dawson, T. Fields, E. May and L. Price of Argonne National Laboratory, plus the invaluable assistance of many students and technicians, and the staff of the Tower-Soudan State Park.

REFERENCES

1. J. Bartelt et al., Phys. Rev. Lett. 50, 651 and 655 (1983).
2. J. Arafune and M. Fukugita, Phys. Rev. Lett. 50, 1901 (1983).
3. W.W.M. Allison and J.H. Cobb, Ann. Rev. Nucl. Part. Sci. 30, 258 (1980).
4. D.M. Ritson, SLAC Report #2950, July, 1982 (unpublished).
5. S.P. Ahlen, T.M. Liss, and G. Tarle, Phys. Rev. Lett. 51, 940 (1983).
6. E.N. Alexeyev et al., Lett. Nuovo Cimento 35, 413 (1982) [Baksan].
7. M.S. Turner, E.N. Parker, and T.J. Bogdan, Phys. Rev. D26, 1296 (1982).

A SEARCH FOR MAGNETIC MONOPOLES IN THE MONT BLANC

NUCLEON DECAY APPARATUS

Michael J. Price

E.F. Division
CERN
Geneva, Switzerland

The Mont Blanc nucleon decay experiment, NUSEX,[1] is situated half way along the road tunnel under Mont Blanc linking France and Italy and is at a depth of approximately 5000 m.w.e. The apparatus has been taking data for the past 16 months. Results on the contained events observed have already been reported.[2,3] In this paper we describe the functioning of the magnetic monopole trigger that has recently been incorporated into the apparatus.

The detector consists of 135 horizontal iron plates ($350 \times 350 \times 1$ cm^3) interleaved with limited streamer tubes. The streamer tubes have wire spacings of 1 cm and, since the cathodes have a reasonably high surface resistivity, the wire signals can induce voltages on strips placed parallel and orthogonal to the wires. The signal from each strip passes into an amplifier-discriminator and then into a one-shot. All one-shot outputs of one plane are connected in parallel to form a fast-OR signal and these fast-ORs are then combined logically to provide the nucleon decay trigger. When a trigger is produced (e.g. if four contiguous planes are hit) the one-shot outputs are loaded into parallel-in serial-out shift registers. This digital information is then read out by CAMAC processors under the control of a PDP11/60 computer to be stored on tape.

The use of the apparatus as a device for detecting through-going magnetic monopoles was suggested by the fact that the streamer tubes, by virtue of their "digital" nature, should be sensitive to a single ion-electron pair. The gas mixture used (29% Ar, 48% CO_2, 23% n-pentane) is such that a minimum ionizing particle should produce of the order of 30 ion-electron pairs in passing through a limited streamer tube. Thus, in principle, the tubes should be capable of detecting particles with an ionziation rate 1/30 that of a minimum ionizing particle.

Recently there has been much speculation on the ionization mechanisms of slowly moving magnetic monopoles. Ahlen and Tarle[4] have concluded that detection methods using ionization techniques would only be sensitive for monopole velocities greater than about $10^{-3}c$ and that for lower velocities (down to $6 \times 10^{-4}c$) scintillators should offer significant advantages. The recent work of Drell et al.[5] however, has suggested that the magnetic nature of the monopole might have a predominant role in the ionization process (at least for simple atoms such as hydrogen and helium). In view of the uncertainty in the details of the ionization mechanism, especially for the complicated gas mixture used, it appeared reasonable to design the monopole tracking electronics to be sensitive down to monopole velocities of about $10^{-4}c$ corresponding to the velocity of the earth in the solar system.

When using the apparatus as a magnetic monopole detector the fast-OR signals from successive planes are ORed together to form a single signal from groups of eight planes. These groups then serve as the basic detection units of the magnetic monopole detector. Two successive groups of eight planes are in turn ORed together to form a trigger signal from a block of 16 planes. This configuration was chosen to reduce the spurious trigger rate to a reasonable level. There is a noise rate of 1 Hz per wire due to radioactivity in the iron, however, radioactivity in the surrounding rock increases this rate significantly in the upper planes of the apparatus. Thus, for the downward going monopole trigger, it is necessary to leave the first two groups of eight planes unconnected. The general principles of the trigger systems are shown schematically in Fig. 1. Taking the downward going monopole trigger as an example: if a hit is recorded in the first group of eight planes a gate G_1 is opened for a duration of 7 μs and the time of arrival of this trigger pulse and of subsequent pulses arriving during the open time of G_1 are recorded in scalar S_1. A trigger signal from the second group of eight planes arriving during the open period of G_1 causes the opening of a 7 μs gate G_2. Again, signals coming from the second group during this time are recorded by scalar S_2. During the time for which G_2 is open any signal from the next block (now of 16 planes) causes the opening of gate G_3 for a time of 15 μs. The times of arrival of the trigger signal and of any other signal from this block during this time are recorded by scalars S_3 and S_4 corresponding to the two groups of eight planes making up the block. This sequence is repeated through the apparatus and a signal recorded in the last block of 16 planes constitutes the magnetic monopole trigger. If an event trigger occurs the times recorded on the scalars are read out by the computer and recorded on magnetic tape.

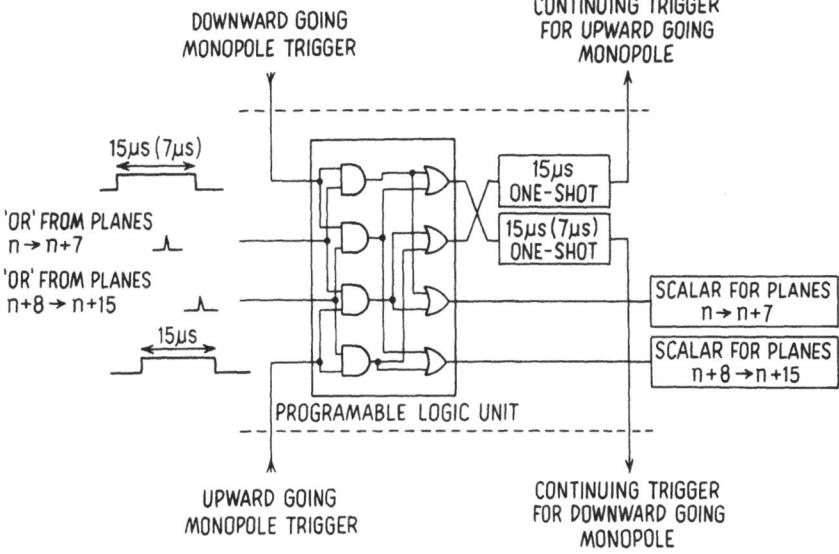

Fig. 1. One logic unit of the upward and downward going monopole trigger.

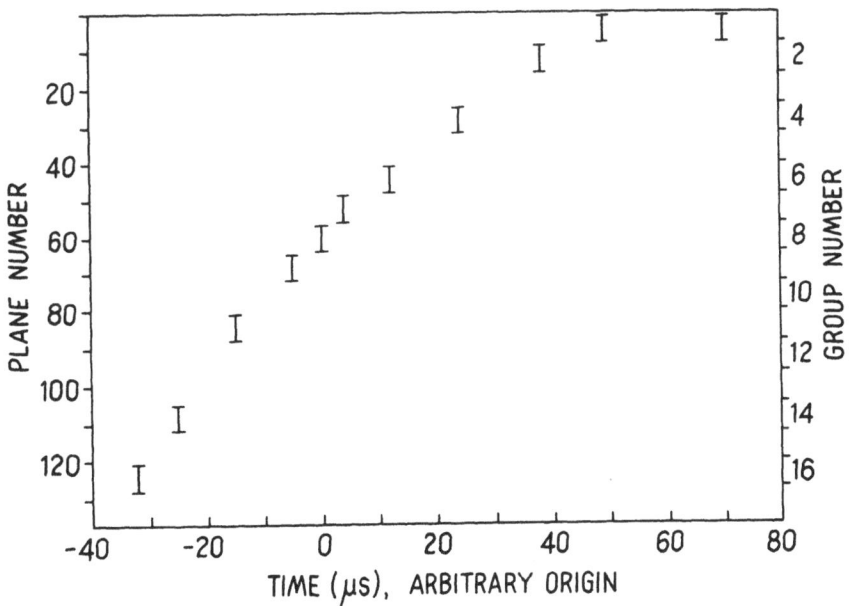

Fig. 2. Hit position versus time for a configuration producing a monopole trigger.

To determine whether the trigger in fact corresponds to the
passage of a monopole the positions of the hits (known to within a
group of eight planes) are plotted against their arrival times: a
straight line would then correspond to a potential magnetic
monopole as shown in Fig. 2.

The efficiency of the trigger can be expressed as a function
of the ionization rate (assuming that 30 electron-ion pairs are
produced per centimeter by a minimum ionizing particle) or,
alternatively, as a function of the mean number of hits per block
of 16 planes. The trigger, in its present form, is not equally
efficient for all ionization rates. If a particle passing through
the apparatus produced an average of n hits per block of 16 planes
then, from Poisson statistics, the probability of obtaining one or
more hits per block would be $(1-\exp(-n))$ and the probability of
obtaining a trigger from a downward going particle would be
$(1-\exp(-n))^7(1-\exp(-n/2))^2$. The efficiency of this trigger as a
function of n (the probability that a particle producing an
average of n hits/block of 16 planes gives a trigger) is shown in
Fig. 3.

In evaluating the significance of a candidate it is clearly
important to know the rate at which noise can simulate a
monopole's passage. Figure 4 shows the expected frequency of
spurious events (i.e. events in which signals from the various
groups satisfy the trigger requirement and are aligned in time) as
a function of the mean number of hits per plane. Comparing this
curve with that in Fig. 3 shows that the probability of noise
simulating a real event is negligible for ionization rates which
give a detection probability of better than 10%.

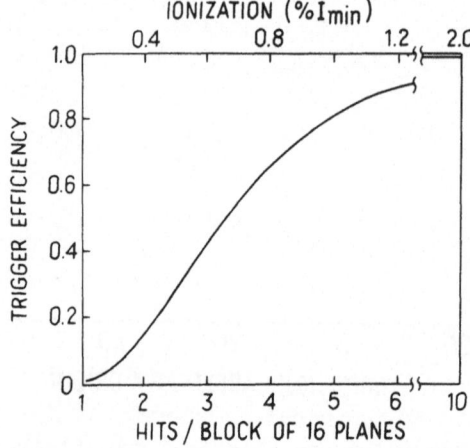

Fig. 3 Trigger efficiency as a function of ionization.

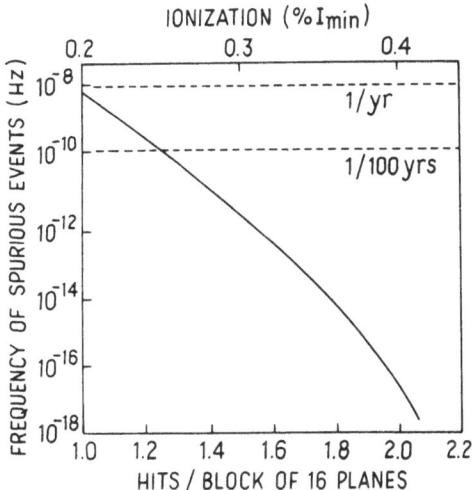

Fig. 4 Frequency of "events" due to noise as a function of ionization.

The acceptance of the apparatus has been calculated to be 9.5 m^2sr^2 per monopole direction. The data accumulated until mid October 1983 permit us to place an upper limit on the cosmic magnetic monopole flux of $F_M \leqslant 1.3 \times 10^{-12}$ $cm^{-2}sr^{-1}s^{-1}$ at the 90% confidence level.

REFERENCES

1. The members of the NUSEX collaboration are: (1) Laboratori Nazionali di Frascati, INFN, Italy; G. Battistoni, P. Campana, V. Chiarella, E. Iarocci, G.P. Murtas, G. Nicoletti, L. Satta, L. Trasatti; (2) Dipartimento di Fisica dell'Universita and INFN, Milano, Italy; E. Bellotti, P. Negri, A. Pullia, E. Fiorini, C. Liguori, S. Ragazzi, M. Rollier, L. Zanotti; (3) Istituto di Cosmogeofisica del CNR and Istituto di Fisica Generale dell'Universita di Torino, Italy; G. Bologna, C. Castagnoli, B. D'Ettorre Piazzioli, P. Galeotti, G. Mannocchi, P. Picchi, O. Saavedra; (4) CERN, Geneva, Switzerland; D.C. Cundy, M.J. Price.
2. G. Battistoni et al., Phys. Lett. 118B 461 (1982).
3. G. Battistoni et al., To be published in Phys. Lett.
4. S.P. Ahlen and G. Tarle, Phys. Rev. 27D 688 (1983).
5. S.D. Drell, N.M. Kroll, M.J. Mueller, S.J. Parke and M.A. Ruderman, Phys. Rev. Lett. 50 644 (1983).

MONOPOLE SEARCH WITH THE FREJUS TUNNEL NUCLEON LIFETIME EXPERIMENT

Paul Eschstruth

Laboratoire de l'Accelerateur Lineaire
Orsay, France

The nucleon lifetime experiment[1] now under construction in the "Laboratoire Souterrain de Modane", France, by the French-German AOPSW Collaboration is well suited for use as a wide aperture GUT monopole detector.

The 1000 ton instrument, of which about a quarter will be installed by the end of 1983, consists of a fine-grain flash-tube calorimeter and a trigger system using Geiger tubes. Its characteristics can be summarized as follows:

Dimensions: 6 m × 6 m × 13 m for 1000 tons.

Fine-grain calorimeter:

- 1000 flash-tube planes each having 1024 6 mm × 6 m tubes
- 10^6 electronics channels
- High voltage pulse: 10 kV/cm
- Gas mixture: Ne 70%, He 30%

Trigger system[2]:

- 124 Geiger tube planes each having 352 16 mm × 6 m tubes
- 45 000 electronics channels
- Wire diameter 100 microns, spacers each 2 m
- Operating voltage 1450 V
- Gas mixture: Argon + 2% ethyl alcohol + 0.05% Freon
- Maximum time jitter 300 ns (drift time), typical pulse length 50 µs.

Figure 1 shows the complete detector in the laboratory cavity. Figure 2 represents the detector with an expanded horizontal scale to show its modular construction.   Each module contains a Geiger tube trigger plane and four pairs of flash tube planes.   The orientation of the flash tubes is alternately horizontal and vertical by pairs, while that of the Geiger tubes changes from one module to the next.

Fast monopoles can be seen in the experiment without a special trigger system.   The trigger electronics for the nucleon lifetime experiment sums clipped signals from groups of adjacent Geiger planes requiring, subject to certain additional conditions which will not be described here, the coincidence of a given number of tubes within the group.   For example, the group size might be set to 5 and the threshold on the number of tubes to 4. For this case, with an effective clipped signal width of 400 ns and a 300 ns drift time, good detection efficiency is obtained for β down to roughly 5×10^{-3}.   Some part of the monopole track would be seen with the flash tubes, sensitive to particles passing during a period of several microseconds before the high voltage pulse.   The identification of such a track as belonging to a monopole would come from timing information which will be provided by a 124 channel multi-hit "logic analyzer" connected to the Geiger planes. A sample will be stored every 50 ns over a 50 μs

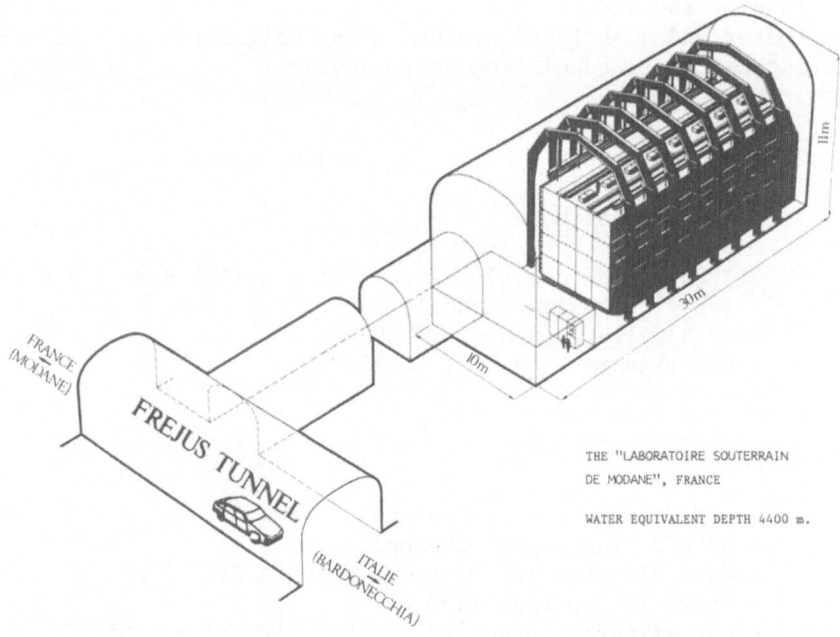

THE "LABORATOIRE SOUTERRAIN
DE MODANE", FRANCE

WATER EQUIVALENT DEPTH 4400 m.

Fig. 1 The complete detector in the laboratory cavity.

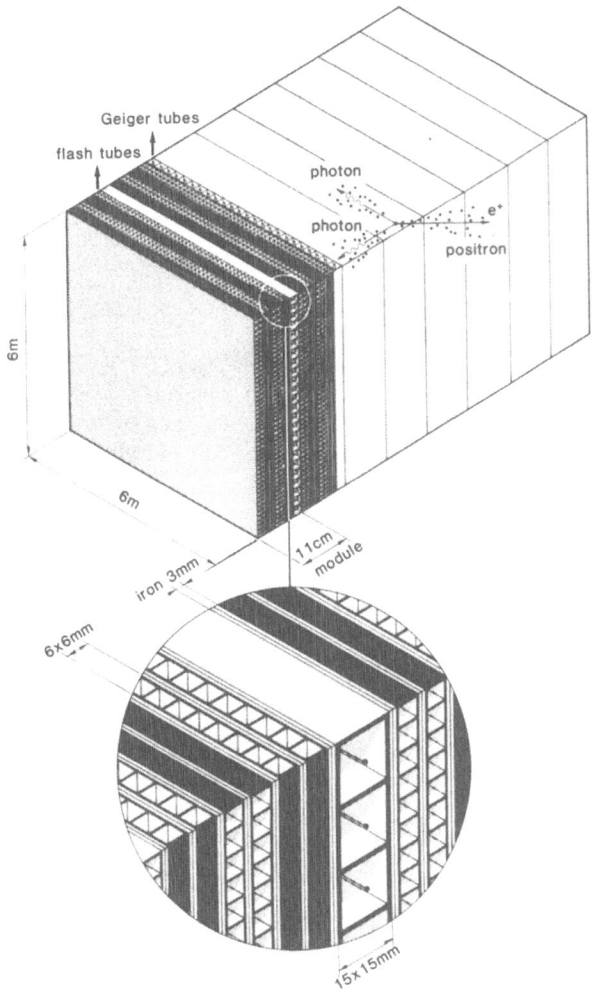

Fig. 2 Detector construction shown schematically, including a
simulated nucleon decay event (not to scale).

period, indicating for each plane the arrival of each new Geiger
tube signal. This system will be used to identify decaying
positive muons (2.2 μs mean life) in nucleon decay event candidates.

The width of the Geiger signal (40 μs minimum) is such that
the strobe for readout can be delayed to allow for following the
track of a slow particle through the entire detector length for β
down to about 1×10^{-3}.

To extend the sensitivity of the experiment to lower β, a special pre-trigger is being constructed to provide a signal for Geiger tube readout when a monopole candidate configuration occurs. The planes will be grouped by, for example 32, each group providing information to a "pile-up" circuit which will furnish a constantly updated count of the number of tubes hit during the previous 30 to 50 μs. When this number reaches a pre-determined threshold, a pre-trigger signal is generated which will, after a suitable delay:

- load the Geiger tube shift registers for possible subsequent readout
- switch the time measurement system to readout mode
- cause a fast on-line processor to look for a linear relationship between plane number and time making it possible to distinguish a monopole candidate from random pile-up.

If the processor accepts the time data, the Geiger information is read out for off line analysis.

Random pile-up will produce triggers about once per second for 8 hits required over 30 μs in 32 planes. This rate is quite sensitive to the threshold setting. Lowering the threshold increases the detector sensitivity for slow particles by reducing the number of planes that must be crossed. It also helps in detecting particles having a reduced probability of ionizing. The threshold chosen will be the lowest compatible with the capabilities of the on line processor.

Numerous gas mixtures allow for operation in the Geiger mode. The one being used is non-inflammable, as imposed by tunnel regulations. Adiabatic excitation of Argon by a slow monopole is a possibility[3] as is subsequent ionization of the alcohol through the Penning effect. Whether or not these effects are confirmed, we note that the high gain nature of the Geiger mode implies good sensitivity to low ionization levels.

Excluding considerations relating to the energy loss of slow monopoles, we can summarize as follows the characteristics of the Geiger planes as a monopole detector:

- a surface area-solid angle product of 1000 m^2 sr;
- physical dimensions and signal characteristics allowing for a β range from 10^{-4} to 10^{-1};
- good track reconstruction with 16 mm tube spacing in 124 6 m × 6 m planes.

ACKNOWLEDGEMENT

The French-German AOPSW collaboration consists of Physikalisches Institut, Aachen; Accelerateur Lineaire, Orsay; Ecole Polytechnique, Palaiseau; Centre d'Etudes Nucleaire, Saclay; and Universitat Gesamthochschule, Wuppertal.

REFERENCES

1. P. Bareyre et al., Proposal to Study the Instability of the Nucleon Using a Calorimetric Detector (1980, unpublished).
2. B. Dudelzak et al., Geiger Tube Planes as a Trigger for the Nucleon Lifetime Experiment in the Frejus Tunnel, Report LAL (Orsay) 83/04 (to be published in Nuclear Instruments and Methods).
3. S.D. Drell et al., Phys. Rev. Lett. 50, 644 (1983).

A MONOPOLE SEARCH USING AN ACCELERATOR DETECTOR

Philip L. Connolly

Japan/USA Neutrino Collaboration
Brookhaven National Laboratory
Upton, Long Island, New York 11973

A neutrino detector at the Brookhaven AGS has been used to investigate the feasibility of using an already constructed apparatus for GUT monopole searches. A flux limit (90% CL) of, $F_m < 5.2 \times 10^{-12}$ cm^{-2} s^{-1} sr^{-1}, was found. The limitations of such an approach are discussed.

This collaboration is primarily interested in the elastic scattering of $\nu(\bar{\nu})$ on electrons and protons and the design of the apparatus is determined by these goals. The main part of the detector consists of 112 identical modules. Each module consists of a plane of liquid scintillator cells for event time, and 2 planes of proportional drift tubes for xy position measurement. In addition there is a gamma catcher for downstream electromagnetic shower measurement and a muon spectrometer for normalization and beam studies.

The apparatus is read out by 4 PDP LSI-11/3 linked to a PDP 11-34. Further apparatus details and early results can be found in a recent publication.[1]

Only the liquid scintillator planes are used in this study. The scintillator is a mixture of NE235A and mineral oil enclosed in extruded acrylic tubes 7.9 cm deep and 25 cm high. Sixteen of these cells are assembled in a vertical wall with an active area of 4.22×4.09 m^2. The horizontal distance between scintillator planes is 17 cm. The overall size is comparable to existing dedicated monopole detectors. Each cell is viewed at each end by a 2" Amperex 2212A phototube. Discriminator thresholds were set so that the apparatus is sensitive to 1/3 × minimum ionizing particle. Times are measured with an accuracy of a few ns.

For neutrino beam running, the apparatus is run in a completely triggerless mode. A gate (10 μs) is opened slightly before arrival of the neutrino beam and all detector elements with data above threshold are read out at the end gate. There is a large set of calibration and monitoring routines which can be run in the period between beam pulses. One of these is called the Horizontal Cosmic Ray (HCR) trigger. In this mode the calorimeter planes are divided into groups of 4 successive planes. In each group, at least 3 of the 4 planes must have a cell in which both PMT tubes fired in coincidence (150 ns.). Super groups of 8, 12, 16, etc. planes can be formed. A gate is opened and if the required trigger is found in time coincidence the data are read out. If the trigger is not satisfied another gate is automatically initiated. This HCR trigger is normally satisfied by single relativistic cosmic ray muons moving in a direction roughly parallel to the neutrino beam direction. On-line analysis of the read out (PDT's included) from this trigger provides a very useful monitor of the efficiency and performance of the detector.

We have modified this trigger (using a super group of 32 planes) to search for low velocity ($\beta_m < 0.2$) penetrating particles. To allow for traversal of slow particles, the gate width was increased to 40 μs and the signals from each plane were latched so that the overall coincidence could be triggered. A veto circuit was included to exclude triggers caused by a single fast particle traversing the super group.

The trigger rate was then dominated by two (or more) independent fast muons. If there were but two fast muons, one of them must have traversed at least 1/2 of the required range. A routine was inserted into the 11/34 program which histogrammed the times of the individual hits. If the number of hits in a single bin exceeded that expected from a particle traversing 1/2 of the super group, the trigger was rejected. No dead time was caused by this routine.

Surviving triggers were written to tape for off-line analysis. Most of these were cases in which the time of arrival was such that the times were histogrammed into 2 adjacent bins and so evaded the on-line software filter. The off-line program required an explanation for each trigger in terms of known phenomenon, i.e., the 32 planes were hit by a small number of tracks each of which had a narrow time width. Any event which did not satisfy this criterion was examined in detail. Most of these were triggers in which most of the planes were hit by a few fast tracks and the remainder of the trigger requirement was satisfied by noise. No interesting candidates were found.

In the velocity range $10^{-3} < \beta_m < 0.2$ and for $\Omega = 14.5 \text{ m}^2$ sr, $t = 3 \times 10^6$ s, we find a flux limit (90 % Confidence Level) of

$$F_m < 5.2 \times 10^{-12} \ cm^{-2} \ sr^{-1} \ s^{-1} \quad .$$

We have shown that a large accelerator detector can be operated for a monopole search. The lower velocity limit in this study was constrained by the gate width which could be readily extended. Using all the apparatus could contribute an additional factor of ≈ 3 but a major limitation is the small solid angle. We are currently investigating methods to make the detector effectively isotropic.

ACKNOWLEDGEMENTS

This research has been supported in part by the U.S. Department of Energy under Contract No. DE-AC02-76CH00016.

REFERENCES

1. L.A. Ahrens, S.H. Aronson, P.L. Connolly, T.E. Erickson, B.G. Gibbard, M. Montag, M.J. Murtagh, S.J. Murtagh, S. Terada, A.M. Thorndike, W.G. Walker, P.J. Wanderer, D.H. White (BNL); J.L. Callas, D. Cutts, R.S. Dulude, B.W. Hughlock, R.E. Lanou, J.T. Massimo, T. Shinkawa (Brown); K. Amako and S. Kabe (KEK); Y. Nagashima, Y. Suzuki, S. Tatsumi (Osaka); K. Abe, E.W. Beier, D.C. Doughty, L.S. Durkin, R.S. Galik, S.M. Heagy, M. Hurley, A.K. Mann, F.M. Newcomer, R. Van Berg, H.H. Williams, T. York (Penn); P. Clark, D. Hedin, M.D. Marx, E. Stern (SUNY/SB); T. Miyachi, (Tokyo), Measurement of the Cross Section of $\nu_\mu + e^- \rightarrow \nu_\mu + e^-$, Phys. Rev. Lett. <u>51</u>, 1514(1983).

EXPERIMENTAL SEARCHES FOR MAGNETIC MONOPOLES AT PARTICLE COLLIDERS

Michael J. Price

E.F. Division
CERN
Geneva, Switzerland

The concept of magnetic monopoles is almost as old as the study of magnetism itself. However, only since 1931, the year of publication of Dirac's[1] classic paper, have monopoles had a fundamental role in modern physics. Of late, Grand Unified Theories have been able to provide predictions of the mass of the magnetic monopole, a parameter missing from Dirac's analysis. Although recent investigations have concentrated on the detection of very heavy GUT monopoles there seem to be no serious theoretical constraints forbidding the existence of monopoles with lower mass. The experimental searches for such light monopoles described here were carried out at three particle colliders with large center of mass energies: the CERN ISR, the CERN SPS collider and PETRA at DESY.[2,3,4] The maximum center of mass energies and particle types were respectively 60 GeV (pp and $p\bar{p}$), 540 GeV ($p\bar{p}$) and 40 GeV (e^+e^-).

Light magnetic monopoles produced in colliders would have high velocities and the search technique is based on the fact that such monopoles would have a high ionization loss. The predicted magnetic charge of the monopole is

$$g = nhc/2e = en/2\alpha = 137ne/2 \tag{1}$$

where the symbols have their usual significance and n is an integer. A magnetic monopole travelling at velocity βc would produce a circulating electric field $(137/2)\beta$ times stronger than the field due to a singly (electrically) charged particle. The expected ionization rate is of the order of $9(n\beta)^2$ GeV g^{-1} cm^2.

In the environment of a particle collider many electrically charged particles can be expected to pass through a monopole detector which should, ideally, be insensitive to them. Some plastic detectors possess sufficient insensitivity; in these detectors a trail of damage on the microscopic level produced by the passage of a highly ionizing particle can be subsequently revealed by suitable chemical treatment. The commercially available plastic foils Kapton and Makrofol have ionization thresholds of 5 GeV g^{-1} cm^2 and 2.5 GeV $g^{-1}cm^2$ respectively and are thus insensitive to minimum ionizing particles. The detectors themselves take the form of several sheets of plastic foils (to avoid accidental coincidences). The thickness of the individual foils is determined by the fact that they should be sufficiently thin to permit easy development but should be thick enough to prevent heavily ionizing end-of-range protons from simulating a signal. Foil thicknesses of the order of 50-100 μm are used which satisfy both these criteria.

As noted above, magnetic monopoles with very large velocities would be highly ionizing and in all three experiments efforts are made to minimize the amount of material between the production region and the detectors themselves. In each experiment Kapton foils are used directly inside the vacuum chamber of the collider; extensive tests had demonstrated the excellent high temperature and degassing properties of this material. In all three experiments magnetic fields are present: 1.5 T at the ISR, 0.7 T at the SPS collider and 1.7 T at PETRA. Such fields serve to separate any $g\bar{g}$ pairs produced and, in the case of the ISR experiment, would actually direct one of the monopoles toward the detector.

In the ISR experiment the foils are located in a tube attached to the main vacuum chamber. The foils, which are arranged in three pairs, are situated some 70 cm from the intersection zone. The SPS experiment is located at the same intersection zone as the UA1 experiment. The internal detectors take the form of three cylinders made of double layers of kapton foil. These cylinders, of length 1 m and diameter 12 cm, are placed 2 m apart straddling the intersection region. There are also two external sets of Kapton foils. In the PETRA experiment the detector consists of a cylinder wound from seven layers of Kapton and again placed around the intersection region.

The Kapton foils were developed in NaOCl solution at 90°C for 2 hours while the Makrofol foils, used in part of the ISR experiment, were treated in an NaOH solution at 50°C for 8 hours. The principle of this etching procedure is that attack of the material occurs at a much higher rate along the track of the heavily ionizing particle than in the bulk of the material. This differential attack eventually results in the formation of a hole along the trail of the particle. The bulk thickness of the foils is reduced by about 25% by this process. Holes in the plastic foils are located by placing them between two electrolyte soaked sponges connected in a circuit containing a microammeter. Calibration of this device using foils exposed to fission fragments had established that a current of 1 μA corresponded to a hole of 4 μm diameter. No holes were seen which could be interpreted as being due to magnetic monopoles.

One of the major difficulties in calculating limits of production cross section in the pp and p̄p experiments is that the detection efficiency depends on the production model. For the ISR results it is assumed (following previous authors[5],[6]) that the monopoles are produced isotropically in the reaction $p + p \rightarrow p + p + g + \bar{g}$ and that the available energy is divided among the outgoing particles proportionally to the relativistic phase space. In Fig. 1 the production cross section limit is plotted as a function of monopole mass. A more reasonable hypothesis may be that monopole production is not due directly to a proton-proton interaction but is similar to a Drell-Yan process $q\bar{q} \rightarrow g\bar{g}$ (the anti-quark being supplied from the sea). This hypothesis is considered in the analysis of the p̄p results (here, of course, the anti-quark comes from the anti-proton). The x-distribution of quarks in the proton and anti-quarks in the anti-proton is taken to be $(1-x^3)$ as indicated by lepton production experiments. If x_1 refers to the quark and x_2 to the anti-quark then monopoles can only be produced when $x_1 x_2 > (2M_g)^2$.

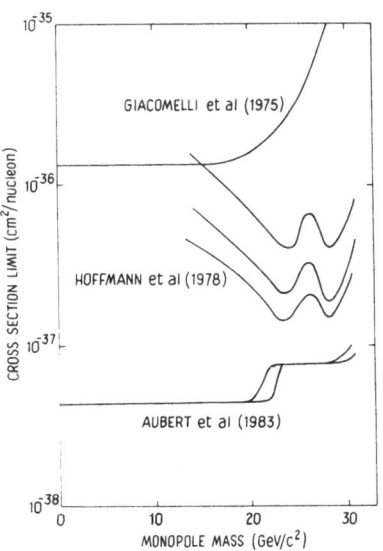

Fig. 1. Limits on monopole production cross section (95% C.L.) as a function of monopole mass in the ISR experiment.

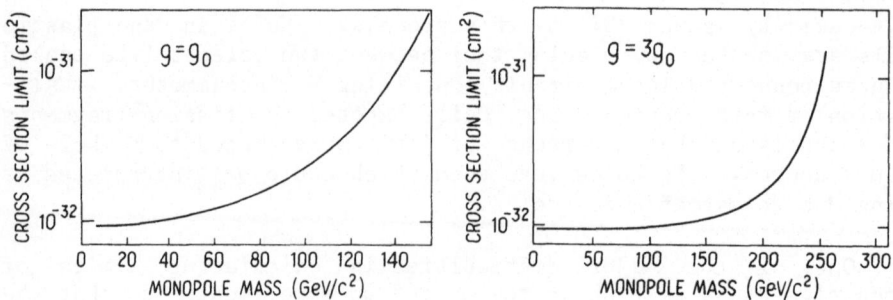

Fig. 2. Limits on monopole production cross section (90% C.L.)
         as a function of monopole mass in the SPS experiment.

It is assumed that production is isotropic in the center of mass
system of the quark-anti-quark pair and that, on account of the
x-distribution, the produced monopoles will suffer Lorenz boosts
in the laboratory system. The results for the SPS collider
experiment are shown in Fig. 2.

     In the PETRA experiment isotropic monopole production was
assumed which, for a center of mass energy $\geqslant$ 34 GeV gives an upper
limit of monopole production cross section in the reaction
$e^+e^- \rightarrow g\bar{g}$ of 4×10^{-38} cm^2 (95% C.L.). This value can be compared
to the limit of 9×10^{-37} cm^2 obtained in a similar experiment at
SLAC.

REFERENCES

1. P.A.M. Dirac, Proc. Roy. Soc. London, A133 60 (1931).
2. B. Aubert, P. Musset, M. Price and J.P. Vialle, (to be
    published. B. Aubert, P. Musset, M. Price and J.P. Vialle,
    Phys. Lett. 120B 465 (1983).
3. B. Aubert, P. Musset, M. Price and J.P. Vialle, Phys. Lett.
    120B 465 (1983).
4. P. Musset, M. Price and E. Lohrmann, Phys. Lett. 128B 333
    (1983).
5. G. Giacomelli, A.M. Rossi, G. Vannini, A. Bussiere, G. Baroni,
    S. Diliberto, S. Petrera and G. Romano, Nuovo Cimento, 28A 21
    (1975).
6. G. Hoffmann, G. Kantardjan, S. Diliberto, F. Medde, G. Romano
    and G. Rosa, Lettere al Nuovo Cimento, 23 357 (1978).
7. K. Kinoshita, P.B. Price and D. Fryberger, Phys. Rev. Lett. 48
    77 (1982).

IONIZATION DETECTOR FOR CABRERA'S MONOPOLE EXPERIMENTS

W. Peter Trower

Physics Department
Virginia Polytechnic Institute and State University
Blacksburg, Virginia 24061

Twenty-one months have elapsed since Cabrera detected a solitary current change in a ring detector consistent with a passing magnetic monopole of Dirac charge.[1] No similar event has since been seen in either his now-retired or new-improved detector.[2] The previously detected smaller current changes, which could be interpreted as near-miss events, occur in only one of the three rings, and then with a reduced frequency of once every other week.[3] Further, no other experiment has produced evidence for monopoles. So, if monopoles exist they are rarer than we thought in spring 1982.

Although time has tarnished its credibility, Cabrera's candidate event stands technically unimpuned. However, since it is the sole evidence that monopoles may exist in the universe, we are instrumenting for a possible, even if increasingly improbable, recurrence. If an ionizing cosmic ray is associated with an induction event, then the monopole myth is strengthened, especially if the ionization is remarkable in yield or temporal distribution. If no ionization is seen in coincidence with near-miss or ring-passing induction signals, then the interpretation of the Cabrera phenomena as a monopole in passage is considerably less compelling.

If the Cabrera candidate event is assumed to be caused by a cosmic ray magnetic monopole, the magnetic charge-direction (either north-up:south-down or south-up:north-down) will unambiguously be known when his coil is calibrated. Ring-passing and near-miss events measure the local monopole abundance. Cabrera's present triaxial experiment and proposed SQUID-read shield experiment can provide no new information about monopole properties,

625

only better information about the magnetic charge magnitude, the local abundance, and the charge-direction. To learn more about the monopole requires using other physical processes whose monopole effects are less reliably predicted than induction.

We are building an ionization detector to surround Cabrera's magnetic monopole experiments. The detector will have single photon sensitivity and will provide several hundred milliseconds of data upon sensing a trigger signal from Cabrera's SQUID. The operation of Cabrera's experiment requires that our detector produce little RF noise, be remotely managed, and fit into his Dewar well inside a granite loaded concrete block. Thus, we choose scintillator for the detector material.

If another induction signal is seen in Cabrera's experiment, our detector will record any attendant ionization signals. No ionization within a few milliseconds of the SQUID trigger will bring into question, variously, the dE/dx estimates, the induction signal authenticity, and/or the predicted monopole velocity. If ionization is identified, then the individual scintillator pulses will give rough speed information even for small ionization yields. Accompanying electrically charged particles will be detected and will show if the induction event was solitary or associated with a shower. The monopole velocity can be determined in direction and estimated in magnitude, from the time-of-flight between two traversed scintillators. Knowing the direction removes the charge-direction ambiguity thus revealing the sign of a causal magnetic charge. Detailed measurements of the ionization characteristics at the single photon level could be used to theoretically retrodict the ionization process. A monopole will excite longer-lived M1 atomic transitions in contrast with the usual faster E1 transitions, and might be measurable if the total ionization yield is sufficient.[4]

Our ionization detector, sketched in Fig. 1, is constructed in three assemblies (segments); roof(9), barrel(12), and plug(1). The scintillator, a Bicron BC412, is 1" thick and 6" wide, where the roof is 54" long and the barrel is 72" long. Segments are attached to solid plexiglass light pipes and then to photomultiplier tubes. These magnetically shielded 3" Hamamatsu R1307-07 tubes have high quantum(> 28%) and collection (> 90%) efficiencies, a clearly distinguishable single photon peak, a low potassium glass envelope, low dark current (< 600 counts/second), and an 11 nanosecond rise time. The 19" diameter plug will be coupled to a 20" Hamamatsu R1449 tube by a silicon-rubber light pipe. A single LED feeds fiber optics to each segment. These pulses are used for calibration and timing.

The electronic chains are built around modules, developed by Nanometric Systems for the Fermilab flash chamber wall. The

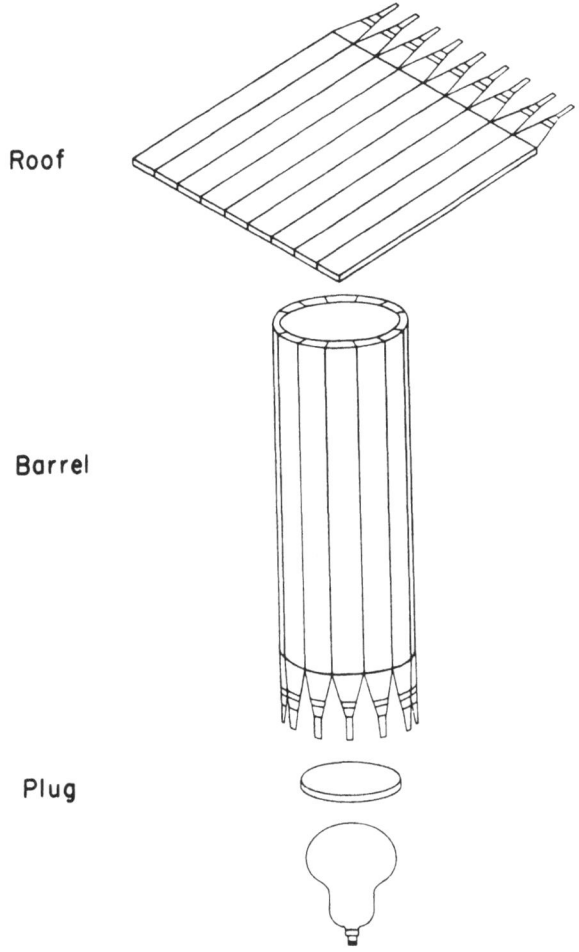

Roof

Barrel

Plug

Fig. 1 Schematic representation of ionization detector.

discriminator input is the photomultiplier anode signal and output
is a TTL adjustable 10-75 nanoseconds wide pulse. These output
pulses, interrogated at 20 MHz by a N-291-B control module will
cause the time of occurrence to be stored by segment in a 4-64K
bit-word wrap-around memory shared by 3 channels of a N-290-B
memory module. A trigger from the SQUID freezes all the memories
and initiates a computer read of their contents.

    The detector will function as a total ionization energy
cosmic ray shower counter, on a segment by segment basis. At
discriminator thresholds set below the one photoelectron peak, the
memory wrap-around time is ~ 1/3 second: Cabrera's SHE 30 MHz
SQUID has signal/noise ~ 1 when sampled at 100 Hz which determines
its 1/20 second rise time.

The summed individual photomultiplier dynode signals are resistively mixed. The resulting output signal is digitized by a TRW 8-bit FLASH analogue-to-digital converter when any discriminator is triggered. The time of occurrence of each non-zero bit is stored in its memory channel, if this summed signal is produced by more than a single photoelectron.

The experiment is controlled by a PDP11/23+ computer operating under RSX using CAMAC for data taking and testing. The experiment control software is being implemented, monitoring methods developed, and automated analysis strategies defined. Daily operation of the detector must proceed with minimum intervention by the resident low temperature physicists, and so the monitoring and data acquisition must be automated and remotely accessable.

Ordinary cosmic rays produce photoelectrons in a single clock cycle while a monopole would do so over several intervals. The dark current is random in occurrence and therefore easily identified. We will make the background rejection semi-automatically, after we study the patterns from our detector.

ACKNOWLEDGEMENTS

Our detector is being implemented in cooperation with Blas Cabrera; in technical collaboration with Tom Nunamaker, Dale Schutt (electronics), and Norm Hill(mechanical); with the support of the Virginia Tech Physics Department shops; and with the assistance of Bill McKinney. This work is supported in part by grants from the NSF (PHY 82-16478) and the Jeffress Trust.

REFERENCES

1. B. Cabrera, Phys. Rev. Lett. 48, 1378 (1982).
2. B. Cabrera, M. Taber, R. Gardner, and J. Bourg, Phys. Rev. Lett. 51, 1933 (1982); and B. Cabrera, in this volume.
3. R.A. Carrigan, Jr., and W.P. Trower, Nature (London) 305, 673 (1983).
4. S.P. Ahlen and G. Tarle, Phys. Rev. D27, 688 (1983).

REVIEW OF MONOPOLE DETECTION BY IONIZATION/EXCITATION TECHNIQUES

Steven P. Ahlen

Physics Department
Indiana University
Bloomington, Indiana 47405

INTRODUCTION

The application of conventional particle detection techniques to searches for GUT monopoles has achieved a remarkable degree of respectability considering the pessimistic attitudes in this regard just two years ago. During this two year period a great deal of progress has been achieved in theoretical work regarding the interaction of monopoles with matter and with the use of this work in interpreting results from cosmic ray telescopes which have been set up to look for monopoles.   In this review I will briefly summarize the results of the theoretical work, of the implications this has on detection techniques and on results from experiments searching for monopoles.

INTERACTION OF MONOPOLES WITH MATTER

When considering the penetration of slow particles in matter it is important to distinguish the energy lost to ionization or electronic excitation of atoms and molecules ("electronic energy loss") from that lost to kinetic energy of recoiling atoms ("atomic or nuclear energy loss").   This is in contrast to the stopping power of fast electrically or magnetically charged particles[1] ($\beta \gg \alpha$) for which electronic energy loss predominates by a large factor. Similarly, whereas single, simple expressions accurately describe the energy loss of fast projectiles[1] in arbitrary media, one must be careful to consider the details of the stopping medium in analyzing the energy loss of slow pro-jectiles.   For example, the approach adopted by Ahlen and Kinoshita[2] is adequate for the analysis of the electronic stopping

power of GUT monopoles in conductors and in materials whose
electronic properties can be approximated by a Fermi gas (e.g. in
semiconductors and insulators with band gaps small compared to the
kinematic limit of energy transfer from a heavy projectile to an
electron). However it is not appropriate for analysis of the
excitation of simple atoms such as hydrogen and helium. For these
types of systems, the elegant calculations of Drell et al.[3] (in
which detailed analysis of energy level mixing is carried out) are
required.

In considering the range of velocities over which a detector
will be sensitive to GUT monopoles it is useful to consider two
types of threshold velocities obtainable from simple kinematics
arguments. The least stringent threshold is that determined by
overall energy conservation between an atom and a monopole. This
is the threshold that would apply in principle to gas excitation
or ionization counters. If E is the energy difference between the
first excited electronic state and the ground state, and if v is
the relative velocity between the monopole and the atom then
excitation is possible only if $(1/2)Mv^2 > E$ where M is the atomic
mass. For helium this corresponds to $\beta = v/c > 1.03 \times 10^{-4}$, for
neon to $\beta > 4.8 \times 10^{-5}$, for argon to $\beta > 2.9 \times 10^{-5}$, for krypton
to $\beta > 1.9 \times 10^{-5}$ and for xenon to $\beta > 1.4 \times 10^{-5}$. However, as
noted by Kroll et al.,[4] it is unlikely that heavy noble gas atoms
could be excited for $\beta < 10^{-4}$ due to diamagnetic repulsion.

The most stringent velocity threshold for excitation or
ionization is obtained by considering two body collisions between
the monopole and an electron. If v_e is the characteristic
electron velocity and if $v \ll v_e$ is the monopole velocity then
excitation is possible if $2mvv_e > E$ where m is the electron mass.
This is the effective threshold velocity observed for the K-shell
excitation of heavy atoms by hydrogen or helium projectiles and
this is due to the fact that the perturbation of an electron bound
to a heavy nucleus by a light nucleus is too weak to induce level
mixings. Thus the threshold is not determined by the atomic mass
but by the electron mass. In general, it can be shown that if a
projectile's interaction with electrons can be regarded as a weak
perturbation, it is legitimate to consider binary interactions
between the electron and the projectile, taking into account the
binding of the electron by imposing on it a distribution of
velocities $f(v_e)$ given by the Fourier transform of its bound state
wave function. It has been argued that this should be true for
organic scintillators[5] for which a monopole velocity threshold of
$\sim 5 \times 10^{-4}$ c is obtained. It is important to note that this is a
conservative threshold and that if it can be demonstrated that
level mixing enhancements occur in aromatic molecules,[3] or that
anomalous high velocity components of pi-electron velocity distributions exist in aromatic molecules,[6] then organic scintillators would be able to detect monopoles with $\beta < 5 \times 10^{-4}$.

In addition to electronic exciation and ionization, other types of material excitation mechanisms have been proposed to search for monopoles. For example it is well known[7] that in a wide variety of minerals and glasses a linear trail of crystal defects can be etched to macroscopic dimensions with hydrofluoric acid. Such tracks have been produced in mica exposure to low energy heavy ions which manifest their presence by displacing mica atoms from their equilibrium lattice sites by means of nuclear scattering. It has been argued[8] that monopoles will capture ^{27}Al nuclei as they pass through the earth's crust and it has been shown[9] that an Al-monopole composite would leave tracks in mica from $3 \times 10^{-4} < \beta < 2 \times 10^{-3}$. Impressive limits on monopole flux have been obtained in this way.[9]

Another form of material excitation which has been proposed[10] to search for monopoles is to look for the thermoacoustic pulse generated in a conductor by the electronic energy loss. Since the electronic energy loss in conductors is proportional to velocity with no threshold[2] such a detector would, in principle, be superior to other excitation techniques in its ability to detect monopoles at any velocity. In this sense it would rival induction detectors and one might expect the cost per area of acoustic detectors to be small compared to induction devices. However, the signal to noise properties of acoustic detectors will have to be significantly improved over existing prototypes[10] before their promise can be realized.

Finally, Datta[11] has discussed the intriguing idea that monopoles might generate magnon shock waves or "magnetic Cerenkov radiation" at velocities $\sim 10^{-4}$ c and that such waves may be detectable. More work needs to be done in this area before an operational detector based on this idea can become available.

COMPLETED EXPERIMENTS

To date there have been four classes of experiments which have been performed to search for GUT monopoles by means of detector excitation. We display the results of the more sensitive experiments in Fig. 1. Also shown are Cabrera's latest limit[12] and indirect limits based on Galactic magnetic field arguments.[13] All limits in the figure[14] correspond to 63% confidence levels. The lower velocity sensitivity may correspond either to a timing threshold, to a pulse discriminator threshold or to a physical threshold. The latter two thresholds have been evaluated for organic scintillator experiments by utilizing the calculation of Ref. 5. This results in a threshold of 5×10^{-4} c for acrylic based scintillators and 6×10^{-4} c for polyvinyltoluene based scintillators. For those experiments which utilized argon proportional counters the conservative estimate for a physical

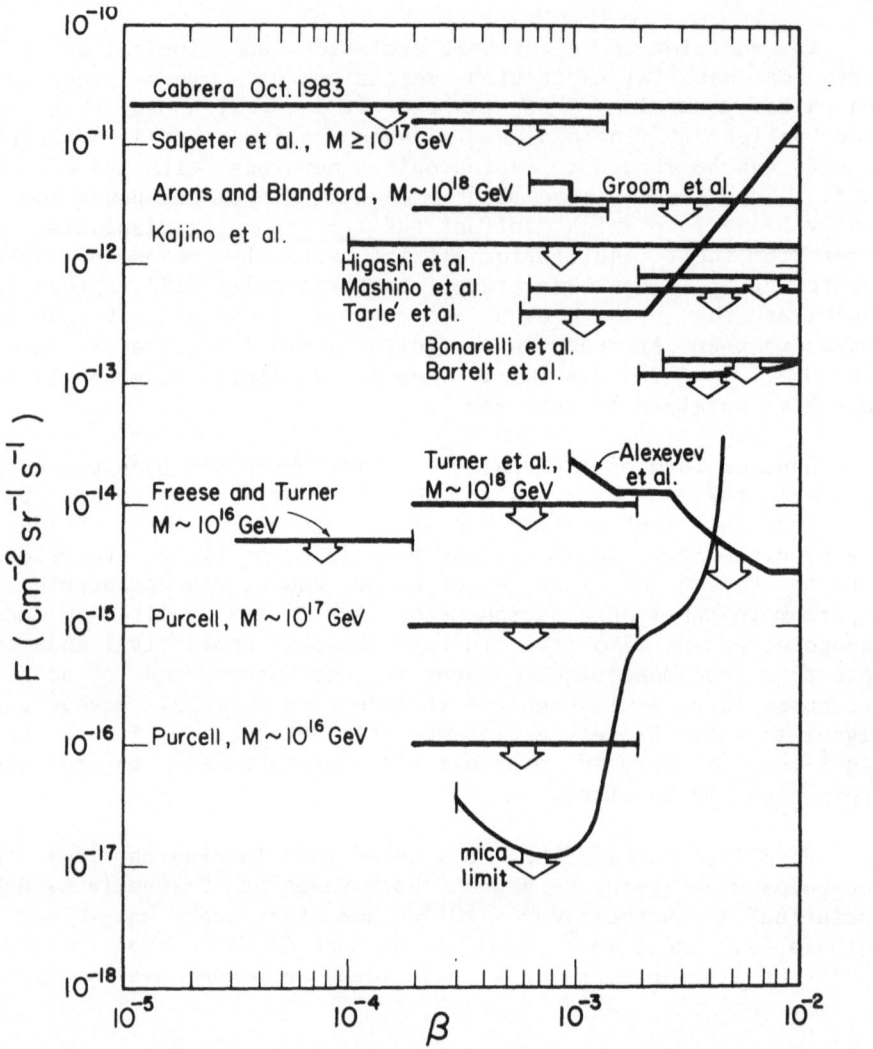

Fig. 1 Limits on monopole flux (63% C.L.), see discussion in text.

threshold is obtained by assuming the binary encounter approximation, which yields a threshold of $\sim 2 \times 10^{-3}$ c. This could be
reduced to lower velocities if level mixing enhancements or
unusually large electron velocity components can be demonstrated
for argon. For the experiment of Kajino et al., which utilized a
He-CH$_4$ gas proportional counter, a threshold of 10^{-4} c was assumed
in accordance with the theoretical work of Drell et al.[3] Note
that in this experiment, He is not primarily ionized. Level
mixing only permits excitation of He which is subsequently

converted to free electrons via the transfer of He excitation to CH_4 ionization. Finally, the results of Ref. 9 are labeled the mica limit. The low velocity sensitivity is determined by diamagnetic repulsion which prevents monopoles from capturing nuclei for velocities smaller than $\sim 3 \times 10^{-4}$ c. The high velocity sensitivity of the mica experiment is due to a reduction of nuclear stopping power with increasing velocity. Note that the zig-zag of the mica limit at $\sim 2 \times 10^{-3}$ c is due to the fact that Al-monopole composites "turn off" due to inadequate stopping power while composites of monopoles with heavier, although rarer nuclei (e.g. Mn) continue to contribute to the limit.

The curve labeled Alexeyev et al. in Fig. 1 is the monopole limit based on the results from the Baksan neutrino detector. This experiment has a geometrical factor of 1800 m^2 sr for monopole velocities. The two levels correspond to two trigger modes corresponding to different ranges of monopole velocities.

It is important to note that the only experiment capable of ruling out monopoles at a level below the "Parker" limit is the mica result. Since the Parker limit is roughly the same as the flux of GUT monopoles required to close the universe, it is seen that insofar as the mica limit can be relied upon, monopoles cannot be the long sought missing mass. In this regard I refer the reader to Ref. 9 where the reliability of the mica experiment is scrutinized. It is argued that the only way to avoid the mica limit is if the monopoles enter the earth with a positive electric charge. This would be the case if monopoles were created as positive dyons or if they managed to bind a proton prior to reaching the earth. This latter scenario could occur through radiative capture of protons in the early universe or through an Auger capture mechanism[15] involving interstellar hydrogen. As either of these possibilities have not been ruled out, the mica result cannot be regarded as compelling. It is interesting to note that experiments based on the He-CH_4 technique are subject to the same limitations as the mica technique since it has been shown[4] that the energy loss of positive dyons in He is small and is of a form that is difficult to detect.

CONCLUSION

It is argued elsewhere[16,17] that the most probable velocities of GUT monopoles relative to the earth would range from $\sim 10^{-3}$ c to 5×10^{-3} c. If this is so then the mica technique has most likely excluded monopoles as the missing mass of the universe, provided monopoles do not carry positive electric charge as they approach the earth. If this is the case, no direct experimental search has yet addressed the issue. In fact, if one accepts the Parker limit, no experimental search for a magnetic monopole component of the cosmic rays has yet been carried out. In order to carry out such an experiment, one requires an apparatus with a geometrical factor of at least 10^4 m^2 sr running for at least one year. In particular, if one wishes to rule out at the 95% confidence level the hypothesis that monopoles having mass of 10^{17} GeV are the missing matter of the universe, it would require an approximately flat detector of area 10^4 m^2 running for two years. This corresponds to a collecting power 50 times larger than the largest·cosmic neutrino experiments in operation (Baksan) or soon to be in operation (Homestake). However, there is good reason to expect that such an apparatus consisting of either liquid or plastic scintillator would have no difficulty identifying a GUT monopole, regardless of its electric charge configuration.[5]

REFERENCES

1. S.P. Ahlen, Rev. Mod. Phys. 52, 121 (1980).
2. S.P. Ahlen and K. Kinoshita, Phys. Rev. D26, 2347 (1982).
3. S. Drell, N. Kroll, M. Mueller, S. Parke, and M. Ruderman, Phys. Rev. Lett. 50, 644 (1983).
4. N.M. Kroll, S.J. Parke, V. Ganapathi and S.D. Drell, in this volume.
5. S.P. Ahlen and G. Tarle, Phys. Rev. D27, 688 (9183).
6. D.M. Ritson, SLAC Report No. SLAC-Pub-2950 (1982).
7. R.L. Fleischer, P. B. Price and R.M. Walker, Nuclear Tracks in Solids: Principles and Applications, University of California Press, Berkeley (1975).
8. C. Goebel, in this volume.
9. S.P. Ahlen, P.B. Price, S. Guo and R.L. Fleischer, in this volume.
10. G. Liu, in this volume.
11. T. Datta, in this volume.
12. B. Cabrera, M. Taber, R. Gardner and J. Bourg, Phys. Rev. Lett. 51, 1933 (1983).
13. M.S. Turner, E.N. Parker and T.J. Bogdan, Phys. Rev. D26, 1296 (1983); E.M. Purcell in Magnetic Monopoles, ed., R.A. Carrigan and W.P. Trower (Plenum, New York, 1983), p. 141; J. Arons and R.D. Blandford, Phys. Rev. Lett. 50, 544 (1983); E.E. Salpeter, S.L. Shapiro and I. Wasserman, Phys. Rev. Lett. 49, 1114 (1982); S. Dimopoulos, S.L. Glashow, E.M. Purcell and F. Wilczek, Nature 298, 824 (1982); K. Freese and M.S. Turner, Phys. Lett. 123B, 293 (1983).
14. Alexeyev et al. Proc. 18th Inter. Cosmic Ray Conf., Bangalore, India 5, 52 (1983); D.E. Groom, E.C. Loh, H.N. Nelson and D.M. Ritson, Phys. Rev. Lett. 50, 573 (1980); J. Bartelt et al. Phys. Rev. Lett. 50, 655 (1983); F. Kajino et al., Proc. 18th Inter. Cosmic Ray Conf., Bangalore, India 5, 56 (1983); S. Higashi, R. Bonarelli et al., Phys. Lett. 126B. 137 (1983); G. Tarle, S.P. Ahlen and T.M. Liss, Phys. Rev.

Lett. 52, 90 (1984); T. Mashimo et al., Phys. Lett. 128B, 327 (1983).

15. L. Bracci and G. Fiorentini, Phys. Lett. 123B, 493 (1983).
16. M.S. Turner, in this volume.
17. E.N. Parker, in this volume.

CONFERENCE HIGHLIGHTS AND SUMMATION - EXPERIMENTAL

Giorgio Giacomelli

CERN
Geneva, Switzerland

and

University of Bologna
INFN, Sezione di Bologna
Bologna, Italy

ABSTRACT

The experimental results presented at the "Monopole '83" workshop are reviewed and discussed.

INTRODUCTION

Before 1931 the subject of magnetic monopoles had received little attention in isolated discussions concerning the symmetries of Maxwell's equations, the analysis of the pole-electron system and the magnetic content of matter.

In 1931 Dirac introduced the magnetic monopole in order to explain the quantization of the electric charge.[1] From 1931 until 1974 a number of experimental searches were performed in cosmic rays, at accelerators and in bulk matter in order to find what we may now call the "classical monopole," characterized by a large magnetic charge and a relatively small mass.[2] During the same period, a slightly larger number of theoretical and phenomenological papers tried to resolve some specific problems, like the energy loss of monopoles, the meaning of the "tail" of the monopole, etc.

In 1974 't Hooft and A.M. Polyakov proved that all gauge

637

theories in which the electromagnetic group U(1) is a subgroup of a larger compact group, predicted the existence of magnetic monopoles with large masses.   Since 1974 the number of mathematical and theoretical papers increased considerably; the increase was even greater after 1979, when the main groundworks on Grand Unified Theories (GUT) of electroweak and strong interactions were laid down.[3]

The first specific conference on magnetic monopoles, held in Trieste at the end of 1981, was dominated by mathematical developments, though the discussions ranged from physics to mathematics, astrophysics, cosmology and experiments.[3]

Experiments searching for GUT (or cosmic) monopoles were somewhat slow to start.   But the number of searches increased considerably after the 1982 Stanford candidate event.[4]

At the Monopole Workshop held in 1982 at Wingspread, Racine, Wisconsin, USA, many new experimental and theoretical results were presented.   An even larger number of papers were presented here in Ann Arbor.   It is interesting to note that the experiments have grown in complexity and that all proton decay experiments are searching for the monopole catalysis of proton decay.   The analyses of all types of energy losses of monopoles in matter have become more detailed.   There have been many discussions of theories involving gravity and the connections with astrophysics and cosmology have become more numerous.

PROPERTIES OF COSMIC MONOPOLES

Let us review briefly the main properties of magnetic monopoles pointing out the recent developments.

Charge and Mass

The magnetic charge g is given by the Dirac relation

$$g = ng_D = hcn/4\pi e = 68.5 \, en, \tag{1}$$

where e is the basic electric charge and n = 1,2,3,...   Last year everybody emphasized the values n = 1, e = 1 and the mass value $m_M \sim 10^{16}$ GeV.   At this workshop we were told by theorists that things could be more varied and that one could have several possibilities.[5]

$$\begin{aligned}
g &= g_D, & 10^{16} &< m_M < 10^{19} \text{ GeV}, \\
g &= g_D, & m_M &\sim 10^4 \text{ GeV}, \\
g &= 2g_D, & 10^{10} &< m_M < 10^{16} \text{ GeV}.
\end{aligned} \tag{2}$$

Large masses were emphasized in order to explain the persistence of galactic magnetic fields. Moreover, the Kaluza-Klein theories,[6] which could be the basis for bringing gravity into a unifying picture, may predict $m_M > 10^{19}$ GeV and very small radii, so that one could ask if the monopole could be a black hole.

In the following I shall emphasize the results for $m_M \sim 10^{16}$ GeV and $g = g_D$.

Size and Structure

The magnetic pole is pictured as having:

- a core with a radius $r_c \sim 10^{-29}$ cm;
- a region up to $r \sim 10^{-16}$ cm, where virtual W^+, W^- and Z^0 may be present;
- a confinement region with $r_{con} \sim 1$ fermi;
- a fermion-antifermion condensate region up to $r_f \sim 1/m_f$;
- for r larger than few fermis the monopole behaves as a point Dirac monopole, which generates a magnetic field $B = g/r^2$.

Flux in Cosmic Rays

Monopoles should have been produced at the phase transition, which occurred at the cosmic time $t \sim 10^{-35}$ s after the Big Bang, when the unifying gauge symmetry group broke down into smaller subgroups, one of which was U(1). The estimates of the monopole production rates in the simplest GUT models are too large; estimates based on modified GUT models, in particular in "inflationary" models, may lead to very small rates. One may conclude that present production models are of little guidance, since the estimated production rates may range from zero to very large values.[5]

The constraint in the monopole flux obtained from the condition that the monopole mass density be smaller than the critical mass density, that is the density which would close the universe, is $F < 5 \times 10^{-5}$ poles cm^{-2} s^{-1} sr^{-1} if one assumes $m_M \sim 10^{16}$ GeV and that the monopole density in the universe is uniform. The upper limit becomes 10^5 times larger if the monopoles are assumed to be clumped in galaxies like the other types of matter.

The bounds obtained from the existence of magnetic fields in most celestial bodies, in particular in our galaxy, lead to stronger constraints on the monopole flux: from the galactic field one has: $F < 10^{-15}$ cm^{-2} s^{-1} sr^{-1} if $m_M < 10^{17}$ GeV (Parker bound). The limit which may be obtained from the existence of a magnetic field in the local supercluster is more doubtful, but could be an order of magnitude smaller.

We have heard that magnetic fields are present in essentially all celestial bodies and that from all of them one can obtain limits on monopole fluxes. It seems that these magnetic fields originate from a continuous dynamo mechanism, determined by the large scale fluid motion; for instance, for our galaxy the magnetic galactic field is stretched in the azimuthal direction along the spiral arms and it is due to the non-uniform rotation of the galaxy. This generates the field with a time scale approximately equal to the rotation period of the galaxy ($\tau \sim 10^8$ years). Though this mechanism seems plausible, there are still uncertainties and the picture is not completely settled. For instance, a primordial magnetic field could give some effect.

One could ask where could the monopoles be. One could ask if they could contribute to the dark matter, in particular to that which surrounds the galaxies and form their halos. It has to be kept well in mind that $\sim 90\%$ of the matter in the universe is not visible. One could also ask if the monopoles cluster like ordinary matter on planetary systems, thus yielding local enhancements. It seems that this is possible, but that the enhancement over the galactic flux could at most be an order of magnitude (and probably only a factor of 2-3). I recall that last year it was thought that the enhancement could be much larger.

On the basis of what has been said above one can expect that on earth could arrive a flux of cosmic poles as sketched in Fig. 1. Note that one could have high velocity poles ($\beta > 3 \times 10^{-3}$, extragalactic flux), poles with $\beta \sim 3 \times 10^{-3}$ bound in the local supercluster, poles with $\beta \sim 10^{-3}$ bound in the galaxy, poles with $\beta \sim 10^{-4}$ bound in the solar system, etc. The flux should increase with decreasing pole velocity, down to $\beta \sim 3 \times 10^{-5}$, which is the escape velocity of the earth. Clearly, the graph is only indicative. Compared to the ideas of one year ago[4] it would seem that the local flux of low velocity monopoles has decreased, while the flux of high energy poles has increased. We have also heard of poles which could enter in some magnetic flux tubes in neutron stars and be accelerated up to $\beta \sim 0.1$.[7]

Monopolonium

Some monopoles may be trapped in a pole-antipole system, which because of the large pole mass, would have a very long lifetime. Not much new has been added to this subject. It has been restated that one could have of the order of 10^3 pole-antipole annihilations per year on a volume of (1 light year)^3. These annihilations would lead to fantastic events, with thousands or millions of secondary particles, which would contain all the known and all the unknown particles of high energy physics.[8]

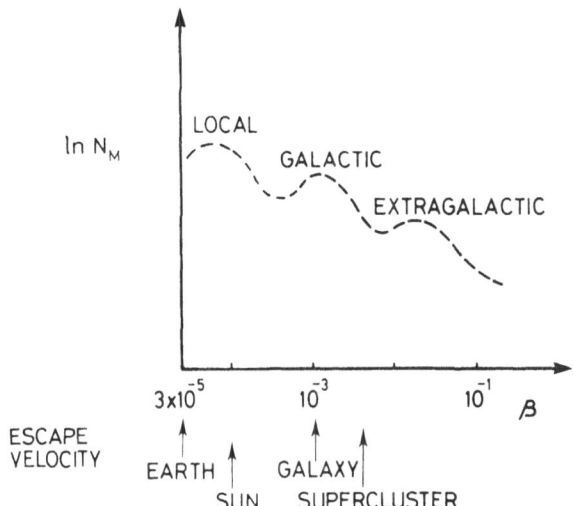

Fig. 1   Sketch of the possibly expected flux of cosmic monopoles
versus their β. The various peaks correspond to poles
trapped locally (to the sun and the earth), to poles
trapped in the galaxy and to extragalactic poles.

Capture of an Atomic Nucleus

The long range electromagnetic interaction of a magnetic
charge with the dipole magnetic moment of a nucleus may yield a
pole-nucleus bound system (plus the emission of energy). G.
Fiorentini[9] has estimated that these "monopolic atoms" have
binding energies in the range 1–100 KeV, with typical linear sizes
of the order of 20 fermi. C. Goebel[10] has estimated that the
capture cross section of the Al^{27} nucleus by a monopole with
$\beta \sim 10^{-3}$ could be $\sim$ 0.3 mb. This corresponds to a mean free path
in earth of $\sim$ 5 km, considering an average abundance of Al in the
earth's crust. A sizeable number of poles moving in the galaxy
should have a proton attached. When traversing a few kilometers
of earth a sizeable number of poles, which do not already have an
attached nucleus, should attach aluminum nuclei.

The formation of "monopolic atoms" could affect the effective
cross section of the catalysis of proton decay by monopoles.

DETECTION OF COSMIC MONOPOLES

We shall now discuss the results of the searches performed with various methods of detection: induction techniques, scintillation counters, proportional chambers, plastic detectors and indirect methods. The knowledge of the energy loss of slow monopoles is crucial for most methods of detection; it is also controversial for poles with $\beta < 10^{-3}$.[11-13]

Superconducting Induction Devices

A monopole passing through a superconducting coil induces a change of magnetic flux connected with the coil and thus a change in the electric current in the superconducting coil. This method of detection is ideal since it is based solely on the long range electromagnetic interaction between the moving magnetic charge and the macroscopic quantum state of the superconducting ring. The method is independent of the monopole mass, velocity and electric charge.

After the pioneering work of the Berkeley group,[14] the Stanford Group operated successfully in 1982 a single coil of 20 cm^2 area.[15] Now several experimental groups have in operation what may be called second generation experiments, characterized by areas one order of magnitude larger, coincidence arrangements and sophisticated procedures for eliminating spurious events.[16] Examples of these detectors are the 3-axis detector at Stanford (Fig. 2), the seventh order gradiometer at IBM (Fig. 3), and the "macrame" arrangment at Chicago (Fig. 4). It may be worth recalling that the fraction of signal collected is limited not by detector size, but by the size of each loop. Figure 5 shows the chart recording of the currents in the three loops of the Stanford detector. It also shows the recording of an accelerometer, which should help in eliminating possible "candidates", in reality caused by mechanical disturbances.

Experimental results have been presented at this workshop by several groups, as listed in Table 1. After the candidate event reported in 1982 by the Stanford group, no other candidate was observed. The present level is $\sim 1/60$th of that reported by Stanford in 1982.

Energy Losses of Monopoles

At high velocities, $\beta > 10^{-2}$, the interaction of poles with matter is well understood: A pole of magnetic charge g behaves as an equivalent electric charge $e_{eq} = g\beta$. The energy loss, due to impact ionization, is described by the Bethe-Bloch formula. For liquid hydrogen the result is curve c of Fig. 6.[13,17-18]

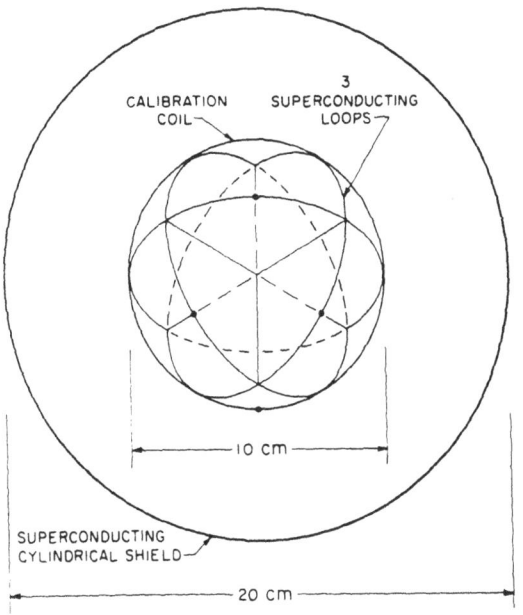

Fig. 2 Schematic top view of the Stanford three-loop superconducting monopole detector.[16]

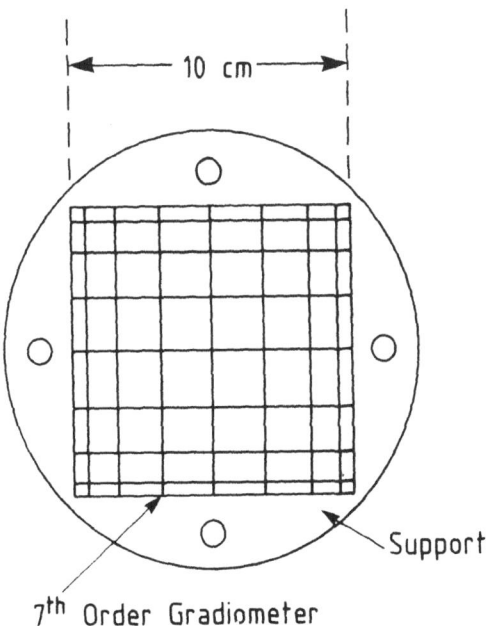

Fig. 3 Schematic top view of IBM 7th order superconducting
gradiometer.[16]

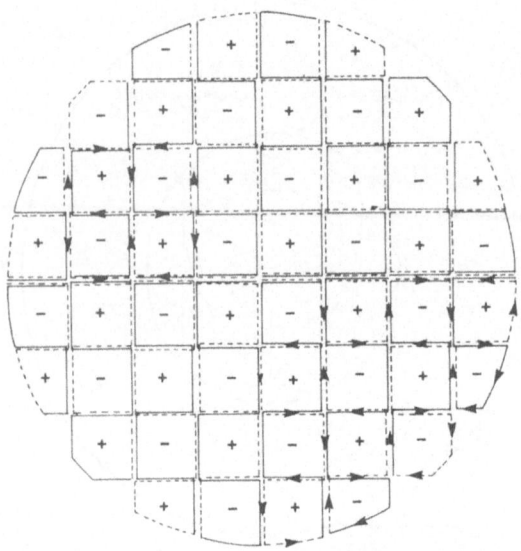

Fig. 4 "Macrame" superconducting structure of Chicago coil.[16]

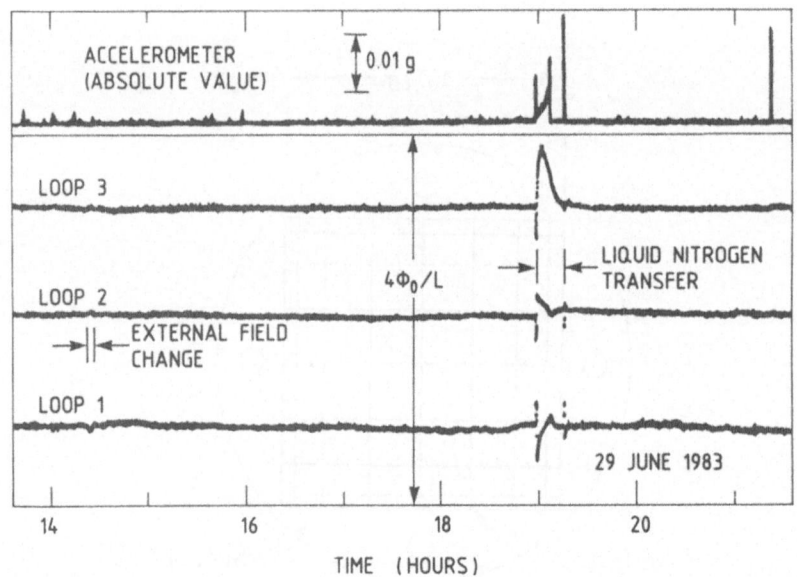

Fig. 5 Chart recording of the loops and of the accelerometer of
        Stanford-2 detector.[16] Note the disturbances generated at
        the time of the liquid nitrogen transfer in the outer
        cryostat.

Table 1. List of experiments searching for cosmic monopoles using superconducting induction devices. The table gives for each group the main feature of the apparatus, the effective area for which one has a 4π solid angle and the flux upper limit (90% confidence level; the first value corresponds to one event). The overall combined upper limit is 1.5×10^{-11} cm^{-2} s^{-1} sr^{-1}.

Group	Main Feature	Area ($cm^2/4\pi$)	Flux Limit (10^{-11} cm^{-2} s^{-1} sr^{-1})
Stanford 1	Single coil	10	61
Stanford 2	3 axis coils	71	2
Chicago-FNAL-Mich.	2 coils	700	7
IBM - 1	Gradiometer	25	51
IBM - 2	Gradiometer	1000	--
Kobe	2 coils	25	200
IC	2 coils	300	--
NBS	Backgr. studies	--	--

For smaller velocities, $10^{-4} < \beta < 10^{-3}$, the energy losses are mainly due to adiabatic excitation of the atoms, arising from the interaction of the atomic electrons with the monopole magnetic field. In this case crossing of the atomic levels may occur (Fig. 7). The net effect is that the passage of a monopole may leave atoms in an excited state. This is the "Drell" effect, which until now has been computed only for atomic hydrogen and helium.[17] The curve b in Fig. 6 shows the energy loss due to the Drell effect in hydrogen. The effect may be used for practical detection either by observing the photons emitted in the de-excitation of the excited atom or by observing the ionization caused by the energy transfer from the excited atoms to complex molecules with a small ionization potential (Penning effect). If the monopole has attached a charged particle (or if it is a dyon) the Drell effect could be reduced.

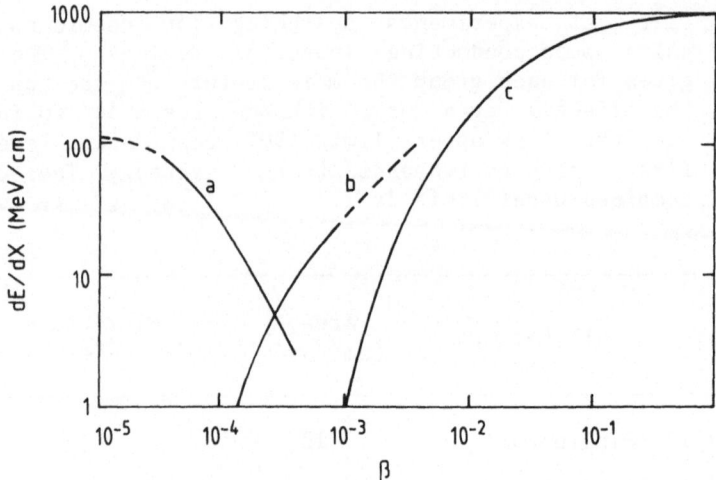

Fig. 6 The energy loss (in MeV/cm) of monopoles in liquid
hydrogen as function of β.[17-18]   Curve (a) corresponds to
elastic monopole-hydrogen atom scattering;[13] curve (b)
corresponds to adiabatic interaction with level crossings[17]; and curve (c) describes the ionization loss.[18]
The dashed parts of the curves correspond to velocity
ranges where the approximations used in the calculations
may break down.

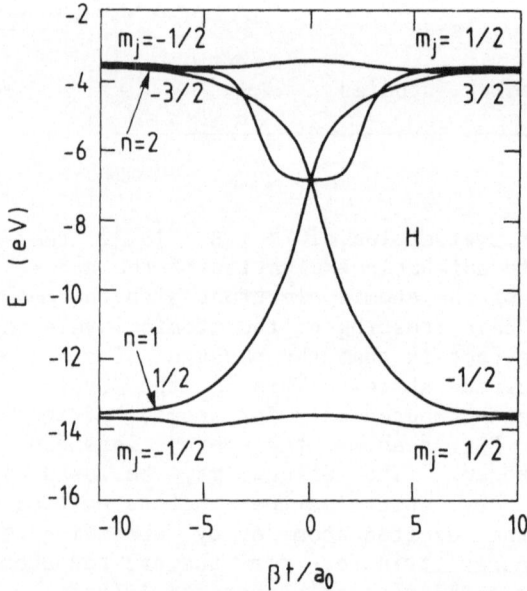

Fig. 7 The energy levels for atomic hydrogen before (left),
during (center) and after (right) the passage of a
magnetic monopole.[17]

The Drell mechanism is effective as long as the monopole-atom collision energy exceeds the spacing of atomic levels. For smaller energies, that is for smaller β, the energy loss is due to elastic monopole-atom collisions arising from the long range magnetic charge-magnetic dipole interaction. The energy is eventually dissipated into heat. Curve a of Fig. 6 shows the energy loss due to elastic collisions in liquid hydrogen.[13]

For materials more complex than liquid hydrogen one expects energy losses which are qualitatively the same, but which should interest lower velocities. Thus the curves b and c of Fig. 6 should shift toward lower β, by even an order of magnitude in β.[11-12]

Ahlen et al.[11,19-20] have pointed out that when using scintillators one cannot limit the considerations to the ionization energy loss, but one should consider the photon yield and the saturation effect in solids. They have obtained the photon yield shown in Fig. 8 for a pole, a dyon and a pole +Al nucleus traversing a plastic scintillator. The curves show that there is an effective threshold at $\beta \sim 5 \times 10^{-4}$. The calculations of other authors[12] would indicate a sensitivity down to $\beta \sim 10^{-4}$. The curves of Fig. 8 represent in any case a lower bound on the light yield, since there must be other effects, like those discussed above due to the magnetic charge-magnetic dipole interaction, which may effectively shift the β-threshold.

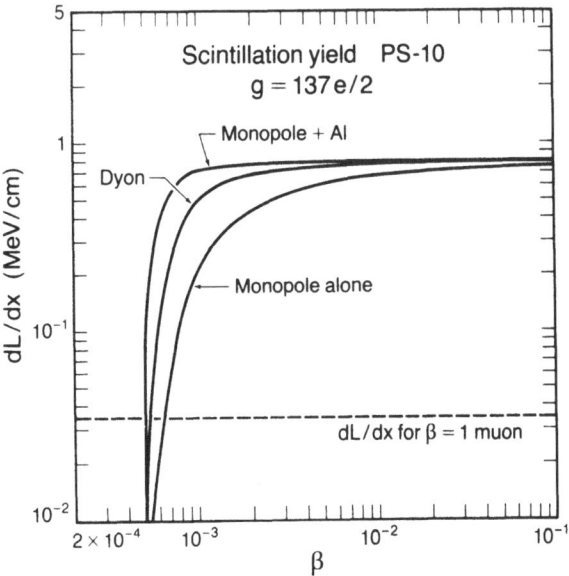

Fig. 8 Scintillation light yield for monopoles in an acrylic scintillator as a function of β.[14-19]

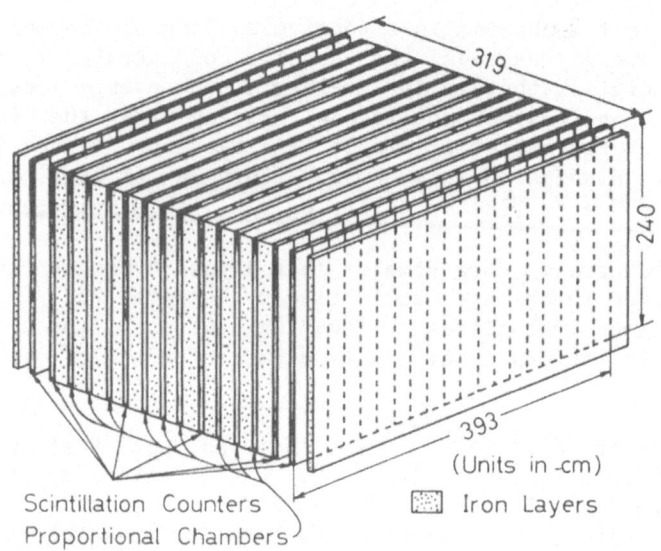

Fig. 9 Layout of the new Tokyo detector, which uses the Drell and
        Penning effects (see text).[21]

Detection with Scintillation Counters and Proportional Chambers

Table 2 gives a comprehensive summary of the electronics
experiments and of the upper limits achieved for the β-range
covered.[4,20]

Besides the early experiments performed with two or more
layers of scintillation counters and/or proportional chambers in
the last year, a number of new methods were developed and applied
to cosmic monopole searches. These methods will now be briefly
discussed (Table 2 gives the relevant parameters also for these
experiments).

The Tokyo Group[21] used a stack of scintillation counters and
of proportional chambers, employing 90% gaseous helium and 10%
CH_4, Fig. 9. In the proportional chambers the monopoles could
excite the helium atoms via the Drell mechanism discussed in the
previous section

$$He \xrightarrow[pole]{} He^*. \tag{3}$$

Then, by the Penning effect, the excitation energy of the excited
He^* is transferred into ionization of the CH_4 molecule

$$He^* + CH_4 \rightarrow He + CH_4^+ + e^-. \tag{4}$$

Table 2. List of experiments which searched or are searching for a flux of cosmic monopoles with scintillation counters, proportional tubes and track-etch detectors.

LABORATORY	LOCATION	DETECTOR	ΩS (M² SR)	dE/dx (MINIMUM)	β-RANGE	FLUX UPPER LIMIT (CM⁻² S⁻¹ SR⁻¹)
1. BNL	BUILDING	PROPORTIONAL	1.9	2.0	3×10^{-4} - 1.2×10^{-3}	3.4×10^{-11}
2. BOLOGNA	BUILDING	SCINTILLATORS	10-36	10-25	10^{-3} - 0.6	3.4×10^{-13}
3. TOKYO	BUILDING	SCINTILLATORS	1.1	1.2	10^{-2} - 10^{-1}	1.5×10^{-11}
"	BUILDING	SCINTILLATORS	1.4	0.025	2×10^{-4} - 5×10^{-3}	1.5×10^{-11}
"	KAMIOKA MINE	SCINTILLATORS	22.0	0.2	6×10^{-4} - 1	1.5×10^{-12}
4. UTAH-STANFORD	MAYFLOWER MINE	SCINTILLATORS	2.7	0.12	1.4×10^{-4} - 3×10^{-2}	8.1×10^{-12}
5. MINNESOTA-ARGONNE	SOUDAN MINE	PROPORTIONAL	71.6	0.5	4×10^{-4} - 3×10^{-2}	4.1×10^{-13}
6. USSR	BAKSAN MINE	SCINTILLATORS	1800.0	0.25	4×10^{-3} - 5×10^{-2}	1.5×10^{-14}
7. INDIA-JAPAN	KOLAR MINE	PROPORTIONAL	218.0	2.5	2×10^{-3} - 0.9	3.5×10^{-14}
8. MONT BLANC	TUNNEL	STREAMER TUBES	12.0	0.02	3×10^{-4} - 0.5	1.9×10^{-12}
9. BNL-BROWN-KEK	NEUTRINO BEAM	DRIFT + SCINT.	14.5	0.3	10^{-3} - 0.2	5.2×10^{-12}
10. TOKYO	BUILDING	SCINT. + GAS	24.7	-	$> 2\times10^{-4}$	1.6×10^{-12}
11. BERKELEY-INDIANA	BUILDING	SCINTILLATOR	17.5	1.2	6×10^{-4} - 2.1×10^{-3}	4.1×10^{-13}
12. BERKELEY	SURFACE	CR39	150.0	Z/β>30	0.02 - 1	1.5×10^{-13}
13. KITAMI	BUILDING	NITROCELLULOSE	1000.0	Z/β>60	0.04 - 1	5.2×10^{-15}

Therefore, one may obtain an effective ionization also for low velocity monopoles, in the $10^{-4} < \beta < 10^{-3}$ range.

The Berkeley-Indiana Group[19] used a single thick slab (7.6 cm) of scintillator. Relativistic charged particles traverse the detector in a time much shorter than the response time of the detector system (~ 40 ns). A GUT pole travelling with $\beta = 10^{-3}$ takes at least 250 ns to traverse the scintillator. The signature of a monopole is thus given by an anomalously wide pulse with a constant pulse height in time. With this arrangement the experimenters should have reached the β limit given by ionization energy losses.

The Mont Blanc detector for proton decay[22] is a cubic detector of 3.5 m/side, made of 134 layers of limited streamer tubes, separated by 1 cm thick iron absorbers. The detector is capable of determining the path of an eventual monopole with a transverse precision of ~ 1 cm^2. The identification may be made by time-of-flight through the whole detector. The authors' point out that because of the many samplings involved the apparatus is capable of detecting monopoles which ionize 1/100 of I_{min} because of the Landau tail in the energy loss distribution.

Figure 10 shows a compilation of upper limits (at the 90% confidence level) for a flux of cosmic GUT poles plotted versus the β of the monopole.

Detection with Track-Etch Detectors

Plastic detectors may be considered as threshold devices with thresholds which depend on the material and on the method of chemical etching. Approximate thresholds for monopoles are: CR 39: $\beta > 0.02$; nitrocellulose: $\beta > 0.04$; lexan (makrofol E): $\beta > 0.3$; mica: β n > 2. These detectors may be exposed for long times (longer than one year). The passage of a heavily ionizing particle produces a number of defects along its path. Since the plastics are good insulators these defects last for long times. The plastics are then chemically etched: The passage of a charged particle may appear as a hole in the plastic if it was heavily etched, or as a cone on each side of the plastic if the etching was moderate.

A Berkeley Group[23] exposed ~ 15 m^2 of several layers of CR 39 for about one year at ground level. They quoted an upper limit $F < 1.5 \times 10^{-13}$ cm^{-2} s^{-1} sr^{-1} for $\beta > 0.02$.

A Japanese Group[24] exposed 100 m^2 of nitrocellulose sheets for 3.3 years at ground level at Kitami, Hokkaido. The experiment was originally designed to search for "classical monopoles." It was modular, with each unit consisting of a stack of a pair of

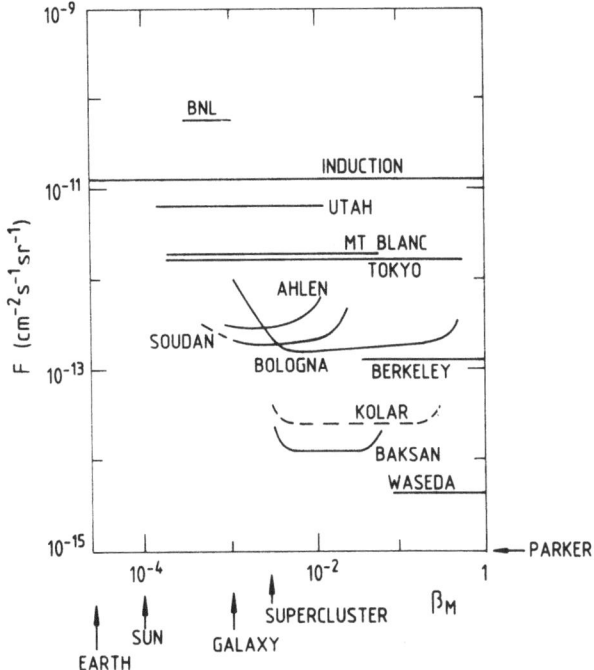

Fig. 10 Compilation of upper limits for a flux of cosmic GUT
monopoles plotted versus the β of the monopoles. The
Stanford-1 Experiment corresponds to one candidate event;
the other experiments are upper limits at the 90%
confidence level. Most limits were obtained with scintillation or gas tube detectors (Tables 1 and 2). The
Berkeley Experiment was performed with CR 39 plastics,
the Kitami Experiment with nitrocellulose sheets.

nitrocellulose sheets, a pair of polycarbonate sheets and an X-ray
film. Only the nitrocellulose is useful for the detection of
relatively slow poles. The authors quoted an upper limit
$F < 5.2 \times 10^{-15}$ cm^{-2} s^{-1} sr^{-1} for poles with $\beta > 0.04$.

Catalysis of Proton Decay

It was suggested in 1980 that a GUT monopole in proximity of
hadronic matter could catalyze baryon number violating processes
such as

$$p + M \rightarrow M + e^+ + \text{mesons}. \tag{5}$$

It was thought that the cross section would be very small,
comparable to the geometrical cross section of the core
($\sim 10^{-58}$ cm^2), where may be found the X-mesons which mediate the
$\Delta B \neq 0$ interactions. In 1982, V.A. Rubakov[25-26] and C.G. Callan[27]

suggested that the cross section could be comparable to the cross
section of ordinary strong interactions ($\sigma \sim 10^{-26}$ cm^2) because
the monopole should be surrounded by a condensate of fermion-
antifermion pairs.   This possibility has stirred up considerable
theoretical interest and many controversies.   At this time one
should in fact consider the cross section to be uncertain by
several orders of magnitude.

   If the $\Delta B \neq 0$ cross section for monopole catalysis of the
proton decay would be large, then a monopole would trigger a chain
of baryon "decays" along its passage through a large detector,
such as those designed to study baryon decay.   The mean free path
λ between two successive monopole-induced proton decays would be
for slow monopoles

$$\lambda(\mathrm{m}) \ = \ \frac{43}{\rho(\mathrm{g\ cm}^{-3})} \ \frac{\beta}{\sigma}, \qquad\qquad\qquad (6)$$

where σ is the cross section in units of the typical strong
interaction cross section ($\sigma_0 = 4 \times 10^{-26}$ cm^2), βc is the velocity
of the pole and ρ the density (in g cm^{-3}) of the material
traversed.   Note that in Eq. (6) one has assumed that $\sigma_{\mathrm{cat}} \sim 1/\beta$.

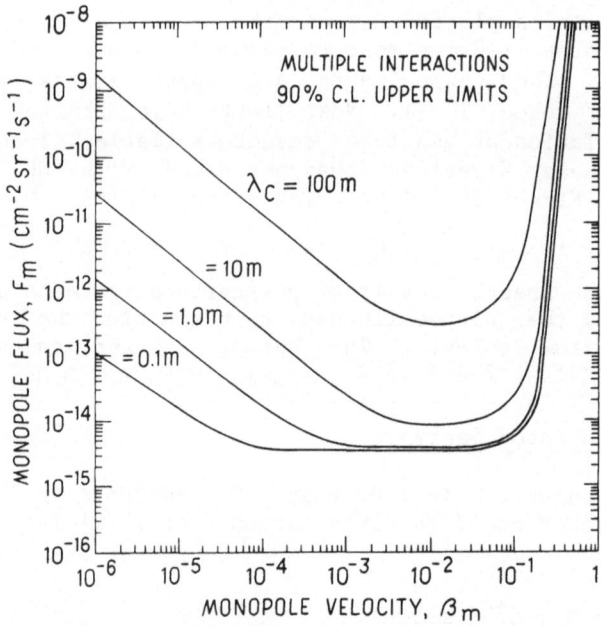

Fig. 11   Upper limits (90% CL) on the monopole flux versus
          monopole velocity for the multiple catalysis of proton
          decay in the IMB water Cherenkov detector for several
          values of the catalysis cross section.[28-29]

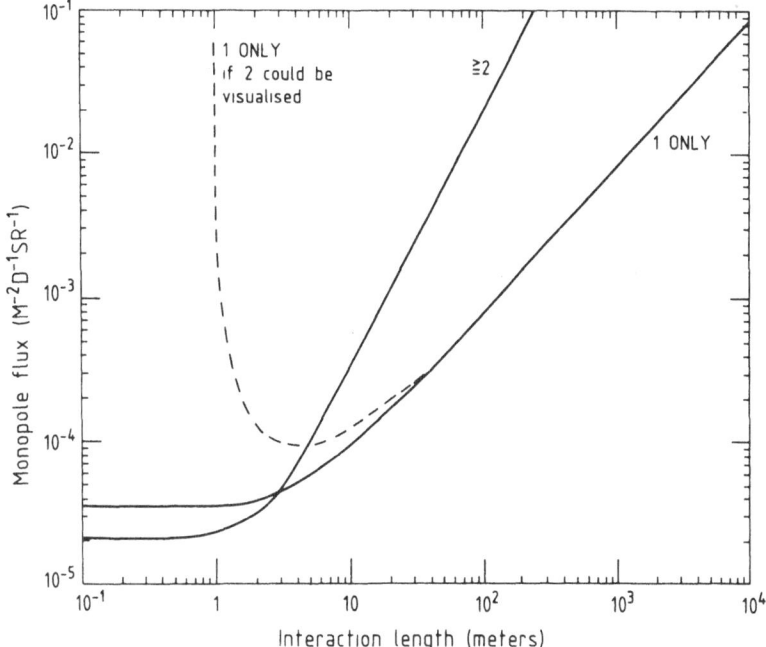

Fig. 12 Upper limits (90% CL) on the monopole flux for the single and multiple monopole catalysis of proton decays in the Mont Blanc Experiment.[22]

While one year ago there were only some rough upper limits from the analysis of bubble chamber pictures and indirect information, now most of the large scale proton decay experiments have reported upper limits for the monopole catalysis of proton decay. Table 3 lists some of the main features of these experiments together with the measured upper limits.[28] Some limits are shown in more detail in Figs. 11 and 12.

The Aachen-Hawaii-Tokyo Group performed a quick experiment using a water Cherenkov counter filled with 12 t of water. From a 12-day run they obtained an upper limit flux of 2×10^{-12} cm^{-2} s^{-1} sr^{-1} valid for $5 \times 10^{-4} < \beta < 5 \times 10^{-2}$.[28]

The Mont Blanc proton decay detector, which was already mentioned in the previous section, has an average density of 3 g cm^{-3}. It is located under Mont Blanc at a depth of 5000 m of water equivalent.[22]

The Irvine-Michigan-Brookhaven (IMB) water Cherenkov detector is a parallelepiped of $17 \times 22.5 \times 18$ m^3, viewed by 2048 photomultipliers. It is located at a depth of 600 m ($\sim$ 2000 m of water equivalent) in the Morton salt mine near Cleveland, Ohio.[29]

Table 3. Some of the main features of the large proton decay experiments and upper limits on the monopole flux. These are at the 90% confidence level and assume $\sigma_{cat}= 10$ mb. The upper limits from the non-observation of a string of interactions (>2) are the most significant; for these the table gives the β-range covered. The upper limits from single interactions are limited by the general background from neutrino interactions.

COLLABORATION	LOCATION	DEPTH (M OF H_2O)	DIMENSION (M^3)	EFFECTIVE SURFACE (M^2)	EFFECTIVE HEIGHT (M)	RUN TIME (YEARS)	UPPER LIMITS FOR $\sigma_{cat} = 10MB$ ($CM^{-2}\ S^{-1}\ SR^{-1}$) FOR > 2 INTERACTIONS	APPROXIMATE β-RANGE FOR > 2 INTERACTIONS
TATA-OSAKA TOKYO	KOLAR GOLD MINE	7600	$6 \times 4 \times 3.7$	30	2.5	2.4	3×10^{-12}	$10^{-3} - 10^{-1}$
FRASCATI-MILAN TORINO	MONT BLANC TUNNEL	5000	$(3.5)^3$	18	2.3	1.2	2.3×10^{-14}	$10^{-3} - 4\times10^{-2}$
IRVINE-MICHIGAN BROOKHAVEN	MORTON SALT MINE	1500	$17 \times 22.5 \times 18$	550	12.8	0.6	3.6×10^{-15}	$6\times10^{-4} - 10^{-1}$
TOKYO-LICEPP	KAMIOKA MINE	2700	15 (DIA.) $\times 16$	218	10.3	0.2	3.8×10^{-14}	$10^{-4} - 4\times10^{-3}$
ANL-MINNESOTA	SOUDAN MINE	1800	$2.9 \times 2.9 \times 1.9$	10	1.7	0.8	1.5×10^{-13}	$> 10^{-3}$

The Soudan-1 prototype consists of horizontal layers of 3456 proportional tubes each 4 cm in diameter, held in a matrix of taconite (iron loaded concrete). The average density of the detector is 1.6 g cm^{-3}; the detector is 2.9 × 2.9 × 1.9 m^3 and weighs 31 t.[28]

The Tokyo water Cherenkov counter is a cylinder of 15 m diameter and 16 m height. They use very large photomultipliers which were specifically designed for the experiment.[28]

The Tata-Osaka-Tokyo detector is composed of 34 layers of proportional counters with 1.2 cm iron plates between the layers. The counters are 10 × 10 cm^2 area by 4 or 6 m length. The detector is a 6 × 4 × 3.7 m^3 parallelepiped with a total weight of 140 t and an average density of 1.6 g cm^{-3}.[28]

Stringent indirect limits on catalysis were obtained by various authors considering the catalysis of nuclear matter within neutron stars. The upper limits were obtained by looking at the general X-ray background and at the X-ray emission for some particular neutron star. The limits (on monopole flux times catalysis cross section times the β of the monopole) are of the order F σ β < 10^{-50} s^{-1} sr^{-1}.[7,30] This value could possibly be modified by unknown properties and structure of the neutron stars.

OTHER METHODS OF DETECTION

Searches in Bulk Matter

As already stated, magnetic monopoles could be trapped in ferromagnetic domains by an image force of the order of 10 eV/Å. Cosmic poles of low velocity (β < 10^{-4}) could have stopped in ferromagnetic materials at the surface of the earth, after traversing it. The probability for stopping is very small. On the other hand some ferromagnetic deposits were formed hundreds of millions of years ago, so that accumulation over long times may have occurred.

The Kobe Group has performed a search for relic monopoles trapped in iron sand using several tens of kilograms of material formed between 10^7 and 10^8 years ago.[31] The sand was heated above the Curie point at which temperature the material stops being ferromagnetic. The poles, which were trapped in the material, would leave it, would fall towards the earth and would be detected in a superconducting induction coil through which they would pass. The Kobe Group placed the upper limit of 2 × 10^{-6} poles per g of ore. It is difficult to extract from this an upper limit on the monopole flux: It is estimated to be of the order of 10^{-13} poles

cm^{-2} s^{-1} sr^{-1} for poles with $\beta < 10^{-4}$.

A Wisconsin Group is proposing an experiment of this type on a large scale using the ancient iron ore processed in a steel mill in Wisconsin.

Searches for Ancient Tracks in Mica

Mica as a track-etch detector has a high threshold (it would not detect even relativistic monopoles with n = 1). On the other hand, it should detect the passage of a "monopolic atom," when the attached nucleus is for instance aluminum, if the speed of the system is of the order of 10^{-3} c, where ionization losses of charged particles are largest.

Ahlen et al.[20] have taken a piece of mica from a mine 5 km deep in Brazil. The age of the mica was estimated to be $\sim 4.6 \times 10^{8}$ years. The authors emphasize that the damage in the mica is caused by nuclear stopping power. After etching in hydrofluoric acid they scanned with an optical microscope 14 cm^{2} of mica. They estimated an upper limit $F < 2 \times 10^{-17}$ cm^{-2} s^{-1} sr^{-1} for poles with $4 \times 10^{-4} < \beta < 2 \times 10^{-3}$. These limits are obtained assuming that the poles may attach an Al nucleus and that the mean free path for Al attachment in the earth's crust be ~ 5 km. It is not clear what would happen if the monopoles would have already attached a light nucleus, like a proton.

Other Detectors

With a detailed calculation De Rujula has shown that a monopole traversing a metal produces a "thermo-acoustic" pulse, whose amplitude is linear in the monopole's velocity[32] (Fig. 13). If the metal is superconducting, there is a novel additional "magneto-acoustic" source whose amplitude is β-independent. He sketched a "sonic antenna" that could respond directionally to monopoles (and even to normal particles of cosmic rays, and to more exotic objects, Fig. 14).

More conventional acoustic detectors applied to metal slabs at room temperature were discussed by Barish.[33] Though not yet operational, acoustic and thermo-acoustic detectors show promise for the future detection of low β monopoles. Induction coils at room temperature have been discussed by Price.[34] These detectors are about to become operational.

EXPERIMENTS AT ACCELERATORS

At present the main emphasis of monopole searches concerns

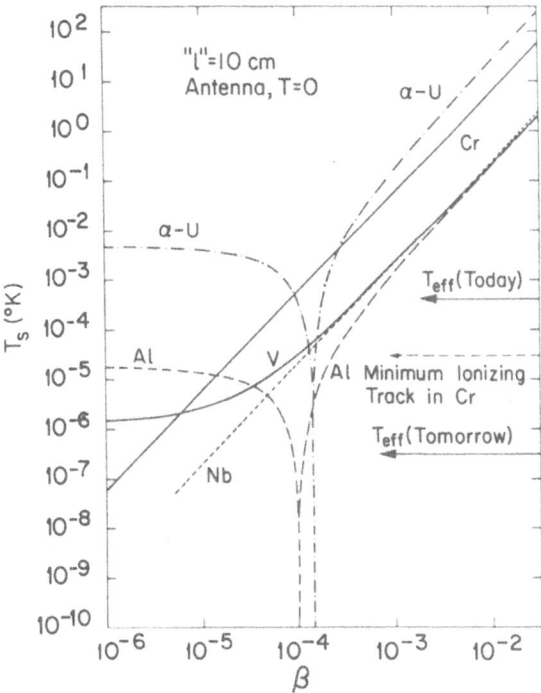

Fig. 13 Signal temperature per eigenmode for a variety of
         materials at very low temperatures vs. monopole velocity.

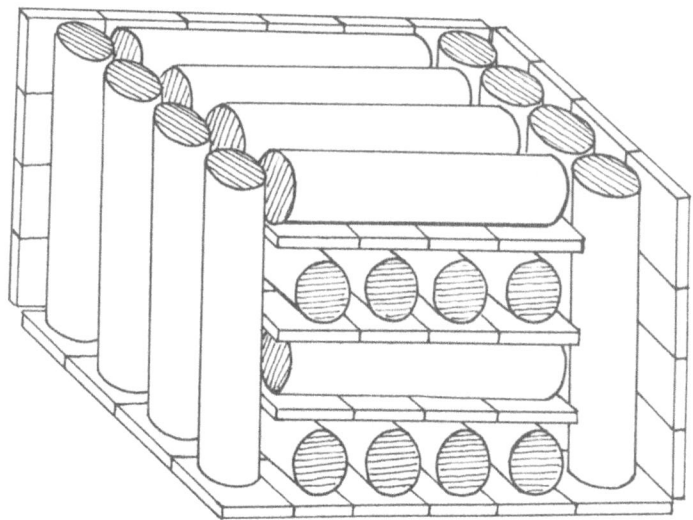

Fig. 14 A possible geometry for a direction-sensitive gravita-
         tional wave antenna with the capability of detecting
         monopoles.[32]

cosmic GUT monopoles. Nevertheless, given the global uncertainty in the field, searches are still made at the highest energy accelerators for what may be called "classical" monopoles. In practice one hypothesizes that monopoles have relatively low masses so that they can be produced in reactions of the type

$$p + p \rightarrow p + p + g + \bar{g},$$

$$\bar{p} + p \rightarrow g + \bar{g}, \qquad\qquad\qquad\qquad\qquad\qquad (7)$$

$$e^+ + e^- \rightarrow g + \bar{g},$$

where $\bar{g}$ is an antimonopole.

Recent searches have been performed at the e^+e^- storage rings PEP[35] and PETRA[36] and the $\bar{p}p$ Collider at CERN.[37] In these experiments the CR39 plastic detectors were used. Sheets of this material surrounded an intersection region. Heavily ionizing relativistic monopoles should have crossed some of the plastic sheets, where they should have left defects. When properly developed, the sheets should show holes along monopole tracks. The experiments at the e^+e^- storage rings placed an upper limit cross section of $\sim 10^{-37}$ cm^2, which is about three orders of magnitude smaller than the QED cross section for point particles. Thus these experiments would exclude poles with masses up to about 16 GeV. The experiment at the CERN $\bar{p}p$ Collider, using kapton foils inside the vacuum chambers and CR39 outside, established an upper limit of $\sigma < 3 \times 10^{-32}$ cm^2 for monopole masses up to 150 GeV.

Figure 15 summarizes, as a function of the monopole mass, the production cross section upper limits (at the 95% C.L.). Solid lines refer to "direct" measurements (such as those discussed above); dashed lines refer to "indirect" measurements (where monopoles should have been stopped and trapped in ferromagnetic materials; later they would have been extracted and accelerated by strong magnetic fields and detected).

CONCLUSIONS AND OUTLOOK

At this workshop we have learned that the list of what monopoles can do has become longer. They may catalyze proton decay, induce nuclear fission of heavy elements, induce β decay, attach nuclei, destroy magnetic fields, etc.; one should probably also mention the possible "induction of lightning" and the "problem of perpetual motion" according to Peter Peregrinus in 1269! We have also learned that the monopole could be so massive that it could be a minuscule black hole. Moreover, monopoles could be part of the dark matter in the universe; alternatively they could be so few as to be undetectable.

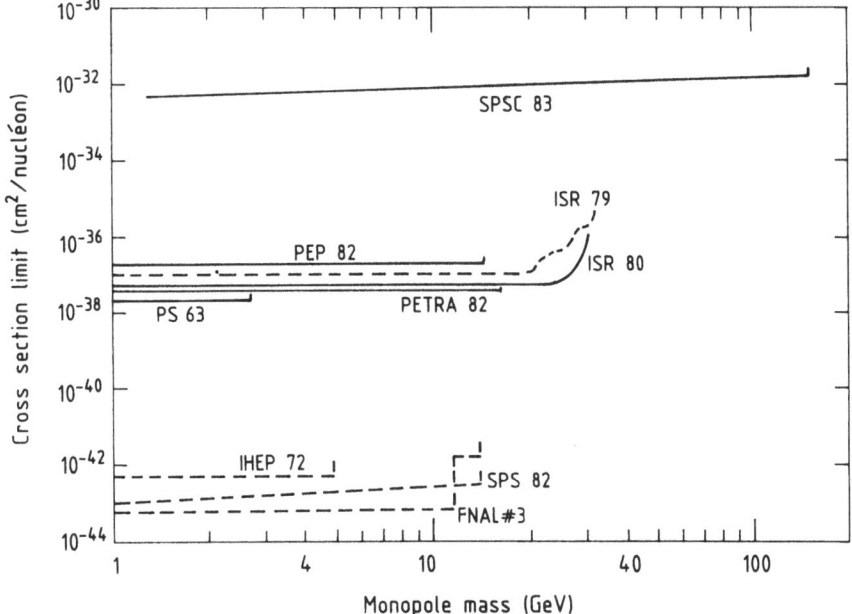

Fig. 15 Compilation of upper limits for classical magnetic monopole production at high energy accelerators plotted
versus monopole mass. Solid and dashed lines refer to
"direct" and "indirect" measurements. The new limits
from the SPS $p\bar{p}$ collider ($\sigma < 3 \times 10^{-32}$ cm^2) extend up to
150 GeV.

It is clear that the field of magnetic monopoles has evolved
into a fascinating interdisciplinary field of physics, with
implications in fundamental theories, in particle physics and in
astrophysics and cosmology. In certain aspects it represents a
connection between physics and cosmology.

The theoretical and phenomenological understanding of monopoles has improved considerably in the last few years. But new
possibilities have opened up. I refer in particular to the
various differing predictions of the monopole mass and of the
monopole production rates in the early universe. Therefore,
theoretical guidance to experiments is not really adequate.

From the experimental point of view, one clearly observes the
trend towards larger and costlier experiments. Moreover, it has
been pointed out by several people, in particular by Frisch, that
in the searches for rare events it is normal to get a candidate,
which is difficult to reject. This forces the experimenters to

use at the same time, as least in large layouts, more than one technique in order to obtain redundancy and gain "convincingness".[16,38]

The present trend towards larger experiments may be summarized as follows:

Induction experiments: Some groups are planning layouts with $1-10$ m^2 coils.

Electronics experiments: The present largest experiments are the following: Baksan ($S\Omega = 1800$ m^2sr), Mont Blanc liquid scintillator detector (700 m^2sr), Homestake (1300 m^2sr), Texas A&M (300 m^2sr), Frejus (1000 m^2sr). There are plans for layouts with ~ 1000 m^2 active detectors at Stanford, Michigan and Italy (Gran Sasso).[39-40]

Plastic detectors: In the Kamioka mine the Japanese are installing 1000 m^2 of CR39.[40]

Catalysis of proton decay: All proton decay experiments have installed new electronics in order to be able to detect a string of catalyzed proton decays.

But what if one finds nothing? To overcome this question several of the larger experiments are planning to add "worthy byproducts," like detection of multimuon events (muon bundles), neutrinos from supernovae explosions, etc. But what the field of monopoles really requires would be some real monopoles!

ACKNOWLEDGEMENTS

I would like to acknowledge many colleagues for discussions and for sending material before publication. In particular, I would like to thank Drs. B. Cabrera, P. Capiluppi, R.A. Carrigan Jr., D. Cline, A. De Rujula, G. Fiorentini, M. Koshiba, G. Mandrioli, P. Musset, A.M. Rossi and J. Stone.

REFERENCES

1. P.A.M. Dirac, Quantized Singularities in the Electromagnetic Field, Proc. Roy. Soc. 133, 60 (1931).
2. G. Giacomelli, Searches for Missing Particles, Invited paper at the 1978 Singapore Meeting on "Frontiers of Physics," Proceedings of the Conference (1978).
3. G. Giacomelli, Review of the Experimental Status (Past and Future) of Monopole Searches, Proceedings of the Conference on "Monopoles in quantum field theory," Trieste (1981).

4. G. Giacomelli, Experimental Status of Monopoles, Proceedings of the 1982 Wingspread Workshop (1982).

5. E.J. Weinberg, Monopoles and Grand Unification, in this volume.

6. M.L. Perry, Monopoles in Kaluza-Klein theories, in this volume.

7. E.W. Kolb, Experimental Limits on Monopole Catalysis, in this volume.

8. C.T. Hill, Monopolonium, Nucl. Phys. B $\underline{229}$, 469 (1983).

9. G. Fiorentini, Binding of Magnetic Monopoles and Atomic Nuclei, in this volume.

10. C. Goebel, Vacuum Anti-Shielding of Magnetic Charge, in this volume; Binding of Monopoles to Nuclei, in this volume.

11. S.P. Ahlen and G. Tarle, Can grand unification monopoles be detected with plastic scintillators?, Phys. Rev. D $\underline{27}$, 688 (1983).

12. D.M. Ritson, Magnetic monopole energy losses, SLAC-PUB-2950 (1982).

13. L. Bracci et al., On the energy loss of very slowly moving magnetic monopoles, IFUP-TH 83/26 (1983).

14. L.W. Alvarez et al., A magnetic monopole detector utilizing superconducting elements, Rev. Sci. Instr. $\underline{4}$, 326 (1971).

15. B. Cabrera, First results from a superconducting detector for moving magnetic monopoles, Phys. Rev. Lett. $\underline{48}$, 1378 (1982).

16. H. Frisch, Monopole Detection by Induction Techniques, in this volume.

17. S.D. Drell et al., Energy Loss of Slowly Moving Magnetic Monopoles in Matter, Phys. Rev. Lett. $\underline{50}$, 644 (1983).

18. S. Geer and W.G. Scott, Calculation of the Energy Loss for Slow Monopoles in Atomic Hydrogen, CERN pp Note (1981).

19. G. Tarle et al., First Results From a Sea Level Search for Supermassive Magnetic Monopoles, in this volume.

20. S. Ahlen, Monopole detection by Ionization/Excitation Techniques, in this volume.

21. F. Kajino et al., A Scintillator-Proportional Counter Search for Monopoles, in this volume.

22. G. Battistoni et al., Nucleon Stability, Magnetic Monopoles and Atmospheric Neutrinos in the Mont Blanc Experiment, CERN/EP 83-147 (1983).

23. P.B. Buford-Price, Searches for Exotic Particles, Proceedings of the Wingspread workshop on magnetic monopoles (1982).

24. T. Doke et al., Search for Massive Magnetic Monopoles by Plastic Track Detectors, Phys. Lett B $\underline{129}$, 370 (1983).

25. V.A. Rubakov, Superheavy Magnetic Monopoles and Decay of the Proton, JETP Lett. $\underline{33}$, 644 (1981).

26. V.A. Rubakov, Adler-Bell-Jackiv Anomaly and Fermion Number Breaking in the Presence of a Magnetic Monopole, Nucl. Phys. B $\underline{203}$, 311 (1982).

27. C.G. Callan Jr., Dyon-Fermion Dynamics, Phys. Rev. D $\underline{26}$, 2058 (1982); Monopole Catalysis Cross Sections, in this volume.

28. S. Errede, Experimental Limits on Monopole Catalysis, in this volume.
29. S. Errede et al., Experimental limits on Magnetic Monopole Catalysis of Nucleon Decay, Phys. Rev. Lett. 51, 245 (1983).
30. K. Freese, Monopoles in Pulsar 1929.10, in this volume.
31. T. Ebisu and T. Watanabe, Search for Superheavy Monopoles in 65 kg of Iron Magnetic Sand with a SQUID Fluxmeter, Journal Phys. for Japan 52, 2617 (1983).
32. C. Bernard, A. De Rujula and B. Lautrup, Sonic Search for Monopoles, Gravitational Waves and Newtorites, CERN/TH 3694 (1983).
33. B. Barish, The Physics of Monopole Detection, in this volume.
34. M.J. Price, The Detection of Cosmic Monopoles Using a Room Temperature Coil, CERN/EF 83-2 (1983).
35. K. Kinoshita et al., Search for Highly Ionizing Particles in e^+e^- Collisions at $\sqrt{s} = 29$ GeV, Phys. Rev. Lett 48, 77 (1982).
36. P. Musset et al., Search For Magnetic Monopoles in e^+e^- Collisions at 34 GeV c.m. Energy, Phys. Lett. B 128, 333 (1983).
37. B. Aubert et al., Search for Magnetic Monopoles in $\bar{p}p$ Interactions at 540 GeV c.m. Energy, Phys. Lett. B 120, 465 (1983).
38. G. Charpak, Detectors for Rare Events, in this volume.
39. D. Ritson et al., Helium Proportional Chamber Monopole Detector, in this volume.
40. J. van der Velde, AMMANDA, An Advanced Monopole, Muon and Neutrino Detector Array, in this volume.
41. K. Kavagoe et al., A Search for Magnetic Monopoles with 1000 m^2 of Plastic Track Detector--Status Report.

MONOPOLES IN 1983

John Preskill

Lauritsen Laboratory of Physics
California Institute of Technology
Pasadena, California 91125

INTRODUCTION

Interest in magnetic monopoles is higher in 1983 than in any previous year in recorded history. To document this claim, I have plotted in Fig. 1a interest in monopoles as a function of time over the past eleven years. "Interest" is defined as the number of high energy physics preprints written per year with the words "monopole(s)" or "dyon(s)" in the title.[1] While this accounting surely does not include all papers written about monopoles, I expect it to accurately reflect historical trends. (The 1983 total was obtained by renormalizing the number of papers written through the end of August.)

Prior to 1974, a plot of interest in monopoles against time reveals only background noise. A signal turned on in that year, which peaked in 1976, then declined slowly. The current resurgence began in 1980.

Each of the two peaks in Fig. 1a is due to several convergent factors, as is better appreciated if we divide interest into several categories. Figure 1b is a plot of experimental interest. (This category includes all papers on the physics of monopole detection.) The Price[2] and Cabrera[3] peaks are clearly visible. Experimental interest responds quickly when there is a reason, but tends to drop to the background noise level when the reason goes away.

Theorists, on the other hand, can maintain interest without a reason. Figure 1c is a plot of interest in the general theory of monopoles in nonabelian gauge theories (excluding the interaction

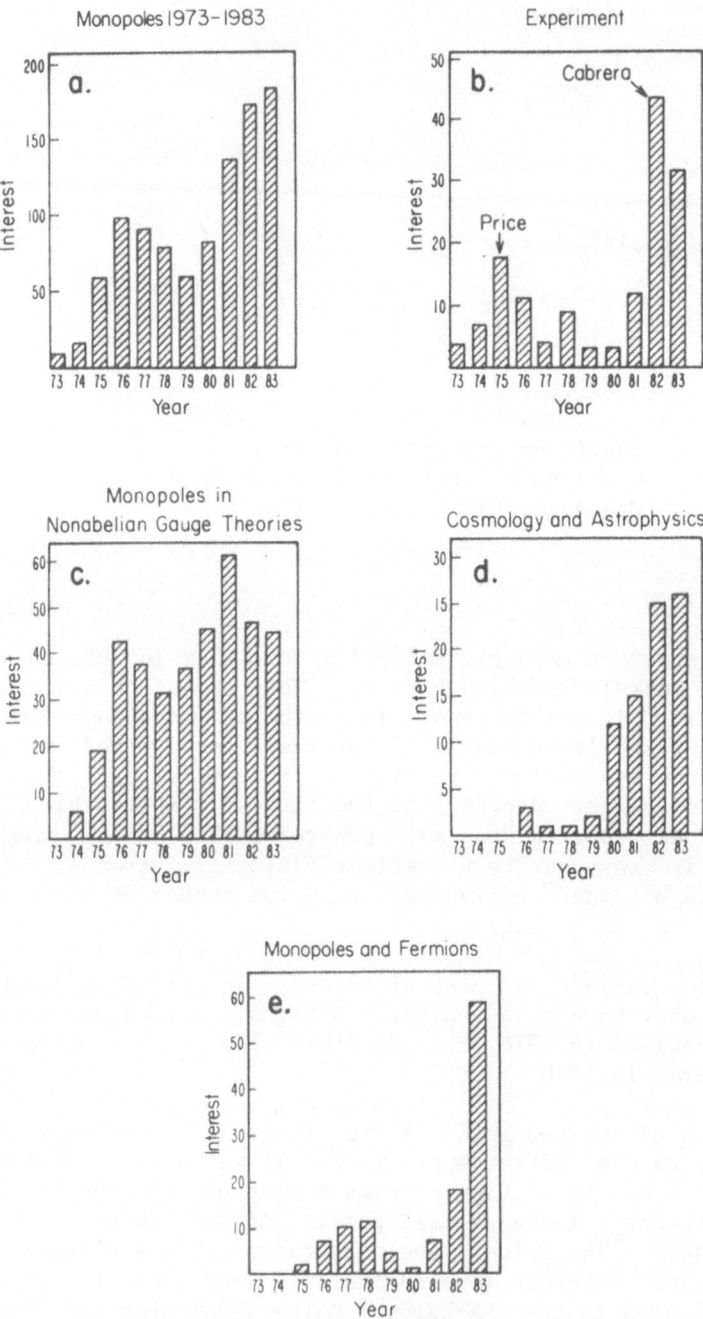

Fig. 1   Interest in magnetic monopoles as a function of time. (a)
overall, (b) experimental, (c) field theory, (d) cosmology
and astrophysics, (e) interactions with fermions.

of fermions and monopoles, which is treated as a separate category). The signal turned on with the pioneering papers of 't Hooft[4] and Polyakov[5], and has held steady since 1976. Of course, a continuous influx of new theoretical discoveries was needed to sustain this signal.

One sees from Fig. 1d that theoretical interest in the role of monopoles in cosmology and astrophysics turned on in 1980, and was already substantial before receiving a big boost from the Cabrera[3] event in 1982.

The most spectacular recent trend is revealed by Fig. 1e, in which interest in monopole-fermion interactions and scattering off monopoles is plotted. This subject attracted transitory interest in the late 1970's, but became a major industry only during the past year, following several seminal papers[6-9] which appeared in 1981 and early 1982.

One concludes from the above historical data that magnetic monopoles are interesting. One reason monopoles are interesting is that they can tell us something about particle physics at extremely high energies.

The existence of magnetic monopoles is a very general consequence of the (grand) unification of the fundamental inter-actions. Any model of particle physics in which the standard low-energy gauge group $SU(3)_{color} \times [SU(2) \times U(1)]_{electroweak}$ is embedded in a semisimple gauge group which is spontaneously broken at a large mass scale M necessarily contains magnetic monopoles. (For a review, see Ref. 10-11.) The size and mass of the monopole are determined by the symmetry breaking scale M; typical pro-perties of a monopole in a unified model are:

Charge: $g = 1/2e$ (Dirac Charge),
Size: $r \sim (eM)^{-1}$,
Mass: $m \sim (4\pi/e)M$.

Obviously, if the magnetic monopole exists, its mass m is of great intrinsic interest; from a measurement of m we would learn, at least in order of magnitude, the fundamental symmetry breaking scale at which electrodynamics becomes truly united with the other particle interactions.

Can the monopole mass m be predicted? Yes, but only if one adopts the "desert hypothesis;" that is, assumes that no un-expected new interactions or particles appear between present-day energies (of order 100 GeV) and the unification scale M. From the desert hypothesis follows the prediction $M \sim 10^{15}$ GeV, so that $r \sim 10^{-28}$ cm and $m \sim 10^{16}$ GeV. But one should remember that the desert hypothesis could easily be wrong, even if the general idea of grand unification is correct. It is sensible to keep an open mind about the monopole mass.

The possibility that magnetic monopoles exist has excited a broad range of theoretical activity. In this summary talk, I review only some of the most recent theoretical developments concerning the implications of magnetic monopoles in cosmology, astrophysics, and particle physics, which have been discussed at this conference. Rather than attempt to give an exhaustive review, I will mention only particular topics which seem to me to be especially important or interesting.

MONOPOLES AND COSMOLOGY

The existence of magnetic monopoles is a general consequence of grand unification of the fundamental interactions. But it is one thing to say that monopoles exist, and quite another to say that we have a reasonable chance of observing one. If monopoles are extremely heavy, as expected, then any monopoles around today must have been produced in the very early universe. Thus, the monopole abundance, like the helium abundance and the baryon abundance, has become a central issue in cosmology, an issue which has exerted a healthy influence on the development of cosmology during the past few years. (See Ref. 12-13 for a review).

It is natural to ask whether the monopole abundance can be predicted. It can be, but, unfortunately, there are many predictions. Here are two predictions for the number density of monopoles $n_{monopole}$:

$$n_{monopole} \sim (m/10^{16})n_{baryon} \tag{1}$$

$$n_{monopole} \sim 0 \tag{2}$$

Equation (1) is the prediction[14] of a standard big-bang cosmology in which monopoles are copiously produced in the phase transition that takes place when the temperature is $T \sim M$. Since Eq. (1) implies that the mass density of the universe is dominated by monopoles by some 16 orders of magnitude, and that the universe is about 10^3 years old, this prediction is sometimes called the "monopole problem." Equation (2) is the prediction[13,15] of a typical inflationary cosmology[16], in which any monopoles produced during the phase transition are subsequently "inflated away," and monopole production is negligible both during the inflationary epoch and after the universe reheats. This prediction will cause no problems until a monopole is observed. But it is apparent that cosmological considerations leave us with no definite expectation for the monopole abundance.

There have been a few recent developments concerning the cosmological monopole abundance. An old suggestion[17] for suppressing the monopole abundance within the context of the standard cosmology, is that the universe may have entered a superconducting phase as it cooled. Since a superconductor tries to expel magnetic flux, monopole-antimonopole pairs would become connected together by flux tubes in this phase, and annihilate rapidly. As the universe cooled further, it might have eventually returned to a normal nonsuperconducting phase, but only after the monopole abundance had been drastically reduced.

Previously it had been believed that the monopole abundance would be quickly reduced to a negligible value in this superconductor scenario, but E. Weinberg[19] has recently suggested that a potentially interesting number of monopoles could survive until the universe re-enters the normal phase. It is crucial to consider the correlations between monopoles and antimonopoles. These correlations are actually very strong, but Weinberg suggests that, nevertheless, some monopoles get confused; unable to decide which antimonopole to pair up with, they get left behind when the pairing occurs. It thus seems possible that the superconductor scenario predicts a detectable, or perhaps even excessively large, abundance of monopoles.

A far more appealing solution to the monopole problem is offered by cosmological inflation[16]; this solution is more appealing because inflation solves many other cosmological problems as well. In the inflationary scenario, the universe undergoes exponential expansion, driven by an effective cosmological constant, after the phase transition in which magnetic monopoles are produced. After many e-foldings of inflation, the monopole abundance is reduced to a negligible value. Eventually the effective cosmological constant which drove the inflation becomes rapidly thermalized, and the universe reheats. Its subsequent evolution is then described by the standard cosmological model.

Within the context of the inflationary scenario, there are at least two mechanisms for producing monopoles which ought to be considered. Naive estimates have been done of the rate of thermal production of monopoles both during[20] and after[14-15] the reheating of the universe. These estimates should probably be re-examined more carefully. It is also possible for monopoles to be produced by the (Hawking) quantum fluctuations during inflation. Preliminary estimates[21] indicate that the abundance of monopoles produced by Hawking fluctuations is linked in an interesting way to the magnitude of the mass density fluctuations in the early universe. It appears that the monopole abundance due to Hawking fluctuations is guaranteed to be unobservably small if the density fluctuations are as small as required by the observed isotropy of the cosmic microwave background. But this issue also deserves a closer look.

MONOPOLES AND ASTROPHYSICS

Although cosmological considerations provide us with no definite prediction for the monopole abundance, the inflationary universe scenario offers the possibility that the monopole abundance is both small enough to be acceptable and large enough to be detectable. In particular, an observable monopole abundance might have been generated during or after reheating. So theoretical cosmology should not discourage an experimenter from looking for monopoles.

But various theoretical limits on the flux of magnetic monopoles in cosmic rays can be derived by considering the effect of monopoles on astrophysical processes[22]. These limits provide valuable guidance for the prospective monopole hunter.

One stringent limit, due to Parker[23], is obtained by noting that monopoles in our galaxy will be accelerated by the galactic magnetic field, and will thus dissipate the energy stored in the field. By demanding that the field energy is not substantially depleted in the time required to regenerate the field, one obtains

$$F \lesssim 10^{-15} \text{ cm}^{-2} \text{ sr}^{-1} \text{ s}^{-1}, \tag{3}$$

if gravitational effects on the trajectory of the monopoles are ignored. Equation (3) is discouraging; it says that less than one event per year should be expected in a detector which covers a football field.

However, if the monopole mass is $m > 10^{17}$ GeV, monopoles accelerated by the galactic magnetic field will not necessarily escape from the galaxy, and the Parker limit is less restrictive[24]; it becomes

$$F \lesssim 10^{-15} \text{ cm}^{-2} \text{ sr}^{-1} \text{ s}^{-1}(m/10^{17} \text{ GeV}), \quad m \gtrsim 10^{17} \text{ GeV}. \tag{4}$$

For $m \gtrsim 3 \times 10^{19}$ GeV, a better constraint than the Parker limit is obtained by demanding that the mass due to monopoles in our galaxy not exceed the total mass of the galaxy.[13,24]

$$F \lesssim 3 \times 10^{-13} \text{ cm}^{-2} \text{ sr}^{-1} \text{ s}^{-1} (3 \times 10^{19} \text{ GeV/m}). \tag{5}$$

(Monopoles with $m > 10^{17}$ GeV will acquire velocities of order 10^{-3} c from gravitational effects.) The flux limits of Eqs. (4) and (5) are shown together in Fig. 2.

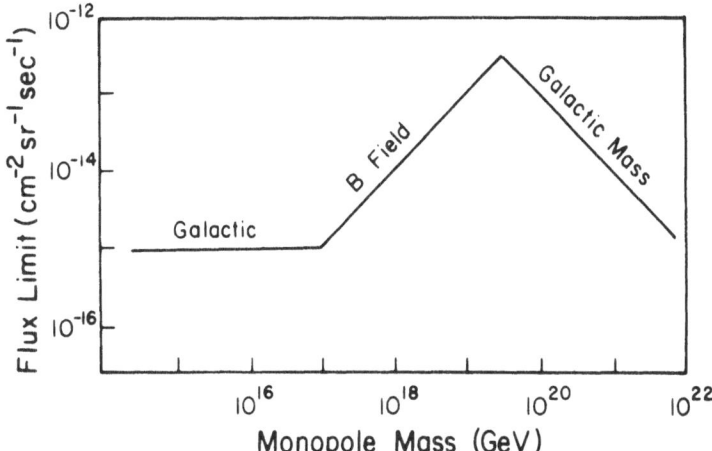

Fig. 2 Astrophysical limits on the monopole flux as a function of
        monopole mass.   "Galactic B field" labels the limit based
        on   the   energetics   of   the   galactic   magnetic   field.
        "Galactic mass" labels the limit based on the total mass
        of the galaxy.

There are two important lessons.   One is that Parker's
reasoning does not exclude the possibility that magnetic monopoles
make up the dark matter of galactic halos, if m $\gtrsim$ 3 × 10^{19} GeV.
The other is that the galactic monopole flux could be a few orders
of magnitude larger than the quoted limit, Eq. (3), so monopole
search experiments which can place bounds on the flux better than
F $\lesssim$ 3 × 10^{-13} cm^{-2} sr^{-1} s^{-1} provide some valuable information.
The above remarks seem especially noteworthy when we consider that
Kaluza-Klein theories actually predict the existence of monopoles
in the mass range where the flux limits are weakest.   (See the
following Section).

At the last monopole conference[25], there was much discussion
of the possibility of evading the Parker limit. One speculation[26],
that the local flux of magnetic monopoles in the solar system
greatly exceeds the ambient flux in the galaxy, seems implausible
on   purely kinematic grounds.[27]   Another, that the observed
galactic magnetic field is due to magnetic charge density fluctuations[24,28], rather than persistent currents, is more difficult to
analyze, but appears to face various problems. For example, it is
hard to understand how such fluctuations could have been established to begin with.

The most powerful limits on the monopole flux are obtained by considering the astrophysical consequences[29-30] of the catalysis by monopoles of nucleon decay.[8-9] In particular, monopoles incident on a neutron star or white dwarf would be captured inside the star, and their catalytic action would heat the star and raise its luminosity. Thus, from observational limits on the luminosity of such stars, we can derive limits on the product of the monopole flux F and the cross section times relative velocity $\sigma\beta$ for catalysis of nucleon decay. Very conservatively, the bound obtained in this way from neutron star luminosities, is[29-30]

$$F \lesssim 10^{-22} \text{ cm}^{-2} \text{ sr}^{-1} \text{ s}^{-1} (\sigma\beta/10^{-27} \text{ cm}^2)^{-1} \, , \tag{6}$$

about seven orders of magnitude more stringent than the Parker limit, if catalysis occurs at a strong interaction rate.

How seriously should we take this bound? In evaluating it, we must first of all decide whether it is plausible that, if magnetic monopoles do exist, they catalyze nucleon decay at a strong interaction rate, as suggested by Callan[9] and Rubakov.[8] It is certainly possible to doubt the existence of the Callan-Rubakov effect. We expect it only if the new interactions associated with the monopole core violate baryon number, and one can construct models for which this is not true.[31]

I believe, however, that it is quite natural to expect the monopoles of a unified gauge theory to exhibit the Callan-Rubakov effect, even though this prediction is not completely general. We have good reasons for suspecting that baryon number is not exactly conserved (the matter-antimatter asymmetry of the universe, the baryon-number anomaly of the standard model), and therefore there is no particular reason to expect the new physics associated with the monopole core to be baryon-number conserving. Monopole hunters should nevertheless bear in mind that the existence of monopoles which fail to catalyze nucleon decay cannot be ruled out.

If the Callan-Rubakov process does occur, the catalysis cross section cannot be calculated accurately, but it seems reasonable to estimate that it is a roughly geometrical strong interaction cross section, if the relative velocity of nucleon and monopole is of order c, as in a neutron star. Thus, the bound, Eq. (6), is discouraging indeed. An even more stringent limit can be obtained if capture of monopoles by the main-sequence progenitor of the neutron star is taken into account. This stronger limit is more mass sensitive however. Planck-mass monopoles, for example, would rarely be captured by main sequence stars.

Once captured by a neutron star, a monopole must be accelerated to a velocity of order c to escape. Harvey[30] has helpfully suggested a possible mechanism for ejecting monopoles from neutron stars, which could considerably weaken the bound, Eq. (6).

It is generally believed that the core of a neutron star is a type II superconductor in which Cooper pairs of protons have condensed. Because the superconducting core expels magnetic flux, monopoles entering the star will eventually come to rest at the surface of the core. Typically, many magnetic flux tubes will have been trapped in the core when it went superconducting, and a monopole floating on the surface of a core will occasionally encounter the opening of a tube, and drop in, penetrating the core.[32]

It is conceivable that, deep within the core, there is an inner core in which charged pions condense. This pion condensate would also be a type II superconductor, but its flux tubes would carry considerably higher energy per unit length than the flux tubes in the proton superconductor. Also, the flux quantum in the pion condensate would be the Dirac magnetic charge carried by the monopole, rather than half the Dirac charge as in the proton superconductor. Thus, two flux tubes in the proton superconductor would coalesce at the surface of the pion condensate, and a monopole drifting down one of them would be rapidly accelerated upon entering the pion condensate, the sizable magnetic field energy stored in the flux tube being efficiently converted to monopole kinetic energy.[30] The monopole could be accelerated to a relativistic velocity, and ejected from the star! (See Fig. 3.) Hence, if we take the fullest advantage of our ignorance concerning the interiors of neutron stars, it is possible that the very stringent bound, Eq. (6), can be evaded.

We are left with limits based on the luminosities of white dwarfs[33]; the interiors of white dwarfs are less exotic and better understood than those of neutron stars. But these limits are rather mass-sensitive; Planck-mass monopoles, for example, would not be captured by white dwarfs.

We conclude then, that the astrophysical arguments based on monopole catalysis which have been used to limit the monopole flux are very suggestive, but not completely compelling. While it seems unlikely, I think it is possible that there is an observable flux of monopoles which catalyze nucleon decay at a strong interaction rate. Terrestrial searches for such monopoles are not purposeless.

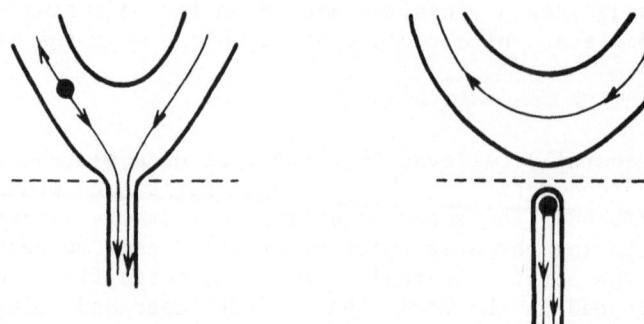

Fig. 3 Magnetic monopole in a flux tube near the boundary between
        a proton pair condensate and a charged pion condensate.
        (a) In the proton pair condensate, the magnetic flux in
        the tube reverses direction at the monopole, and there is
        no net magnetic force on the monopole.   (b) In the pion
        condensate, the flux tube terminates on the monopole, and
        the monopole accelerates rapidly.

MONOPOLES AND FIELD THEORY

     The theory of magnetic monopoles is a triumph of modern
quantum field theory.   Among the remarkable discoveries of the
past ten years are the following:

- Magnetic monopoles can arise as topologically stable solitons in spontaneously broken nonabelian gauge field
  theories.[4-5]

- Electrically charged dyons arise as quantum mechanical
  excitations of these monopoles.[34]

- A composite of a bosonic monopole and an elementary boson
  can be a fermion.[35]

- Monopoles can carry a nonabelian (color) magnetic field.[36]

- The color magnetic field of a monopole is screened if color
  electric fields are confined.[37,11]

- Quark confinement in quantum chromodynamics can be regarded
  as a consequence of the condensation of monopoles in the
  vacuum state.[38]

- Dyons can carry the anomalous electric charge $Q = e\theta/2\pi$.[39]

- Monopoles can carry fractional fermion number.[40]

I will not discuss further any of these "classical" contributions to monopole theory. Instead, I confine my attention to more recent developments.

Kaluza-Klein Monopoles

More than sixty years ago, Kaluza[41] proposed a novel way of unifying gravitation with other gauge interactions. His idea is that spacetime is not really 4-dimensional, but $(4 + n)$-dimensional, that $(4 + n)$-dimensional spacetime is endowed with a metric satisfying a $(4 + n)$-dimensional generalization of Einstein's equations, and that n dimensions have become "spontaneously compactified" with radii of order the Planck length. At energies much less than the Planck mass, the microscopic compact dimensions are concealed from view, but a remnant of the underlying $(4 + n)$-dimensional theory survives at low energy; the massless fields in a theory of this type include, in addition to the 4-dimensional metric, spin-one gauge fields associated with the isometry group G of the compact n-dimensional manifold M. These gauge fields are components of the $(4 + n)$-dimensional metric which have managed to avoid acquiring masses upon compactification.

Models of particle interactions based on the concept of extra spacetime dimensionality still endure today, and are known as Kaluza-Klein theories.

It is natural to expect that a Kaluza-Klein theory unifying electromagnetism with gravitation will, like other unified gauge theories, contain magnetic monopoles as static topologically stable soliton solutions to the classical field equations. This expectation has been verified, and the solution has been explicitly constructed, for the simplest Kaluza-Klein theory (that originally considered by Kaluza), in which $n = 1$, the compact manifold is the circle $M = S^1$, and the isometry group is $G = U(1)$.[42]

In general, a Kaluza-Klein soliton is a metric satisfying the $(4 + n)$-dimensional field equations which approaches the vacuum solution at spatial infinity. The behavior of the solution at spatial infinity thus defines an M bundle over S^2, the boundary of three-dimensional space. If this bundle is topologically nontrivial, the soliton will be stable. In the 5-dimensional theory, the S^1 bundles over S^2 are labeled by an integer, which, it turns out, can be identified as the magnetic charge. The Dirac quantization condition arises, as we might expect, from the requirement that a singularity of the metric be a harmless coordinate singularity.

Generally, one expects a Kaluza-Klein monopole to have a mass of order $(1/e)M_{Planck}$. In the 5-dimensional theory, it has been found that[42]

$$M_{monopole} = \frac{1}{4} \alpha^{-1/2} M_{Planck} \sim 5 \times 10^{19} \text{ GeV}. \qquad (7)$$

As remarked in the previous section this is a quite interesting monopole mass experimentally.

The monopole of the 5-dimensional theory has some unusual properties. The 4-dimensional constant-time slices of both the monopole and antimonopole solutions have handles; therefore, a monopole-antimonopole pair has a different topology than the vacuum, and cannot annihilate classically. Quantum gravity effects, which include fluctuations in topology, would presumably allow the pair to annihilate.

Also, the interactions of monopoles with other monopoles, antimonopoles, and test particles are rather strange in the 5-dimensional Kaluza-Klein theory.[42] But these strange interactions appear to be consequences of the existence of a massless scalar particle in this theory, which would be absent in a more realistic theory.

The Kaluza-Klein monopoles appear to deserve further study. The explicit monopole solution has been constructed in only the simplest Kaluza-Klein theory. Dyonic excitations of Kaluza-Klein monopoles do not seem to arise in the usual way, because there are no charged fields excited in the monopole core. It will also be interesting to study further the interactions of fermions with Kaluza-Klein monopoles.[43] We must keep in mind that quantum gravity could exert an important effect on these interactions. Since quantum gravity (e.g., virtual black holes) is not expected to conserve baryon number, it is possible that the Callan-Rubakov effect, if it occurs at all, arises in a much different way for Kaluza-Klein monopoles than for ordinary grand-unified monopoles.

Stable Multimonopoles

Until recently, static multimonopole solutions[44] had been constructed only in the "Prasad-Sommerfield limit,"[45] in which the Higgs fields which acquire vacuum expectation values and spontaneously break the gauge symmetry are exactly massless. In this limit, the Coulomb repulsion of the monopoles is exactly cancelled by the attractive force due to the exchange of the massless scalars, and the monopoles are noninteracting. It was expected that for any nonzero Higgs mass, monopoles would repel, and no stable bound states of monopoles would exist.

But it was recently discovered that stable multimonopole solutions do in fact exist.[46] The key observation is that monopoles with nonabelian long range magnetic fields can, in some cases, orient their magnetic charges in orthogonal directions in group space, and reduce their Coulomb repulsion to zero. In the Prasad-Sommerfield limit, the scalar-exchange force would also vanish for this relative charge orientation. But the scalar-exchange force is typically due to Higgs scalars in several different representations of the unbroken gauge group; when the charges are oriented so that there is no Coulomb repulsion, some of the Higgs scalars generate an attractive force and some a repulsive force. By choosing a Higgs scalar which generates an attractive force to be much lighter than the others, we can obtain an attractive interaction between the monopoles over a large range of separations. The result is a stable two-monopole bound state.

The above remarks apply to the SU(5) monopole if we recall that at separations much less than M_W^{-1}, the monopole charges are free to choose orientations anywhere in SU(3) × SU(2) × U(1). Binding occurs if the SU(3) × SU(2) singlet member of the Higgs 24 multiplet is much lighter than the others. The existence of bound states of three, four, and six monopoles has also been demonstrated.[46]

Further work on the properties of these multimonopoles is desirable. In particular, their interactions with fermions should be investigated. Since some multimonopoles will not be spherically symmetric, and since W, Z fields are excited inside the multimonopoles, these interactions could have qualitatively new features.

Demise of Global Color

The dyonic charge excitations of the 't Hooft-Polyakov monopole arise when the global charge rotator degree of freedom of the monopole is quantized semiclassically. It was expected that semiclassical quantization of a monopole with a nonabelian "chromomagnetic" field would similarly give rise to a spectrum of "chromodyon" excitations in definite representations of the global color group. But early attempts to calculate the chromodyon spectrum of the SU(5) monopole uncovered some puzzles.[47]

Recently, it was discovered[48-49] that the only dyonic excitations of a chromomagnetic monopole which exist are those associated with the color rotations of the monopole which commute with its long range field; these "color hypercharge" excitations do not form complete color multiplets. Their failure to do so was clarified by the subsequent discovery[50] that a global color rotation of a monopole cannot even be implemented if the rotation does not commute with the monopole charge.

The surprising statement that global color rotations of a chromomagnetic monopole cannot be defined is better understood if we recall that global color is not a symmetry at all in the usual sense; physical states are gauge-invariant. Global color is important in monopole physics only as a means of keeping track of the collective coordinates of a monopole solution, for the purpose of semiclassical quantization.[51]

What has been found is that color rotations which act nontrivially on the long range chromomagnetic field are not acceptable as collective coordinates. Perhaps this is not really so surprising. The excitations associated with these rotations cannot be supported by the monopole core; they propagate to spatial infinity along the lines of magnetic force.

MONOPOLES AND FERMIONS

The most spectacular recent developments in the theory of magnetic monopoles concern the interactions of monopoles and fermions. This subject has a complex history of which I will give a brief overview.

The recent developments have woven together several independent lines of inquiry into the properties of monopoles. First, an analysis[52] of the quantum mechanics of a charged spin-1/2 particle in the field of a point monopole led Goldhaber[53] to observe that this problem is inherently ambiguous; the behavior of an electron scattering off a point monopole cannot be uniquely determined until a boundary condition is chosen for the electron wave function at the location of the magnetic pole.

Later, Dokos and Tomaras[54] pointed out that a magnetic monopole in a grand unified theory can catalyze processes which change baryon number. They noted that the dyonic excitations of the SU(5) monopole have baryon-number-violating couplings, and that a collision which excites the dyon degree of freedom need not conserve baryon number. But they believed that the lightest dyon was split from the monopole ground state by an amount of order $\alpha M_X \gtrsim 10^{13}$ GeV, so that the catalysis cross section, suppressed by a huge energy denominator, was extremely small.

Meanwhile, Witten[39] discovered that the monopole ground state carries the anomalous electric charge $Q = e\theta/2\pi$, where θ, the "vacuum angle,"[55] is an arbitrary parameter defined modulo 2π. Witten's discovery, like Goldhaber's, suggested that the dynamics of a fermion coupled to a monopole is rather subtle, for it is known that, because of the axial anomaly,[56] the angle θ becomes unobservable if there are massless charged fermions.[55] The question arose, what happens to the anomalous charge of the monopole as m_e, the electron mass, approaches zero?

This issue was clarified by Blaer et al.,[6] Wilczek,[7] Rubakov,[8] and Callan,[9] who concluded that the electric charge of the monopole is smeared out over a region with radius of order m_e^{-1}; thus, the anomalous charge disappears in the limit $m_e = 0$. It follows that the dyon is actually split from the monopole ground state by a tiny amount of order αm_e, and is easily excited. Moreover, Wilczek[7] and Rubakov[8] emphasized that, because of the axial anomaly,[56] the monopole is not an eigenstate of chirality or baryon number. This observation opened the possibility of a large chirality-violating, and, perhaps, baryon-number-violating cross section in monopole-fermion scattering.

Callan[9] had meanwhile developed a different perspective on the catalysis process. Besson[57] and he recognized that the boundary condition needed to completely specify the physics of a fermion scattering off a point monopole could be obtained by considering the case of a nonsingular monopole, and then taking the limit of vanishing monopole core size. This procedure led Callan to the remarkable conclusion that the physics of the monopole core need not "decouple" from low energy fermion-monopole scattering. In particular, baryon-number-violating interactions inside the core can induce baryon-number-changing scattering processes with a cross section unsuppressed by the exceedingly small core size.

Thus, two different explanations of the catalysis phenomenon emerged. One is that the catalysis of baryon-number-changing reactions by monopoles is a consequence of an anomaly afflicting the baryon number current. The other is that catalysis is a consequence of a baryon-number-violating boundary condition reflecting the physics of the monopole core. The two explanations are sometimes confused, but cannot, in general, be regarded as complementary descriptions of the same phenomenon.[58-60] By the "Callan-Rubakov effect," one usually means symmetry violation arising in the form of a boundary condition at the monopole core.

To begin to understand the Callan-Rubakov phenomenon, recall that if a particle of electric charge e moves in the field of a point monopole with magnetic charge g, the electromagnetic field carries angular momentum

$$\vec{J}_{em} = e\, g\, \hat{r} , \tag{8}$$

where $\hat{r}$ is the unit vector pointing toward the charged particle from the monopole.[11] If the charged particle were to pass through the monopole, this contribution to the angular momentum would change discontinuously. Therefore, conservation of angular momentum forbids the particle to pass through the pole, unless its charge or intrinsic spin can change instantaneously as it does so.

The above remark has a quantum mechanical counterpart, as we see[52-53] by solving the Dirac equation for a massless electron in the field of a point monopole with eg = 1/2. The J = 0 solutions to the Dirac equations are peculiar; the positive helicity solution is purely an outgoing wave, and the negative helicity solution is purely an incoming wave. (For a positron, the helicities of the solutions are reversed.) Both solutions are singular at the origin, the location of the pole, and the Dirac equation itself provides no criterion for matching up the incoming and outgoing solutions. The Hamiltonian defined by the Dirac equation is therefore not self-adjoint; probability is not conserved unless the Hamiltonian is supplemented by a boundary condition which relates the incoming and outgoing waves.

This peculiar behavior is not too hard to understand. An electron in the monopole field has $\vec{J}_{em} = -1/2\ \hat{r}$. Hence, an incoming (outgoing) electron must have negative (positive) helicity to be in a state with $\vec{J} = \vec{J}_{em} + \vec{\sigma} = 0$. For a positron, $\vec{J}_{em}$ has the opposite sign, and the helicities are reversed.

The boundary condition at the origin determines the fate of a left-handed electron which scatters off a monopole in the J = 0 partial wave. But there are only two options; it becomes either a right-handed electron or a left-handed positron, because those are the only available outgoing modes with J = 0. The boundary condition must therefore either violate chirality (which is otherwise a good symmetry of the Hamiltonian) or require the monopole to absorb electric charge. If the charge-conserving boundary condition is chosen, then the chirality-changing J = 0 cross section will saturate the unitarity limit.

That boundary condition must be specified to determine the final state of an electron scattering off a point monopole is the key to the Callan-Rubakov effect. What makes the effect profoundly disturbing is that it seems to violate a fundamental principle of quantum field theory, the decoupling principle,[61] which asserts that the effects of very short distance physics must be power suppressed at low energy. The decoupling principle leads one to expect that the amplitude for monopole-fermion scattering at energies much less than the inverse size of the monopole core will not depend on the structure of the core, except for power corrections which vanish as the size of the core goes to zero. This expectation fails because of the ambiguity in monopole-fermion scattering when the monopole is pointlike; information about the structure of the core survives at low energy in the form of a boundary condition which exerts a strong influence on low-energy physics. In particular, the boundary conditions may violate a symmetry (like baryon number) which would otherwise be a good symmetry of the effective theory describing low-energy physics.

The analysis of the scattering of a low-energy fermion off a nonsingular monopole can be divided into two parts. First, we decide what boundary conditions to impose as the limit of vanishing core size is taken. Then the interaction of a point monopole with fermions satisfying the appropriate boundary conditions is studied. The second step is highly nontrivial. Fermion pair creation effects, which are responsible for smearing out the dyon charge over a region of order the fermion Compton wavelength, must be taken into account as fully as possible. Both Rubakov[8],[62] and Callan[9],[63] suggested that, since only $J = 0$ fermions can penetrate to the core of the monopole, the problem can be well-approximated by an effective $(1 + 1)$-dimensional quantum field theory describing the $J = 0$ partial wave. The qualitative features of this $(1 + 1)$-dimensional theory are most easily glimpsed if it is converted into an equivalent "bosonized" theory[64] in which the fermions are represented by solitons. The soliton picture of monopole-fermion scattering is especially convenient when we try to understand the effects of fermion masses.

Returning to the problem of finding the appropriate boundary conditions satisfied by the fermions, let us consider, for concreteness, the case of the SU(5) model with a single generation of fermions.[65] The magnetic charge of the SU(5) monopole is actually a linear combination of ordinary magnetic charge and color magnetic charge. At distances from the monopole center much less than 10^{-13} cm and much greater than the radius of the core, the only fermions which interact with the monopole are those which carry Q, the corresponding combination of electric charge and color electric charge. The right-handed quarks and leptons with nonzero Q are, in an appropriate gauge,

$$Q = 1: \quad e_R^+ \ \bar{d}_{3R} \ u_{1R} \ u_{2R} \ \text{(incoming)} \tag{9}$$

$$Q = -1: \quad d_{3R} \ e_R^- \ \bar{u}_{2R} \ \bar{u}_{1R} \ \text{(outgoing)},$$

where 1, 2, 3 are color indices. The behavior of these fermions in the field of the SU(5) monopole is identical to the behavior of an electron or positron in the field of an ordinary Dirac monopole. The only new feature is that there are four (Dirac) fermions interacting with the monopole, and the boundary condition at the origin causes these fermions to mix in a manner determined by the structure of the core of the monopole.

One can attempt to determine the boundary condition by solving the Dirac equation in the field of the nonsingular SU(5) monopole with finite core radius.[66],[57],[9] The result is that the helicity of the incoming fermion is preserved; incoming and outgoing states are matched up as in Eq. (9). We see that two units of Q are transferred to the monopole, exciting its dyon degree of freedom.

But if we now investigate the consequences of this boundary condition, taking proper account of pair creation effects, we realize that the picture in which the incoming fermion falls to the core and deposits charge there, suggested by the solution to the Dirac equation, is not very accurate. An enormous Coulomb barrier prevents charge from being deposited on the core. It is energetically favored for the charge to be spread out over a region with a radius of order a fermion Compton wavelength. As a result, our original procedure for finding the correct boundary condition is called into question. It seems that a more appropriate boundary condition is one that forbids charge to accumulate at the origin.[67]

Fortunately and remarkably, in the case of the SU(5) monopole we can obtain quite nontrivial information about the scattering process by merely demanding that none of the charges coupling to massless gauge bosons accumulate on the core.[68] This constraint is especially interesting because the W and Z bosons must be regarded as effectively massless at distances from the center of the monopole much less than M_W^{-1}. Since left-handed and right-handed fermions with the same electric charge have different values of $(T_3)_{weak}$, simple chirality violating processes such as

$$e_L^- + M \rightarrow e_R^- + M \tag{10}$$

are forbidden for massless fermions! If, for example, two u-quarks scatter off the monopole, there is only one possible two-fermion final state; the allowed process is

$$u_{1R} \ u_{2R} + M \rightarrow \bar{d}_{3L} \ e_L^+ + M \ . \tag{11}$$

Baryon number violation is forced on us, if we ignore the masses of the fermions. All we need to know about the SU(5) model is that it contains a monopole which couples to the charge Q given by Eq. (9), in a phase with unbroken $SU(3)_c \times SU(2) \times U(1)_{ew}$.

The process, Eq. (11), with two fermions in the initial state, must occur in two steps. It is natural to inquire about the intermediate state produced when u_{1R} scatters off the monopole. What one finds[58-59,69] is rather subtle and mysterious; the intermediate state consists of four "semitons," each with fermion number 1/2. The reaction

$$u_{1R} + M \rightarrow \frac{1}{2} (u_{1L} \ \bar{u}_{2R} \ \bar{d}_{3L} \ e_L^+) + M \tag{12}$$

changes baryon number by $-\frac{1}{2}$ unit.

The semitons are destabilized by fermion mass terms or the effects of the strong QCD interaction. At a distance from the monopole center where these effects become important, the intermediate state in Eq. (12) evolves into a final state with baryon and lepton number differing by an integer from that of the initial state. One possibility is that the semitons in Eq. (12) evolve into u_{1L}; chirality violating processes like Eq. (10) are allowed if the fermions have masses.

The evolution of semitons into final state quarks and leptons can be studied numerically by integrating the classical equations of motion of the bosonized version of the effective $(1 + 1)$-dimensional field theory, in which the fermions are represented by solitons. The classical approximation cannot be justified in detail, but is expected to give qualitatively correct results. Unsurprisingly, the semiton intermediate state is found[59] to evolve into a final state with a different baryon number than the initial state with a probability of order one. It is also found that adding more generations of fermions has no qualitative effect on the baryon-number-changing processes.

The above considerations strongly suggest that the baryon-number-changing cross section for a quark of energy E scattering off a monopole is of order E^{-2}, if E^{-1} is much greater than the radius of the monopole core, and much less than both the Compton wavelength of the quark and the size of a hadron. But we are really interested in the cross section of nucleon decay. There are actually two questions of experimental interest. One is, what is the cross section for capture of nuclei by the monopole? It is presumably large,[70] and the capture rate might conceivably determine the catalysis rate in terrestrial experiments. The other is, what is the cross section for the catalysis process itself? In particular, is it much smaller then the naive guess $\sigma\beta \sim 10^{-27}$ cm^2?

A first step toward answering the latter question can be taken within the context of an approximation in which the nucleon is regarded as a soliton in a nonlinear effective chiral field theory.[71] This approximation should be reasonable at distances large compared to the confinement scale, and so is in a sense complementary to the quark model description of the nucleon. It is found[72] that a chiral soliton which encounters a point monopole can "unwind" and decay, if an appropriate boundary condition is satisfied at the pole. Preliminary calculations thus suggest that the catalysis of nucleon decay by monopoles is not significantly suppressed by nonperturbative strong interaction effects.

Our understanding of the catalysis of nucleon decay by monopoles is still far from complete. For one thing, the production and evolution of semitons could probably be elucidated by further work. Also, the effects of the nondiagonal color and weak charges on the scattering process have not been analyzed.[73]

It might be possible to use the chiral soliton approach to obtain some qualitative information about the velocity dependence and multiplicities in catalyzed nucleon decay, although the corrections to this approximation are expected to be of order one. Any other new ideas about how cross sections might be calculated would be welcome.

CONCLUSIONS

I have attempted to review only a few of the recent developments in theoretical physics which concern magnetic monopoles. Various matters deserving further investigation have been noted, such as monopole production in the inflationary comsology, the properties of Kaluza-Klein monopoles and multimonopole bound states, and the rates of processes catalyzed by monopoles.

The magnetic monopole has proved to be a rich source of theoretical speculation and discovery. On the basis of an eyeball extrapolation of Fig. la, it seems reasonable to conjecture that the monopole will continue to command our attention for some time to come.

ACKNOWLEDGEMENTS

Work supported by the U.S. Department of Energy under contract DEAC 03-81-ER40050.

REFERENCES

1. SPIRES-3, SLAC library data base (1983).
2. P.B. Price et al., Phys. Rev. Lett. 35, 487 (1975).
3. B. Cabrera, Phys. Rev. Lett. 48, 1378 (1982).
4. G. 't Hooft, Nucl. Phys. B79, 276 (1974).
5. A.M. Polyakov, JETP Lett. 20, 194 (1974).
6. A. Blaer, N. Christ, and J. Tang, Phys. Rev. Lett. 47, 1364 (1981).
7. F. Wilczek, Phys. Rev. Lett. 48, 1146 (1982).
8. V. Rubakov, JETP Lett. 33, 644 (1981); Nucl. Phys. B203, 311 (1982).
9. C.G. Callan, Phys. Rev. D25, 2141 (1982); Phys. Rev. D26 2058 (1982).
10. E. Weinberg, these proceedings.
11. S. Coleman, in The Unity of the Fundamental Interactions, A.

Zichichi, New York, Plenum Press (1983).

12. M. Turner, these proceedings.

13. J. Preskill, in The Very Early Universe, ed. G.W. Gibbons, S.W. Hawking, and S.T.C. Siklos, New York, Cambridge (1983).

14. J. Preskill, Phys. Rev. Lett. 43, 1365 (1979).

15. M.S. Turner, Phys. Lett. 115B, 95 (1982).

16. G. Lazarides, Q. Shafi, and W.P. Trower, Phys. Rev. Lett. 49, 1756 (1982).

17. P. Langacker and S.-Y. Pi, Phys. Rev. Lett. 45, 1 (1980).

18. F.A. Bais and P. Langacker, Nucl. Phys. B197, 520 (1982); A.V. Vilenkin, Nucl. Phys. B196, 240 (1982).

19. E. Weinberg, Phys. Lett. 126B, 441 (1983).

20. M.S. Turner, these proceedings; W. Collins and M.S. Turner, Fermi Inst. Preprint 83-41 (1983).

21. M.S. Turner, unpublished; A.S. Goldhaber, A.H. Guth, and S.-Y. Pi, unpublished.

22. M.S. Turner, in Magnetic Monopoles, ed. R.A. Carrigan and W.D. Trower, New York, Plenum (1983); these proceedings.

23. E.N. Parker, Ap. J. 160, 383 (1970).

24. M.S. Turner, E.N. Parker, and T.J. Bogdan, Phys. Rev. D26, 1296 (1982).

25. R.A. Carrigan and W.D. Trower, Magnetic Monopoles, New York, Plenum Press (1983).

26. S. Dimopoulos, S.L. Glashow, E.M. Purcell, and F. Wilczek, Nature 298, 824 (1982).

27. K. Freese and M.S. Turner, Phys. Lett. 123B, 293 (1983).

28. E.E. Salpeter, S.L. Shapiro, and I. Wasserman, Phys. Rev. Lett. 49, 1114 (1982).

29. E.W. Kolb, S.A. Colgate and J. Harvey, Phys. Rev. Lett. 49, 1373 (1982); S. Dimopoulos, J. Preskill, and F. Wilczek, Phys. Lett. 119B, 320 (1982); E.W. Kolb, these proceedings.

30. J. Harvey, these proceedings.

31. S. Dawson and A.N. Schellekens, Phys. Rev. D27, 2119 (1983).

32. J. Harvey, Princeton Preprint Print-82-0868 (1982).

33. K. Freese, these proceedings.

34. B. Julia and Z. Zee, Phys. Rev. D11, 2227 (1975); E. Tomboulis and G. Woo, Nucl. Phys. B107, 221 (1976).

35. R. Jackiw and C. Rebbi, Phys. Rev. Lett. 36, 1116 (1976); P. Hassenfratz and G. 't Hooft, Phys. Rev. Lett. 36, 1119 (1976); A. Goldhaber, Phys. Rev. Lett. 36, 1122 (1976).

36. G. 't Hooft, Nucl. Phys. B105, 538 (1976); E. Corrigan, D. Olive, D. Fairlie and J. Nuyts, Nucl. Phys. B106, 475 (1976).

37. G. 't Hooft, Nucl. Phys. B153, 141 (1979).

38. G. 't Hooft, Nucl. Phys. B190, 455 (1981).

39. E. Witten, Phys. Lett. 86B, 283 (1979).

40. R. Jackiw and C. Rebbi, Phys. Rev. D13, 3398 (1976); J. Goldstone and F. Wilczek, Phys. Rev. Lett. 47, 986 (1981).

41. Th. Kaluza, Sitzungsber. Preuss. Akad. Wiss. Math. Phys. 1921, 966 (1921); O. Klein, Z. Phys. 37, 895 (1926).

42. D. Gross and M. Perry, Nucl. Phys. B226, 29 (1983); R. Sorkin, Phys. Rev. Lett. 51, 87 (1983).

43. P. Nelson, Harvard Preprint HUTP-83/A074 (1983).

44. E. Weinberg, Phys. Rev. D20, 936 (1979); Nucl. Phys. B167, 500 (1980).

45. M. Prasad and C. Sommerfield, Phys. Rev. Lett. 35, 760 (1975); E. Bogomol'nyi, Sov. J. Nucl. Phys. 24, 449 (1976); S. Coleman, S. Parke, A. Neveu, and C. Sommerfield, Phys. Rev. D15, 544 (1977).

46. C.L. Gardner, J.A. Harvey, Princeton Preprint Print-83-1062 (1983).

47. C. Dokos and T. Tomaras, Phys. Rev. D21, 2940 (1980).

48. A. Abouelsaood, Nucl. Phys. B226, 309 (1984); Phys. Lett. 125B, 467 (1983).

49. P. Nelson, Phys. Rev. Lett. 50, 939 (1983).

50. P. Nelson and A. Manohar, Phys. Rev. Lett. 50, 943 (1983); A. Balachandran et al., Phys. Rev. Lett. 50 1553 (1983).

51. P. Nelson and S. Coleman, Harvard Preprint HUTP-83/A067 (1983).

52. Y. Kazama, C.N. Yang, and A.S. Goldhaber, Phys. Rev. D15, 2287 (1977).

53. A.S. Goldhaber, Phys. Rev. D16, 1815 (1977).

54. C. Dokos and T. Tomaras, Phys. Rev. D21, 2940 (1980).

55. G. 't Hooft, Phys. Rev. Lett. 37, 8 (1976); C.G. Callan, R.F. Dashen and D.J. Gross, Phys. Lett. 63B, 334 (1976); R. Jackiw and C. Rebbi, Phys. Rev. Lett. 37, 177 (1976).

56. J.S. Bell and R. Jackiw, Nuovo. Cim. 51, 47 (1969); S.L. Adler, Phys. Rev. 177, 2426 (1969).

57. C. Besson, Ph.D. Thesis, Princeton University (1982).

58. A. Sen, Fermilab-Pub-83/28 (1983); these proceedings.

59. S. Dawson and A.N. Schellekens, Fermilab-Pub-83/43-THY (1983).

60. Z.F. Ezawa and A. Iwazaki, preprint (1983).

61. T. Appelquist and J. Carazzone, Phys. Rev. D11, 2856 (1975).

62. V.A. Rubakov and M.S. Serbryakov, Nuc. Phys. B218, 240 (1983).

63. C.G. Callan, Nucl. Phys. B212, 391 (1983).

64. S. Coleman, Phys. Rev. D11, 2088 (1975); S. Mandelstam, Phys. Rev. D11, 3026 (1974).

65. H. Georgi and S.L. Glashow, Phys. Rev. Lett. 32, 438 (1974).

66. R. Jackiw and C. Rebbi, Phys. Rev. D13, 3398 (1976).

67. Y. Kazama and A. Sen, Fermilab-Pub-83/58 (1983); T.-M. Yan, Cornell Preprint CLNS-83/563 (1983); A.P. Balachandran and J. Schechter, Syracuse Preprint (1983); Z.F. Ezawa and A. Iwazaki, Preprint (1983); H. Yamagishi, Princeton Preprint (1983).

68. B. Grossman, G. Lazarides, and A.I. Sanda, Phys. Rev. D28, 2109 (1983).

69. C.G. Callan, Princeton Preprint 83-0306 (1983).

70. G. Fiorentini, these proceedings.

71. T. Skyrme, Proc. Roy. Soc. A260, 127 (1961); Nucl. Phys. 31,

556 (1962); E. Witten, Nucl. Phys. B223, 433 (1983).

72.   C.G. Callan   and E. Witten,   Princeton Preprint-84-0054
      (1983); C.G. Callan, these proceedings.

73.   N.S. Craigie, W. Nahm, and V.A. Rubakov, Preprint (1983).

PARTICIPANTS

Steven Ahlen, Physics Department, Indiana University, Bloomington IN 47405, USA.

Carl Akerlof, Randall Physics Laboratory, University of Michigan, Ann Arbor MI 48109, USA.

Metin Arik, Physics Department, Bogazici Universitesi, PK 2 Bebek, Istanbul, TURKEY.

F. Alexander Bais, Institute for Theoretical Physics, University of Utrecht, Princetonplein 5, 3508 TA Utrecht, THE NETHERLANDS.

A.P. Balachandran, Physics Department, Syracuse University, Syracuse NY 13210, USA.

Robert Ball, Randall Physics Laboratory, University of Michigan, Ann Arbor MI 48109, USA.

Varouzhan Baluni, Randall Physics Laboratory, University of Michigan, Ann Arbor MI 48109, USA.

Barry Barish, Lauritsen Laboratory, California Institute of Technology, Pasadena CA 91125, USA.

John Bartelt, SLAC, Bin 78, PO Box 4349, Stanford University, Stanford CA 94305, USA.

C. Bemporad, Piazza Toricelli 2, Università di Pisa, Pisa, ITALY.

Alain Blanchard, Observatoire de Paris-Meudon, 92195 Meudon Principal Cedex, FRANCE.

Geoff Blewitt, Physics Department, California Institute of Technology, Pasadena CA 91125, USA.

Blas Cabrera, Physics Department, Stanford University, Stanford CA 94305, USA.

Curtis Callan Jr., Physics Department, Princeton University, Princeton NJ 08594, USA.

Richard Carrigan Jr., Fermilab MS 208, PO Box 500, Batavia IL 60510, USA.

Rosanna Cester, CERN-EP Division, CH-1211 Geneva 23, SWITZERLAND.

Lay Nam Chang, Physics Department, Virginia Polytechnic Institute and State University, Blacksburg VA 24061, USA.

Praveen Chaudhari, IBM Watson Research Center, PO Box 218, Yorktown Heights NY 10598, USA.

John C.C. Chi, IBM Watson Research Center, PO Box 218, Yorktown Heights NY 10598, USA.

John R. Clem, Ames Laboratory, Department of Physics, Iowa State University, Ames IA 50011, USA.

Philip Connolly, Physics Department, Brookhaven National Laboratory, Upton, Long Island NY 11973, USA.

Jerry A. Cowen, Department of Physics, Michigan State University, East Lansing MI 48824, USA.

Neil Craigie, ICTP, PO Box 586, Miramare-Strada Costiera 11, 34110 Trieste, ITALY.

Robin Craven, Fermilab MS 341, PO Box 500, Batavia IL 60510, USA.

Timir Datta, Physics Department, University of South Carolina, Columbia SC 29208, USA.

Sumit Das, Fermilab, PO Box 500, Batavia IL 60510, USA.

Sally Dawson, Lawrence Berkeley Laboratory 50A-3115, University of California, Berkeley CA 94720, USA.

Alvaro DeRujula, CERN-Theory Division, CH-1211 Geneva 23, SWITZERLAND.

A.K. Drukier, Max Planck Institut für Physik und Astrophysik, Föhringer Ring 6, 8000 München 40, FEDERAL REPUBLIC OF GERMANY.

Martin Einhorn, Randall Physics Laboratory, University of Michigan, Ann Arbor MI 48109, USA.

Steven Errede, Randall Physics Laboratory, University of Michigan, Ann Arbor MI 48109, USA.

Paul Eschstruth, LAL-ORSAY, Bâtiment 200-91405 Orsay Cedex, FRANCE.

William Fairbank, Physics Department, Stanford University, Stanford CA 94305, USA.

Daniele Fargion, Istituto di Fisica, Università di Roma, Piazza Aldo Moro 2, I-00815 Rome, ITALY.

Fred Fickett, US Department of Commerce, National Bureau of Standards, 325 Broadway, Boulder CO 80303, USA.

Gianni Fiorentini, Istituto di Fisica, Università di Pisa, Piazza Torricelli 2, I-56100 Pisa, ITALY.

William Ford, Randall Physics Laboratory, University of Michigan, Ann Arbor MI 48109, USA.

Katherine Freese, Astronomy and Astrophysics Center, University of Chicago, 5640 South Ellis Avenue, Chicago IL 60637, USA.

Jean-Marie Frère, Randall Physics Laboratory, University of Michigan, Ann Arbor MI 48109, USA.

Henry Frisch, Enrico Fermi Institute, University of Chicago, 5640 South Ellis Avenue, Chicago IL 60637, USA.

Piero Galeotti, Istituto di Fisica, Università di Torino, Corso M. D'Azeglio 46, I-10125 Torino, ITALY.

Giorgio Giacomelli, Istituto di Fisica, Università di Bologna, Via Irnerio 46, I-40126 Bologna, ITALY.

Charles Goebel, Physics Department, University of Wisconsin, Madison WI 53706, USA.

Alfred S. Goldhaber, Physics Department, State University of New York, Stony Brook NY 11794, USA.

Maurice Goldhaber, High Energy Physics, Brookhaven National Laboratory, Upton, Long Island NY 11973, USA.

Maury Goodman, Department of Physics, Massachusetts Institute of Technology, Cambridge MA 02139, USA.

Bernard Grossman, Department of Physics, Rockefeller University, 1230 York Avenue, New York NY 10021, USA.

H. Richard Gustafson, Randall Physics Laboratory, University of Michigan, Ann Arbor MI 48109, USA.

C.N. Guy, Blackett Laboratory, Imperial College, Prince Consort Road, London SW7 2BZ, UNITED KINGDOM.

Ray Hagstrom, High Energy Physics Division, Argonne National Laboratory, 9700 South Cass Avenue, Argonne IL 60439, USA.

Jeffrey Harvey, Department of Physics, Princeton University, Princeton NJ 08544, USA.

Richard M. Heinz, Physics Department, Indiana University, Swain Hall West 259, Bloomington IN 47405, USA.

Roberto Iengo, ICTP, PO Box 586, Miramare-Strada Costiera 11, 34100 Trieste, ITALY.

Joseph Incandela, University of Chicago, 5640 South Ellis Avenue, Chicago IL 60637, USA.

Keith Jones, Editorial Office of Nuclear Physics, NORDITA, Blegdamsvej 17, DK-2100, Copenhagen, DENMARK.

Lawrence W. Jones, Randall Physics Laboratory, University of Michigan, Ann Arbor MI 48109, USA.

Fumiyoshi Kajino, Physics Department, Virginia Polytechnic Institute and State University Blacksburg VA 24061, USA.

Gordon Kane, Randall Physics Laboratory, University of Michigan, Ann Arbor MI 48109, USA.

Yi Han Kao, Physics Department, State University of New York, Stony Brook NY 11794, USA.

Edwin Kashy, Department of Physics, Michigan State University, East Lansing MI 48824, USA.

Boris Kayser, Division of Physics, National Science Foundation, Washington DC 20550, USA.

Che Ming Ko, Cyclotron Institute, Texas A&M University, College Station TX 77843, USA.

Edward Kolb, Fermilab MS 209, PO Box 500, Batavia IL 60510, USA.

Jean Krisch, Randall Physics Laboratory, University of Michigan, Ann Arbor MI 48109, USA.

Norman M. Kroll, Department of Physics, University of California-San Diego, La Jolla CA 92093, USA.

Moyses Kuchnir, Fermilab MS 341, PO Box 500, Batavia IL 60510, USA.

Gabor Kunstatter, Department of Physics, University of Toronto, Toronto, Ontario M5S 1A7, CANADA.

Charles Lane, Physics Department 256-48, California Institute of Technology, Pasadena CA 91125, USA.

John Learned, Physics Department, University of Hawaii, 2505 Correa Road, Honolulu HI 96822, USA.

Paul H.E. Lee, Atomic Energy of Canada Ltd., Chalk River Laboratories, Chalk River, Ontario K0J 1J0, CANADA.

Ian Leedom, Randall Physics Laboratory, University of Michigan, Ann Arbor MI 48109, USA.

Harry Lipkin, Physics Department, Weizmann Institute of Science, Rehovot 76100, ISRAEL.

Tony M. Liss, Physics Department, University of California, Berkeley CA 94720, USA.

Gang Liu, Physics Department 256-48, California Institute of Technology, Pasadena CA 91125, USA.

David London, Physics Department, University of Chicago, 5640 South Ellis Avenue, Chicago IL 60637, USA.

Michael Longo, Randall Physics Laboratory, University of Michigan, Ann Arbor MI 48109, USA.

John LoSecco, Physics Department, California Institute of Technology, Pasadena CA 91125, USA.

Robert March, Department of Physics, University of Wisconsin, Madison WI 53706, USA.

Alessandro Marini, Laboratorio Nazionali di Frascati, Casella Postale 13, I-00044 Frascati, ITALY.

Paul Musset, CERN-EP Division, CH-1211 Geneva 23, SWITZERLAND.

Frank Nezrick, Fermilab, PO Box 500, Batavia IL 60510, USA.

Jean Nuyts, Université de l'Etat, Avenue Maistriau 19, B-700 Mons, BELGIUM.

Dan O'Conner, Physics Department, University of Hawaii, 2505 Correa Road, Honolulu HI 96822, USA.

Keith Olive, Fermilab MS 209, PO Box 500, Batavia IL 60510, USA.

Stephen J. Parke, Fermilab, PO Box 500, Batavia IL 60510, USA.

E.N. Parker, Enrico Fermi Institute, Department of Astronomy and Astrophysics, University of Chicago, Chicago IL 60637, USA.

Roberto Peccei, Max Planck Institut für Physik und Astrophysik, Föhringer Ring 6, 800 München 40, FEDERAL REPUBLIC OF GERMANY.

Malcolm Perry, Department of Physics, Princeton University, Princeton NJ 08544, USA.

Murray Peshkin, Argonne National Laboratory-203, 9700 South Cass Avenue, Argonne IL 60439, USA.

Earl A. Peterson, Physics Department, University of Minnesota, 116 Church Street S.E., Minneapolis MN 55455, USA.

Pio Pistilli, Istituto di Fisica, Università di Roma, Piazza Aldo Moro 2, I-00815 Rome, ITALY.

John Preskill, Physics Department, California Institute of Technology, MS 452-48, Pasadena CA 91125, USA.

Mike Price, CERN-EF Division, CH-1121 Geneva 23, SWITZERLAND.

Zhao Ming Qiu, Physics Department, Columbia University, 538 West 120 Street, New York NY 10027, USA.

Sumathi Rao, Fermilab-Theory Group, PO Box 500, Batavia IL 60510, USA.

David Ritson, Physics Department, Stanford University, Stanford CA 94305, USA.

Michel Rollier, Istituto di Fisica, Università di Milano, Via Celoria 16, I-20133 Milano, ITALY.

Francesco Ronga, Laboratorio Nazionali di Frascati, Casella Postale 13, I-00044 Frascati, ITALY.

Serge Rudaz, Physics Department, University of Minnesota, 116 Church Street S.E., Minneapolis MN 55455, USA.

Joseph Scanio, Department of Physics, ML 11, University of Cincinnati, Cincinnati OH 45221, USA.

A. Schellekens, Physics Department, State University of New York, Stony Brook NY 11794, USA.

J.C. Schouten, Blackett Laboratory, Imperial College, Prince Consort Road, London SW7 2AZ, UNITED KINGDOM.

David Seckel, Fermilab MS 209, PO Box 500, Batavia IL 60510, USA.

Ashoke Sen, Fermilab MS 106, PO Box 500, Batavia Il 60510, USA.

Richard Seto, Physics Department, Columbia University, 538 West 120 Street, New York NY 10027, USA.

Marco Severi, Istituto di Fisica, Università "La Saphenza", Piazza Aldo Moro 2, I-00185 Rome, ITALY.

Qaisar Shafi, Bartol Research Foundation, University of Delaware, Newark DE 19711, USA.

Joshua D. Silver, Clarendon Laboratory, University of Oxford, Parks Road, Oxford OX1 3PU, UNITED KINGDOM.

Helene Sol, Observatoire de Paris-Meudon, 92195 Meudon Principal Cedex, FRANCE.

Sunil Somalwar, Physics Department, University of Chicago, 5640 South Ellis Avenue, Chicago IL 60637, USA.

Hidenori Sonoda, Physics Department, MS 452-48, California Institute of Technology, Pasadena CA 91125, USA.

Rafael Sorkin, Department of Physics, Syracuse University, Syracuse NY 13210, USA.

Richard Steinberg, Physics Department, University of Pennsylvania, Philadelphia PA 19104, USA.

Paul Steinhardt, Physics Department, University of Pennsylvania, Philadelphia PA 19104, USA.

James Stone, Randall Physics Laboratory, University of Michigan, Ann Arbor MI 48109, USA.

Daniel Stump, Department of Physics, Michigan State University, East Lansing MI 48824, USA.

Lawrence Sulak, Randall Physics Laboratory, University of Michigan, Ann Arbor MI 48109, USA.

Gregory Tarlé, Randall Physics Laboratory, University of Michigan, Ann Arbor, MI 48109, USA.

Claudia Tesche, IBM Watson Research Center, PO Box 218, Yorktown Heights NY 10598, USA.

Yukio Tomozawa, Randall Physics Laboratory, University of Michigan, Ann Arbor MI 48109, USA.

W. Peter Trower, Physics Department, Virginia Polytechnic Institute and State University, Blacksburg VA 24061, USA.

C.C. Tsuei, IBM Watson Research Center, PO Box 218, Yorktown Heights NY 10598, USA.

Michael Turner, Fermilab, PO Box 500, Batavia, IL 60510, USA.

Hans Vanderplicht, Department of Physics, Michigan State University, East Lansing MI 48824, USA.

John van der Velde, Randall Physics Laboratory, University of Michigan, Ann Arbor MI 48109, USA.

Martinus Veltman, Randall Physics Laboratory, University of Michigan, Ann Arbor MI 48109, USA.

John Vergados, School of Physics, University of Ioannina, Ioannina, GREECE.

Thomas Walsh, Physics Department, University of Minnesota, 116 Church Street S.E., Minneapolis MN 55455, USA.

Tadashi Watanabe, Department of Physics, Kobe University, Rokkodai, Kobe 657, JAPAN.

Robert Webb, Physics Department, Texas A&M University, College Station TX 77843, USA.

Michael L. Weber, Randall Physics Laboratory, University of Michigan, Ann Arbor MI 48109, USA.

Erick Weinberg, Physics Department, Columbia University, 538 West 120 Street, New York NY 10027, USA.

C. Wetterich, Institut für theoretische Physik, Universität Bern, Sidlerstrasse 5, CH-3012 Bern, SWITZERLAND.

Tung Mow Yan, Newman Laboratory, Cornell University, Ithaca NY 14853, USA.

Cosmas Zachos, Argonne National Laboratory-362, 9700 South Cass Avenue, Argonne IL 60439, USA.